T0183200

Lecture Notes in Computer Science 9315

Commenced Publication in 1973
Founding and Former Series Editors:
Gerhard Goos, Juris Hartmanis, and Jan van Leeuwen

More information about this series at http://www.springer.com/series/7409

Yo-Sung Ho · Jitao Sang
Yong Man Ro · Junmo Kim
Fei Wu (Eds.)

Advances in Multimedia Information Processing – PCM 2015

16th Pacific-Rim Conference on Multimedia
Gwangju, South Korea, September 16–18, 2015
Proceedings, Part II

 Springer

Editors
Yo-Sung Ho
Gwangju Institute of Science
  and Technology
Gwangju
Korea (Republic of)

Jitao Sang
Chinese Academy of Sciences
Institute of Automation
Beijing
China

Yong Man Ro
KAIST
Daejeon
Korea (Republic of)

Junmo Kim
KAIST
Daejeon
Korea (Republic of)

Fei Wu
College of Computer Science
Zhejiang University
Hangzhou
China

ISSN 0302-9743   ISSN 1611-3349   (electronic)
Lecture Notes in Computer Science
ISBN 978-3-319-24077-0   ISBN 978-3-319-24078-7   (eBook)
DOI 10.1007/978-3-319-24078-7

Library of Congress Control Number: 2015948170

LNCS Sublibrary: SL3 – Information Systems and Applications, incl. Internet/Web, and HCI

Springer Cham Heidelberg New York Dordrecht London

Printed on acid-free paper

Springer International Publishing AG Switzerland is part of Springer Science+Business Media
(www.springer.com)

Preface

We are delighted to welcome readers to the proceedings of the 16th Pacific-Rim Conference on Multimedia (PCM 2015), held in Gwangju, South Korea, September 16–18, 2015. The Pacific-Rim Conference on Multimedia is a leading international conference for researchers and industry practitioners to share and showcase their new ideas, original research results, and engineering development experiences from areas related to multimedia. The 2015 edition of the PCM marked its 16th anniversary. The longevity of the conference would not be possible without the strong support of the research community, and we take this opportunity to thank everyone who has contributed to the growth of the conference in one way or another over the last 16 years.

PCM 2015 was held in Gwangju, South Korea, which is known as one of the most beautiful and democratic cities in the country. The conference venue was Gwangju Institute of Science and Technology (GIST), which is one of the world's top research-oriented universities. Despite its short history of 22 years, GIST has already established its position as an educational institution of huge potential, as it ranked the fourth in the world in citations per faculty in the 2014 QS World University Rankings.

At PCM 2015, we held regular and special sessions of oral and poster presentations. We received 224 paper submissions, covering topics of multimedia content analysis, multimedia signal processing and communications, as well as multimedia applications and services. The submitted papers were reviewed by the Technical Program Committee, consisting of 143 reviewers. Each paper was reviewed by at least two reviewers. The program chairs carefully considered the input and feedback from the reviewers and accepted 138 papers for presentation at the conference. The acceptance rate of 62 % indicates our commitment to ensuring a very high-quality conference. Out of these accepted papers, 68 were presented orally and 70 papers were presented as posters.

PCM 2015 was organized by the Realistic Broadcasting Research Center (RBRC) at Gwangju Institute of Science and Technology (GIST) in South Korea. We gratefully thank the Gwangju Convention and Visitors Bureau for its generous support of PCM 2015.

We are heavily indebted to many individuals for their significant contributions. Firstly, we are very grateful to all the authors who contributed their high-quality research and shared their knowledge with our scientific community. Finally, we wish to thank all Organizing and Program Committee members, reviewers, session chairs, student volunteers, and supporters. Their contributions are much appreciated. We hope you all enjoy the proceedings of the 2015 Conference on Multimedia.

September 2015

Yo-Sung Ho
Jitao Sang
Yong Man Ro
Junmo Kim
Fei Wu

Organization

Organizing Committee

General Chair

Yo-Sung Ho   Gwangju Institute of Science and Technology, South Korea

Program Chairs

Jitao Sang   Chinese Academy of Sciences, China
Yong Man Ro   Korea Advanced Institute of Science and Technology, South Korea

Special Session Chairs

Shang-Hong Lai   National Tsinghua University, Taiwan
Chao Liang   Wuhan University, China
Yue Gao   National University of Singapore, Singapore

Tutorial Chairs

Weisi Lin   Nanyang Technological University, Singapore
Chang-Su Kim   Korea University, South Korea

Demo/Poster Chairs

Xirong Li   Renmin University of China, China
Lu Yang   University of Electronic Science and Technology of China, China

Publication Chairs

Junmo Kim   Korea Advanced Institute of Science and Technology, South Korea
Fei Wu   Zhejiang University, China

Publicity Chairs

Chin-Kuan Ho   Multimedia University, Malaysia
Gangyi Jiang   Ningbo University, China
Sam Kwong   City University of Hong Kong, Hong Kong
Yoshikazu Miyanaga   Hokkaido University, Japan
Daranee Hormdee   Khon Kaen University, Thailand

Thanh-Sach Le   Ho Chi Minh City University, Vietnam
Ki Ryong Kwon   Pukyong National University, South Korea

Web Chair

Eunsang Ko   Gwangju Institute of Science and Technology, South Korea

Registration Chairs

Young-Ki Jung   Honam University, South Korea
Youngho Lee   Mokpo National University, South Korea

Local Arrangement Chairs

Young Chul Kim   Chonnam National University, South Korea
Pankoo Kim   Chosun University, South Korea

Technical Program Committee

Sungjun Bae	Shoko Imaizumi	Youngho Lee
Hang Bo	Byeungwoo Jeon	Haiwei Lei
Xiaochun Cao	Zhong Ji	Donghong Li
Kosin Chamnongthai	Yu-Gang Jiang	Guanyi Li
Wen-Huang Cheng	Jian Jin	Haojie Li
Nam Ik Cho	Xin Jin	Houqiang Li
Jae Young Choi	Zhi Jin	Leida Li
Wei-Ta Chu	SoonHeung Jung	Liang Li
Peng Cui	YongJu Jung	Songnan Li
Wesley De Neve	Yun-Suk Kang	Xirong Li
Cheng Deng	Hisakazu Kikuchi	Yongbo Li
Weisheng Dong	Byung-Gyu Kim	Chunyu Lin
Yao-Chung Fan	Changik Kim	Weisi Lin
Yuming Fang	ChangKi Kim	Weifeng Liu
Sheng Fang	Chang-Su Kim	Bo Liu
Toshiaki Fujii	Hakil Kim	Qiegen Liu
Masaaki Fujiyoshi	Hyoungseop Kim	Qiong Liu
Yue Gao	Jaegon Kim	Wei Liu
Yanlei Gu	Min H. Kim	Xianglong Liu
Shijie Hao	Seon Joo Kim	Yebin Liu
Lihuo He	Su Young Kwak	Dongyuan Lu
Ran He	Shang-Hong Lai	Yadong Mu
Min Chul Hong	Duy-Dinh Le	Shogo Muramatsu
Richang Hong	Chan-Su Lee	Chong-Wah Ngo
Dekun Hu	Sang-Beom Lee	Byung Tae Oh
Min-Chun Hu	Sanghoon Lee	Lei Pan
Ruimin Hu	Sangkeun Lee	Yanwei Pang
Lei Huang	Seokhan Lee	Jinah Park

**New Media Representation and Transmission Technologies for Emerging
UHD Services**

Special Poster Sessions

Contents – Part I

Multimedia Applications and Services

Video Coding and Processing

Multimedia Representation Learning

Regular Poster Session

Visual Understanding and Recognition on Big Data

**Coding and Reconstruction of Multimedia Data with Spatial-Temporal
Information**

3D Image/Video Processing and Applications

Motion and Depth Assisted Workload Prediction for Parallel View Synthesis

Zhanqi Liu, Xin Jin$^{(\boxtimes)}$, and Qionghai Dai

Shenzhen Key Lab of Broadband Network and Multimedia,
Graduate School at Shenzhen, Tsinghua University, Shenzhen, China
liu-zql3@mails.tsinghua.edu.cn,
jin.xin@sz.tsinghua.edu.cn, qhdai@tsinghua.edu.cn

Abstract. In this paper, a parallel system together with a real-time workload balancing algorithm is proposed for view synthesis on multi-core platforms. First, a numerical relationship between the number of holes after warping and the work-load of view synthesis is derived based on correlation analysis for the texture regions. Then, according to the difference between the adjacent depth maps, a novel model is proposed to predict the synthesis workload accurately. Experimental results show that the workload difference among cores is reduced obviously and higher speedup ratio is achieved with negligible quality degradation by applying the proposed workload balancing system.

Keywords: Parallel view synthesis · Depth image based rendering · Workload balance · FTV

1 Introduction

Free viewpoint TV (FTV) allows viewers to watch a 3D scene from multiple perspectives without wearing glasses, which brings immerse visual feelings to people [1]. For this reason, FTV has been known as the future of 3DTV and has attracted great attention both from industry and academy. Considering the capacity limit of transmission cable, the Moving Picture Experts Group (MPEG) has adopted a system architecture for FTV that performs depth estimation at the sender and view synthesis at the receiver with the data format Multiview Video plus Depth (MVD) [2].

The MVD based architecture can reduce the data size to be transferred, while big computational burden is also introduced at the receiver for generating the virtual views using depth image-based rendering (DIBR) [3]. In order to obtain virtual views with good quality, DIBR introduces Warping, Blending, Hole Filling and Boundary Noise Re-moving [3] to map the reference viewpoints to the virtual viewpoint and to correct the pixel values accordingly. However, the process presents high computational complexity, which restricts the application of FTV especially on portable devices, such as smart phones, tablet PCs etc.

Recently, some works have been proposed for fast view synthesis based on GPU [4, 5]. However, for multi-core systems, which are widely applied in portable devices, to the best of our knowledge, no parallel view synthesis algorithm has been proposed yet. In addition, the efficiency of parallelism is mainly constrained by the execution

Y.-S. Ho et al. (Eds.): PCM 2015, Part II, LNCS 9315, pp. 3–13, 2015.
DOI: 10.1007/978-3-319-24078-7_1

time of the core with the heaviest workload [6]. Representative workload prediction methods, like PAST [7] and moving average (MA) [7], have been widely applied to balance the workload of general applications [8]. PAST predicts the workload of the current time interval according to that of the previous. MA predicts the workload in the next interval using the average of the workload in the previous intervals [7]. Nevertheless, all these methods are not dedicated to view synthesis whose workload is highly affected by video contents, and none of them utilizes the characteristics contained in the texture or depth images. Therefore, developing a parallel view synthesis system with a workload balancing algorithm targeted to view synthesis is fundamentally needed by FTV for its immigration to the portable devices.

Consequently, a parallel view synthesis system with a real-time workload balancing algorithm is proposed in this paper. First, a mathematical model between the number of holes in the warped depth maps and the workload of view synthesis is built. Then, a novel depth-assistant prediction method for the number of holes is designed. Based on the difference of depth maps between two neighboring frames, which reflects the movement of objects in video content, the number of holes in current frame is predicted from the previous frame. The predicted number of holes will be used in the cost function, which decides the locations for partitioning the frame so as to balance the workload. Experimental results demonstrate that the proposed workload prediction method can balance workload among cores and the speed of view synthesis is accelerated obviously with the proposed algorithm.

The rest of the paper is organized as follows. In Sect. 2, the proposed parallel view synthesis system is introduced. In Sect. 3, the real-time workload balancing algorithm is described in detail. The experimental results are provided in Sect. 4 with conclusions in Sect. 5.

2 The Proposed Parallel Structure

The architecture of the proposed parallel view synthesis system with workload balancing, which consists of Depth Maps Process, Feature Detection, Workload prediction, Partition Location Decision and Parallel View Synthesis, is depicted in Fig. 1.

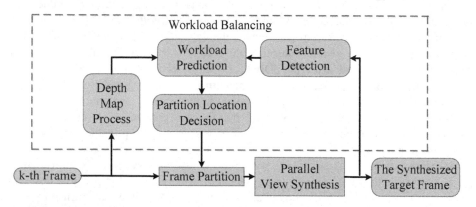

Fig. 1. The proposed parallel view synthesis system

Before partitioning the k-th frame for parallel view synthesis, the distribution of workload around the whole frame should be predicted first. Assuming the k-th frame of reference view is waiting to do view synthesis, and the previous frame has completed its view synthesis work in which the total number of holes in the warped depth maps are recorded in *Feature Detection*. In *Depth Maps Process*, the summation of depth values in each line is recorded. In *Workload Prediction*, the workload of the k-th frame is predicted based on the number of holes in the previous two frame and the difference of depth maps among the current frame and the previous two frames. Then, a cost function will be minimized to make a decision on the partition locations of the k-th frame in *Partition Location Decision*. The target of the function is to balance the synthesis workload among the partitions. Finally, *Parallel View Synthesis* is executed by synthesizing one partition on one core and merging the synthesized partitions to form the k-th frame at the virtual viewpoint for output.

3 Adaptive Workload Balancing

3.1 Definition of Workload

To evaluate which process consumes the major workload during view synthesis, execution time is retrieved for the major processes in 1-D view synthesis. Forward Warping, Merging, Hole Filling [10] and an improvement in depth maps are evaluated, while Boundary Noise Removal (BNR) is turned off since it doesn't provide a stable performance in objective quality. It is observed that Forward Warping and Merging occupy around 94 % of the total execution time of view synthesis. The rest of time is composed of Hole Filling and a temporal improvement in the depth map. Thus, the workload of view synthesis for a frame, Ψ can be approximated by

$$\Psi \approx \Psi_F + \Psi_M, \tag{1}$$

where Ψ_F and Ψ_M are workload of Forward Warping and Merging, respectively.

Forward Warping warps the reference partition to the virtual viewpoint pixel by pixel according to the 3D warping formula [3]. So, its workload is proportional to partition size, which is given by

$$\Psi_F = 2\alpha_1 WH, \tag{2}$$

where α_1 denotes the warping workload per pixel; W and H denote the width and height of frame, respectively. Multiplying by 2 represents warping from the left and right views simultaneously.

In Merging, the target viewpoint is generated by blending the two images warped from the left and right reference views pixel by pixel. Since it processes the pixel in the non-hole regions, the workload of Merging, Ψ_M, is proportional to the total number of pixels in the non-hole regions. So, it is given by

$$\Psi_M \approx \alpha_2(\lambda 2WH - \Theta), \tag{3}$$

where α_2 denotes the blending workload of a pixel in non-hole regions; Θ denotes the sum of hole pixels in two warped depth maps. λ equals to 1 or 2 corresponding to mapping to integer pixel or half pixel precision, respectively.

Generally, the normalized workload of a task is proportional to the execution time of a task [9]. Since α_1 and α_2 are algorithm and platform dependent, they are retrieved and updated dynamically with view synthesis for a stable result on a specific platform. Therefore, α_1 and α_2 are given by

$$\alpha_1 \approx \frac{\sum\limits_{N} T_F}{N \cdot 2WH}, \quad \alpha_2 \approx \frac{\sum\limits_{N} T_M}{\sum\limits_{N} (\lambda 2WH - \Theta)}, \tag{4}$$

where N is the total number of processed frames, T_F and T_M are the execution time of Warping and Merging during processing a frame, respectively.

Finally, the workload of view synthesis for a frame can be computed by

$$\Psi \approx 2\alpha_1 WH + \alpha_2(\lambda 2WH - \Theta). \tag{5}$$

3.2   Workload Prediction

According to the definition of workload in Eq. (5), the only variable is Θ, which represents the number of holes in a frame. So, predicting the workload can be approximated by predicting the number of holes.

After warping to the targeted view, we have found that the number of holes in each frame is strongly related to the sum of depth values (denoted by Ω) in corresponding depth maps. Figure 2 shows Θ and Ω retrieved during view synthesis in each frame for

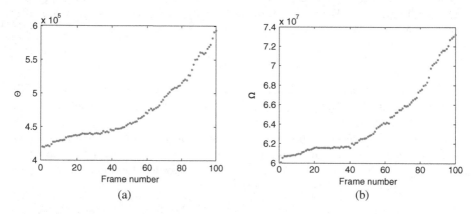

Fig. 2. (a) The total number of holes after warping in each frame; (b) The summation of depth values in each frame

sequence "Champagne" [13]. Investigating the best fit to the data points, a linear function is found by the blue line as shown in Fig. 3 (the green dots is the retrieved real data). The same phenomenon has also been observed in other sequences.

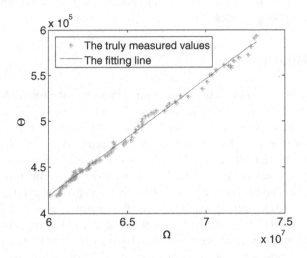

Fig. 3. Data retrieved for Θ and Ω with a linear fitting result

From the physical sense, it is easily to interpret. When the objects in video content moves towards the camera, the depth values will grow large. Then, the depth difference between the object edges and the background will become big, which will generates more holes in the warped view.

Therefore, the depth maps can be utilized to assist the prediction of holes number. Based on the linear relationship between Θ and Ω, we can use the variation ratio of Ω to approximate the variation ratio of Θ. Therefore, the prediction values to the number of holes Θ_k in the k-th frame is given by

$$\overset{\Lambda}{\underset{k}{\Theta}} = \Theta_{k-1} + \frac{\Omega_k - \Omega_{k-1}}{\Omega_{k-1}} \cdot \Theta_{k-1}. \tag{6}$$

3.3   Partition Location Decision

According to our experiment, partitioning the frame horizontally introduces an average of 0.4 dB increment in PSNR to the synthesized views relative to partitioning vertically. Therefore, horizontal splitting is chosen here. After the prediction of the number of holes, the line-wise workload of the k-th frame can be calculated. A cost function is defined to find the best partition locations of the k-th frame for workload balancing as follows:

$$h_m = \arg\min_{h_m \in (1,H)} \left| \sum_{i=1}^{h_m} \psi_{k,i} - m \cdot \frac{\Psi_k}{n} \right|, \ (m=1, 2, \ldots, n-1), \tag{7}$$

where n is the number of cores, $\psi_{k,i}$ is the predicted workload at the line i of the k-th frame. Ψ_k is the whole predicted workload of the k-th frame.

4   Experimental Results

The proposed parallel system is integrated into reference software, VSRS 3.5 [11], to test its performance. The experiments are conducted on a PC with an Intel®Core™ i5-3470 3.2 GHz quad-core processor and 12 GB RAM under 64-bit Windows 7. 1-D mode view synthesis with half-pixel precision is selected for VSRS, while other options are set to the default values [11]. The single-thread VSRS (denoted by *VSRS*) has been chosen as benchmark. Both extrapolation (*EXTRA*) and interpolation (*INTER*) are evaluated under 2-thread ($n = 2$) and 4-thread ($n = 4$) parallel synthesis cases. Four sequences (the first 100 frames) with representative features, as listed in Table 1, are selected for testing. 9(10, 8) represents view 9 was synthesized by view 10 and 8. For simplicity, the name of each sequence is abbreviated by the first three characters in the following. The proposed parallel system without workload balancing is denoted by *Propw/oWB* and the proposed parallel system with workload balancing is denoted by *Propw/WB*.

The performance of workload balancing is measured by the average reduction ratio in the normalized maximum workload difference among the cores, which is given by

Table 1. Test sequences and viewpoints

Sequences	Resolution	*INTER*	*EXTRA*
Bookarrival [12]	1024 × 768	9(10, 8)	8(10)
Champagne [13]	1280 × 960	40(39, 41)	41(39)
Newspaper [12]	1024 × 768	5(4, 6)	4(6)
PoznanStreet [14]	1920 × 1088	4(3, 5)	5(3)

Table 2. Prediction error ratio (%)

Seq.	PAST [7]	MA [7]	Proposed	Proposed vs PAST	Proposed vs MA
Boo.	1.10	1.30	1.10	0.00	15.40
Cha.	0.40	0.60	0.30	25.00	50.00
New.	3.00	3.40	2.90	3.30	14.70
Poz.	0.40	0.50	0.30	25.00	40.00
Ave.	1.23	1.45	1.15	13.33	30.03

$$WDRR = 1 - \frac{1}{N}\sum_{N}\frac{\left(\max\limits_{i,j\in[1,n]}\left|W_{i,\text{WB}} - W_{j,\text{WB}}\right|\right)\Big/\left(\max\limits_{i\in[1,n]}W_{i,\text{WB}}\right)}{\left(\max\limits_{i,j\in[1,n]}\left|W_{i,n\text{WB}} - W_{j,n\text{WB}}\right|\right)\Big/\left(\max\limits_{i\in[1,n]}W_{i,n\text{WB}}\right)}, \qquad (8)$$

where $W_{i,nWB}$ and $W_{i,WB}$ are the workload of the i-th partition processed without and with workload balancing, respectively. So, the larger $WDRR$ is, the larger the improvement in workload balancing among the cores is achieved.

4.1 The Performance for Workload Prediction

In order to test the performance of the proposed prediction method in Eq. (6), we compare it with PAST [7] and MA [7] (using the previous two frames) in Table 2. The prediction error ratio (ER) is defined by

$$ER = \frac{1}{N}\sum_{N}\frac{\left|\Theta - \overset{\wedge}{\Theta}\right|}{\Theta} \qquad (9)$$

From the table, we can see that the proposed mode has less error ratio in all test sequences, which means a more accurate prediction than PAST and MA. On average, the prediction error ratio of our method is 1.15 %, which is smaller than that of PAST (1.23 %) and MA (1.45 %). In all, the proposed workload prediction method reduces the error ratio by 13.33 % and 30.03 % compared with PAST and MA, respectively.

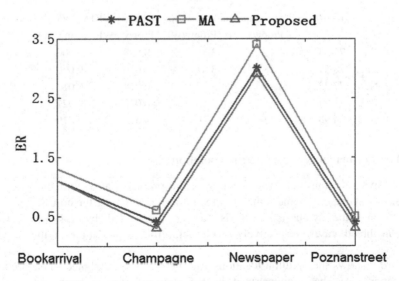

Fig. 4. The prediction error ratio for different methods

Figure 4 displays the prediction error ratio for these three methods visually. It is obviously that the error ratio line of our method is under other two lines, so the proposed depth assistant method outperforms the representative workload prediction method PAST and MA.

4.2 The Performance for View Extrapolation

Tables 3 and 4 compare the performance for view extrapolation. As shown in the table, relative to *Propw/oWB*, the proposed workload prediction reduces the workload difference among the cores by 86.25 % and 74.90 % for 2-thread and 4-thread cases, respectively. It further accelerates the synthesis speed of *Propw/oWB* by 10.1 % and 8.0 % at maximum for 2-thread and 4-thread cases, respectively. Speedup approaching to the theoretical maximum given by Amdahl's law [15] is achieved with negligible quality degradation.

Table 3. Performance comparison for n = 2.

Seq.	*WDRR* (%)	Speedup Ratio vs. *VSRS*		ΔPSNR (dB) vs. *VSRS*	
		Propw/oWB	Propw/WB	Propw/oWB	Propw/WB
Boo.	81.13	1.79	1.88	-0.002	0.000
Cha.	90.02	1.81	1.86	0.000	0.000
New.	86.97	1.78	1.86	-0.009	-0.008
Poz.	86.88	1.79	1.97	-0.010	-0.010
Ave.	**86.25**	1.79	**1.89**	-0.005	**-0.005**

Table 4. Performance comparison for n = 4.

Seq.	*WDRR* (%)	Speedup Ratio vs. *VSRS*		ΔPSNR (dB) vs. *VSRS*	
		Propw/oWB	Propw/WB	Propw/oWB	Propw/WB
Boo.	72.54	3.13	3.38	-0.008	-0.006
Cha.	78.51	3.18	3.31	0.000	0.000
New.	77.32	3.14	3.24	-0.014	-0.012
Poz.	71.24	3.07	3.25	-0.023	-0.021
Ave.	**74.90**	3.13	**3.30**	-0.011	**-0.010**

4.3 The Performance for View Interpolation

Tables 5 and 6 demonstrate the efficiency of the proposed algorithm for view interpolation. On average, it reduces the workload difference among the cores by 90.52 %/ 78.96 % and the synthesis speed is further accelerated by 7.4 %/6.5 % for 2-thread/4-thread cases, respectively, with negligible degradation in the synthesis quality.

Figure 5 shows the partition locations and the workload assigned to each core for "Newspaper". As that demonstrated in the figure, with the proposed workload

Table 5. Performance comparison for n = 2.

Seq.	WDRR (%)	Speedup Ratio vs. *VSRS*		ΔPSNR (dB) vs. *VSRS*	
		Propw/oWB	Propw/WB	Propw/oWB	Propw/WB
Boo.	86.82	1.82	1.88	-0.026	-0.012
Cha.	92.13	1.79	1.86	0.000	0.004
New.	87.37	1.71	1.85	0.001	-0.007
Poz.	95.75	1.73	1.96	-0.002	0.000
Ave.	**90.52**	1.76	**1.89**	-0.007	**-0.004**

Table 6. Performance comparison for n = 4

Seq.	WDRR (%)	Speedup Ratio vs. *VSRS*		ΔPSNR (dB) vs. *VSRS*	
		Propw/oWB	Propw/WB	Propw/oWB	Propw/WB
Boo.	72.42	3.22	3.35	0.002	0.002
Cha.	81.95	3.13	3.30	0.001	0.004
New.	76.17	3.09	3.29	0.000	-0.010
Poz.	85.29	2.97	3.26	-0.014	-0.016
Ave.	**78.96**	3.10	**3.30**	-0.003	**-0.005**

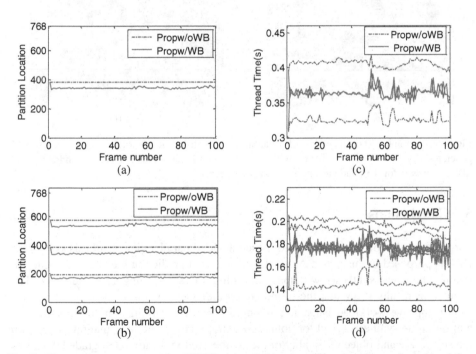

Fig. 5. Newspaper sequence under *INTER*: (a) and (b) are partition location and workload for 2-thread case, respectively; (c) and (d) are partition location and workload for 4-thread case, respectively

prediction method, the partition locations will be adjusted adaptively and the workload among the cores is balanced during the view synthesis process.

4.4   The Subjective Quality

Figure 6 compares the subjective quality for the view synthesized by VSRS and Propw/WB for "PoznanStreet" (a maximal drop in the objective quality). As shown in the figure, no quality degradation can be observed visually.

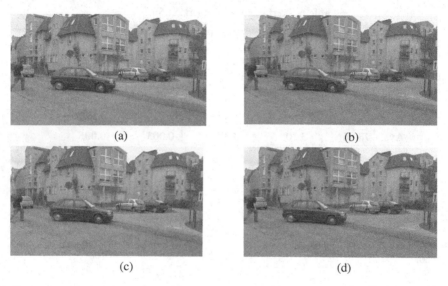

(a)   (b)

(c)   (d)

Fig. 6. (a) Frame 100 of the original PoznanStreet view 4; Frame 100 of the synthesized view 4 generated by: (b) VSRS; (c) Propw/WB for 2-thread case under *INTER* synthesis; and (d) Propw/WB for 4-thread case under *INTER* synthesis;

5   Conclusions

Targeting multi-core platforms, a workload balanced parallel view synthesis system for FTV is proposed in this paper. A novel motion-based prediction method is designed to predict the view synthesis workload according to the warping characteristics. The simulation results show that the proposed parallel view synthesis system can significantly improve workload balancing among the cores and brings an additional acceleration in the synthesis speed for both view extrapolation and interpolation. Also, both the subjective and objective quality of the synthesized view are not degraded, which is very beneficial to fast view synthesis on multi-core platforms.

Acknowledgment. This work was supported by project of NSFC 61371138, China.

References

1. Tanimoto, M., Tehrani, M.P., Fujii, T., Yendo, T.: Free-Viewpoint TV. In: Signal Processing Magazine. IEEE, vol. 28, no. 3, pp. 67–76 (2011)
2. Tech, G., Wegner, K., Chen, Y., Yea, S.: 3D-HEVC test model 5. In: JCT-3 V Doc. JTC3 V-E1005, 5th Meeting, Vienna (2013)
3. Fehn, C.: A 3D-TV approach using depth-image-based rendering (DIBR). In: Proceedings of VIIP, Benalmádena, vol. 3, pp. 84–88 (2003)
4. Jung, J.I., Ho, Y.S.: Parallel view synthesis programming for free viewpoint television. In: 2012 IEEE International Conference on Signal Processing, Communication and Computing, Hongkong, pp. 88–91 (2012)
5. Shin, H.C., Kim, Y.J., Park, H., Park, J.I.: Fast view synthesis using GPU for 3D display. IEEE Trans. Consum. Electron. 54(4), 2068–2076 (2008)
6. Sihn, K.H., Baik, H., Kim, J.T., Bae, S., Song, H.J.: Novel approaches to parallel H.264 decoder on symmetric multicore systems. In: IEEE International Conference on Acoustics, Speech and Signal Processing, Taipei, pp.2017–2020 (2009)
7. Pering, T., Burd, T., Brodersen, R.: The simulation and evaluation of dynamic voltage scaling algorithms. In: International Symposium on Low Power Electronics and Design, Monterey, pp. 76–81 (1998)
8. Jin, X., Goto, S.: Hilbert transform based workload prediction and dynamic frequency scaling for power efficient video encoding. IEEE Trans. CAD Integr. Circ. Syst. 31(5), 649–661 (2012)
9. Sinha, A., Chandrakasan, A.P.: Dynamic voltage scheduling using adaptive filtering of workload traces. In: Fourteenth International Conference on VLSI Design, pp. 221–226. IEEE, Bangalore (2001)
10. Sung, Y.S., Hosik, S., Ju, J.Y., Man, R.Y.: Inter-view consistent hole filling in view extrapolation for multi-view image generation. In: IEEE International Conference on Image Processing, Paris, pp. 2883–2887 (2014)
11. View Synthesis Reference Software (VSRS 3.5) in Technical report ISO/IEC JTC1/SC29/WG11, Guangzhou (2010)
12. 3D Video Databse in Mobile 3DTV. http://sp.cs.tut.fi/mobile3dtv/stereo-video/
13. MPEG-FTV Test Sequence. http://www.tanimoto.nuee.nagoyau.ac.jp/mpeg/
14. Rusanovskyy, D., Müller, K., Vetro, A.: Common test conditions of 3DV core experiments. In: JCT-3 V Doc. JCT3 V-D1100, 4th meeting, Stockholm (2012)
15. Rodgers, D.P.: Improvements in multiprocessor system design. ACM SIGARCH Comput. Architect. News Arch. 13(3), 225–231 (1985)

Graph Cuts Stereo Matching Based on Patch-Match and Ground Control Points Constraint

Xiaoshui Huang[1,2]($\boxtimes$), Chun Yuan[2], and Jian Zhang[1]

[1] University of Technology Sydney, Sydney, Australia
[2] Graduate School at Shenzhen, Tsinghua University, Shenzhen, China
xiaoshui.huang@student.uts.edu.au

Abstract. Stereo matching methods based on Patch-Match obtain good results on complex texture regions but show poor ability on low texture regions. In this paper, a new method that integrates Patch-Match and graph cuts (GC) is proposed in order to achieve good results in both complex and low texture regions. A label is randomly assigned for each pixel and the label is optimized through propagation process. All these labels constitute a label space for each iteration in GC. Also, a Ground Control Points (GCPs) constraint term is added to the GC to overcome the disadvantages of Patch-Match stereo in low texture regions. The proposed method has the advantage of the spatial propagation of Patch-Match and the global property of GC. The results of experiments are tested on the Middlebury evaluation system and outperform all the other PatchMatch based methods.

Keywords: Patch-Match · Graph cuts · GCP · Stereo matching · Low texture

1 Introduction

Depth reconstruction is a technique that estimates depth from images. It is a baseline technique for building 3D objects and environments in stereo vision, but conducting reconstruction from several cameras is always an challenging problem. The performance of the depth reconstruction, including such elements as accuracy and efficiency, will directly impact the quality of computer vision, graphics and multimedia applications. One solution to the depth reconstruction problem is to conduct stereo matching from image pairs for depth estimation. This is an attractive research topic which has been applied to many applications.

The existing stereo matching methods are divided into two categories: global algorithms and local algorithms [12]. Global methods [7,13,15] consider the problem as an energy function that is solved by optimization methods. Although such methods have been shown to achieve precise disparity estimation in discontinuous regions, many challenging problems remains. Local methods [17,18] estimate disparity at a given centre pixel using colour or intensity values in the

© Springer International Publishing Switzerland 2015
Y.-S. Ho et al. (Eds.): PCM 2015, Part II, LNCS 9315, pp. 14–23, 2015.
DOI: 10.1007/978-3-319-24078-7_2

user-defined window. However, these algorithms have weaknesses in respect of disparity discontinuity and low texture areas, because they assume that the area inside the user-defined window is locally frontal-parallel.

Significant progress has recently been made in accuracy of stereo vision. One breakthrough that has obtained good accuracy results is a global method using GC. Another is the use of 3D models [2,3,11], in which a local 3D disparity plane is estimated for each pixel and accurate photo consistency between matching pixels is measured even with large matching windows. Sub-pixel precision is achieved with this method.

Even though GC obtains good results to some extent, it can only obtain disparity accuracy in pixel level accuracy. How to integrate such a 3D model with GC to achieve sub-pixel precision accuracy is a significant research topic. While stereo with standard 1D discrete disparity labels [6,14] is directly solved by discrete optimizers such as GC [4,7] and belief propagation(BP) [8], such approaches cannot be directly used for continuous 3D labels due to the infinite label space $(a, b, c) \in R^3$.

Many papers related to this research topic has been published [2,9,13]. Patch-Match [3] is a successful method used in [2,13] to efficiently infer an accurate 3D disparity plane using spatial propagation; each pixel's candidate plane is randomly selected, refined and then propagated to neighboring pixels [3]. In [2], Patch-Match is combined with BP to create an efficient stereo matching method, PMBP, for pairwise Markov random fields(MRF). However, BP is a MRF optimization method which is considered to be a sequential optimizer, in which each node is improved individually while other conditions are kept in the current state. In contrast, GC improves all nodes simultaneously by accounting for interactions across nodes, and this global property helps GC to avoid local minima [15]. A local shared labels strategy is proposed in [16] to combine GC and Patch-Match. This method has high computation complexity because of its still large label space, though it achieves better results than PMBP [2].

Patch-Match achieves great performance in complex texture regions, but low texture regions remain a challenge for these 3D model methods because of their local propagation strategy. Methods based on GCPs have recently achieved great improvements in low texture regions, but it is a challenging problem to acquire high quality GCPs. Occlusions are another challenge for the accurate computation of visual correspondence. Occluded pixels are visible in only one image, with no corresponding pixel in other images. Non-local (NL) algorithm [17] uses non-local strategy to detect these occlusion pixels, but it needs to compute a separate disparity map for left and right image.

In this paper, we focus on the accuracy of depth estimation in low texture, and our work also concentrates on complex texture regions, such as regions full of slant and edge areas, which is also a significant challenge in stereo matching. We propose a new label space computation method that effectively uses the global strengths of graph cut and local depth continuous strength of Patch-Match. Our method demonstrates great ability to handle complex texture areas as well as low texture areas. As discussed above, GC and Patch-Match cannot be directly combined, so a powerful method is required to effectively decide the label space for GC in infinite R3 sub-pixel space. Our new method has the ability to use both strengths and obtains good results at relatively high speed. At the same time, GCPs are used to improve the accuracy of disparity in low texture and occlusion areas. We propose a new GCP term called soft GCPs, which is unlike other GCP methods. It varies on different kind of GCPs.

Our contributions comprise a new method that incorporates spatial propagation of Patch-Match into GC-based global optimization and a new GCP constraint term that is used to guide the process of matching in low texture and occlusion regions. The major contributions of our method are as follows:

(1) A new label space computation method is proposed. We use Patch-Match to randomly select the disparity label of each pixel and compare with its 8 neighbors; the optimal label that produces minimum pixel dissimilarity is the current label. It is quite simple, very effective and obtains relatively good results. It has the advantages of Patch-Match and effectively combines Patch-Match and GC by considering a small number of labels in such large R3 infinite label space. The new label space computation method is reasonable because depth is locally continuous in most part of the images.
(2) A new GCPs constraint term is added to GC. We define the GCPs constraint term when a pixel is GCP, or we omit the term when the pixel is not GCP. This strategy is effective because it does not need a huge amount of computation on the GCP of every pixel. The GCPs constraint term is considered to be a soft constraint.It has the ability to improve energy minimization to achieve good disparity results.

2   Proposed Method

This section describes the proposed stereo matching method. Given two input images I_L and I_R, our purpose is to estimate the disparity maps of both images.

2.1   Problem Formulation

Sufficient label space is prerequisite for conducting Graph cuts(GC). As the main contribution of this paper, we develop a effective label selection algorithm to combine Patch-Match and GC. Initially, the optimal label for each pixel is selected in a local 3×3 window size, then the whole range of pixel label forms the label space of the whole image. Due to propagation steps in obtaining optimal label for each pixel, our label computation method has the advantages of obtaining more effective label space and maintaining good matcher.

Existing global methods (i.e., graph cuts) have achieved remarkable results, but the capability of the traditional Graph cuts(GC) stereo model is limited. To reduce ambiguities in matching, additional information is required to formulate an accurate model. In this paper, the GCPs constraints(G) are integrated as additional regularization terms to obtain a precise disparity estimation(D) for a scene with low texture characteristics. According to Bayes rule, the posterior probability over D given G and D_L is

$$p(D|D_L, G) = \frac{p(D_L, G|D)p(D)}{p(D_L, G)} \tag{1}$$

where D_L is the initial disparity map conducted by Patch-Match. Due to the initial disparity map D_L and G are constant during each iteration and independent, $p(D_L, G)$ is constant. Therefore,

$$p(D|D_L, G) \propto p(D_L, G|D)p(D) \propto p(D_L|D)p(G|D)p(D) \qquad (2)$$

Maximizing this posterior is equivalent to minimizing its negative log likelihood [7], therefore disparity map (D) is obtained by minimizing the following energy function:

$$E(D) = E_d + E_s + E_G \qquad (3)$$

2.2   Disparity Map Computation

Label Space Computation. The method of computing label space is divided into two steps: random selection and propagation. First, we randomly select the disparity label l_i of each pixel. After selection, we form the initial label space $L = (l_1, l_2, ..., l_i, ...l_n)$, where n is the number of pixels. Second, we optimize the initial label at each pixel, using its neighbors. Our goal in the second step is to form the optimal label space $S(S \subset L)$,

$$S = \cup_{i=1...k}\{s_i\} \qquad (4)$$

where k is the dimension of S and s_i is the optimal label for p_i. s_i is obtained by selecting the optimal label that has the minimum ρ (through formulation (10)) compared with its 8-connected neighbors and its own last label,

$$s_i = l(\min_{l_j \in N_i} \{\rho(p_i, p_i')\}) \qquad (5)$$

where N_i is the 3×3 labels centred in pixel p_i, p_i' is the projection of p_i with disparity l_j, $l(\cdot)$ is the label in N_i that has minimum ρ.

From formulation (4) and (5), we can obviously find that S is usually smaller than L, which is rational in reality. The new method obtains label space by propagation which is using the continuous property of neighbours' labels.

GC Settings Data Term. To measure photo-consistencies, we use a data term that was recently proposed by [18]. In the term, we use the 3D model to represent disparity. A pixels disparity d_i can be represented as $d_i = a_i x + b_i y + c_i$. Therefore, the objective of finding the best disparity for each pixel is to search its disparity plane $f_p = (a_i, b_i, c_i) \in R^3$. We find the best f_p though an energy method (e.g., GC) by minimizing $E(f_p)$. With the disparity f_p, we project the pixel $q = (q_x, q_y)^T$ in the left image to the right image by using the projection function:

$$\phi_{pq} = q \quad (a_p q_x \mid b_p q_y \mid c_p, 0)^T \qquad (6)$$

Here, we only suppose it has x-axis disparity, because we use the Middlebury dataset to test our algorithm. If there is y-axis disparity, the $\phi_{pq} = q - (0, a_p q_x + b_p q_y + c_p)^T$; if there is both axis disparity, this function becomes

$$\phi_{pq} = q - (a_{px} q_x + b_{px} q_y + c_{px}, a_{py} q_x + b_{py} q_y + c_{py})^T \qquad (7)$$

The data term of pixel p in the left image is defined as:

$$E_d = \sum_{q \in W_p} \omega(p, q) \rho(q, q') \qquad (8)$$

Where W_p denotes a window centred on pixel p, pixel q' is the projection of pixel q, $q' = \phi_{pq}$ .The weight implements the adaptive support window proposed in [18], and defined as

$$\omega(p, q) = e^{-\|I^q - I^p\|_1 / \gamma} \qquad (9)$$

The function $\rho(q, q')$ represents the pixel dissimilarity degree between q and q'. It is calculated by the following formulation,

$$\rho(q, q') = (1 - \alpha) \cdot \| I_q - I_{q'} \|_1 + \alpha \| \nabla I_q - \nabla I_{q'} \|_1 \qquad (10)$$

where I is the intensity of the pixel and ∇I is the gradient of the pixel. In our experiment, we use Middlebury datasets, so there is only disparity on x-axis, thus $\nabla I = \nabla_x I$. If there is disparity on the y-axis, $\nabla I = \nabla_y I$; if on both axes, we define $\nabla I = sqrt(\nabla_x I \cdot \nabla_x I + \nabla_y I \cdot \nabla_y I)$.

Smooth Term. For the smooth term, a second-order regularization, curvature-based term [11] is used. It defined as

$$E_s(f_p, f_q) = max(\omega_{pq}, \varepsilon) min(S(f_p, f_q), \tau) \qquad (11)$$

where ω_{pq} is a parameter which is defined in (9), it help to have a more robust smooth term. $S(f_p, f_q)$ estimates the penalty of the discontinuity between label f_p and label f_q in the same pixel. The penalty function is defined as

$$S(f_p, f_q) = |d_p(f_p) - d_p(f_q)| + |d_q(f_p) - d_q(f_q)| \qquad (12)$$

GCP Constraint Term. GCPs are usually referred as highly reliable matched points that are used to guide the local structure of pixels. They improve the performance of GC by dealing with the problem of matching ambiguities in repetitive or low-texture areas. Because obtaining precision GCPs is a difficult task, GCPs constraint term is considered as a soft constraint. The GCPs constraint term is defined as

$$E_G = \lambda_G \sum_{p \in I_L} \lambda_{sg} \Psi(f_p, \hat{f}_p) \qquad (13)$$

where $\Psi(f_p, \hat{f}_p)$ penalizes the disparity assignment that diverges from the Patch-Match. λ_G is a parameter that controls the GCPs constraint term and λ_{sg} controls GCPs term when the current pixel is GCP and when is not GCP,

$$\lambda_{sg} = \begin{cases} 0, & \text{if the pixel is not GCP} \qquad (14) \\ 1, & \text{if the pixel is GCP} \qquad (15) \end{cases}$$

In this paper, our robust penalty function $\Psi(f_p, \hat{f}_p)$ is derived from the Total Variance model [1] as

$$\Psi(a,b) = -In(1-\eta)exp(\frac{-\mid a-b \mid}{r} + \eta) \qquad (16)$$

where r and η are parameters that control the sharpness and upper-bound of the robust function.

We use the methods in [9,17] to obtain GCPs, and the GCPs in I_L are obtained from several local stereo-matching algorithms. These methods include the left-right check procedure, which helps to find pixels which are occluded.

GC Optimization. We use the Fusion-move [8,11]method for optimization, which is an extension of the α-expansion algorithm [6]. In Fusion-move, an arbitrary values for each pixel is allowed instead of a fixed value for each optimization, and each optimization can be conducted by parallelization.

Optimization is conducted iteratively. The label space is computed using our label space computation method described in Sect. 3.2. The fusion-move method is then used to optimize the energy formulation (3). After each iteration, we update the last label space into a new one, which becomes the initialization of the next iteration. The optimization is continued until the energy converges or iteration times reach the maximum.

3 Experiments

The proposed method is evaluated on the Middlebury datasets and an explanation of the results is given. In addition, we compare the performance of our algorithm with GCPs and without GCPs. The performance of the proposed method is analysed.

Experiment Setting. The following settings is used throughout the experiments. We use a PC with an I5 CPU (2.50 GHz × 4 cores) and 16G memory. The parameters of our algorithm are set as $(\alpha, \tau, \varepsilon, \lambda_G, \gamma, \eta) = (0.9, 1.0, 0.01, 15, 2, 0.005)$. All labels of last label in formulation (5) is set into 0 when the first iteration of GC.

3.1 Evaluation on the Middlebury Datasets

Table 1 is our selected algorithms in the Middlebury evaluation database, there are all Patch-Match(PM) related methods ([7,16,17]). Our algorithm achieves

Table 1. Quantitative comparison of methods on the Middlebury datasets.

Algorithm	Avg. Rank	Tsukuba all	Venus all	Teddy all	Cones all	Average error
GC-LSL [13]	51.4	2.73	0.36	3.77	7.37	4.19
PMBP [2]	47.6	2.21	0.49	8.57	6.64	4.46
PM-PM [16]	42.2	3.23	0.30	8.27	6.43	4.52
PMF [10]	44.5	2.04	0.49	5.87	6.80	4.06
PM [3]	47.7	2.33	0.39	8.16	7.80	4.59
PM-Huber [5]	48.8	0.22	5.56	5.56	6.69	4.56
Ours	**40.4**	2.19	**0.21**	**3.11**	**4.32**	**3.73**

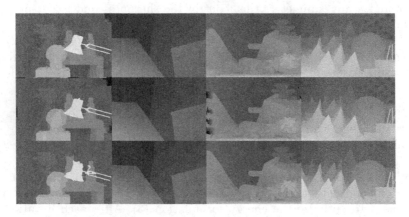

Fig. 1. The results of disparity maps of GC-LSL(row 1), PMBP(row 2), and the proposed method(row 3). The datasets are Tsukuba, Venus, Teddy, Cones, which is displayed in a row for each method.

the current best average error (3.73) and average rank (40.4) of all the PM based methods on threshold 1, and the proposed method ranked better than the others. In addition, our algorithm gets best results on the Venus, Teddy and Cones datasets.

We are closely compare our method to the related methods GC-LSL [13] and PMBP [2]. From the Table 1, we can see that our results achieve better accuracy compared to GC-LSL and PMBP on the four datatasets on threshold 1. In threshold 0.75, our algorithm can also obtain better results on the last three datasets(Venus, Teddy, Cones) than the others(The results can be seen on supplement materials). Because of our GCPs constraint term, low texture regions obtain guided information when they are finding matching pixels. Also, our new space computing method combine the strengths of PatchMatch with GC. The both strengths contribute the improvement of accuracy. Figure 1 illustrates the results evaluated on the Middlebury system. Row 3 shows our disparity results, from which we can clearly see that our results looks smoother and less artificial.

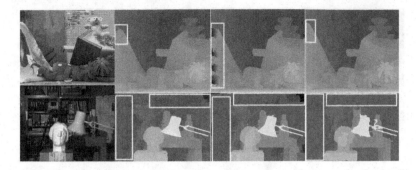

Fig. 2. Example regions that better than other methods. GC-LSL (col 1), PMBP (col 2), and the proposed method (col 3).

Figure 2 is the detailed comparison with GC-LSL and PMBP on Teddy and Tsukuba datasets. From the comparison results, we can see that our results (col 3) in the red rectangular region clearly obtain better results, demonstrating that our method achieves good results in both complex texture regions(row 1) and low texture regions(row 2).

The proposed method achieves better accuracy primarily because we find one optimal label space in each iteration of GC. It is a new way to compute label space, which illustrates that it is effective to conduct expansion steps in GC. Also, an additional GCP constraint term is added to our energy formulation which contributes to the improvements in low texture regions. As we discussed above, our algorithm can not obtain the best results compared to our related methods under threshold 0.75 in the first dataset of Middlebury (Tsukuba). It is because our GCPs are computed using some fast pixel level accuracy methods, while existed sub-pixel level accuracy methods have much slow speed and are not robust to get a same disparity in sub-pixel level accuracy. Such GCPs may have inaccurate points in sub-pixel level. It influences the accuracy. In real application, if we can use high accuracy device to obtain sparse high level accuracy GCPs, this drawback will be overcome.

3.2 Disparity Results with and Without GCPs

Our method contains two main contributions to stereo matching. One is a new graph cut combination with Path-Match, and the other is the introduction of GCPs into our new global method. In this section, we test them separately.

Figure 3 shows the disparity result of our algorithm both with and without considering GCPs. From the results, we can see that GCPs contributes significantly to improvement in the accuracy of the final disparity map. Table 2 gives the quantitative evaluation on Middlebury evaluation system. From the table, we can see that the disparity error of Teddy without considering GCPs is 6.3, while the disparity error considering the GCPs constraint term guide the

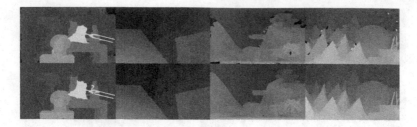

Fig. 3. Disparity map without GCPs(row 1) and with GCPs(row 2).

Table 2. Quantitative comparison of disparity with and without GCPs

Algorithm	Avg. Rank	Tsukuba			Venus			Teddy			Cones			Average error
		ncc	all	disc	ncc	all	disc	ncc	all	disc	ncc	all	disc	
without GCPs	47.6	2.09	2.37	9.90	0.31	0.35	2.63	2.88	6.3	8.6	2.97	5.75	6.2	5.37
with GCPs	25.6	1.63	2.17	8.71	0.15	0.19	2.13	1.91	2.29	5.47	1.32	2.02	3.6	2.64

disparity computation is 2.29, which contributes to the improved results. The performance in other three Middlebury datasets (Tsukuba, Venus and Cones) has different extend improvement in its disparity results.

3.3 Performance of Proposed Method

Graph cuts generate the highest computation cost step in our algorithm. The proposed algorithm obtains higher computation speed than GC-LSL, because more effective label space is obtained for the GC. The proposed method iterates six times in each experiment which costs about 15 min. From their papers [2,13], we can find that, GC-LSL requires about 70 min when CPUs are used and PMBP need 16 min either.

4   Conclusion

A new method stereo matching method is proposed that obtains accuracy results in both the complex texture regions and low texture regions. The new algorithm integrates the Patch-Match stereo method into the graph cuts optimization method. In addition, a new GCPs constraint term is added to energy formulation of GC to obtain more accurate results in low texture regions.

Acknowledgement. This work is supported by the High Technology Development Program of China(863 Program), under Grant No.2011AA01A205, National Significant Science and Technology Projects of China, under Grant No.2013ZX01039001-002-003; by the NSFC project under Grant Nos. U1433112, and 61170253.

References

1. Barnes, C., Shechtman, E., Goldman, D.B., Finkelstein, A.: The generalized patch-match correspondence algorithm. In: Daniilidis, K., Maragos, P., Paragios, N. (eds.) ECCV 2010, Part III. LNCS, vol. 6313, pp. 29–43. Springer, Heidelberg (2010)
2. Besse, F., Rother, C., Fitzgibbon, A., Kautz, J.: PMBP: patchmatch belief propagation for correspondence field estimation. Int. J. Comput. Vis. **110**, 1–12 (2013)
3. Bleyer, M., Rhemann, C., Rother, C.: Patchmatch stereo-stereo matching with slanted support windows. In: BMVC, vol. 11, pp. 1–11 (2011)
4. Boykov, Y., Kolmogorov, V.: An experimental comparison of min-cut/max-flow algorithms for energy minimization in vision. IEEE Trans. Pattern Anal. Mach. Intell. **26**(9), 1124–1137 (2004)
5. Heise, P., Klose, S., Jensen, B., Knoll, A.: Pm-huber: Patchmatch with huber regularization for stereo matching. In: IEEE International Conference on Computer Vision (ICCV), pp. 2360–2367. IEEE (2013)
6. Kolmogorov, V., Zabih, R.: Computing visual correspondence with occlusions using graph cuts. In: IEEE International Conference on Computer Vision (ICCV), vol. 2, pp. 508–515. IEEE (2001)
7. Kolmogorov, V., Zabin, R.: What energy functions can be minimized via graph cuts? IEEE Trans. Pattern Anal. Mach. Intell. **26**(2), 147–159 (2004)
8. Lempitsky, V., Rother, C., Blake, A.: Logcut - efficient graph cut optimization for markov random fields. In: International Conference on Computer Vision, pp. 1–8 (2007)
9. Liu, J., Li, C., Mei, F., Wang, Z.: 3D entity-based stereo matching with ground control points and joint second-order smoothness prior. Vis. Comput. **31**, 1–17 (2014)
10. Lu, J., Yang, H., Min, D., Do, M.N.: Patchmatch filter: efficient edge-aware filtering meets randomized search for fast correspondence field estimation. In: IEEE Conference on Computer Vision and Pattern Recognition (CVPR), pp. 1854–1861. IEEE (2013)
11. Olsson, C., Ulén, J., Boykov, Y.: In defense of 3D-label stereo. In: IEEE Conference on Computer Vision and Pattern Recognition (CVPR), pp. 1730–1737. IEEE (2013)
12. Scharstein, D., Szeliski, R.: A taxonomy and evaluation of dense two-frame stereo correspondence algorithms. Int. J. Comput. Vis. **47**(1–3), 7–42 (2002)
13. Taniai, T., Matsushita, Y., Naemura, T.: Graph cut based continuous stereo matching using locally shared labels. In: IEEE Conference on Computer Vision and Pattern Recognition (CVPR), pp. 1613–1620, June 2014
14. Wang, L., Yang, R.: Global stereo matching leveraged by sparse ground control points. In: IEEE Conference on Computer Vision and Pattern Recognition (CVPR), pp. 3033–3040. IEEE (2011)
15. Woodford, O., Torr, P., Reid, I., Fitzgibbon, A.: Global stereo reconstruction under second-order smoothness priors. IEEE Trans. Pattern Anal. Mach. Intell. **31**(12), 2115–2128 (2009)
16. Xu, S., Zhang, F., He, X., Shen, X., Zhang, X.: PM-PM: patchmatch with potts model for object segmentation and stereo matching. IEEE Trans. Image Process. **24**(7), 2182–2196 (2015)
17. Yang, Q.: A non-local cost aggregation method for stereo matching. In: IEEE Conference on Computer Vision and Pattern Recognition (CVPR), pp. 1402–1409, June 2012
18. Yoon, K.J., Kweon, I.S.: Adaptive support-weight approach for correspondence search. IEEE Trans. Pattern Anal. Mach. Intell. **28**(4), 650–656 (2006)

Synthesized Views Distortion Model Based Rate Control in 3D-HEVC

Songchao Tan[1(✉)], Siwei Ma[2], Shanshe Wang[2], and Wen Gao[2]

[1] School of Computer Science and Technology,
Dalian University of Technology, Dalian 116024, China
sctan@jdl.ac.cn
[2] Institute of Digital Media, Peking University, Beijing 100871, China

Abstract. In this paper, we propose a synthesized views distortion model based rate control algorithm for the high efficiency video coding (HEVC) based 3D video compression standard. The major contributions of the paper include the following two aspects. Firstly, we investigate the distortion dependency between the synthesized views and the coded views including texture video and depth maps. Then we propose a synthesized views distortion model for 3D-HEVC, and based on the distortion model an efficient joint bit allocation scheme is proposed. Experimental results show that the proposed rate control algorithm achieves better performance on both the coded texture views and synthesized views. The maximum overall (including all coded texture views and all synthesized views) performance improvement can be up to 14.4 % and the average BD-rate gain is 6.9 %. Moreover, it can accurately control the bitrate to satisfy the total bitrate constraint.

Keywords: Rate control · Bit allocation · 3D video coding

1 Introduction

The 3D extension of High Efficiency Video Coding [1] (3D-HEVC) was developed by the Joint Collaborative Team on 3D Video Coding Extension Development (JCT-3 V) led by ISO/IEC MPEG and ITU-T VCEG. 3DV video coding aims at coding the visual information of a 3D scene that usually contains multi-view texture data and its corresponding depth information [2]. In 3D-HEVC, one view is selected as a base view which is coded independently of the other views to provide backward compatibility to HEVC decoders. Other views (termed as dependent views) are coded with inter-view prediction using the base view or other reference views to reduce the redundancy between views.

As an important module of video encoder, rate control (RC) is employed to regulate the bit rate meanwhile to guarantee a good video quality. For each video coding standard, rate control is always a hot research topic and many different rate control schemes for different video coding standards have been proposed, such as quadratic model for H.264 [3] and URQ model [4], R-lambda [5] and rate-GOP [6] for HEVC. However for 3D-HEVC, rate control becomes more complicated. Different from the traditional 2D video coding standards, 3D-HEVC utilizes the depth map to generate the

© Springer International Publishing Switzerland 2015
Y.-S. Ho et al. (Eds.): PCM 2015, Part II, LNCS 9315, pp. 24–32, 2015.
DOI: 10.1007/978-3-319-24078-7_3

synthesized virtual views. The coding quality of synthesized views depends on the quality of texture video and depth maps [7]. As such, it is important to balance the bit allocation for texture video and depth maps to get better quality of synthesized views. And many other techniques are adopted to improve the coding performance. All these techniques bring great challenges to establish an accurate rate distortion (R-D) model and bit allocation scheme in rate control for 3D-HEVC.

In the literature, several rate control schemes for 3D-HEVC are proposed. Two representative rate control methods, URQ and R-lambda have been proposed for 3D-HEVC in [8,9], which are the extension of the methods in HEVC. In order to get better performance, depth maps-based inter-views MAD prediction is proposed to improve the prediction accuracy of the to-be-generated bits for the current unit. However, there is still large room for R-D performance improvement. In our previous work [10], we proposed an adaptive rate control scheme for 3D-HEVC. The algorithm performance including bit rate mismatch and R-D performance is significantly improved compared to the above two algorithms.

In this paper, we further propose a novel rate control scheme based on the synthesized views distortion model in 3D-HEVC. Firstly, we investigate the distortion dependency between the synthesized views and the input texture video and depth maps, and formulate a distortion model for synthesized views. Secondly, based on this model, the bit allocation scheme for texture video and depth maps is formulated as an optimization problem.

The rest of this paper is organized as follows. In Sect. 2, the R-D characteristics of both the coded views and synthesized views are investigated. In Sect. 2.1, the R-D model for coded texture views is proposed. A view synthesis distortion model to characterize the distortion dependency of the texture video and the depth maps on the synthesized virtual views is investigated in Sect. 2.2. In Sect. 2.3, an effective joint bit allocation based rate control scheme is designed for 3D-HEVC. In Sect. 3, the experimental results are given to demonstrate the efficiency of the proposed RC algorithm. Finally, Sect. 4 concludes this paper.

2   Rate and Distortion Analysis in 3D-HEVC

As illustrated in Fig. 1, the texture videos are captured by synchronizing the multiple camera arrays. The associated depth maps are also generated for virtual view synthesis. At the encoder, texture video and depth maps are encoded using 3D-HEVC. At the client side, the arbitrary virtual views are synthesized from the decoded texture video and depth maps. Then the decoded texture video and synthesized views would be presented for viewing at receiver side. Therefore, the quality of coded texture views and the virtual synthesized views needs to be optimized as follows

$$min(D_v + D_c),$$
$$s.t. R_d + R_t \leq R_c, \tag{1}$$

where R_t and R_d are the bit rate of texture video and depth maps, respectively. D_c and D_v are the distortion of texture video and synthesized views respectively.

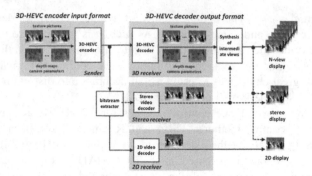

Fig. 1. General framework of the 3D-HEVC [1] system

In order to model the expression in (1), we need to investigate the rate and distortion (R-D) relationship for coded texture views and synthesized views, respectively.

2.1   R-D Model for the Coded Texture Views

To obtain the R-D characteristics of the coded views, we encode the original texture video with 4 quantization parameters (QP) (25, 30, 35, and 40). As an example, the R-D curve of the texture distortion and the bit rate of test sequence '*Newspaper_CC*' are illustrated in Fig. 2.

It can be observed that power functions can be used to fit the R-D points of texture video well.

$$D_t(R_t) \cong \alpha_c R_t^{-\beta_c}, \tag{2}$$

where D_t is the distortion of the coded texture views. R_t is the bit rate of texture views. α_c and β_c are model parameters.

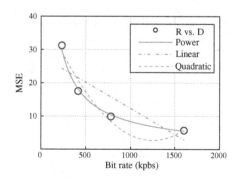

Fig. 2. The relationship between distortion and bit rate of coded texture views, '*Newspaper_CC*'.

2.2   R-D Model for Synthesized Views

To find the best bit budget between the texture and depth, we also need to establish the relationship of bit rate and the synthesized views distortion. We investigate the synthesized views quality influence on the bit rate of texture video (R_t) and the bit rate of depth map (R_d).

In Fig. 3, the quality influence of texture and depth on synthesized views is investigated by changing the texture quantization parameter Q^T from 5 to 45 meanwhile fixing the depth quantization parameter Q^D at 24, 34, 39, 44 and 49 respectively. The quality of synthesized views (D_s) is measured in term of MSE. As shown in Fig. 3, once R_t/R_d is determined, the D_s - R_t/D_s - R_d relationship can be approximated as power expression.

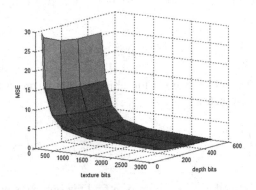

Fig. 3. The R-D surface of synthesized views distortion and bit rate. '*kendo*'.

The D_s - R_t relationship as

$$D_s(R_t) = \alpha_t R_t^{-\beta_t}. \tag{3}$$

And the D_s - R_d relationship as

$$D_s(R_d) = \alpha_d R_d^{-\beta_d}. \tag{4}$$

Therefore from (3) to (4), we get the distortion model for the synthesized views as follows,

$$D_s = \alpha_t R_t^{-\beta_t} + \alpha_d R_d^{-\beta_d}, \tag{5}$$

where D_s is the distortion of synthesized views. R_t and R_d are the bits for texture and depth. α_t, β_t, α_d and β_d are the model parameters.

2.3 A Joint Bit Allocation Based RC Scheme for 3D-HEVC

Rate control for 3D-HEVC needs to solve the bit allocation on texture/depth level, view level and frame level. The optimum bit allocation problem is to effectively distribute the bit budget between texture and depth so that the minimum views synthesis and coded views distortion are achieved. Based on the proposed coded texture views and synthesized views R-D model (5), we formulate the overall quality based optimum bit allocation as

$$\left(R_t^{opt}, R_d^{opt}\right) = \arg\min\left(\alpha_t R_t^{-\beta_t} + \alpha_c R_t^{-\beta_c} + \alpha_d R_d^{-\beta_d}\right),$$

$$s.t. R_t + R_d \leq R_c. \tag{6}$$

ζ is used to represent the proportional relationship between R_t and R_d, defined as

$$\zeta = \frac{R_t^{opt}}{R_d^{opt}}. \tag{7}$$

Therefore, from (6) and (7), we get the objective optimization function with only one variable ζ, as shown below,

$$(\zeta) = \arg\min\left(\alpha_t\left(\frac{\zeta}{1+\zeta}R_c\right)^{-\beta_t} + \alpha_c\left(\frac{\zeta}{1+\zeta}R_c\right)^{-\beta_c} + \alpha_d\left(\frac{1}{1+\zeta}R_c\right)^{-\beta_d}\right). \tag{8}$$

Many optimization methods can be used to find the optimal solution of (8). In this paper, Newton iterative method is used to get the approximate optimal value. The target bit rate for the texture and depth can be expressed as follows

$$R_t = R_c \cdot \frac{\zeta}{1+\zeta}, \tag{9}$$

$$R_d = R_c \cdot \frac{1}{1+\zeta}. \tag{10}$$

In order to estimate these parameters, we first encode the frames in the first GOP. Then the model parameters are calculated by the least square error method.

Based on the optimal target bit rate for the texture and depth, the bit rate ratio between the different views can be further determined by the statistical analysis. In this paper, we use anchor's bits ratio between the base view and the dependent views to allocate the bits for different views.

After allocating the target bit rate for texture/depth level and view level, the target bit rate needs to be allocated for the different frames. The frame level bit allocation is proposed in our previous work [7] as follows

$$R_{n,i} = \begin{cases} R_n^{remain} \cdot \phi & I\,frame \\ R_n^{remain} \cdot w_i \cdot (1 - \phi) & others \end{cases}, \tag{11}$$

$$R_n^{remain} = \frac{bit\,rate}{framerate} \cdot N_n + \left(R_{n-1}^{remain} - R_{n-1}^{actual}\right) / N_{rest}^G, \tag{12}$$

where $R_{n,i}$ is the target bits for i^{th} frame in n^{th} GOP. R_n^{remain} and R_n^{actual} are the target and actual bits in n^{th} GOP. N_n is the numbers of n th GOP's frames. N_{rest}^G is the number of the rest GOP which is not coded. ϕ is a proportion of the I frame in a GOP which is recommended to be 0.4 and 0.25 respectively for the first and the rest GOPs based on experiments. w_i is the weight of the frames in RA hierarchical structure getting from experience.

$$w_i = \begin{cases} 0.07 & if\,(POC\%8 == 0) \\ 0.056 & if\,(POC\%8 == 4) \\ 0.0454 & if\,(POC\%4 == 2) \\ 0.035 & else \end{cases}, \tag{13}$$

where POC denotes Picture Order Count and represents an output order of the pictures in the video stream.

When overflow or underflow occurs, the difference between the target bits and the actual bits in a GOP will be distributed to the rest GOPs averagely.

Trade-off between the output bit rate (R) and the quality (D) of the compressed video are determined by the quantization step size (Qs), which is indexed by quantization parameter (Q). The R-Qs and D-Qs model have been studied extensively for the previous video coding standards such as H.264/AVC and HEVC. Here we use a linear model which is proposed in our previous work [7] as follows

$$R = \alpha \times X/QP, \tag{14}$$

where α is the model parameter. R is the coding rate. QP is the quantization parameter. X is the complexity estimation for the current picture which is computed as following.

$$X = \left(\sum_{i=0}^{n} (w_i \times SAD_i) / \sum_{i=0}^{n-1} (w_i \times SAD_i) \right)^{1-\lambda} \times R_{n-1} \times QP_{n-1}, \tag{15}$$

where n is the current frame number. QP_{n-1} is the quantization parameter of the $(n-1)$ th frame. R_{n-1} is the actual bits of the $(n-1)$ th frame. w_i is defined as:

$$w_i = 0.5^{n-i} / \sum_{i=0}^{n} 0.5^{n-i} \tag{16}$$

3 Experimental Results

To evaluate the proposed 3D-HEVC rate control algorithm, the proposed algorithm is integrated into the reference software HTM10.0. In order to evaluate the performance of the proposed RC algorithm and R-lambda algorithm is utilized for comparison. We have tested our algorithm on all of eight sequences defined in the CTCs (1024 × 768 and 1920 × 1088). Each sequence is composed of three views: the left, the center (coded first) and the right view. After coding, six synthesized views were rendered.

3.1 Control Accuracy

To evaluate the accuracy of the bit rate control, the following measurement is adopted.

$$Error = \frac{|R_{actual} - R_{target}|}{R_{target}} \times 100\%, \tag{17}$$

where $Error$ is the bits error. R_{target} and R_{actual} are the number of target bits and the actual output bits, respectively.

As illustrated in Table 1, it can be seen that the proposed RC algorithm achieves smaller mismatch between target bits and actual output bits. That is because the frame level bit allocation proposed in our previous work [10] is designed more suitable for I-SLICE instead of relying on overflow/underflow handling strategy.

Table 1. Proposed Algorithm R-D Performance Compared With Anchor And R-lambda

	Anchor VS Proposed (BD-Rate)			R-lambda VS Proposed (BD-Rate)			Bits Error (%)	
	Texture PSNR / Texture bits	Syn views PSNR / Total bits	All views PSNR / Total bits	Texture PSNR / Texture bits	Syn views PSNR / Total bits	All views PSNR / Total bits	R-lambda	Proposed
Balloons	2.2 %	2.1 %	2.0 %	-3.8 %	-5.3 %	-5.2 %	1.47 %	0.28 %
Kendo	2.5 %	2.3 %	2.2 %	-4.3 %	-5.7 %	-5.6 %	0.58 %	0.37 %
Newspaper_CC	3.8 %	3.0 %	3.0 %	-2.6 %	-3.4 %	-3.3 %	2.60 %	0.54 %
GT_Fly	0.1 %	-0.8 %	-0.5 %	-8.4 %	-9.5 %	-9.8 %	1.45 %	1.46 %
Pozan_Hall2	4.7 %	3.3 %	3.5 %	-12.0 %	-14.0 %	-14.4 %	2.88 %	0.62 %
Poznan_Street	5.0 %	3.3 %	3.6 %	-4.1 %	-4.6 %	-4.8 %	6.01 %	3.88 %
Undo_Dancer	4.9 %	3.4 %	3.6 %	-6.4 %	-7.7 %	-7.8 %	0.98 %	0.96 %
Shark	3.5 %	2.1 %	2.5 %	-3.7 %	-4.7 %	-4.6 %	1.43 %	1.63 %
1024*768	2.8 %	2.5 %	2.4 %	-3.6 %	-4.8 %	-4.7 %	1.55 %	0.40 %
1920*1088	3.6 %	2.3 %	2.5 %	-6.9 %	-8.1 %	-8.3 %	2.55 %	1.71 %
average	**3.3 %**	**2.4 %**	**2.5 %**	**-5.7 %**	**-6.9 %**	**-6.9 %**	2.18 %	**1.22 %**

3.2 R-D Performance

In order to objectively evaluate the performance of the proposed RC algorithm, R-lambda algorithm proposed in [9] is utilized for comparison. In [9], the target bit rate of each texture video is set as corresponding bit rate in HTM anchor and depth maps are coded with fixed QP as the same as anchor. In the proposed algorithm, the target bit rate for all coded views' bit rate (including three texture videos and three depth maps) is set as anchor's total bit rate.

As illustrated in Table 1 and Fig. 4, we can see that the proposed algorithm shows much better R-D performance than R-lambda for both coded texture views and synthesized views. Based on the proposed synthesized views distortion model, the optimal bit allocation for the texture and depth is achieved. The maximum performance improvement for all views (including coded texture views and synthesized views) can be up to 14.4 % and the average BD-rate gain is 6.9 %.

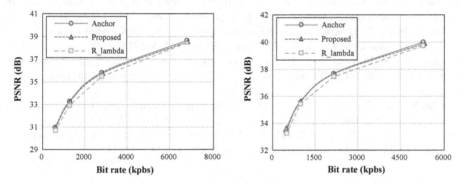

Fig. 4. The R-D curves for all views of the proposed RC scheme compared with anchor and R-lambda '*Undo_Dancer*' and '*GT_Fly*'

Furthermore, two R-D curves are shown in Fig. 4. It can be observed the proposed RC algorithm shows much better R-D performance than R-lambda model for both high bit rate and low bit rate.

4 Conclusions

This paper has presented a synthesized views distortion model based joint bit allocation and rate control method to achieve the best overall quality for 3D-HEVC. The distortion dependency is investigated between the coded views and the synthesized views. The proposed bit allocation method is classified into three levels, namely texture/depth level, view level and frame level. Experimental are conducted on different video sequences and the results show that the proposed method can achieve much better R-D performance than other algorithms for 3D-HEVC.

References

1. Bross, B., Han, W.-J., Sullivan, G.J., Ohm, J.-R., Wiegand, T.: High efficiency video coding (HEVC) text specification draft 8. JCTVC-J1003, Stockholm, July 2012
2. Kauff, P., Atzpadin, N., Fehn, C., Müller, M., Schreer, O., Smolic, A., Tanger, R.: Depth Map Creation and Image Based Rendering for Advanced 3DTV Services Providing Interoperability and Scalability. Signal Processing: Image Communication, Special Issue on 3DTV, pp. 217–234, February
3. Ma, S., Gao, W., Lu, Y.: Rate-distortion analysis for H.264/AVC video coding and its application to rate control. IEEE Trans. Circ. Syst. Video Technol. **15**(12), 1533–1544 (2005)
4. Choi, H., Nam, J., Yoo, J., Sim, D., Bajić, I.V.: Rate control based on unified RQ model for HEVC. JCT-VC of ITU-T SG16 WP3 and ISO/IEC JTC1/SC29/WG11, JCT-VC H0213 (m23088), San José, CA, USA, February 2012
5. Li, B., Li, H., Li, L., Zhang, J.: lambda Domain Rate Control Algorithm for High Efficiency Video Coding. IEEE Trans. Image Process. **23**(9), 3841–3854 (2014)
6. Wang, S., Ma, S., Wang, S., Zhao, D., Gao, W.: Rate-GOP based rate control for high efficiency video coding. IEEE J. Sel. Top. Sign. Process. **7**(6), 1101–1111 (2013)
7. Ma, S., Wang, S., Gao, W.: Low complexity adaptive view synthesis optimization in HEVC based 3D video coding. IEEE Trans. Multimedia **16**(1), 266–271 (2014)
8. Lim, W., Sim, D., Bajić, I.V.: JCT3 V – Improvement of the rate control for 3D multi-view video coding. ISO/IEC JTC1/SC29/WG11, JCT3 V-C0090, Geneva, Switzerland, January 2013
9. Lim, W., Sim, D., Bajić, I.V.: JCT3 V –The rate control schemes for 3D multi-view video coding. ISO/IEC JTC1/SC29/WG11, JCT3 V-D0111, Incheon, KR, April 2013
10. Tan, S., Si, J., Ma, S., Wang, S., Gao, W.: Adaptive Frame Level Rate Control in 3D-HEVC. Visual Communication and Image Processing, Malta, December 2014

Efficient Depth Map Upsampling Method Using Standard Deviation

Su-Min Hong$^{(\boxtimes)}$ and Yo-Sung Ho

School of Information and Communications,
Gwangju Institute of Science and Technology (GIST), 123 Cheomdan-Gwagiro,
Buk-Gu, Gwangju 500-712, Republic of Korea
{sumin,hoyo}@gist.ac.kr

Abstract. In this paper, we present an adaptive multi-lateral filtering method to increase depth map resolution. Joint bilateral upsampling (JBU) increases the resolution of a depth image considering the photometric property of corresponding high-resolution color image. The JBU uses both a spatial weighting function and a color weighting function evaluated on the data values. However, JBU causes a texture copying problem. Standard deviation is a measure that is used to quantify the amount of variation or dispersion of a set of data in image. Therefore, it includes an edge information in the each kernel. In the proposed method, we decrease the texture copying problem of the upsampled depth map by using adaptive weighting functions are chosen by the edge information. Experimental results show that the proposed method outperformed compared to the other depth upsampling approaches in terms of bad pixel rate.

Keywords: Depth map upsampling · Joint bilateral upsampling

1 Introduction

Accurate and high resolution depth sensing is an issue of the importance in a number of applications, including 3DTV, image based rendering, view synthesis, among many others. Various methods for acquisition of depth information have been researched and can be classified into two types: a passive sensor based method and an active sensor-based method. Passive depth sensing uses multi-view images to calculate depth information [1]. In the active sensor-based methods, depth information is obtained from the object directly using physical sensors, such as infrared ray (IR) sensors [2]. Recently, KINECT depth camera and Time-of-Flight (ToF) range camera became a popular alternative for dense depth sensing. Although ToF cameras can capture depth information for object in real time, but are noisy and subject to low resolutions. For example, the current ToF camera 'Mesa Imaging SR4000' provides a 176×144 depth map up to 30 frames per second. For the actual utilization, we need the same resolution of the depth image and color image [3]. Therefore, an efficient depth map upsampling method is necessary.

Several techniques have been developed to solve problems of the depth map captured by ToF cameras. Gaussian filtering and other smoothing filtering have bad

© Springer International Publishing Switzerland 2015
Y.-S. Ho et al. (Eds.): PCM 2015, Part II, LNCS 9315, pp. 33–40, 2015.
DOI: 10.1007/978-3-319-24078-7_4

results in depth discontinuous region. Therefore, many depth map upsampling methods based on edge preserving.

A joint bilateral upsampling (JBU) removes the over smooth at the depth discontinuity region by adding additional information [4]. The information is an original color image used in depth estimation. JBU uses the two weighting functions with respect to photometric similarity of the neighbor pixels. However, there are two problems on JBU: using the color image as a guide cause edge blurring some regions and texture copy from the color image to the depth map. To solve these problems, noise-aware filter for depth upsampling (NAFDU) is adaptively apply the color weighting function based on edge information [5]. Also, Markov random field (MRF) based method is proposed in [6]. MRF depth map upsampling defined the probability model by MRF using color and depth information, and the model is optimized by conjugate gradient to provide upsampled depth map.

In this paper, we propose a depth map upsampling method that uses the edge information to decide the weighting functions. To get the edge information, we calculate the standard deviation value in each kernel. Using this value, we apply the blending function for depth map upsampling. Blending function defines the how much contributes to the each weighting functions. By applying blending function to the depth map upsampling, we obtain the high resolution depth map without texture copying of color guide image. In addition, the proposed upsampling method makes high-resolution depth map while protecting edge region.

2   Related Work

Low resolution of depth map provided by ToF cameras cause the problems in some 3D applications making. Using the color image information for the upsampling is a valid approach to enhance and improve the ToF data. Kopf et al. presented joint bilateral upsampling (JBU), a modified version of the joint bilateral filter for the depth map upsampling considers guide color image information [4]. In the JBU, the upsampled depth map is made via the Gaussian weighted sum of neighbors within a filter kernel. Assume that there are input low resolution depth map I, output high resolution depth map $\tilde{S}$ and high resolution color image $\tilde{I}$. Formally, the depth value $\tilde{S}_p$ at p in an upsampled depth map $\tilde{S}$ is computed by JBU as Eq. (1).

$$\tilde{S}_p = \frac{1}{k_p} \sum_{q_\downarrow \in \Omega} I_{q_\downarrow} f(\|p_\downarrow - q_\downarrow\|) g(\|\tilde{I}_p - \tilde{I}_q\|) \tag{1}$$

where $p_\downarrow$ and $q_\downarrow$ denote the pixel coordinate in a low resolution depth map and p and q are denote the corresponding coordinates in the high resolution color image. Ω represent the spatial neighborhood around p, and k_p is a normalizing factor. In this equation f and g respectively represent the spatial filter and range filter that have Gaussian distribution, and $\|\cdot\|$ is an Euclidean distance operator. The final goal of JBU is upsample of low resolution depth map while protecting the edge information.

Although the JBU can upsampled the depth map with protect the edge region, it has texture copy from the color image to the upsampling result. Texture copying problem is caused by the different discontinuity information between the color and depth data. Also, it is found in if depth sensor contains a lot of random noise. To solve this problem, the Noise-Aware Filter for Depth Upsampling, shortly NAFDU, was proposed [5]. NAFDU is a kind of adaptive multilateral filter that blends two range filters of color and depth information. Upsampled depth value $\tilde{S}_p$ is computed by NAFDU as Eq. (2).

$$\tilde{S}_p = \frac{1}{k_p} \sum\nolimits_{q_\downarrow \in \Omega} I_{q_\downarrow} f(\|p_\downarrow - q_\downarrow\|) [\alpha(\Delta\Omega) g(\|\tilde{I}_p - \tilde{I}_q\|) + (1 - \alpha(\Delta\Omega)) h(\|I_{p_\downarrow} - I_{q_\downarrow}\|)]$$

(2)

where α denotes the blending function and $\Delta\Omega$ represent the depth difference between the minimum and maximum depth values in the each filter kernel. Blending function of NAFDU is defined by Eq. (3).

$$\alpha(\Delta\Omega) = \frac{1}{1 + e^{-\varepsilon(\Delta\Omega - \tau)}}$$

(3)

The blending function is dependent to the depth value difference ($\Delta\Omega$). So, if the $\Delta\Omega$ is increased, α is close to 1 and NAFDU works like JBU. Otherwise, NAFDU use the depth information in the range filter, so it works like bilateral upsampling.

3 Efficient Depth Map Upsampling Using Edge Information

The proposed method modifies the conventional NAFDU. First of all, we compute the standard deviation in filter kernel. Based on standard deviation value, we estimate the edge region in the image. According to the edge information, we apply the blending function for adaptive weighting. For the edge region, our proposed method works like JBU. Otherwise, proposed method applies the bilateral upsampling using the median filtered depth map. Figure 1 shows the overall structure of our proposed method.

3.1 Edge Region

The goal of depth map upsampling is to estimate a high quality and high resolution depth map. It is difficult to expect a good result of depth map upsampling without considering the edge region. In our algorithm, we use the standard deviation to get the edge information. In statistics, the standard deviation (σ) is a measure that is used to quantify the amount of variation or dispersion of a set of data values. A standard deviation close to 0 indicates that the data points tend to be very close to the mean of the set, while a high standard deviation indicates that the data points are spread out over a wider range of values [7]. Standard deviation in the filter kernel is defined by Eq. (4).

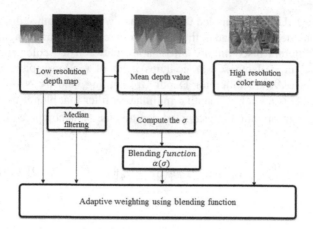

Fig. 1. Overall structure of the proposed method

$$\sigma_p = \sqrt{\frac{\sum_{i=1}^{n}\left(q_i - M_p\right)^2}{n}} \qquad (4)$$

where p denote the center pixel in the filter kernel and q represent the neighbor pixels in the filter kernel. n is the number of pixels in the filter kernel. Also, we can get the mean value in the filter kernel using Eq. (5).

$$M_p = \frac{1}{n}\sum_{i=1}^{n} q_i \qquad (5)$$

The standard deviation is a measure of how spreads out pixel values are. In addition, it can represent the dispersion of a set of pixel values from the filter kernel. So, we can decide which region is the edge by using standard deviation.

3.2   Blending Function

Our proposed method use the adaptive weighting functions based on edge information. Blending function is given by Eq. (6).

$$\alpha(x) = \frac{1}{1 + e^{-\varepsilon(x-\tau)}} , x = \frac{\sigma}{\sigma_{max}} \qquad (6)$$

blending function is the distribution of the 0 to 1 [5]. Here x is the ratio between the result of Eq. (4). and the maximum standard deviation in each kernel. The largest standard deviation possible will be found where the largest possible gaps are. So, we can calculate the largest standard variation very easy. The maximum standard deviation is calculated as in Eq. (7) [8].

$$\sigma_{max} = \sqrt{(M - \min(\Omega))(\max(\Omega) - M)} \tag{7}$$

where Ω represent the spatial neighborhood around p. M indicates the mean in the Ω and min, max represents the minimum depth value and maximum depth value in the Ω. The standard deviation distribution is normalized in 0 to 1 by consider maximum standard deviation. Also, it can represent how much edge information is included in each kernel. If standard deviation is very small in some region then the x have almost zero. Otherwise, x is the close to one when the standard deviation is almost same with maximum standard deviation. So, we can determine the edge in the filter kernel by using the x value (Fig. 2).

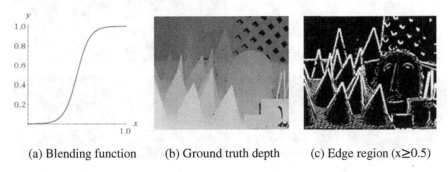

(a) Blending function   (b) Ground truth depth   (c) Edge region (x≥0.5)

Fig. 2. Blending function and Edge region

3.3 Adaptive Multilateral Upsampling

Assume that there are input low resolution depth map I, output high resolution depth map $\tilde{S}$, high resolution color image $\tilde{I}$ and median filtered depth map M (in case of input depth map captured by ToF camera). After blending function is determined, upsampled depth value $\tilde{S}_p$ is defined by Eq. (8).

$$\tilde{S}_p = \frac{1}{k_p} \sum_{q_\downarrow \in \Omega} I_{q_\downarrow} f\left(\|p_\downarrow - q_\downarrow\|\right) \left[\alpha(x)g\left(\|\tilde{I}_p - \tilde{I}_q\|\right) + (1 - \alpha(x))h\left(\|M_{p_\downarrow} - M_{q_\downarrow}\|\right)\right] \tag{8}$$

Here p and q denote the pixel coordinate in a high resolution color image and $p_\downarrow$ and $q_\downarrow$ are denote the corresponding coordinates in the low resolution depth map. Ω is a spatial neighborhood around target pixel. f, g and h are all Gaussian functions, and $\alpha(x)$ is a blending function.

If the x is increased, α is close to 1 and proposed method upsample of low resolution depth map while protecting the edge information. Otherwise, If the x is decreased, α is close to 0 and proposed method use the median filtered depth map information. Noise in ToF cameras is due to systematic errors and other sources impacting the measurements [9]. For example, surfaces with low infrared reflectance properties result in error region of the depth map. So, input depth map is median filtered to reduce the amount of noise.

4 Experimental Results

To evaluate the performance of the proposed upsampler, we used four test image sets Cones, Teddy, Tsukuba and Venus provided by Middlebury stereo were used [10].

These data sets are omposed of color images and ground-truth depth maps as hown in Fig. 3. So, we use the original depth map information in the range filter h. The ground truth depth maps were downsampled by factors of 2, 4 and 8 using the nearest

Fig. 3. Image sets for experimet

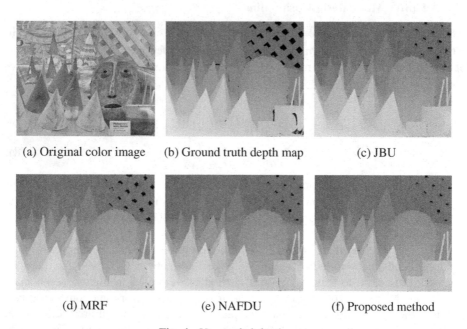

(a) Original color image   (b) Ground truth depth map   (c) JBU

(d) MRF   (e) NAFDU   (f) Proposed method

Fig. 4. Upsampled depth maps

Table 1. Performance comparison

Dataset	Scale	JBU	MRF	NAFDU	Proposed
Cones	2×	2.50	2.84	2.49	1.92
	4×	4.31	4.12	4.42	3.84
	8×	10.01	10.02	10.02	10.01
Teddy	2×	4.41	4.81	4.44	4.09
	4×	6.12	6.48	6.09	6.01
	8×	11.12	11.21	11.19	**11.09**
Tsukuba	2×	3.41	4.71	3.44	2.92
	4×	5.31	5.39	5.31	5.12
	8×	8.84	8.69	8.94	8.21
Venus	2×	0.83	1.12	0.93	0.26
	4×	0.92	1.2	0.91	0.76
	8×	3.81	4.05	3.80	3.73

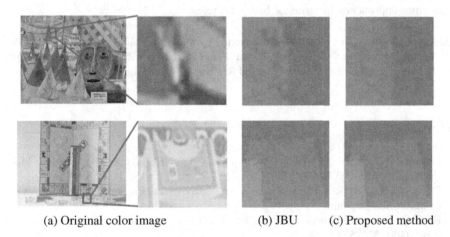

   (a) Original color image   (b) JBU   (c) Proposed method

Fig. 5. Enlarged texture copying area

neighbor method. In the experiment, the length of one side in the local window for the blending function, the length of one side in the local window for the filtering, variance of the spatial filter σ_s, variance of the range filter σ_r were set to factor 2: 7, 7, 1, 3, factor 4: 9, 7, 3, 8 and factor 8: 17, 11, 5, 10.

Also, ε and τ of the blending function were set to 50 and 0.3 in this test. The proposed method is compared with the joint bilateral upsampling (JBU) [4], the noise-aware filter for depth upsampling (NAFDU) [5] and the MRF-based depth upsampling [6]. Figure 4 shows the upsampled depth maps of Cones of the scaling factor 4.

Although the JBU can reconstruct depth edges, it has texture copying problem. So, proposed method use the adaptive weighting functions based on edge information (Fig. 5).

For an objective evaluation of the depth maps, the bad pixel rate (BPR), whose absolute difference is greater than 1, was used.

Table 1. show the result of the BPR(%) comparison. According to the results, the proposed method outperforms the conventional algorithms in terms of the BPR for all the scaling factors and all the tested images.

5   Conclusion

In this paper, we have proposed a multilateral depth map upsampling method for low resolution depth maps. The proposed upsampler is based on edge information, and we define the edge region that considers standard deviation in the image. After that, our proposed method use the adaptive weighting functions based on blending function. It is an efficient method because it increases depth map resolution while protecting the edge region. In addition this method can solve the texture copying problem. From the experimental results, the proposed method outperformed compared to the other depth upsampling approaches in terms of bad pixel rate.

Acknowledgements. This research was supported by the 'Cross-Ministry Giga KOREA Project' of the Ministry of Science, ICT and Future Planning, Republic of Korea(ROK). [GK15C0100, Development of Interactive and Realistic Massive Giga-Content Technology]

References

1. Okutomi, M., Kanade, T.: A multiple-baseline stereo. IEEE Trans. Pattern Anal. Mach. Intell. **15**(4), 353–363 (1993)
2. C.S. Swiss Ranger SR-2. The Swiss center for electronics and microtechnology. http://www.csem.ch/fs/imaging.htm
3. Fehn, C., Barre, R., Pastoor, S.: Interactive 3-D TV concepts and key technologies. Proc. IEEE **94**(3), 524–538 (2006)
4. Kopf, J., Cohen, M.F., Lischinski, D., Uyttendaele, M.: Joint bilateral upsampling. ACM Trans. Graph. **26**(3), 1–6 (2007)
5. Chan, D., Buisman, H., Theobalt, C., Thrun, S.: A noise-aware filter for real-time depth upsampling. In: ECCV Workshop on Multi-camera and Multi- modal Sensor Fusion Algorithms and Applications, pp. 1–12 (2008)
6. Diebel, J., Thrun, S.: An application of markov random fields to range sensing. In: Proceedings of Advances in Neural Information Processing Systems, pp. 291–298 (2006)
7. http://en.wikipedia.org/wiki/Standard_deviation
8. Baek, E., Ho, Y.: Gaussian alpha blending for natural image synthesis. In: Proceedings of Korean Institute Smart Media, vol. 4(1), pp. 138-141(2015)
9. Schwarz, L.A., Mkhistaryan, A., Mateus, D., Navab, N.: Human skeleton tracking from depth data using geodesic distances and optical flow. Image Vis. Comput. **20**(1), 217–226 (2012)
10. Scharstein, D., Szeliski, R.: A taxonomy and evaluation of dense two-frame stereo correspondence algorithms. Int. J. Comput. Vis. **47**(13), 7–42 (2002)

Orthogonal and Smooth Subspace Based on Sparse Coding for Image Classification

Fushuang Dai[1,2], Yao Zhao[1,2(✉)], Dongxia Chang[1,2],
and Chunyu Lin[2]

[1] Institute of Information Science,
Beijing Jiaotong University, Beijing 100044, China
yzhao@bjtu.edu.cn
[2] Beijing Key Laboratory of Advanced Information Science
and Network Technology, Beijing, China

Abstract. Many real-world problems usually deal with high-dimensional data, such as images, videos, text, web documents and so on. In fact, the classification algorithms used to process these high-dimensional data often suffer from the low accuracy and high computational complexity. Therefore, we propose a framework of transforming images from a high-dimensional image space to a low-dimensional target image space, based on learning an orthogonal smooth subspace for the SIFT sparse codes (SC-OSS). It is a two stage framework for subspace learning. Firstly, a sparse coding followed by spatial pyramid max pooling is used to get the image representation. Then, the image descriptor is mapped into an orthonormal and smooth subspace to classify images in low dimension. The proposed algorithm adds the orthogonality and a Laplacian smoothing penalty to constrain the projective function coefficient to be orthogonal and spatially smooth. The experimental results on the public datasets have shown that the proposed algorithm outperforms other subspace methods.

Keywords: Image classification · Orthogonal and smooth subspace · Sparse coding · Max pooling

1 Introduction

Image classification is one of the most fundamental problems in computer vision and pattern recognition. The primary objective of image classification is to assign one or more category labels to an image. It has been applied in a variety of fields such as video surveillance, image and video retrieval, web content analysis and so on. In image classification, image representation plays a very important role. In the past, many representation methods based on color, texture, shape, etc. were proposed. For the images of single color, single texture and single shape, the representations use these simple features are very excellent for image classification. Moreover, they have a small amount of computation and low computational complexity. However, in real life, a large number of images contain various and colorful content, in this case, the above-mentioned representations are far not enough to do image classification. In recent years, the Scale Invariant Feature Transform (SIFT) [1] descriptor has been

© Springer International Publishing Switzerland 2015
Y.-S. Ho et al. (Eds.): PCM 2015, Part II, LNCS 9315, pp. 41–50, 2015.
DOI: 10.1007/978-3-319-24078-7_5

widely used. SIFT descriptor is the local feature of the image, it keeps invariant to rotate, scale zoom, brightness variation, and a certain stability to the change of the viewing angle, affine transformations and the noise.

However, due to SIFT descriptor' high dimension, it leads to the "curse of dimensionality" [2]. So, the work of pre-processing for SIFT descriptor is quite important. In recent years, the bag of visual words (BoW) model [3] is widely applied for image analysis. But, there are two main recognized drawbacks [4] existing in BoW model: (1) In dictionary, the approach selects randomly visual words, which severely limits the descriptive power of image representation. (2) The model loses the spatial information of raw image. To overcome the first problem, sparse coding succinctly represents data vectors with several basis vectors in the dictionary, which requires only a relatively small number of bases to represent the signal. There are two basic requirements in sparse coding. One is that sparse coding feature is similar to the original as possible. Another is that coding coefficient is sparse. This shows that sparse coding not only can well retain the original features, but also greatly simplifies the complexity of the subsequent calculation. By overcoming the second problem of the BoW model, a spatial pyramid method is proposed [4, 5], which extended the BoW model by partitioning the image into sub-regions and computing histograms of local features.

From a space perspective, image classification can be considered as a classifier design problem. In this circumstance, an image of $n \times m$ pixels can be represented by a $n \times m$-dimensional vector. Then, the researchers focus on designing a classifier which can classify images in the $n \times m$-dimensional space effectively. In fact, the linear low-dimensional subspace [6] can significantly improve the performance of image classification. Principle Component Analysis (PCA) [7] and Linear Discriminate Analysis (LDA) [8] are very prominent subspace methods and they have got promising accuracies in image classification. However, they failed to consider the specific structures of images and cannot fully explore the spatial information. Hence, the orthogonal smooth subspace [9] is proposed. However, the algorithm uses the color features to classify. For single colorful images with very obvious target, the algorithm gets good classification performance. But, for various colorful target images e.g. in each class, the colors of images are various, and there is interference of other objects in the background of the target image, such as these images in Caltech101, this algorithm will lose its superior classification performance.

Based on the above-mentioned analysis, considering the low accuracy and high computational complexity of high-dimensional data in the classification algorithms, an orthogonal smooth subspace for the SIFT sparse codes (SC-OSS) was proposed. In the new algorithm, we transform images from a high-dimensional image space to a low-dimensional space. In the new algorithm, sparse coding followed by spatial pyramid max pooling is used to get the image representation. Then, the obtained representation is mapped into an orthonormal and smooth subspace in low dimension. Experiments will show that our approach get higher classification accuracy on the public datasets.

The rest of the paper is organized as follows. In Sect. 2, we describe and analyze sparse codes of SIFT descriptor using spatial pyramid max pooling. Section 3 presents the orthogonal smooth subspace learning model. Section 4 explains the parameters setting. Section 5 shows some experimental results, and Sect. 6 concludes the paper.

2 SIFT Sparse Codes Using Spatial Pyramid Max Pooling

2.1 The Sparse Codes

Let $\mathbf{X}$ be a set of SIFT appearance descriptors, $\mathbf{X} = [\mathbf{x}_1, \mathbf{x}_2, \ldots, \mathbf{x}_M]^T \in \mathbf{R}^{M \times D}$. The sparse coding method mainly solves the following problem

$$\min_{\mathbf{u}, \mathbf{v}} \sum_{m=1}^{M} \|\mathbf{x}_m - \mathbf{u}_m \mathbf{V}\|^2 + \lambda |\mathbf{u}_m| \tag{1}$$

where $\mathbf{V} = [\mathbf{v}_1, \ldots, \mathbf{v}_k]^T$ is the codebook, K is the size of the codebook. Normally, the codebook $\mathbf{V}$ is an overcomplete basis set, i.e.,$K > D$. $\mathbf{U} = [\mathbf{u}_1, \ldots, \mathbf{u}_M]^T$ is the coding matrix, $\mathbf{u}_m$ is constrainedL_1-norm regularization, i.e., after the optimization, the only nonzero element in $\mathbf{u}_m$ denotes the vector $\mathbf{x}_m$.

The problem (1) is not convex with respect to both variables $\mathbf{U}$ and $\mathbf{V}$. Thus, we cannot directly get the global optima for the problem. So, the way is to solve (1) iteratively by alternatively optimizing over $\mathbf{V}$ or $\mathbf{U}$ while fixing the other. Fixing $\mathbf{V}$, we can solve the following optimization by optimizing over each coefficient $\mathbf{u}_m$ individually

$$\min_{\mathbf{u}_m} \|\mathbf{x}_m - \mathbf{u}_m \mathbf{V}\|^2 + \lambda |\mathbf{u}_m| \tag{2}$$

The optimization (2) is well known as Lasso in the Statistics, and it is essentially a linear regression problem with L_1-norm regularization on the coefficients. Fixing $\mathbf{U}$, we can solve the optimization (3) by the Lagrange dual as used in [10]. The optimization (3) is a least square problem with quadratic constraints, where typically applied a unit L_2-norm constraint on $\mathbf{v}_k$ to avoid trivial solutions.

$$\min_{\mathbf{V}} \|\mathbf{X} - \mathbf{U}\mathbf{V}\|_F^2$$
$$s.t. \|\mathbf{v}_k\| \leq 1, \forall k = 1, 2, \ldots K \tag{3}$$

2.2 Spatial Pyramid Max Pooling

In spatial pyramid model, the feature will be partitioned into $2^l \times 2^l$ blocks in different scales. Here, we construct two levels spatial pyramid, i.e., $l = 0, 1$. From sparse coding, we already get the coefficient $\mathbf{U}$. Then, in each level of spatial pyramid model, the max pooling function [13] is used to compute the image features:

$$r_j = \max\{|u_{1j}|, |u_{2j}|, \ldots, |u_{Mj}|\} \tag{4}$$

where r_j is the j-th element of the image feature, u_{ij} is the element of j-th column and i-th row of the coding matrix $\mathbf{U}$, and M is the number of local descriptors in this region.

3 Orthogonal and Smooth Subspace Method

From the above framework of sparse coding followed by spatial pyramid max pooling, we get a $5 \times K$-dimensional feature space. And it is a high dimensional space. The performance of image classification can be improved significantly in a linear low-dimensional subspace. So we propose an orthogonal smooth subspace for the SIFT sparse codes (SC-OSS). A more detailed description of the proposed algorithm is shown in Fig. 1. Firstly, we extract SIFT feature to get the local descriptor. Then, sparse coding and max pooling is applied to get the high dimensional description. At last, the orthonormal and smooth subspace method is proposed to obtain the low dimensional description of the image.

In the following, we will project each image feature into a low dimensional subspace by $\mathbf{y}_i = \mathbf{W}^T \mathbf{x}_i$ after getting the $5 \times K$-dimensional vector. $\mathbf{W} \in \mathbf{R}^{(n_1 \times n_2) \times d}$ is the transformation matrix and d is the dimensionality of the subspace.

A popular image classification aim is to find a set of centers for which the within-cluster spread is small; the between-clustering spread is large in some sense. Therefore, the optimization objection used of our paper is:

$$\max \ \text{trace}\left(\mathbf{S}_w^{-1}\mathbf{S}_b\right) \tag{5}$$

where $\mathbf{S}_w$ is the within-class scatter matrix and $\mathbf{S}_b$ is the between-class scatter matrix. $\mathbf{S}_w$ measures how compact or tight the classes are and it is defined as

$$\mathbf{S}_w = \sum_{j=1}^{C} \sum_{i=1}^{q_j} (\mathbf{y}_i - \mathbf{m}_j)(\mathbf{y}_i - \mathbf{m}_j)^T \tag{6}$$

where $\mathbf{m}_j = \frac{1}{q_j} \sum_{i=1}^{q_j} \mathbf{y}_i$ ($j = 1, 2, \ldots, C$), q_j is the number of samples in class j. $\mathbf{S}_b$ measures how scatter the class centers are from the sample mean and it is given by

$$\mathbf{S}_b = \sum_{j=1}^{C} q_j (\mathbf{m}_j - \mathbf{m})(\mathbf{m}_j - \mathbf{m})^T \tag{7}$$

where $\mathbf{m} = \frac{1}{n} \sum_{i=1}^{n} \mathbf{y}_i$. By making the substitution $\mathbf{y}_i = \mathbf{W}^T \mathbf{x}_i$ into (6) and (7), we can get

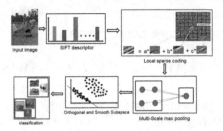

Fig. 1. The flow-chart of the proposed SC-OSS algorithm

$$S_w = \sum_{j=1}^{C} \sum_{i=1}^{q_j} \left(\mathbf{W}^\mathsf{T}\mathbf{x}_i - \mathbf{W}^\mathsf{T}\mathbf{m}_j\right)\left(\mathbf{W}^\mathsf{T}\mathbf{x}_i - \mathbf{W}^\mathsf{T}\mathbf{m}_j\right)^\mathsf{T} = \mathbf{W}^\mathsf{T} \sum_{j=1}^{C} \sum_{i=1}^{q_j} \left(\mathbf{x}_i - \mathbf{m}_j\right)\left(\mathbf{x}_i - \mathbf{m}_j\right)^\mathsf{T}\mathbf{W}$$
$$= \mathbf{W}^\mathsf{T}\mathbf{S}_{wx}\mathbf{W}$$

$$(8)$$

$$S_b = \sum_{j=1}^{C} q_j\left(\mathbf{W}^\mathsf{T}\mathbf{m}_j - \mathbf{W}^\mathsf{T}\mathbf{m}\right)\left(\mathbf{W}^\mathsf{T}\mathbf{m}_j - \mathbf{W}^\mathsf{T}\mathbf{m}_j\right)^\mathsf{T}$$
$$= \mathbf{W}^\mathsf{T} \sum_{j=1}^{C} q_j\left(\mathbf{m}_{jx} - \mathbf{m}_x\right)\left(\mathbf{m}_{jx} - \mathbf{m}_x\right)^\mathsf{T}\mathbf{W} = \mathbf{W}^\mathsf{T}\mathbf{S}_{bx}\mathbf{W}$$

$$(9)$$

Then we can rewrite (5) as

$$\mathrm{maxtrace}\left(\left(\mathbf{W}^\mathsf{T}\mathbf{S}_{wx}\mathbf{W}\right)^{-1}\left(\mathbf{W}^\mathsf{T}\mathbf{S}_{bx}\mathbf{W}\right)\right) \tag{10}$$

where $\mathbf{S}_{wx}$ and $\mathbf{S}_{bx}$ are the within-class scatter matrix and between-class scatter matrix in the original image feature space. Here, we add the orthogonal constraints, i.e., $\mathbf{W}^\mathsf{T}\mathbf{W} = \mathbf{I}$. It guarantees that the embedding coefficients are invariant only when one multiples the transformation matrix by an orthogonal matrix. And then, we use the 2-D discretized Laplacian smoothing term, which has been successfully used in [11]. It is a $n_1 n_2 \times n_1 n_2$ matrix:

$$\mathbf{\Delta} = \mathbf{D}_1 \otimes \mathbf{I}_2 + \mathbf{I}_1 \otimes \mathbf{D}_2 \tag{11}$$

where $\mathbf{I}_j$ is the $n_j \times n_j$ identity matrix for $j = 1, 2$. $\otimes$ is the Kronecker product. $\mathbf{D}_j$ is

$$\mathbf{D}_j = \frac{1}{h_j^2}\begin{pmatrix} -1 & 1 & & & & 0 \\ 1 & -2 & 1 & & & \\ & \cdots & \cdots & \cdots & & \\ & & & 1 & -2 & 1 \\ 0 & & & & 1 & -1 \end{pmatrix} \tag{12}$$

where $h_j = \frac{1}{n_j}$, for $j = 1, 2$.

Therefore, the objective function can be written as:

$$\arg\max_{\mathbf{W}^T\mathbf{W}=\mathbf{I}} \text{trace}\left(\left(\mathbf{W}^T\left(\mathbf{S}_{wx} + \lambda\mathbf{\Delta}^T\mathbf{\Delta}\right)\mathbf{W}\right)^{-1} \times \left(\mathbf{W}^T\mathbf{S}_{bx}\mathbf{W}\right)\right) \qquad (13)$$

Where λ is the balance parameter and $\mathbf{I}$ is an identity matrix. Use the Lagrangian method to solve the following problem.

$$\Phi(\mathbf{W}) = \text{trace}\left(\left(\mathbf{W}^T\left(\mathbf{S}_{wx} + \lambda\mathbf{\Delta}^T\mathbf{\Delta}\right)\mathbf{W}\right)^{-1} \times \left(\mathbf{W}^T\mathbf{S}_{bx}\mathbf{W}\right)\right) - \text{trace}\left(\Gamma\left(\mathbf{W}^T\mathbf{W} - \mathbf{I}\right)\right)$$

$$(14)$$

Let $\mathbf{A} = \mathbf{S}_{wx} + \lambda\mathbf{\Delta}^T\mathbf{\Delta}$, $\mathbf{B} = \mathbf{S}_{bx}$. Differentiating $\Phi(\mathbf{W})$ with respect to $\mathbf{W}$ and set it to be zero. Then, we have

$$\left(\mathbf{W}^T\mathbf{A}\mathbf{W}\right)^{-1}\left(\mathbf{A}\mathbf{W}\left(\mathbf{W}^T\mathbf{A}\mathbf{W}\right)^{-1}\mathbf{W}^T\mathbf{B}\mathbf{W} - \mathbf{B}\mathbf{W}\right) - \mathbf{W}\Gamma = \mathbf{0} \qquad (15)$$

Multiple both sides of Eq. (15) by $\mathbf{W}^T$, we can get $\mathbf{W}^T\mathbf{W}\Gamma = \mathbf{0}$. Note that there always exists $\mathbf{W}^T\mathbf{W} = \mathbf{I}$. Therefore, we can deduce $\Gamma = \mathbf{0}$. Hence,

$$\mathbf{A}\mathbf{W}\left(\mathbf{W}^T\mathbf{A}\mathbf{W}\right)^{-1}\mathbf{W}^T\mathbf{B}\mathbf{W} - \mathbf{B}\mathbf{W} = \mathbf{0} \qquad (16)$$

Let $\mathbf{D} = \mathbf{W}^T\mathbf{A}\mathbf{W}$, $\mathbf{E} = \mathbf{W}^T\mathbf{B}\mathbf{W}$. Substituting $\mathbf{D}$ and $\mathbf{E}$ into (16), we can get

$$\mathbf{B}\mathbf{W} = \mathbf{A}\mathbf{W}\mathbf{D}^{-1}\mathbf{E} \qquad (17)$$

Since $\mathbf{D}$ and $\mathbf{E}$ are symmetric, there always exists a nonsingular matrix $\mathbf{P}$, satisfying $\mathbf{P}^T\mathbf{D}\mathbf{P} = \mathbf{I}$ and $\mathbf{P}^T\mathbf{E}\mathbf{P} = \mathbf{\Lambda}$, where $\mathbf{\Lambda}$ is a diagonal matrix. Hence, $\mathbf{D} = \left(\mathbf{P}^T\right)^{-1}\mathbf{P}^{-1}$, $\mathbf{E} = \left(\mathbf{P}^T\right)^{-1}\mathbf{\Lambda}\mathbf{P}^{-1}$. By making the substitution $\mathbf{D}$ and $\mathbf{E}$ into (17), we can get

$$\mathbf{B}\mathbf{W}\mathbf{P} = \mathbf{A}\mathbf{W}\mathbf{P}\mathbf{\Lambda} \qquad (18)$$

Let $\mathbf{V} = \mathbf{W}\mathbf{P}$, then $\mathbf{W} = \mathbf{V}\mathbf{P}^{-1}$. Therefore, $\mathbf{V}$ is the generalized Eigen matrix of the matrix pairs $\mathbf{B}$ and $\mathbf{A}$. Substituting $\mathbf{W} = \mathbf{V}\mathbf{P}^{-1}$ into $\mathbf{W}^T\mathbf{W} = \mathbf{I}$, we have $\left(\mathbf{P}^{-1}\right)^T\mathbf{V}^T\mathbf{V}\mathbf{P}^{-1} = \mathbf{I}$. Perform the following eigenvalue decomposition of the matrix $\mathbf{V}^T\mathbf{V}$: $\mathbf{V}^T\mathbf{V} = \mathbf{U}\mathbf{\Sigma}\mathbf{U}^T$, then

$$\left(\mathbf{P}^{-1}\right)^T\mathbf{U}\mathbf{\Sigma}\mathbf{U}^T\mathbf{P}^{-1} = \mathbf{I} \qquad (19)$$

Thus, we can get $\mathbf{P}^{-1} = \mathbf{U}\mathbf{\Sigma}^{-1/2}$. Therefore, we can get the transformation matrix

$$W = VP^{-1} = VU\Sigma^{-1/2} \tag{20}$$

After getting the low dimensional feature y_i, the Nearest Neighbor is used to classify images.

4 Parameters Selecting

For the orthogonal smooth subspace, we get the balance parameter λ by another parameter α, which is irrelevant to numerical scale.

$$\lambda = \alpha \max(\text{diag}(S_{wx})) \tag{21}$$

Here, we determine $\alpha = 0.001$ by cross validation based on previous work [9].

Another parameter is the dimensionality: d. The optimal dimensionality of subspace is equal to $\text{rank}(S_{bx})$. In real applications, we know that $\text{rank}(S_{bx}) = C - 1$. Thus, we also empirically choose $d = C - 1$ in the following experiments.

5 Experiments

For the purpose of testing the performance of our SC-OSS algorithm, experiments are conducted on the Caltech 101 data set (only the twenty maximum data sets are used), the Coil20 data set and the Flickr 13 animal images data set selected from a subset of NUS-WIDE data sets. In order to verify the classification performance of the SC-OSS, we compare it with LDA [8], OLDA [12], OSSL [9], KSPM [4], ScSPM [5] and LLC [14].

Firstly, we compare the performance of the algorithms with different number of training images and the experiments are repeated for 10 times. The average classification accuracy is shown in Tables 1, 2 and 3.

From Table 1, we observe that SC-OSS performs best among all the methods. It outperforms OSSL by more than 20 % and even outperforms LDA and OLDA by a large margin. And, it also achieves higher accuracy than ScSPM, KSPM and LLC. From Tables 2 and 3, the SC-OSS algorithm again achieves much better performance

Table 1. Classification rate (%) comparison on twenty maximum datasets of Caltech101 dataset (Mean ± Std-Dev%).

Algorithms	15training	30training	45training	60training
LDA	20.5 ± 1.8	26.2 ± 4.1	32.8 ± 2.4	32.8 ± 2.0
OLDA	27.6 ± 3.8	34.7 ± 3.0	36.0 ± 1.2	39.2 ± 4.4
OSSL	43.0 ± 1.9	52.2 ± 1.5	57.1 ± 0.7	61.1 ± 1.0
KSPM	68.3 ± 1.5	76.4 ± 2.1	80.5 ± 1.0	81.5 ± 1.8
ScSPM	**76.3 ± 1.5**	80.0 ± 0.7	82.4 ± 0.6	83.9 ± 0.8
LLC	68.9 ± 1.8	74.8 ± 0.9	77.8 ± 0.8	80.3 ± 0.8
SC-OSS	73.1 ± 1.5	**81.8 ± 1.0**	**87.2 ± 0.8**	**90.2 ± 0.7**

Table 2. Classification rate (%) comparison on Coil20 dataset (Mean ± Std-Dev%).

Algorithms	*5training*	10training
LDA	43.4 ± 8.0	45.4 ± 11.9
OLDA	55.7 ± 6.0	60.3 ± 4.1
OSSL	84.5 ± 2.0	89.7 ± 0.8
KSPM	76.2 ± 2.0	80.5 ± 0.9
ScSPM	89.2 ± 1.4	94.2 ± 1.6
LLC	88.8 ± 0.6	93.8 ± 0.8
SC-OSS	**89.4 ± 0.8**	**96.9 ± 0.8**

Table 3. Classification rate (%) comparison on Flickr 13 animal images dataset (Mean ± Std-Dev%).

Algorithms	*50training*	75training	100training
LDA	9.7 ± 0.5	10.2 ± 0.6	9.1 ± 0.5
OLDA	10.1 ± 0.8	10.6 ± 0.7	9.2 ± 0.5
OSSL	10.8 ± 0.7	10.4 ± 0.9	11.2 ± 0.6
KSPM	26.9 ± 0.8	28.8 ± 0.6	28.7 ± 0.6
ScSPM	34.5 ± 0.8	36.8 ± 1.1	37.4 ± 0.8
LLC	**34.8 ± 1.0**	36.8 ± 1.2	38.0 ± 1.4
SC-OSS	30.9 ± 0.8	**37.2 ± 0.8**	**40.7 ± 1.2**

than listing algorithms. Meanwhile, from experimental standard derivation, we generally find that our method standard derivations are relatively smaller than another algorithms, it means that our method is more stable to training samples.

In the following, we compare the max pooling with other pooling methods, namely, the mean of absolute values (Abs) and the square root of mean squared statistics (Sqrt). The other two pooling functions are defined as

$$\text{Abs}: \quad z_j = \frac{1}{M} \sum_{i=1}^{M} |u_{ij}| \tag{22}$$

$$\text{Sqrt}: \quad z_j = \sqrt{\frac{1}{M} \sum_{i=1}^{M} u_{ij}^2} \tag{23}$$

As shown in Table 4, the performance of max pooling is the best, probably due to its robustness to local spatial variations. We also investigate the effects of codebook

Table 4. The performance comparison using different pooling methods on ten maximum datasets of Caltech 101 dataset (Mean ± Std-Dev%).

Pooling	Max	Abs	Sqrt
Caltech101	**88.7 ± 0.8**	78.9 ± 1.3	84.3 ± 1.2

sizes. Here, we design three sizes: 256, 512 and 1024. As shown in Table 5, we find that the performance for our method keeps increasing when the codebook size goes up to 1024.

Table 5. The effects of codebook size on our method on ten maximum datasets of Caltech 101 dataset (Mean ± Std-Dev%).

Codebook size	256	512	1024
15training	80.8 ± 1.2	81.1 ± 1.5	**81.5 ± 1.1**
30training	84.5 ± 1.3	87.1 ± 1.5	**88.7 ± 0.8**
45training	85.9 ± 1.0	88.5 ± 0.6	**90.1 ± 0.8**
60training	88.1 ± 0.7	89.7 ± 0.8	**91.6 ± 0.5**

6 Conclusion

In this paper, we propose a framework of transforming images from a high-dimensional image space to a low-dimensional target image space, based on learning an orthogonal smooth subspace for the SIFT sparse codes (SC-OSS). By sparse coding followed by spatial pyramid max pooling, we get the excellent image representation. Then, the high-dimensional representation is mapped into a low dimension subspace to classify images. The low dimension subspace adds the orthogonality and a Laplacian smoothing penalty to constrain the projective function coefficient to be orthogonal and spatially smooth. So, comparing with other methods, our method performs better on the public datasets. Moreover, in terms of the sparse coding, it also has much value to be worth studying. Further research of this proposed method is a valuable direction.

Acknowledgement. This paper was supported in part by National Natural Science Foundation of China (61210006, 61100141), Program for Changjiang Scholars and Innovative Research Team in University (IRT201206), the Fundamental Research Funds for the Central Universities of China (2013JBM021).

References

1. Lowe, D.G.: Distinctive image features from scale-invariant keypoints. Int. J. Comput. Vis. **60**(2), 91–110 (2004)
2. Bai, E.: Big data: The curse of dimensionality in modeling. In: Control Conference, pp.6–13. IEEE press, Chinese (2014)
3. Lou, X., Huang, D., Fan, L., et al.: An image classification algorithm based on bag of visual words and multi-kernel learning. J. Multimedia **9**(2), 269–277 (2014)
4. Lazebnik, S., Schmid, C., Ponce, J.: Beyond bags of features: spatial pyramid matching for recognizing natural scene categories. IEEE Comput. Soc. Conf. Comput. Vis. Pattern Recogn. **2**, 2169–2178 (2006)

5. Yang, J., Yu, K., Gong, Y., et al.: Linear spatial pyramid matching using sparse coding for image classification. In: IEEE Conference on Computer Vision and Pattern Recognition, pp. 1794–1801 (2009)
6. Yan, Y., Zhang, Y.: Discriminant projection embedding for face and palmprint recognition. Neurocomputing **17**, 3534–3543 (2008)
7. Zhou, J., Jin, Z., Yang, J.: Multiscale saliency detection using principle component analysis. In: The 2012 International Joint Conference on Neural Networks (IJCNN), pp. 1–6. IEEE press, Brisbane (2012)
8. Duda, R.O., Hart, P.E., Stork, D.G.: Pattern Classification, 2nd edn. Wiley-Inter science, Hoboken, NJ (2000)
9. Hou, C., Nie, F., Zhang, C., et al.: Learning an orthogonal and smooth subspace for image classification. Signal Process. Lett. **16**(4), 303–306 (2009). IEEE press
10. Lee, H., Battle, A., Raina, R., et al.: Efficient sparse coding algorithms. In: Advances in neural information processing systems, pp.801–808 (2006)
11. Liu, F., Liu, X.: Locality enhanced spectral embedding and spatially smooth spectral regression for face recognition. In: Information and Automation, pp.299–303. IEEE press, Shenyang (2012)
12. Ye, J.: Characterization of a family of algorithms for generalized discriminant analysis on undersampled problems. J. Mach. Learn. Res. **6**, 483–502 (2005)
13. Serre, T., Wolf, L., Poggio, T.: Object recognition with features inspired by visual cortex. Comput. Vis. Pattern Recogn. **2**, 994–1000 (2005)
14. Wang, J., Yang, J., Yu, K., et al.: Locality-constrained linear coding for image classification. In: IEEE Conference on Computer Vision and Pattern Recognition, pp.3360–3367 (2010)

Video/Image Quality Assessment and Processing

Sparse Representation Based Image Quality Assessment with Adaptive Sub-dictionary Selection

Leida Li[1(✉)], Hao Cai[1], Yabin Zhang[2], and Jiansheng Qian[1]

[1] School of Information and Electrical Engineering,
China University of Mining and Technology, Xuzhou 221116, China
`readerll04@hotmail.com`
[2] School of Computer Engineering, Nanyang Technological University,
Singapore 639798, Singapore

Abstract. This paper presents a sparse representation based image Quality metric with Adaptive Sub-Dictionaries (QASD). An overcomplete dictionary is first learned using natural images. A reference image block is represented using the overcomplete dictionary, and the used basis vectors are employed to form an undercomplete sub-dictionary. Then the corresponding distorted image block is represented using all the basis vectors in the sub-dictionary. The sparse coefficients are used to generate two feature maps, based on which a local quality map is generated. With the consideration that sparse features are insensitive to weak distortions and image quality is affected by various factors, image gradient, color and luminance are integrated as auxiliary features. Finally, a sparse-feature-based weighting map is proposed to conduct the pooling, producing an overall quality score. Experiments on public image databases demonstrate the advantages of the proposed method.

Keywords: Image quality assessment · Sparse representation · Dictionary learning · Adaptive sub-dictionary

1 Introduction

Modern image quality assessment (IQA) aims to build computational models for measuring image quality, and meantime keep consistent with human perception [1]. The current IQA metrics can be classified into full-reference (FR), reduced-reference (RR) and no-reference (NR) approaches, which require full, partial and no information of an undistorted reference image during the evaluation of image quality.

Mean squared error (MSE) and its variant peak signal-to-noise-ratio (PSNR) are the most commonly used FR-IQA metrics. However, they have been criticized for not correlating well with human perception. Recent years have witnessed remarkable advances in IQA. Wang et al. [2] addressed the milestone work structural similarity (SSIM), which was based on the assumption that the perception of image quality was mainly determined by structural degradations. A multi-scale version of SSIM, i.e., MS-SSIM, was later developed to account for the influence of varying viewing conditions on the perceived quality [3]. Sheikh et al. [4] measured image quality using

© Springer International Publishing Switzerland 2015
Y.-S. Ho et al. (Eds.): PCM 2015, Part II, LNCS 9315, pp. 53–62, 2015.
DOI: 10.1007/978-3-319-24078-7_6

information fidelity criterion (IFC) based on natural scene statistics. The fidelity criterion was obtained by measuring the mutual information shared between the reference and distorted images. Later, IFC was improved to the visual information fidelity (VIF) metric with enhanced performance [5]. In [6], Larson and Chandler claimed that the human visual system (HVS) adopted different strategies in viewing low-quality and high-quality images. These two strategies were combined to produce the most apparent distortion (MAD) model. Zhang et al. [7] employed phase congruency to represent image local structures and proposed the feature similarity (FSIM) model. Liu et al. [8] employed gradient similarity (GSM) to represent structure and contrast changes simultaneously. Luminance change was then integrated to produce the overall quality score. Xue et al. [9] proposed the gradient magnitude similarity deviation (GMSD) model, where the standard deviation of gradient magnitude similarity map was computed as the image quality score.

Most of the current IQA models are based on low-level features. However, the HVS tends to extract high-level features for image understanding. From this perspective, low-level features may not be sufficient for IQA, and high-level features can benefit quality assessment. Recently, dictionary based sparse representation has been shown effective in capturing underlying image structures. Dictionaries learned from natural images consist of basis vectors that behave similarly to the simple cells in mammalian primary visual cortex [10], which is believed to be able to extract higher-level features. As a result, sparse representation can be explored for advanced IQA. Recently, attempts have been done to use sparse representation in IQA [11–13]. In this paper, we propose a sparse representation based image Quality metric with an Adaptive Sub-Dictionary selection strategy (QASD). An overcomplete dictionary is first learned using natural images. A reference image block is represented using this dictionary. Then the relevant basis vectors are used to construct an adaptive sub-dictionary, and the corresponding distorted image block is represented using the basis vectors in the sub-dictionary. With the sparse coefficients, two feature maps are generated. A local quality map is then generated by computing the similarity of the feature maps. Considering that sparse features are insensitive to weak distortions and image quality is affected by many different factors, gradient, color and luminance are integrated as auxiliary features. Since the feature maps also characterize the visual importance of local regions, they are also used to generate a weighting map, based on which the features are pooled to generate an overall quality score. Experiments demonstrate that the proposed method produces the state-of-the-art performance.

2 Dictionary Based Sparse Representation

Dictionary based sparse representation is to approximate a signal with a small number of basis vectors in a dictionary. With an overcomplete dictionary $\mathbf{D} \in \mathbf{R}^{n \times K}$ ($n < K$), which consists of K n-dimension basis vectors $\{\mathbf{d}_i\}_{i=1}^{K}$, a given signal $\mathbf{y}$ can be represented as a linear combination of the basis vectors:

$$\mathbf{y} = \mathbf{D}\mathbf{x} = \sum\nolimits_{i=1}^{K} x_i \mathbf{d}_i, \quad \text{s.t.} \quad \|\mathbf{y} - \mathbf{D}\mathbf{x}\|_2 \leq \epsilon, \tag{1}$$

where $\mathbf{x} \in \mathbf{R}^K$ denotes the representation vector, and $\|\cdot\|_2$ is the l_2 norm.

In the sparse representation of a signal, we always expect to approximate it using as few basis vectors as possible. Therefore, sparse representation can be converted into the following optimization problem:

$$\min_{\mathbf{x}} \|\mathbf{x}\|_0, \quad \text{s.t.} \quad \|\mathbf{y} - \mathbf{D}\mathbf{x}\|_2 \leq \epsilon, \tag{2}$$

Where $\|\cdot\|_0$ denotes the l_0 norm, which counts the number of nonzero elements.

Overcomplete dictionaries are commonly used for sparse representation. Generally, a large number of natural image patches are used to train this kind of dictionaries. The process of dictionary learning is to find an optimal set of basis vectors that can produce the best approximation of each input signal. In this paper, we employ Lee's algorithm [14] to train the dictionary. To obtain the sparse coefficients, pursuit algorithms are usually employed, which can select the optimal basis vectors to approximate a signal with a preset threshold of approximation error, while satisfying a sparsity constraint. In this paper, the orthogonal matching pursuit (OMP) algorithm [15] is employed to obtain the sparse coefficients.

It has been demonstrated that overcomplete dictionaries learned from natural images can capture the underlying structures of images. Furthermore, the basis vectors in a dictionary behave similarly to the simple cells in the primary visual cortex [10]. As a result, sparse representation naturally meets the requirements of perceptual IQA.

3 Proposed Image Quality Model

The flowchart of the proposed metric is illustrated in Fig. 1. QASD consists of two components, namely sparse feature similarity and auxiliary feature similarity. Sparse feature similarity is the main component of QASD, where higher-level features are extracted for measuring the structure distortions. Auxiliary features are used to compensate for the insensitiveness of sparse features to weak distortions, and also to account for the impacts of other factors in IQA.

3.1 Adaptive Sub-dictionary Selection

In the proposed method, an overcomplete dictionary is employed to represent the reference and distorted images block-wisely. For fair comparison, the reference-distorted image block-pair should be represented using the same basis vectors. However, in sparse representation, the basis vectors are automatically selected to produce the minimal approximation error. This indicates that if the overcomplete dictionary is directly used to represent both the reference and distorted image blocks, different basis vectors will be used. In this case, the produced sparse coefficients are not fair to compare, and image distortions cannot be correctly captured.

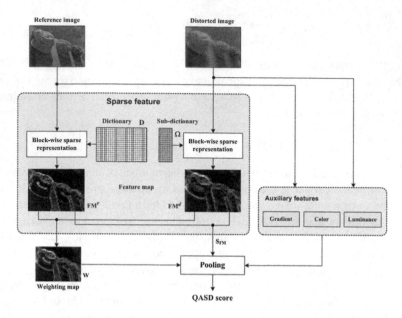

Fig. 1. Flowchart of the proposed QASD metric.

In order to ensure that the same basis vectors are used to represent each reference-distorted block pair, an adaptive sub-dictionary (ASD) selection strategy is proposed, which is illustrated in Fig. 2. For a reference image block $\mathbf{y}^r$, it is represented using the overcomplete dictionary $\mathbf{D}$ with the OMP algorithm, producing the representation vector $\mathbf{x}^r$ and the corresponding used basis vectors $\{\mathbf{d}_j\}_{j=1}^{L}$, L is the sparsity. Then all the L basis vectors are used to constitute an undercomplete sub-dictionary $\mathbf{\Omega}$. The corresponding distorted image block $\mathbf{y}^d$ is then represented using all the basis vectors in $\mathbf{\Omega}$, producing the representation vector $\mathbf{x}^d$. By doing so, the

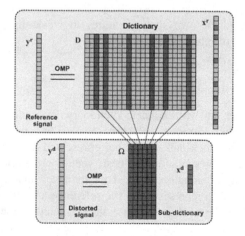

Fig. 2. Illustration of adaptive sub-dictionary selection.

reference-distorted block-pairs are represented using the same basis vectors, so the difference between $\mathbf{x}^r$ and $\mathbf{x}^d$ can correctly captures the distortions.

3.2 Sparse Feature Similarity

Sparse feature similarity is used to extract higher-level features for measuring image distortions, which we believe is important for IQA. Given a reference image $\mathbf{F}_1$ and a distorted image $\mathbf{F}_2$, both of size $M \times N$, they are first partitioned into nonoverlapping blocks of size $R \times R$, which are denoted by $\left\{\mathbf{y}_{ij}^r\right\}$ and $\left\{\mathbf{y}_{ij}^d\right\}$, where $i = 1, 2, \cdots, \lfloor \frac{M}{R} \rfloor$, $j = 1, 2, \cdots, \frac{N}{R}$, $\lfloor \cdot \rfloor$ is the floor operation. For a reference image block $\mathbf{y}_{ij}^r$, it is represented using the overcomplete dictionary $\mathbf{D}$:

$$\mathbf{y}_{ij}^r = \mathbf{D}\mathbf{x}_{ij}^r, \text{ s.t. } \left\| \mathbf{y}_{ij}^r - \mathbf{D}\mathbf{x}_{ij}^r \right\|_2 \leq \epsilon, \tag{3}$$

Where $\mathbf{x}_{ij}^r$ is the representation vector with sparsity L. Then all L basis vectors used in the representation of $\mathbf{y}_{ij}^r$ are used to constitute a sub-dictionary $\mathbf{\Omega}_{ij}$. Then the distorted block $\mathbf{y}_{ij}^d$ is represented using all the basis vectors in the sub-dictionary $\mathbf{\Omega}_{ij}$:

$$\mathbf{y}_{ij}^d = \mathbf{\Omega}\mathbf{x}_{ij}^d, \text{ s.t. } \left\| \mathbf{y}_{ij}^d - \mathbf{\Omega}\mathbf{x}_{ij}^d \right\|_2 \leq \epsilon. \tag{4}$$

With the representation vectors of the reference and distorted image blocks, i.e., $\left\{x_{ij}^r\right\}$ and $\left\{x_{ij}^d\right\}$, we propose to construct two feature maps. Then the similarity between the feature maps is utilized to represent the structure distortions. In this paper, the feature maps are computed as follows:

$$\mathbf{FM}^r(i,j) = \sqrt{\left\langle \mathbf{x}_{ij}^r, \mathbf{x}_{ij}^r \right\rangle}, \ \mathbf{FM}^d(i,j) = \sqrt{\left\langle \mathbf{x}_{ij}^d, \mathbf{x}_{ij}^d \right\rangle}, \tag{5}$$

Where $\mathbf{FM}^r$ and $\mathbf{FM}^d$ denote the feature maps for the reference and distorted images, respectively, and $\langle \cdot, \cdot \rangle$ denotes inner product.

Figure 3 shows a reference image and its three distorted versions, together with their feature maps. It is observed from the figure that the feature maps are sensitive to distortions. This indicates that the feature maps can be used to measure the distortions in an image, based on which image quality can be evaluated.

With the feature maps $\mathbf{FM}^r$ and $\mathbf{FM}^d$, sparse feature similarity is computed as:

$$\mathbf{S}_{\mathrm{FM}}(i,j) = \frac{2\mathbf{FM}^r(i,j)\mathbf{FM}^d(i,j) + c_1}{[\mathbf{FM}^r(i,j)]^2 + [\mathbf{FM}^d(i,j)]^2 + c_1}, \tag{8}$$

where $i = 1, 2, \ldots, M, j = 1, 2, \ldots, N$, c_1 is a constant used to ensure numerical stability. The similarity of the feature maps is computed pixel-wisely, so $\mathbf{S}_{\mathrm{FM}}$ serves as a local qualitymap, which is indicative of local distortions in the image.

Fig. 3. Images and feature maps. (a) Reference image; (b) Gaussian blurred image; (c) JPEG compressed image; (d) Sparse sampling and reconstruction; (e)-(h) are feature maps of (a)-(d).

3.3 Auxiliary Feature Similarity

The sparse features extracted based on a learned overcomplete dictionary can capture the underlying structures in images and provide higher-level clues for image analysis. This also indicates that they are insensitive to weak distortions in images. In addition, image quality is affected by several other factors, including contrast, color and luminance. Therefore, sparse feature alone may not work well. Therefore, auxiliary features are integrated, including gradient, color and luminance.

Gradient can capture the structure changes caused by weak distortions, so gradient acts as a complementary feature for measuring structure distortions. In this paper, gradient is computed in the Y component of YCbCr color space, a model widely used in image processing. For a reference image and a distorted image, their Y, Cb, Cr compoents are denoted by $\mathbf{Y}^r$, $\mathbf{Cb}^r$, $\mathbf{Cr}^r$ and $\mathbf{Y}^d$, $\mathbf{Cb}^d$, $\mathbf{Cr}^d$, respectively. The gradients of $\mathbf{Y}^r$ and $\mathbf{Y}^d$ are first computed using the Scharr operator [7], which are denoted by $\mathbf{G}^r$ and $\mathbf{G}^d$, respectively. Then the local gradient similarity is computed as:

$$\mathbf{S}_G(i,j) = \frac{2\mathbf{G}^r(i,j)\mathbf{G}^d(i,j) + c_2}{[\mathbf{G}^r(i,j)]^2 + [\mathbf{G}^d(i,j)]^2 + c_2}. \tag{9}$$

Similarly, local color similarity is computed in Cb and Cr channels as follows:

$$\mathbf{S}_C(i,j) = \frac{2\mathbf{Cb}^r(i,j)\mathbf{Cb}^d(i,j) + c_3}{[\mathbf{Cb}^r(i,j)]^2 + [\mathbf{Cb}^d(i,j)]^2 + c_3} \cdot \frac{2\mathbf{Cr}^r(i,j)\mathbf{Cr}^d(i,j) + c_3}{[\mathbf{Cr}^r(i,j)]^2 + [\mathbf{Cr}^d(i,j)]^2 + c_3}. \tag{10}$$

Luminance is the final feature. In this paper, luminance similarity is computed using the method in [16], and the luminance similarity score is denoted by Q_L.

3.4 Sparse-Feature-Based Pooling

So for, we have obtained the sparse feature similarity map $\mathbf{S}_{FM}$, gradient similarity map $\mathbf{S}_G$, color similarity map $\mathbf{S}_C$ and luminance similarity score Q_L. The final step is to combine them to generate an overall quality score. Furthermore, in order to adapt to the characteristics of the HVS, a perceptual weighting is needed. In this paper, a weighting map is proposed using the feature maps as follows:

$$\mathbf{W}(i,j) = \max\big(\mathbf{FM}^r(i,j), \mathbf{FM}^d(i,j)\big). \tag{11}$$

Then the three similarity maps are weighted to produce three scores, i.e., Q_{FM}, Q_G and Q_C. Then the final image quality score is obtained by:

$$Q = Q_{FM} \cdot (Q_G)^\alpha \cdot (Q_C)^\beta \cdot (Q_L)^\gamma, \tag{12}$$

where α, β, γ are used to adjust the relative importance of gradient, color and luminance. In this paper, they are experimentally set to $\alpha = 0.25$, $\beta = 0.03$, $\gamma = 0.65$.

4 Experimental Results and Discussions

We evaluate the performance of the proposed method on three popular image quality databases, i.e., LIVE, CSIQ and TID2013. There are 779, 886 and 3000 images in these databases, respectively. In LIVE and CSIQ, the subjective quality is measured using Difference Mean Opinion Score (DMOS), and in TID2013 MOS is used.

Three criterions are employed to evaluate the performances, i.e., Pearson linear correlation coefficient (PLCC), Spearman rank order correlation coefficient (SRCC) and root mean squared error (RMSE). Before computing these criteria, a five-parameter logistic mapping conducted between the subjective and predicted scores [7]. In implementation, the images are divided into 8×8 blocks for sparse representation. A group of ten natural images are used to train the overcomplete dictionary of size 64×256. Specifically, 10000 randomly selected 8×8 patches are adopted to train the dictionary using Lee's dictionary learning method [14]. Figure 4 shows the training images and the overcomplete dictionary. In experiments, the sparsity L is set to 2.

Figure 5 shows the scatter plots of the subjective scores against the computed QASD scores on the three databases. It is observed that the QASD scores are highly consistent with subjective evaluations.

Table 1 summarizes the experimental results of QASD and the state-of-the-art IQA metrics, including SSIM [2], MS-SSIM [3], VIF [5], MAD [6], FSIM [7], GSM [8], GMSD [9] and SFF [16]. In each database, the best three results are marked in boldface. It is observed form the table that QASD outperforms all the other metrics in TID2013, which is the largest database. In LIVE and CSIQ, although QASD does not performance the best, its performance is quite similar to the best metrics, which are MAD and SFF, respectively.

In this paper, a new sparse-feature-based weighting function is proposed to conduct the pooling. To evaluate the effectiveness of the new weighting function, we conduct

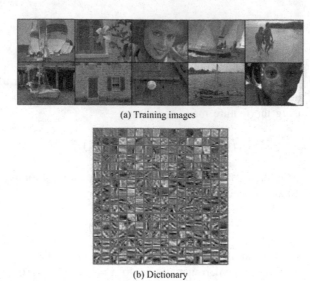

(a) Training images

(b) Dictionary

Fig. 4. Training images and the overcomplete dictionary.

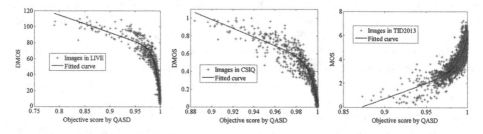

Fig. 5. Scatter plots of subjective scores against predicted scores by QASD.

Table 1. Experimental results of QASD and state-of-the-art metrics.

Database	Criterion	SSIM	MS-SSIM	VIF	MAD	FSIM	GSM	GMSD	SFF	QASD
TID2013	PLCC	0.7895	0.8329	0.7720	0.8267	**0.8589**	0.8463	0.8553	**0.8706**	**0.8894**
	SRCC	0.7417	0.7859	0.6677	**0.8083**	0.8022	0.7946	0.8044	**0.8513**	**0.8657**
	RMSE	0.7608	0.6861	0.7880	0.6976	**0.6349**	0.6603	0.6423	**0.6099**	**0.5667**
LIVE	PLCC	0.9212	0.9489	**0.9604**	**0.9675**	0.9597	0.9512	0.9595	**0.9632**	0.9602
	SRCC	0.9226	0.9513	0.9636	**0.9669**	0.9634	0.9561	0.9603	**0.9649**	**0.9646**
	RMSE	10.6320	8.6184	**7.6137**	**6.9073**	7.6742	8.4326	7.6937	**7.3460**	7.6334
CSIQ	PLCC	0.8579	0.8991	0.9277	**0.9500**	0.9118	0.8964	**0.9541**	**0.9643**	0.9466
	SRCC	0.8719	0.9133	0.9195	0.9466	0.9240	0.9108	**0.9570**	**0.9627**	**0.9516**
	RMSE	0.1349	0.1149	0.0980	**0.0820**	0.1078	0.1164	**0.0786**	**0.0695**	0.0846

an experiment of QASD without the sparse-feature-based weighting. The performances of the proposed method with/without the weighting function are summarized in Table 2. It can be seen from the table that for all databases, integrating sparse-feature-based weighting improves the performances, in terms of both prediction accuracy and monotonicity.

Table 2. Performance of QASD with/without sparse-feature-based weighting.

Weighting	LIVE		CSIQ		TID2013	
	PLCC	SRCC	PLCC	SRCC	PLCC	SRCC
YES	0.9602	0.9646	0.9466	0.9516	0.8894	0.8657
NO	0.9546	0.9574	0.9409	0.9465	0.8844	0.8618

5 Conclusion

In this paper, we have presented a new image quality metric based on sparse representation. The contributions of this paper are as follows. First, an adaptive sub-dictionary selection method is proposed. In order to compare the reference-distorted image block-pairs, they have to be represented using the samebasis vectors in the dictionary. The proposed method solves this problem by constructing a sub-dictionary using the basis vectors that have been adopted for representing the reference block. Then the distorted block is represented using all basis vectors in the sub-dictionary. Second, feature maps are constructed based on the sparse coefficients. The feature maps can capture the distortions in an image, and they are used to generate a local quality map between the reference and distorted images. Third, the feature maps of the reference anddistorted images are combined to generate a weighting map, which is used to adapt to the characteristics of the human visual system and produce an overall quality score. With the consideration that the sparse features are not sensitive to weak distortions and image quality is affected by different factors, gradient,color and luminance are integrated as auxiliary features for more efficient quality assessment. We have done experiments on three public image quality databases to evaluate the performance of the proposed method. The experimental results show that our metric can produce quality scores highly consistent with subjective evaluations, and it performs consistently well across databases. We have also compared our method with eight popular image quality metrics, and the results demonstrate that our method achieves state-of-the-art performance.

Color is important in the perception of image quality. In this paper, color is processed separately. As future work, we will try to process color distortions in a holistic way [17], which we believe is beneficial for more advanced quality assessment.

Acknowledgements. This work is supported by National Natural Science Foundation of China (61379143), Fundamental Research Funds for the Central Universities (2015QNA66), and the S&T Program of Xuzhou City (XM13B119).

References

1. Lin, W.S., Jay Kuo, C.-C.: Perceptual visual quality metrics: a survey. J. Vis. Commun. Image Represent. **22**(4), 297–312 (2011)
2. Wang, Z., Bovik, A.C., Sheikh, H.R., Simoncelli, E.P.: Image quality assessment: from error visibility to structural similarity. IEEE Trans. Image Process. **13**(4), 600–612 (2004)
3. Wang, Z., Simoncelli, E.P., Bovik, A.C.: Multi-scale structural similarity for image quality assessment. In: Proceedings of the IEEE Asilomar Conference on Signals, Systems and Computers, pp. 1398–1402 (2003)
4. Sheikh, H.R., Bovik, A.C., de Veciana, G.: An information fidelity criterion for image quality assessment using natural scene statistics. IEEE Trans. Image Process. **14**(12), 2117–2128 (2005)
5. Sheikh, H.R., Bovik, A.C.: Image information and visual quality. IEEE Trans. Image Process. **15**(2), 430–444 (2006)
6. Larson, E.C., Chandler, D.M.: Most apparent distortion: Full-reference image quality assessment and the role of strategy. J. Electron. Imaging **19**(1), 001–006 (2010)
7. Zhang, L., Zhang, L., Mou, X.Q., Zhang, D.: FSIM: a feature similarity index for image quality assessment. IEEE Trans. Image Process. **20**(8), 2378–2386 (2011)
8. Liu, A.M., Lin, W.S., Narwaria, M.: Image quality assessment based on gradient similarity. IEEE Trans. Image Process. **21**(4), 1500–1512 (2012)
9. Xue, W.F., Zhang, L., Mou, X.Q., Bovik, A.C.: Gradient magnitude similarity deviation: a highly efficient perceptual image quality index. IEEE Trans. Image Process. **23**(2), 684–695 (2014)
10. Olshausen, B.A., Field, D.J.: Emergence of simple-cell receptive field properties by learning a sparse code for natural images. Nature **381**(13), 607–609 (1996)
11. Cheng, C., Wang, H.L.: Image quality assessment using sparse representation in ICA domain. In: Asia-Pacific Signal and Information Processing Association Annual Summit and Conference, pp. 1047–1050 (2011)
12. He, L.H., Tao, D.C., Li, X.L., Gao, X.B.: Sparse representation for blind image quality assessment. In: International Conference on Computer Vision and Pattern Recognition (CVPR), pp. 1146–1153 (2012)
13. Guha, T., Nezhadarya, E., Ward, R.K.: Sparse representation-based image quality assessment. Sig. Proc. Image Comm. **29**(10), 1138–1148 (2014)
14. Lee, H., Battle, A., Raina, R., Ng, A.Y.: Efficient sparse coding algorithms. In: Proceedings 19th Annual Conference Neural Information Processing Systems (NIPS), pp. 801–808 (2007)
15. Pati, Y.C., Rezaiifar, R., Krishnaprasad, P.S.: Orthogonal matching pursuit: Recursive function approximation with applications to wavelet decomposition. In: Proceedings 27th Annual Asilomar Conference Signals, Systems, and Computers, pp. 40–44 (1993)
16. Chang, H.W., Yang, H., Gan, Y., Wang, M.H.: Sparse feature fidelity for perceptual image quality assessment. IEEE Trans. Image Process. **22**(10), 4007–4018 (2013)
17. Chen, B.J., Shu, H.Z., Coatrieux, G., Chen, G., Sun, X.M., Coatrieux, J.L.: Color image analysis by quaternion-type moments. J. Math. Imaging Vis. **51**(1), 124–144 (2015)

Single Image Super-Resolution via Iterative Collaborative Representation

Yulun Zhang[1][✉], Yongbing Zhang[1], Jian Zhang[2],
Haoqian Wang[1], and Qionghai Dai[1,3]

[1] Graduate School at Shenzhen, Tsinghua University, Shenzhen 518055, China
zhangy114@mails.tsinghua.edu.cn
[2] Institute of Digital Media, Peking University, Beijing 100871, China
[3] Department of Automation, Tsinghua University, Beijing 100084, China

Abstract. We propose a new model called iterative collaborative representation (ICR) for image super-resolution (SR). Most of popular SR approaches extract low-resolution (LR) features from the given LR image directly to recover its corresponding high-resolution (HR) features. However, they neglect to utilize the reconstructed HR image for further image SR enhancement. Based on this observation, we extract features from the reconstructed HR image to progressively upscale LR image in an iterative way. In the learning phase, we use the reconstructed and the original HR images as inputs to train the mapping models. These mapping models are then used to upscale the original LR images. In the reconstruction phase, mapping models and LR features extracted from the LR and reconstructed image are then used to conduct image SR in each iteration. Experimental results on standard images demonstrate that our ICR obtains state-of-the-art SR performance quantitatively and visually, surpassing recently published leading SR methods.

Keywords: Iterative collaborative representation · Super-resolution

1 Introduction

Single image super-resolution (SR), a classical and important problem in image processing and computer vision, aims at restoring a high-resolution (HR) image from its degraded low-resolution (LR) measurement, while minimizing visual artifacts as far as possible. Recent popular and state-of-the-art SR methods are mostly based on machine learning (ML). These methods either establish the mapping relationships between the LR and HR features from external low- and high-resolution exemplar pairs [1,2,4,6,8,9,12,14,15,17], or learn an end-to-end mapping between the LR and HR images by a deep learning technique [3]. These methods have been obtaining superior and challenging results, but neglect to use the reconstructed image for further SR enhancement.

This work was partially supported by the National Natural Science Foundation of China under Grant 61170195, U1201255, and U1301257.

Y.-S. Ho et al. (Eds.): PCM 2015, Part II, LNCS 9315, pp. 63–73, 2015.
DOI: 10.1007/978-3-319-24078-7_7

The neighbor embedding (NE) based [1,2] and sparse coding (SC) based methods [8,9,14,15] are the main streams and represent the state-of-the-art methods among the ML-based algorithms. These methods often upscale the LR input to the size of the desired HR image using simple interpolation-based methods (e.g., Bicubic interpolation) as a start of SR. In these methods, great efforts have been taken to extract features from the up-scaled LR images and learn the mapping relationships between the features of LR and HR images. Similar to Zeyde's method in [15] LR features are extracted to recover the corresponding HR features and the overlapping reconstructed HR patches are finally averaged to obtain the desired HR images. When we apply the principal component analysis (PCA) algorithm to reduce the dimensionality of the LR features while preserving 99.9 % of the average energy, the dimensionality of the LR features would be reduced largely and far smaller than that of HR features. However, if we extract the same features as LR features above from either the reconstructed or the original HR images and apply the same PCA dimensionality reducing process, the dimensionality of the extracted features will be larger than that of LR features. This inspires us to make an assumption that the reconstructed HR image contains more useful information than the interpolated LR image for high-quality image SR. So, why not utilize the reconstructed HR image as an intermediary result for the final HR image reconstruction?

Motivated by the observation above, in this paper, we propose a simple yet efficient model called iterative collaborative representation (ICR). We make use of the reconstructed HR image as an intermediary result for further image SR in an iterative way. The main difference between our method and the popular

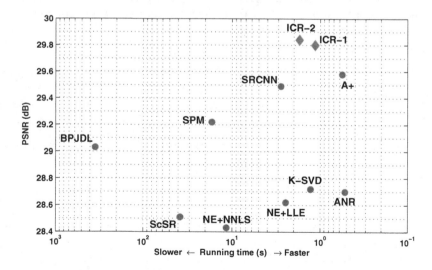

Fig. 1. Average running time (s) versus PSNR (dB) for the tested super-resolution (SR) methods (scaling factor $s = 3$). The proposed ICR (shown in red) obtains the state-of-the-art SR results and maintains competitive speed in comparison to other ML-based methods. More details are shown in Tables 1 and 2 (Color figure online).

ML-based approaches [1–4,6,8,9,12,14,15,17] lies in that we extract LR features from the reconstructed HR image to recover the next better reconstructed HR image. Based on Timofte *et al.*'s work about using collaborative representation [16] for fast image SR [8,9], we learn the projection matrices mapping LR feature spaces to HR feature spaces offline for each iteration in the learning phase. The goal is to further improve the final HR output in an iterative way. As a result, our ICR achieves superior high-quality results compared with the previous most efficient SR methods, such as SRCNN [3] and A+ [9], while maintaining computation efficiency (see Fig. 1) without additional enhancement processes, like back-projection method [5] in Yang *et al.*'s [14] and He *et al.*'s [4] work.

Overall, the contributions of this work can be mainly summarized as follows. **First**, we assume that the reconstructed HR image contains more useful information than that of the original interpolated image generated by simple methods (e.g., Bicubic interpolation). We extract LR features from the reconstructed HR image for further image SR. **Second**, we present an iterative collaborative representation model for image SR. The model would enhance the quality of the SR results significantly by only one iteration. **Third**, we demonstrate that ICR is useful in SR and can achieve high-quality and speed efficiently with little pre/post-processing.

2   Related Work

Neighbor embedding (NE) [1,2], sparse coding (SC) [4,6,8,9,14,15], and deep learning [3] approaches are the three main types of ML-based SR algorithms, which train the mapping relationships between LR and HR features. Chang *et al.* [2] generated HR patches by using a manifold embedding technique. Bevilacqua *et al.* [1] proposed another NE-based SR algorithm using nonnegative neighbor embedding. Yang *et al.* [13,14] proposed a SC-based SR algorithm by assuming that LR and HR features share the same reconstruction coefficients over their corresponding LR and HR dictionaries. This work was further developed by Zeyde *et al.* [15], who utilized PCA and orthogonal matching pursuit (OMP) [10] to reduce the dimensionality of LR features and solve sparse representation respectively. More sophisticated SC formulations were proposed recently, for example, a Bayesian method using a beta process prior was applied to learn the over-complete dictionaries in [4] and a statistical prediction model based on sparse representation was used in [6]. At the same time, fast SR algorithms were proposed by combining dictionary learning [8] or clustering [12] and regression. Timofte *et al.* proposed anchored neighborhood regression (ANR) [8] learning sparse dictionaries and regressors anchored to the dictionary atoms. Yang *et al.* proposed to cluster features into numerous subspaces and created simple mapping functions to generate SR images [12]. Very recently, Dong *et al.* [3] proposed a model super-resolution convolutional neural network (SRCNN) by using a deep learning method. In these methods, the SR procedure started by interpolating the LR input to the size of the desired HR image and did not utilize the reconstructed HR image for further image SR.

Collaborative representation mainly developed by Zhang *et al.* has recently shown an extensive popularity partially due to its powerful performance in pattern classification [16]. By combining dictionary learning and collaborative representation [16], Timofte *et al.* proposed adjusted anchored neighborhood regression (A+) [9] by learning the regressors from the full training material. This work was further improved in [17] by Zhang *et al.*, who employed unified mutual coherence between the dictionary atoms and atoms/samples when training the dictionary and sampling anchored neighbors for image SR. These methods obtained state-of-the-art SR results with fast speed by introducing collaborative representation [16], from which our ICR also benefits.

Based on the fact that there have been a few studies of using the reconstructed image for further SR, we propose the ICR model. In this model, we study the mapping relationships between features of the reconstructed HR image and its corresponding original HR image. The major steps of our ICR are illustrated in Fig. 2. To the best of our knowledge, the image SR problem has not witnessed such an SR framework.

3   Iterative Collaborative Representation for Super-Resolution

3.1   Image SR Based on Collaborative Representation

Timofte *et al.* proposed ANR [8] and A+ [9] for fast image SR by using collaborative representation [16]. Based on these works, we give a brief review of the pipeline of collaborative representation [16] based SR. Let $\left\{ \boldsymbol{y}_L^i \right\}_{i=1}^{N_s}$ and $\left\{ \boldsymbol{y}_H^i \right\}_{i=1}^{N_s}$ be the LR and HR features set extracted from LR/HR training images, the dimensionality of these LR features have been reduced by a PCA projection matrix $\mathbf{P}$. Then the LR dictionary $\boldsymbol{D} = \{ \boldsymbol{d}_k \}_{k=1}^{K}$ is trained in the LR feature space. For each LR dictionary atom $\boldsymbol{d}_k$, we group its LR anchored neighborhood $\boldsymbol{N}_{L,k}$ by searching max nearest training LR features and the corresponding HR features are used to form $\boldsymbol{N}_{H,k}$. We use collaborative representation [16] with l_2-norm regularized least squares regression to obtain the projection matrix for each LR-HR feature neighborhood $\{ \boldsymbol{N}_{L,k}, \boldsymbol{N}_{H,k} \}_{k=1}^{K}$. The problem becomes

$$\boldsymbol{x}_i = arg\ \min_{\boldsymbol{x}_i} \left\| \boldsymbol{y}_L^i - \boldsymbol{N}_{L,k} \cdot \boldsymbol{x}_i \right\|_2^2 + \lambda \left\| \boldsymbol{x}_i \right\|_2^2, \tag{1}$$

where $\boldsymbol{y}_L^i \in \mathbb{R}^{n \times 1}$ is an arbitrary LR feature, $\boldsymbol{N}_{L,k}$ is the LR neighborhood whose corresponding LR dictionary atom $\boldsymbol{d}_k$ is the closet to $\boldsymbol{y}_L^i$, $\boldsymbol{x}_i$ is the coefficient vector of $\boldsymbol{y}_L^i$ over $\boldsymbol{N}_{L,k}$, and $\lambda > 0$ is a weighting parameter. The problem above has a closed-form solution given by

$$\boldsymbol{x}_i = \left(\boldsymbol{N}_{L,k}^T \boldsymbol{N}_{L,k} + \lambda \boldsymbol{I} \right)^{-1} \boldsymbol{N}_{L,k}^T \boldsymbol{y}_L^i, \tag{2}$$

from which we can compute the corresponding projection matrix $\boldsymbol{F}_k$ offline by

$$\boldsymbol{F}_k = \boldsymbol{N}_{H,k} \left(\boldsymbol{N}_{L,k}^T \boldsymbol{N}_{L,k} + \lambda \boldsymbol{I} \right)^{-1} \boldsymbol{N}_{L,k}^T, \quad k = 1, ..., K. \tag{3}$$

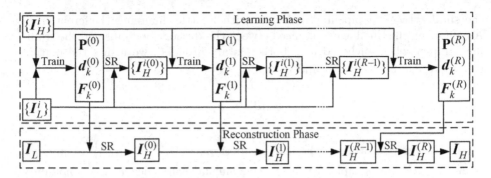

Fig. 2. Pipeline of our iterative collaborative representation framework.

The SR problem can then be solved by calculating for each LR feature y_L^i its nearest dictionary atom d_k, followed by converting y_L^i to its HR feature y_H^i with the stored projection matix F_k via

$$y_H^i = F_k y_L^i. \tag{4}$$

3.2 Learning Phase in ICR

As shown in Fig. 2, the original LR and HR training images are used to obtain the PCA transform matrix $\mathbf{P}^{(0)}$, LR dictionary atoms $\left\{d_k^{(0)}\right\}_{k=1}^{K}$, and their corresponding projection matrices $\left\{F_k^{(0)}\right\}_{k=1}^{K}$ mapping LR feature y_L to its corresponding HR feature y_H. Here, we utilize the same process as that in A+ [9] for the following three reasons: First, the LR features are first- and second-order gradients in the horizontal and vertical directions from the interpolated LR images, as a result, this type of features will be sensitive to the region whose gradients change obviously. Second, the SR results obtained by A+ [9] usually have sharper edges and less artifacts (e.g., ringing artifacts) along the edges than that of most of other state-of-the-art methods (see Figs. 4, 5, 6, and 7). Third, by using one of the published most efficient SR methods as a start point for our implementation, we can demonstrate the effectiveness of our proposed ICR model clearly.

We then utilize A+ [9] to upscale the original LR training images $\left\{I_L^i\right\}_{i=1}^{N_I}$ to $\left\{I_H^{i(0)}\right\}_{1=1}^{N_I}$, whose size is the same as that of their original HR images $\left\{I_H^i\right\}_{i=1}^{N_I}$. In our ICR model, we regard $\left\{I_H^{i(r-1)}\right\}_{r=1}^{R}$ (R is the total number of iteration) as LR images for simplicity. In the r-th iteration of the learning phase, we extract LR and HR feature pairs from $\left\{I_H^{i(r-1)}, I_H^i\right\}_{i=1}^{N_I}$ to learn PCA transform matrix $\mathbf{P}^{(r)}$, LR dictionary atoms $\left\{d_k^{(r)}\right\}_{k=1}^{K}$, and projection matrices $\left\{F_k^{(r)}\right\}_{k=1}^{K}$ using the similar training pipeline as A+ [9] with a little difference. Different from A+ [9] using the Euclidean distance between the dictionary atoms and the training samples when anchored neighborhoods are grouped, we utilize the mutual coherence $c_{k,i}$ between $d_k^{(r)}$ and y_L^i via

$$c_{k,i} = \left| [d_k^{(r)}]^{\mathrm{T}} y_L^i \right|, \tag{5}$$

where $d_k^{(r)}$ is k-th atom in the learned dictionary $D^{(r)} = \left\{ d_k^{(r)} \right\}_{k=1}^{K}$ and y_L^i is an LR training sample. By using the mutual coherence $c_{k,i}$, we not only comply with distance measurements between LR features and dictionary atoms in the reconstruction phase, but also speed up about 25 % in computational time compared with that in A+ [9].

3.3   Reconstruction Phase in ICR

As we learned R relationship projecting models $\left\{ \mathbf{P}^{(r)}, \left\{ d_k^{(r)}, F_k^{(r)} \right\}_{k=1}^{K} \right\}_{r=1}^{R}$ after R iterations in the learning phase, the SR problem can then be efficiently solved in an iterative way (see the reconstruction phase in Fig. 2). For an LR input image I_L, we upscale it to $I_H^{(0)}$ by A+ [9] as the initial input for our ICR model. In the r-th iteration of the reconstruction phase, we extract a set of LR features $\left\{ y_L^{i(r-1)} \right\}$ from the reconstructed HR image $I_H^{(r-1)}$. For each LR feature $y_L^{i(r-1)}$, we search its nearest atom $d_k^{(r)}$ with highest coherence $c_{k,i}$ from the LR dictionary $D^{(r)}$, followed by recovering its HR feature $y_H^{i(r)}$ via (4). The HR patch $p_H^{i(r)}$ is then obtained by adding HR feature $y_H^{i(r)}$ to the corresponding interpolated LR patch p_L^i and we combine these HR patches to a whole HR image $I_H^{(r)}$ by averaging the intensity values over the overlapping regions. Finally, we obtain the R-th SR result $I_H^{(R)}$ as our final HR output I_H.

4   Experimental Results

In this section, a series of experiments are reported to demonstrate the efficiency and robustness of our proposed ICR-based SR algorithm compared with other state-of-the-art methods.

Fig. 3. 8 test images from left to righ: bike, bird, butterfly, flowers, plants, ppt3, starfish, and woman.

4.1   Experimental Settings

In our experiments, for a fair comparison, we use 91 training images proposed by Yang *et al.* [13]. The experiment is conducted on 8 commonly used LR natural images shown in Fig. 3, including humans, animals, and plants. Peak signal-to-noise ratio (PSNR), structural similarity (SSIM) [11], and visual information fidelity (VIF) [7] are employed in our experiments to evaluate the quality of the SR results. All of these evaluation metrics are performed between the luminance channel of the original HR and the reconstructed image. We compare our ICR with Bicubic interpolation and 9 state-of-the-art ML-based SR methods: NE+LLE [2], ScSR [14], K-SVD [15], NE+NNLS [1], ANR [8], BPJDL [4], SPM [6], SRCNN [3], and A+ [9]. The implementations are all from the publicly available codes provided by the authors[1].

We down-sample the original HR images to generate LR images for training and testing on the luminance channel by Bicubic interpolation. We set scaling factor $s = 3$ and the total number of iteration $R = 2$ throughout the paper. Same as A+[9], we set $K = 1024, max = 2048, \lambda = 0.1$ in a set of experimental results.

4.2   Performance

We compare the proposed ICR with other state-of-the-art methods in terms of PSNR, SSIM, and VIF values in Table 1, where the results of ICR-1 and ICR-2 are produced by our ICR model. $R = 1$ and $R = 2$ are set in ICR-1 and ICR-2 respectively. Although ICR does not outperform all of the competing methods for each test image, ICR obtains best results on average in terms of PSNR. On the other hand, ICR achieves the best results for each image in terms of SSIM and VIF. This demonstrates that ICR recovers more structural details than other methods. According to Table 1, a large average PSNR, SSIM, and VIF gains of our ICR-2 over the second best method A+ [9] (regardless of ICR-1) are 0.26 dB, 0.007, and 0.017. Such an improvement is notable, since A+ [9] has been the most efficient dictionary-based SR method so far. This indicates that our ICR is efficient for high-quality image SR by using features extracted from the reconstructed HR image in each iteration. Another amazing phenomenon is that even we conduct only one iteration (i.e., $R = 1$), the SR results by ICR-1 can also outperform the second best method A+ [9] by a large average PSNR, SSIM, and VIF gains. While, the average PSNR, SSIM, and VIF gains of ICR-2 over ICR-1 is very smal. As a result, $R = 1$ or $R = 2$ is an optimal value to balance the quality of SR and computation complexity.

Table 2 shows the running time comparisons of numerous methods with a scaling factor $s = 3$ in the SR phase. We profile the running time of all the methods in a MATLAB 2012a environment using the same machine (3.20 GHz Core(TM) i5-3470 with 16 GB RAM). According to Table 2, ICR ranks the third or fourth place with a very small gap between ICR and the fastest method ANR. And one iteration in our ICR would take very little time while enhancing the SR results significantly.

---

[1] The source code of the proposed ICR will be available after this paper is published.

Table 1. Quantitative comparisons on PSNR (dB), SSIM, and VIF for 3× magnification. For each image, there are three rows: PSNR (dB), SSIM, and VIF. The best result for each image is highlighted.

Images	BI	NE+LLE [2]	ScSR [14]	K-SVD [15]	NE+NNLS [1]	ANR [8]	BPJDL [4]	SPM [6]	SRCNN [3]	A+ [9]	ICR-1	ICR-2
bike	22.83	23.89	23.97	23.89	23.76	23.98	24.23	24.45	24.51	24.68	24.85	**24.92**
	0.704	0.765	0.767	0.765	0.756	0.769	0.781	0.788	0.787	0.798	0.805	**0.808**
	0.310	0.383	0.354	0.382	0.371	0.389	0.404	0.409	0.397	0.422	0.434	**0.437**
bird	32.63	34.54	34.20	34.62	34.16	34.63	34.88	35.15	34.97	35.60	**35.79**	35.68
	0.926	0.948	0.940	0.948	0.943	0.949	0.950	0.952	0.950	0.956	0.958	**0.959**
	0.552	0.659	0.587	0.657	0.640	0.665	0.675	0.668	0.658	0.692	0.704	**0.706**
butter-fly	24.04	25.73	25.69	25.99	25.45	25.89	26.52	26.78	27.59	27.34	27.53	**27.63**
	0.822	0.868	0.862	0.877	0.861	0.870	0.887	0.899	0.901	0.910	0.915	**0.917**
	0.365	0.459	0.416	0.467	0.446	0.468	0.500	0.507	0.516	0.532	0.545	**0.551**
flowers	27.28	28.42	28.33	28.48	28.23	28.52	28.78	28.86	29.02	29.10	29.29	**29.34**
	0.803	0.839	0.831	0.839	0.832	0.841	0.847	0.847	0.849	0.854	0.858	**0.859**
	0.370	0.448	0.401	0.448	0.436	0.453	0.468	0.464	0.462	0.481	0.491	**0.495**
plants	31.30	32.81	32.52	32.79	32.63	32.81	33.13	33.14	33.52	33.79	34.07	**34.16**
	0.872	0.905	0.894	0.904	0.900	0.905	0.910	0.907	0.911	0.921	0.926	**0.927**
	0.476	0.590	0.514	0.582	0.574	0.593	0.605	0.590	0.599	0.634	0.648	**0.655**
ppt3	23.90	25.20	25.32	25.49	25.20	25.28	25.55	25.87	26.32	26.36	26.61	**26.67**
	0.884	0.912	0.908	0.920	0.910	0.913	0.921	0.926	0.929	0.939	0.944	**0.945**
	0.371	0.448	0.420	0.462	0.444	0.452	0.473	0.481	0.482	0.521	0.540	**0.547**
star-fish	27.01	28.15	28.04	28.14	28.04	28.18	28.46	28.79	**29.04**	28.63	28.83	28.88
	0.814	0.852	0.846	0.851	0.847	0.853	0.860	0.865	0.871	0.864	0.870	**0.872**
	0.373	0.447	0.397	0.446	0.438	0.451	0.464	0.470	0.471	0.473	0.483	**0.486**
woman	28.57	30.21	29.99	30.36	30.01	30.30	30.73	30.69	30.92	31.18	31.41	**31.45**
	0.890	0.916	0.904	0.917	0.912	0.917	0.921	0.923	0.924	0.929	0.931	**0.932**
	0.454	0.559	0.481	0.562	0.549	0.562	0.580	0.579	0.571	0.605	0.617	**0.622**
Average	27.19	28.62	28.51	28.72	28.43	28.70	29.03	29.22	29.49	29.58	29.80	**29.84**
	0.839	0.876	0.869	0.878	0.870	0.877	0.885	0.888	0.890	0.896	0.901	**0.903**
	0.409	0.499	0.446	0.501	0.487	0.504	0.521	0.521	0.520	0.545	0.558	**0.562**

Table 2. Running time (seconds) comparisons of different methods

Methods	NE+LLE [2]	ScSR [14]	K-SVD [15]	NE+NNLS [1]	ANR [8]	BPJDL [4]	SPM [6]	SRCNN [3]	A+ [9]	ICR-1	ICR-2
Time	2.47	38.12	1.29	11.59	0.52	349.13	16.88	2.81	0.56	1.15	1.73

4.3 Visual Results

To further demonstrate the effectiveness of ICR, in Figs. 4, 5, 6, and 7 we compare our visual results with that of ANR, BPJDL, SPM, SRCNN, and A+, as they perform the best among all existing methods. As we can see from Figs. 4 and 6, ANR, BPJDL, SPM, and SRCNN would always produce obvious ringing artifacts along the edges and fail to recover detailed textures. A+ can alleviate the blurring and ringing artifacts along the main edges, however, it also generates blurred edges (see Fig. 5) and introduces some unpleasing artifacts (see Fig. 7) when compared to ICR. This also helps to demonstrate that the reconstructed HR image contains more useful information than the interpolated LR image for better SR performance. As can be observed, our ICR produces sharper edges without obvious ringing effects (see Figs. 4 and 7), less blurring effects (see Fig. 5), and finer textures with more details (see Fig. 6).

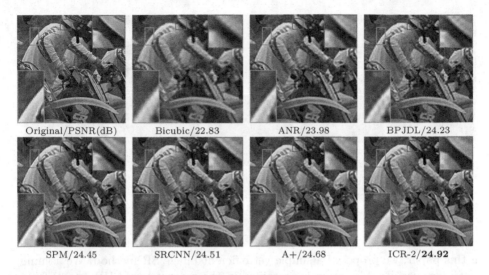

Original/PSNR(dB)   Bicubic/22.83   ANR/23.98   BPJDL/24.23

SPM/24.45   SRCNN/24.51   A+/24.68   ICR-2/**24.92**

Fig. 4. Visual comparisons on **bike** with a scaling factor 3.

Original/PSNR(dB)   Bicubic/31.30   SRCNN/33.52   A+/33.79   ICR-2/**34.16**

Fig. 5. Visual comparisons on **plants** with a scaling factor 3.

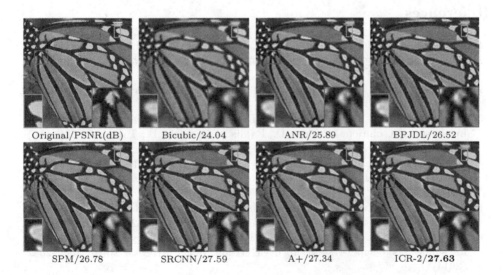

Original/PSNR(dB)   Bicubic/24.04   ANR/25.89   BPJDL/26.52

SPM/26.78   SRCNN/27.59   A+/27.34   ICR-2/**27.63**

Fig. 6. Visual comparisons on **butterfly** with a scaling factor 3.

| Original/PSNR(dB) | Bicubic/28.57 | SRCNN/30.92 | A+/31.18 | ICR-2/**31.45** |

Fig. 7. Visual comparisons on **woman** with a scaling factor 3.

5   Conclusion

In this paper, we proposed a simple yet efficient image SR method. We assume that more useful information is contained in the reconstructed HR image, from which we extract LR features for further image SR. Our ICR model employs dictionary learning and collaborative representation [16] to learn numerous projection matrices offline in each iteration. By using these projection matrices, we first upscale the LR image to obtain the initial HR image, from which we extract LR features to reconstruct the next better HR image. When compared with other state-of-the-art methods, our ICR shows the best performance both in terms of objective evaluation metrics and subjective visual results. We conjecture that additional performance can be gained by using more iterations and different feature extraction strategies.

References

1. Bevilacqua, M., Roumy, A., Guillemot, C., Alberi-Morel, M.L.: Low-complexity single-image super-resolution based on nonnegative neighbor embedding. In: BMVC (Sep 2012)
2. Chang, H., Yeung, D.Y., Xiong, Y.: Super-resolution through neighbor embedding. In: CVPR, pp. 1–6 (2004)
3. Dong, C., Loy, C.C., He, K., Tang, X.: Learning a deep convolutional network for image super-resolution. In: Fleet, D., Pajdla, T., Schiele, B., Tuytelaars, T. (eds.) ECCV 2014, Part IV. LNCS, vol. 8692, pp. 184–199. Springer, Heidelberg (2014)
4. He, L., Qi, H., Zaretzki, R.: Beta process joint dictionary learning for coupled feature spaces with application to single image super-resolution. In: CVPR, pp. 345–352 (2013)
5. Irani, M., Peleg, S.: Motion analysis for image enhancement: Resolution, occlusion, and transparency. J. Vis. Commun. Image Represent. **4**(4), 324–335 (1993)
6. Peleg, T., Elad, M.: A statistical prediction model based on sparse representations for single image super-resolution. IEEE Trans. Image Process. **23**(6), 2569–2582 (2014)
7. Sheikh, H.R., Bovik, A.C.: Image information and visual quality. IEEE Trans. Image Process. **15**(2), 430–444 (2006)

8. Timofte, R., De, V., Gool, L.V.: Anchored neighborhood regression for fast example-based super-resolution. In: ICCV, pp. 1920–1927 (2013)
9. Timofte, R., De Smet, V., Van Gool, L.: A+: adjusted anchored neighborhood regression for fast super-resolution. In: Cremers, D., Reid, I., Saito, H., Yang, M.-H. (eds.) ACCV 2014. LNCS, vol. 9006, pp. 111–126. Springer, Heidelberg (2015)
10. Tropp, J.A., Gilbert, A.C.: Signal recovery from random measurements via orthogonal matching pursuit. IEEE Trans. Inf. Theor. 53(12), 4655–4666 (2007)
11. Wang, Z., Bovik, A.C., Sheikh, H.R., Simoncelli, E.P.: Image quality assessment: from error visibility to structural similarity. IEEE Trans. Image Process. 13(4), 600–612 (2004)
12. Yang, C.Y., Yang, M.H.: Fast direct super-resolution by simple functions. In: ICCV, pp. 561–568 (2013)
13. Yang, J., Wright, J., Huang, T., Ma, Y.: Image super-resolution as sparse representation of raw image patches. In: CVPR, pp. 1–8 (2008)
14. Yang, J., Wright, J., Huang, T.S., Ma, Y.: Image super-resolution via sparse representation. IEEE Trans. Image Process. 19(11), 2861–2873 (2010)
15. Zeyde, R., Elad, M., Protter, M.: On single image scale-up using sparse-representations. In: Proceedings of the 7th International Conference Curves Surfing, pp. 711–730 (2010)
16. Zhang, L., Yang, M., Feng, X.: Sparse representation or collaborative representation: Which helps face recognition? In: ICCV, pp. 471–478 (2011)
17. Zhang, Y., Gu, K., Zhang, Y., Zhang, J., Dai, Q.: Image super-resolution based on dictionary learning and anchored neighborhood regression with mutual incoherence. In: ICIP (2015)

Influence of Spatial Resolution
on State-of-the-Art Saliency Models

Zhaohui Che[(⊠)], Guangtao Zhai, and Xiongkuo Min

Institute of Image Communication and Information Processing,
Shanghai Jiao Tong University, Shanghai, China
{chezhaohuihy,zhaiguangtao,minxiongkuo}@gmail.com

Abstract. Visual attention has been widely investigated and applied in recent decades. Various computation models have been proposed to modeling visual attention, but most researches are conducted under the assumption that given images have few limited spatial resolutions. Spatial resolution is an important feature of image. Image resolution may have some influence on visual attention, and it may also affect the effectiveness of visual attention models. The influence of spatial resolution on saliency models has not been systematically investigated before. In this paper, we discuss two problems related to image resolution and saliency: (1) Most saliency models contain down-sampling function which changes the resolution of original images to lower the computational complexity and keep the formalization of the algorithm. In the first part, we discuss the influence of the down-sampling ratio on the effectiveness of classic saliency models. (2) In the second part, we investigate the effectiveness of saliency models in images of various resolutions. A dataset which provides images and corresponding eye movement data in various spatial resolutions is used in this part. We apply the default rescaling parameters and keep them unchanged. Then we analyze the performance of classic models on 8 resolution levels. In summary, we systematically investigate and analyze problems concerning spatial resolution in the research of saliency modeling. The results of this work can provide a guide to the use of classic models in images of different resolutions and they are helpful to computational complexity optimization.

Keywords: Saliency model · Spatial multi-resolutions · Down-sample · Average gradient

1 Introduction

The research of saliency modeling has been developing for several years, and it aims at detecting regions that human visual system (HVS) is interested in. Dozens of saliency models mentioned in [1] have been proposed and widely applied to computer vision, computer graphics, robotics and other research areas. Spatial resolution is an important feature of image, and the influence of spatial resolution on saliency models has not been systematically investigated before.

Previous works have studied influence of spatial resolution on human fixations. The subjective experiment of Judd [2] focused on the consistency of human fixations in different resolutions and they drew some conclusions about human fixations and

© Springer International Publishing Switzerland 2015
Y.-S. Ho et al. (Eds.): PCM 2015, Part II, LNCS 9315, pp. 74–83, 2015.
DOI: 10.1007/978-3-319-24078-7_8

resolutions: (1) Fixations from low resolution images can predict fixations from high resolution images; (2) Consistency of fixations increases as the resolution rises. But when resolution reaches the critical value, human fixations' consistency remains stable although the resolution increases. The advantages of Judd's work were sufficient dataset and enough observers. But they paid no attention to saliency models' objective performance in images of multi- resolutions. We design objective experiments to evaluate the influence of spatial resolution on twelve state-of-the-art saliency models. Anton Garcia-Diaz [3] also analyzed the performance of their saliency model in multi-resolutions. They controlled spatial resolution by changing human visual field and they just evaluated performance of five saliency models. We control spatial resolution critically and test more models on sufficient datasets.

Most of saliency models adopt default down-sample ratio to lower the computational complexity and keep the formalization of algorithms. A comprehensive analysis and evaluation of the down-sample ratio's influence on different saliency models' performance will be helpful. It provides a guidance to computational complexity minimization and robust modeling. Except for the influence of down-sample ratio, we also measure the performance of state-of-the-art models in images of various resolutions. Moreover, we adopt a novel approach to measure saliency models' robustness to images in various resolutions in this paper. Each model owns best resolution corresponding to its peak performance. In a word, we research two problems here: the influence of down-sample ratio on saliency performance, and saliency models' robustness to images in multi-resolutions.

The rest of this paper is organized as follows: In Sect. 2, we introduce the test datasets and evaluation metrics. Moreover, two experiments related to two problems discussed before will be introduced in detail in this section. Section 3 shows experiments results, and we will highlight our contributions here. We summarize this paper and give future research directions in Sect. 4.

2 Objective Experiment

2.1 Test Models

We choose thirteen state-of-the-art saliency models as the test group in this paper. These models are good at predicting human fixations and all of them provide executable codes refer to [4]. The test group can widely represent most saliency models because it contains bottom-up models and combination of bottom-up and top-down models. Thus the results of this paper are widely applicable. We list all test models in Table 1.

2.2 Metric and Database

We adopt the Area Under Curve of Judd (AUC-Judd) [2] as evaluation metric in this paper. AUC is a popular metric reflecting the accuracy rate about predicting human fixations of saliency models. Each saliency model will generate a saliency map from original image. We can get eye tracking data of original image by eye tracker,

Table 1. Model Introduction

	Model	Author	Year	Frame	Self-Rescaling	Category
1	Itti & Koch [5]	Laurent Itti et al.	1998	BU	N	C
2	Torralba [6]	Antonio Torralba et al.	2003	BU &TD	N	B
3	AIM [7]	Neil D.B & John Tsotsos	2005	BU	Y	B & I
4	GBVS [8]	Jonathan Harel et al.	2006	BU	N	G
5	Image Signature [9]	Xiaodi Hou et al.	2012	BU	Y	S
6	SUN [10]	Lingyun Zhang et al.	2008	BU & TD	Y	B
7	AWS [3]	Anton Garcia-Diaz et al.	2010	BU & TD	Y	O
8	Murray [11]	Naila Murry et al.	2011	BU	N	C
9	RARE2012 [12]	Nicolas Riche et al.	2013	BU	N	I
10	Covsal [13]	Erkut Erdem et al.	2012	BU	Y	O
11	Achanta [14]	Ranhakrishna Achanta et al.	2009	BU	N	S
B	Context Aware [15]	Stas Goferman et al.	2008	BU	N	O

Frame : BU(bottom-up), TD(top-down), BU&TD (combination of BU and TD); **Self-Rescaling** (whether models contain down-sample steps in their algorithms) : Y(yes), N(no); **Category**: C(Cognitive Model), B(Bayesian Model), D(Decision Model), I(Information Theoretic Model), G(Graphical Model), S(Spectral Analysis Model), P(Pattern Classification Model), O(Other Model)

and generate human fixation map. The saliency map will be treated as a binary classifier to divide pixels into positive and negative samples at various thresholds. The true positive rate is the proportion of saliency map values above threshold at fixation locations. The false positive rate is the proportion of saliency map values above threshold at non-fixated pixels. In this metric, the thresholds are sampled from saliency map values. The area of region under the ROC curve is AUC. Higher AUC score indicates that saliency model can predict human fixations more precisely.

We adopt Bruce and Tsotsos' database [7] and MIT Low-resolution database [2] as test datasets in this paper.

2.3   Experiment Design

The frameworks of two experiments are shown in Fig. 1. Figure 1 (a) is corresponding to the first experiment which investigates the influence of down-sample ratio. Figure 1(b) is corresponding to the second experiment which evaluates different saliency models' robustness to images in multi-resolutions.

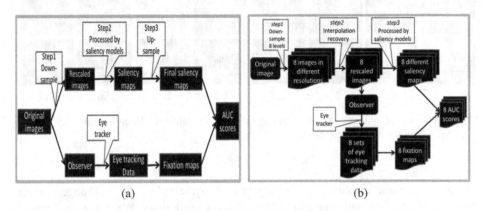

(a)   (b)

Fig. 1. (a): The flow chart of experiment one about influence of down-sample ratio on models' performance; (b): The flow chart of experiment two about different saliency models' robustness to images in different spatial resolutions.

The first experiment aims at evaluating the influence of down-sample ratio. We kept the input images unchanged but changed the down-sample ratio by adjusting parameters in algorithms. It's worth noting that we added a simple down-sample step by a factor of 2 to algorithms without self-rescaling step (refer to self-rescaling option in Table 1). Then we got AUC scores corresponding to different down-sample ratios. Judd's implementation [16] was adopted to calculate AUC score in this experiment. We changed the down-sample ratio (refer to step 1 in Fig. 1(a)) while keeping other parameters constant. Down-sample ratios were regarded as independent variables and AUC scores were dependent variables in this experiment.

We chose MIT Low-resolution database proposed by Judd et al. as test dataset in the second experiment. Two sets of images and their saliency maps from MIT Low-resolution database are shown in Fig. 2. This database was generated as follows: Judd et al. applied a low-pass binomial filter to each feature channel and down-sampled images. Since low-pass filter compressed the range of color channels, Judd et al. rescaled each down-sampled image while maintaining the mean of luminance pixel values unchanged. The down-sampled images were up-sampled by the binomial filter to the original size for visualization. They used the code of Steerable Pyramid Toolbox proposed by E.P Simoncelli [17] in down-sample and up-sample steps. As illustrated in Fig. 1(b), the second experiment aims at quantizing the saliency models' robustness to

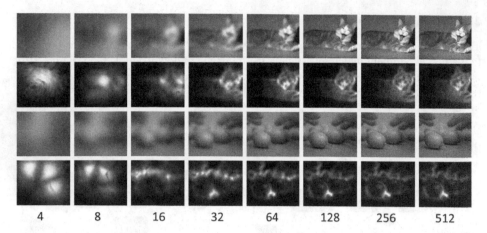

| 4 | 8 | 16 | 32 | 64 | 128 | 256 | 512 |

Fig. 2. Two sets of images and corresponding saliency maps in eight different spatial resolutions from MIT Low-Resolution Database.

images in various resolutions. We kept the down-sample ratio unchanged while inputting a set of stimuli images in eight different resolutions. We got eight fixation maps corresponding to different resolution levels. Finally, we calculated eight different AUC scores. In this experiment, spatial resolutions of input images were independent variables and AUC scores were dependent variables.

2.4   Relationship between Two Experiments

We introduce similarities and differences between two experiments in this part. Both of them have close ties with spatial resolution, and they are tested by the same saliency models on the same metric. But the essential differences are down-sampled images' size and fixation maps. The size of down-sampled images are smaller than original image in the first experiment but the same in the second experiment. Experiment one focuses on the amount of image pixels while experiment two focuses on the degree of image vagueness. In addition, as we can see in the steps of AUC calculation in Fig. 1, fixation maps are the same in the first experiment as down-sample ratios change, but there are eight different fixation maps corresponding to eight resolutions in the second experiment.

A previous research [18] of Wei-Ta Chu et al. supports above discussion. They researched how an image's resolution (pixels) and physical dimensions (inches) affect viewers' appreciation of it. And they found that image size do affects humans' perception of it, besides, the impact of image scaling is highly depended on the image content. Therefore, we design two experiments to research the influence of spatial resolution on both of constant image size and alterable image size.

3   Result and Contribution

The results of two experiments are shown in two line charts in Fig. 3. There are 13 lines in both of two charts, and each line represents a saliency model. Image signature model provides two approaches to generate saliency map in RGB and Lab color space, so we plot two lines to represent approaches in Lab and RGB color space individually. The X-axis in Fig. 3(a) and (b) represents the pixel width of down-sampled images (we use pixel width of down-sampled images to represent down-sample ratios since they are linearly dependent), and the Y-axis represents AUC score accordingly.

3.1   Result of the First Experiment

As we can see in Fig. 3(a), there is a general tendency that performances of most saliency models keep stable in high resolutions and they decrease in medium resolutions, then they decrease rapidly in low resolutions. This tendency looks similar to human visual perception. The images become blurrier as resolution decrease. When resolution is less than 16×16 pixels, down-sampled images will be similar to noisy images. AUC score tend to 0.5 when stimuli are noisy images. Contrary to general tendency, there are three special curves corresponding to AIM, SUN and Torralba model whose AUC score increases slowly as resolution decreases in medium resolution levels. It's worth noting that these three models are all based on Bayesian frame. Itti & Koch and GBVS can just process images larger than 128×128 pixels, so the limit of down-sampled image's pixel width is 128 pixels in these two models. In addition, influence of down-sample ratio is tiny in high resolutions on GBVS and Itti & Koch. Similar situation happens to Torralba model. The AUC scores of RARE2012 and Context-aware keep stable during all resolutions because these two models transform different input images into fixed scale.

We call models with alterable down-sample ratio as steerable models in this paper. On the contrary, models like Context-aware with fixed scale are called as non-steerable models. We draw conclusions above all: (1) Steerable models like Achanta, Signature-Lab, Signature-RGB, AWS, Murray and Covsal follow the general tendency mentioned above; (2) AIM and SUN follow tendency that AUC scores increase during high and middle resolutions until reach a peak, then performances decline rapidly in low resolutions; (3) The performances of non-steerable models such as Context-aware and RARE2012 keep stable during all resolutions; (4) GBVS, Itti & Koch and Torralba have high limit of down-sample ratio. Their performances are stable in high resolution except Torralba whose AUC score increases as resolution decreases.

3.2   Result of the Second Experiment

The result of the second experiment is shown in Fig. 3(b). We find the general tendency in Fig. 3(b) as follows: AUC scores keep stable in high resolutions (512,256pixels) and increase slowly as resolutions decrease in medium levels (128,64 pixels), then they decrease rapidly in low resolutions (32,16,8,4 pixels). When resolutions range from

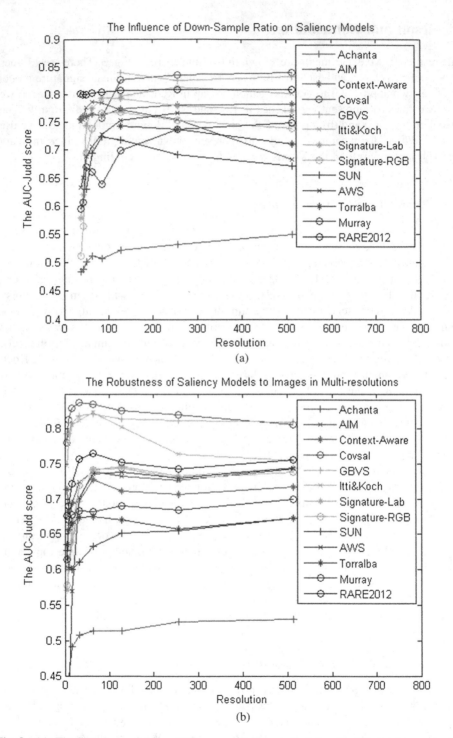

Fig. 3. (a): The line chart reflecting AUC scores of saliency models with different down-sample ratios; (b): The line chart reflecting AUC scores of models in different spatial resolutions;

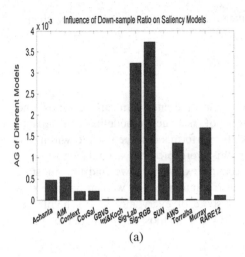 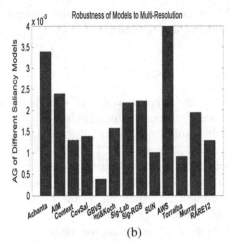

Fig. 4. (a): Average gradient of saliency models with different down-sample ratios; (b): Average gradient of saliency models in different spatial resolutions.

512 to 256 pixels, images still contain enough information so models' performances are almost the same. When resolution drops to 32×32 pixels or less, images are similar to noisy images so that AUC scores decrease quickly even less than 0.5.

We use a simple approach to evaluate the robustness to images in multi-resolutions called as Average Gradient. It's the absolute value of arithmetic average of slopes between every two adjacent break points of line charts shown in Fig. 3(a) and (b). We express this metric by its acronym AG which is defined by Eq. (1). AG reflects the fluctuation degree of AUC score as the resolution varies. Higher AG indicates bigger variation of AUC score, thus meaning that the saliency model is less robust. For example, the line of signature-Lab and signature-RGB models in Fig. 3(a) have big fluctuations and their AG scores are pretty high in Fig. 4(a). On the contrary, the stable line belongs to Achanta model has low AG score. The AG scores of models in two experiments are plotted in two histograms in Fig. 4.

$$AG = \frac{1}{scN - 1} \sum_{i=2}^{scN} |\frac{S_i - S_{i-1}}{R_i - R_{i-1}}| \tag{1}$$

When in Eq. (1), AG is average gradient; scN is total number of resolution levels; S_i is the AUC score in the i-th resolution level; R_i is the pixel width of image in i-th resolution level.

As we can see in Fig. 4, the X-axis represents model title and the Y-axis represents AG score. In the first experiment, the AG scores reflect the influences of down-sample ratios on models' performances. Image signature model in both of Lab and RGB color spaces are influenced dramatically by down-sample ratio as we can see in Fig. 4(a). In the second experiment, the AG scores reflect the models' robustness to images in multi-resolutions. As we can see in Fig. 4(b), GBVS is the most robust saliency model and AWS is the least robust model. The AUC scores of AWS and Achanta are less than

0.5 in low resolutions (16,4 pixels), thus meaning that they are worse than chance model in low resolutions.

3.3 Contribution

We evaluate performances of various models with different down-sample ratios, and provide theoretic foundation to research of saliency modeling in spatial multi-resolutions. We find that most of models' performances keep stable within the scope [128 pixels to 512 pixels]. Moreover, this work provides a guidance to computational complexity minimization. In the second experiment, we find that most of saliency models reach a peak value in medium resolution so that we can find the best resolution corresponding to the highest AUC score of every model. It's helpful to performance improvement. We also find that GBVS is the most robust model to spatial resolution. It's revealing for robust modeling.

4   Conclusion and Discussion

In this paper, we investigate two problems related to spatial resolution in the research of visual attention. One is the influence of the common down-sample operation on the effectiveness of saliency models. Another one is the efficiency of saliency models in images of various resolutions. This paper provides a comprehensive performance evaluation about these two problems. Concerning the first problem, we evaluate the influence of different down-sample ratios on saliency models' performances. We find that AUC scores of most models remain stable in high resolutions and decrease slowly in medium levels then decrease rapidly in low resolutions. As far as the second problem, we evaluate the saliency models' robustness to resolutions. We find that most models get higher AUC scores in medium resolutions but the scores decrease quickly in low resolutions. Besides that, GBVS is the most robust saliency model. The results of this work are useful to computational complexity optimization, performance improvement and robust modeling.

We pay attention to details about rescaling operations (down-sample or up-sample steps) in saliency models. Both of Fig. 1(a) and (b) contain two rescaling operations: step1 and step3 in Fig. 1(a), step1 and step2 in Fig. 1(b). It's still an open topic to research the influence of lost information in rescaling operations on saliency model's performance. In the second experiment, we find that most of models get higher score in medium resolution. We want to explain the reason of this situation and find the every model's best resolution corresponding to peak performance in the future.

Acknowledgement. This work was supported in part by NSFC (61422112, 61331014, 61371146).

References

1. Borji, A., Itti, L.: State-of-the-art in visual attention modeling. IEEE Trans. Pattern Anal. Mach. Intell. **35**(1), 185–207 (2013)
2. Judd, T., Durand, F., Torralba, A.: Fixations on low-resolution images. J. Vision **11**(4), 1–20 (2011)
3. Garcia-Diaz, A., Leboran, V., Fdez-Vidal, X.R., Pardo, X.M.: On the relationship between optical variability, visual saliency, and eye fixations: a computational approach. J. Vision **12**(6), 1–22 (2012)
4. http://saliency.mit.edu
5. Itti, L., Koch, C., Niebur, E.: A Model of saliency-based visual attention for rapid scene analysis. IEEE Trans. Pattern Anal. Mach. Intell. **20**(11), 1254–1259 (1998)
6. Torralba, Antonio, Oliva, Aude, Castelhano, Monica S., Henderson, John M.: Contextual guidance of eye movements and attention in real-world scenes : the role of global features on object search. J. Optical Soc. Am. **20**(7), 28–71 (1998)
7. Bruce, N.D.B., Tsotsos, J.K.: Saliency based on information maximization. In: Proceedings of Advances in Neural Information Processing Systems (2005)
8. Harel, J., Koch, C., Perona, P.: Graph-based visual saliency. Neural Inf. Process. Syst. **19**, 545–552 (2006)
9. Hou, X., Harel, J., Koch, C.: Image signature: highlighting sparse salient regions. IEEE Trans. Pattern Anal. Mach. Intell. **34**(1), 194–201 (2012)
10. Zhang, L., Tong, M.H., Marks, T.K., Shan, H., Gottrell, G.W.: SUN : a Bayesian framework for saliency using natural statistics. J. Vision **8**(32), 1–20 (2008)
11. Murray, N., Vanrell, M., Otazu, X., Alejandro Parraga, C.: Saliency estimation using a non-parametric low-level vision model. In: Proceeding of IEEE Computer Vision and Pattern Recognition (2011)
12. Riche, N., Mancas, M., Duvinage, M., Mibulumukini, M., Gosselin, B.: RARE2012: a multi-scale rarity-based saliency detection with its comparative statistical analysis. Signal Process. Image Commun. **28**(6), 642–658 (2013)
13. Erdem, E., Erdem, A.: Visual saliency estimation by nonlinearly integrating features using region covariances. Journal of Vision **13**(4), 1–20 (2013)
14. Achanta, R., Hemami, S.S., Estrada, F.J., Su, S.: Frequency-Tuned Salient Region Detection. In: Proceedings of the IEEE Conference on Computer Vision and Pattern Recognition (2009)
15. Goferman, S., Zelnik-Manor, L., Tal, A.: Context-Aware Saliency Detection. In: Proceedings of the IEEE Conference on Computer Vision and Pattern Recognition (2010)
16. Judd, T., Ehinger, K., Durand, F., Torralba, A.: Learning to Predict Where Humans Look. In: IEEE 12th International Conference on Computer Vision (2009)
17. Simoncelli, E.P., Freeman, W.T., Adelson, E.H., Heeger, D.J.: Shiftable multi-scale transforms. IEEE Trans. Inf. Theory **38**(2), 587–607 (1992)
18. Chu, W.-T., Chen, Y.-K., Chen, K.-T.: Size does matter: how image size affects aesthetic perception? In: Proceeding of ACM Multimedia (2013)

Depth Map Upsampling via Progressive Manner Based on Probability Maximization

Rongqun Lin[1][✉], Yongbing Zhang[1], Haoqian Wang[1],
Xingzheng Wang[1], and Qionghai Dai[1,2]

[1] Graduate School at Shenzhen, Tsinghua University, Shenzhen 518055, China
linrq14@mails.tsinghua.edu.cn
[2] Department of Automation, Tsinghua University, Beijing 100084, China

Abstract. Depth maps generated by modern depth cameras, such as Kinect or Time of Flight cameras, usually have lower resolution and polluted by noises. To address this problem, a novel depth upsampling method via progressive manner is proposed in this paper. Based on the assumption that HR depth value can be generated from a distribution determined by the ones in its neighborhood, we formulate the depth upsampling as a probability maximization problem. Accordingly, we give a progressive solution, where the result in current iteration is fed into the next to further refine the upsampled depth map. Taking advantage of both local probability distribution assumption and generated result in previous iteration, the proposed method is able to improve the quality of upsampled depth while eliminating noises. We have conducted various experiments, which show an impressive improvement both in subjective and objective evaluations compared with state-of-art methods.

Keywords: Progressive manner · Denoising · Depth map · Upsampling · Probability Maximization

1 Introduction

As an indication of true position in 3D space, depth maps play a more and more important role in a variety of different applications including object reconstruction, medical, 3D television and entertainment. Although there exist several range measuring approaches to capture depth map, it is difficult to acquire depth information accurately and sufficiently. For instance, time of flight (ToF) cameras can use active sensing to capture depth map per-pixel at video frame-rate, and they become easily accessed and popular. However, the main disadvantage of such cameras is that the resolution of generated depth maps is relatively low compared with their associated color image. This is due to chip size limitations and the captured depth maps always contain amounts of acquisition noise. These defects limit the practical applications based on depth information.

This work was partially supported by the National Natural Science Foundation of China under Grant 61170195, U1201255, and U1301257.

Y.-S. Ho et al. (Eds.): PCM 2015, Part II, LNCS 9315, pp. 84–93, 2015.
DOI: 10.1007/978-3-319-24078-7_9

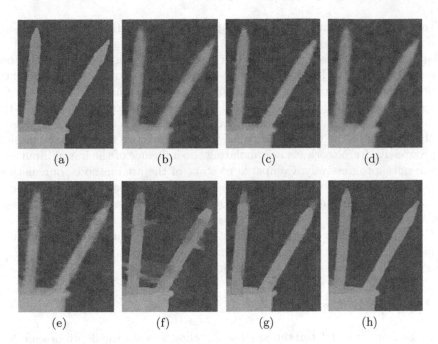

Fig. 1. Depth map upsampling results of art compared in details (upscaled ×4) with noise. (a) Ground truth. (b) Diebel et al. [2]. (c) Yang et al. [11]. (d) Chan et al. [1]. (e) He et al. [5]. (f) Park et al. [7]. (g) Ferstl et al. [3]. (h) Ours. (Zoom in for better view.)

Therefore, depth upsampling and denoising becomes a vital problem to the development of 3D applications [4, 8].

To address the above problem, several approaches are proposed to upsample the depth maps via using additional corresponding color or intensity image as cues. Diebel et al. [2] performed upsampling using a MRF formulation exploiting the fact that discontinuities in depth and intensity image tend to co-align and weights of the smoothness term were computed according to texture derivatives. Yang et al. [11] used a joint bilateral filtering of a depth cost volume and a RGB image in an iterative process. Chan et al. [1] used a noise-aware bilateral filter to eliminate the noise during upsampling the depth map. Park et al. [7] proposed a more complicated approach, which is based on a least-square optimization that combines several weighting factors with nonlocal means filtering, segmentation, image gradients and edge saliency for depth upsamling. Frestl et al. [3] formulated the depth map upsampling as a global energy optimization problem using Total Generalized Variation (TGV) regularization.

While these methods achieve good quality in smooth regions, the major drawback is that their upsampling results contain blurring, over smoothing or texture-copying around thin structures or sharp discontinuities. Moreover, when upsampling a noised LR depth map, these methods produce worse results, since the noise can be propagated to the upsampled regions, especially around sharp edges or thin objects. All of the above methods cannot generate accurate HR depth maps with sharp edges and less noise.

To generate accurate HR depth maps with sharp edges while depressing noise effects, in this paper we propose a depth upsampling algorithm based on probability maximization in a progressive manner. The main contributions of our work are two-fold: (1) Inspired by the work in [10], we provide the mathematic derivations and build our model based on the assumption that the depth of a pixel in HR depth map can be generated from a distribution determined by the depth of pixels in its neighborhood in the same HR depth map. (2) To preserve the sharp edges of the upsampled depth map and remove the noise, we exploit the progressive framework via accumulating the influence of the initial input and remove noise progressively. Compared to state of the art methods, our method is superior in terms of both subjective and objective evaluations. Figure 1 shows that our results can simultaneously remove the noise and achieve sharper edges with less artifacts .

The paper is organized as follows: Sect. 2 details about our approach including our model and the progressive framework. The experimental results are reported in Sect. 3, followed by a conclusion of our work in Sect. 4.

2 Proposed Method

In this section, we will detail the proposed method to solve the depth upsampling and denoising problem simultaneously. Firstly, we derive a probability maximization problem to build our model. After necessary derivations, we analyze our model and show that our model describes the intrinsic properties of the depth map. Secondly, to preserve the sharp edges of the upsampled depth map and remove the noise, we introduce the progressive framework. In each iteration, we use the output of previous iteration to update the input accumulating the influence of the initial input until the output converges to a stable result, which can remove some noise progressively.

2.1 Our Model

Inspired by [10], we assume the depth of a pixel in HR depth map can be generated from a distribution determined by those in its neighborhood in the same HR depth map. It yields the following model for each pixel i.

$$p(d_i^H | d_\Omega^H) \propto \sum_{j \in \Omega} \eta_{ij} exp\{-(d_i^H - d_j^H)^2 / 2\sigma_d^2\}, \tag{1}$$

where Ω is a spatial pixel block centered at pixel i, d_i^H is the value of pixel i in HR depth map. d_Ω^H is the depth of pixels in its neighborhood in the same HR depth map. η_{ij} is the mixture coefficient corresponding statistically to color image and spatial position. And $\eta_{ij} = \alpha_{ij} / \sum_{j \in \Omega} \alpha_{ij}$, where

$$\alpha_{ij} = exp\{-\frac{\sum_{k=1}^{3}(I_i^k - I_j^k)^2}{2\sigma_I^2}\} exp\{-\frac{(p - q)^2}{2\sigma_s^2}\}, \tag{2}$$

with I_i^k being the k_{th} color channel of pixel i in color image. p is the position of pixel i and q is the position of pixel j.

Then we maximize the probability and the optimal d_i^H for each pixel i is yielded as

$$d_i^H = \operatorname*{argmax}_{d_i^H} p(d_i^H | d_\Omega^H) = \operatorname*{argmin}_{d_i^H} (-ln\ p(d_i^H | d_\Omega^H)). \tag{3}$$

And we get

$$\frac{\partial(-ln\ p(d_i^H | d_\Omega^H))}{\partial d_i^H} = \frac{\sum\limits_{j \in \Omega} \eta_{ij} exp\{\frac{-(d_i^H - d_j^H)^2}{2\sigma_d^2}\} \frac{(d_i^H - d_j^H)}{\sigma_d^2}}{\sum\limits_{j \in \Omega} \eta_{ij} exp(\frac{-(d_i^H - d_j^H)^2}{2\sigma_d^2})},$$

$$\propto d_i^H - \frac{\sum\limits_{j \in \Omega} \alpha_{ij} exp\{\frac{-(d_i^H - d_j^H)^2}{2\sigma_d^2}\} d_j^H}{\sum\limits_{j \in \Omega} \alpha_{ij} exp\{\frac{-(d_i^H - d_j^H)^2}{2\sigma_d^2}\}}. \tag{4}$$

Setting Eq. 4 to zero yields a per-pixel constraint for each d_i^H.

However, we meet a **chicken-egg** dilemma. Our goal is to obtain the HR depth map, while d_i^H can be reliably acquired only when d_Ω^H is available. Thus we introduce a roughly estimated HR depth map (eg. Bicubic upsampling result) to break this dilemma. In addition, it can be used to avoid incorrect depth prediction due to depth color inconsistency (some pixels with similar intensity may have different depth, vice versa).

Therefore, we rewrite our model as follow. We write the whole objective function in a matrix form. We denote by d^{in} the input vector, d^{out} the output vector, $\boldsymbol{W}$ the weight matrix. Each matrix element is $\boldsymbol{W}_{ij} = w_{ij}/\sum_{j \in \Omega} w_{ij}$, where w_{ij} is the mixture coefficient corresponding statistically to color image, spatial position and the roughly estimated HR depth map.

$$w_{ij} = exp\{-\frac{\sum\limits_{k=1}^{3}(I_i^k - I_j^k)^2}{2\sigma_I^2}\} exp\{(-\frac{(p - q)^2}{2\sigma_s^2}\} exp\{\frac{-(d_i^{est} - d_j^{est})^2}{2\sigma_d^2}\}, \tag{5}$$

σ_I, σ_s and σ_d adjust the importance of the spatial distance, intensity difference and estimated depth changes.

The objective function with respect to d^{out} is therefore formulated as

$$min(E(d^{out})) = min\{(d^{out} - \boldsymbol{W}d^{in})^T (d^{out} - \boldsymbol{W}d^{in})\}. \tag{6}$$

The global minima can be obtained by solving $d^{out} = \boldsymbol{W}d^{in}$, which is our final practical model.

Our model describes the intrinsic properties of the depth map. Although our model is similar to the combination of the neighborhood smoothness term and NLM regularization in [7], their method takes segmentation, image gradient, edge saliency and non-local means into consideration, which is more complicated. Moreover, the difference between our method and that of [7] is the idea how we get the final HR depth map, which will be detailed in the following part.

2.2 Progressive Framework

In our model mentioned above, we set the parameters appropriately and consequently the w_{ij} of the pixels in the neighborhood almost has slight difference in their values. From a different perspective, pixels in the nearby area often have similar depth.

Algorithm A1. PROPOSED METHOD

1: **Input:** low resolution depth map d^L .
2: **initialization:** Map d^L to high-resolution image as initial input d^{ini}, compute W.
3: **for:** k = 1 : max-step **do**
4:      update the k_{th} input $d^{in}(k)$ according to 7 .
5:      compute $d^{out}(k) = W d^{in}(k)$.
6:      **if**   (k = max-step)
7:          $d^H = d^{out}(k)$.
8:          break.
9:      **end if**
10:     **if** $sum(d^{out}(k) - d^{in}(k)) \leq \varepsilon$, $(\varepsilon = 0.0001)$
11:         $d^H = d^{out}(k)$.
12:         break.
13:     **end if**
14: **end for:**
15: **Output:** high resolution depth map d^H .

If the input contains more information, our model can produce the result once. However, the initial input is the one with high resoltion but has fewer non-zero values in the position we map the low one to high one. It means that in initial depth map there exist many zero values, which is not enough to get the final HR depth map once.

The idea of our method is that the depth of every pixel in HR depth map can be acquired via **accumulating the influence of the initial input** . During the progressive process, **the updated result gets improved and contains more information for the next iteration**. Therefore we propose a progressive framework.

In each step, we update the input as shown in the following.

$$d_i^{in}(k) = \begin{cases} d_i^{ini}, & \text{if } i \in \Psi \\ d_i^{out}(k-1), & \text{otherwise} \end{cases} \tag{7}$$

where $d_i^{in}(k)$ is the depth of pixel i in k_{th} input, $d_i^{out}(k-1)$ is the depth of pixel i in $(k-1)_{th}$ output. We map d^L to high-resolution image as initial input d^{ini},

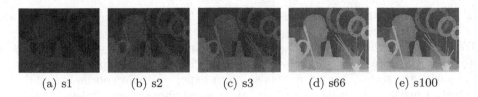

(a) s1   (b) s2   (c) s3   (d) s66   (e) s100

Fig. 2. Progressive process displayed and intermediate results with specified step.

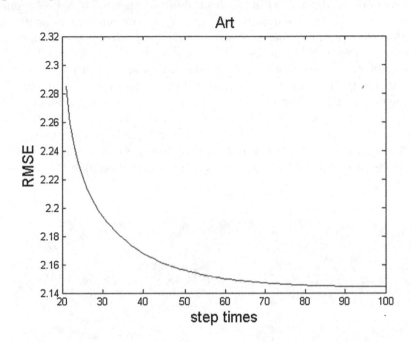

Fig. 3. RMSE in different step(up-scaled ×2)without noise

and the d_i^{ini} is the depth of pixel i in the initial input. Ψ is the non-zero position in the initial input. The replacement makes k_{th} input have the same value as the initial input on the Ψ, which was called **anchor points**. But in the last step, we don't do the replacement, in order to cope with noisy input.

Pixels in the output of first step usually have a small value, since there exist fewer non-zero values in initial input. Thus in Fig. 2, result in the first step seems dark and it gets brighter and brighter during the progressive process. As is shown in Fig. 3, our method is so powerful that the RMSE results converge so quickly. The reason we can deal with the noised situation well is that we can remove some noise in every step and also keep the thin structures like sharp edges. In the last step the output except the anchor points has noiseless and accurate values. We regenerate the values on the anchor points, thus the noise on the anchor points is removed.

3   Experimental Results

We test our method using synthetic examples from Middlebury 2007 datasets [6,9] and the dataset of Frestl et al. [3] for quantitative and qualitative comparisons with the state of the art methods. In our experiments, we normalize all the values of pixels and spatial coordinates and the roughly estimated depth map is obtained by bicubic interpolation from the low resolution depth map.

During our experiments, σ_I influences the importance of guided color image. When it fixes at a very small digit, the results will contain serious artifacts. It means that the color image has excessive influence. σ_s influences the importance of spatial difference. When it fixes at a very small digit, the results will be over-smoothing. σ_d influences the importance of estimated depth changes. When it fixes at a very small digit, the results will have blurings around the edges. Consequenlty, in this paper we set the values of parameters as follows: $\sigma_I = 0.12, \sigma_s = 0.02, \sigma_d = 0.04$. And the size of neighborhood region Ω is 9×9. We use the same setting when upsampling the noised input LR depth map.

To demonstrate the effectiveness of our proposed method, we show our results in two parts: upsample the clean LR depth map and upsample the noised LR depth map.

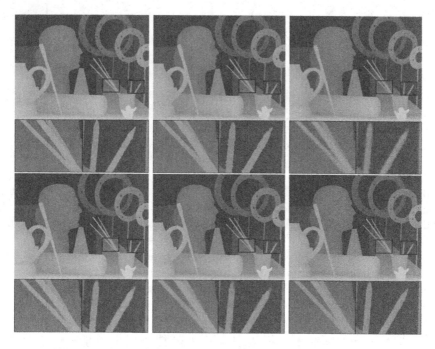

Fig. 4. Visual quality comparisons of ×4 upsampling on the Middlebury *Art* datasets *without noise*. Row 1 Column 1: Ground Truth. Row 1 Column 2: Yang et al. [11]. Row 1 Column 3: He et al. [5]. Row 2 Column 1: Park et al. [7]. Row 2 Column 2: Ferstl et al. [3]. Row 2 Column 3: Ours. (Zoom in for better view.)

3.1   Upsample the Clean LR Depth Map

We downscale the Middlebury 2007 datasets [6,9] by bicubic interpolation method as the clean LR depth map. Especially, we choose three of them to form our test dataset: Art, Book and Moebius, which have clutter depth values and their corresponding color images have complicated textures. We conduct quantitative evaluations on our results and the results provided by Frestl et al. [3].

The quantitative results in terms of the Root Mean Square Error (RMSE) against the ground-truth depth maps are shown in Table 1. Beside the bilinear interpolation, the proposed method is compared with five recent methods: Diebel et al. [2], Yang et al. [11], He et al. [5], Park et al. [7], and Ferstl et al. [3]. The best result for each dataset is highlighted. What can be clearly seen from the numerical results is that our approach is the best compared to other state of the art methods.

A visual comparison for the different methods is given in Fig. 4. For clean inputs, Yang et al. [11], He et al. [5], Park et al. [7], and Ferstl et al. [3] methods introduce some jaggy artifacts along edges because they depend too much on the guide color image. Our method can generate results with sharper edges and less artifacts.

Table 1. RMSE comparisons on Middlebury 2007 datasets **without noise**(upscaled ×2, ×4).

	Art		Books		Moebius	
	×2	×4	×2	×4	×2	×4
Bilinear	2.834	4.147	1.119	1.673	1.016	1.499
Diebel et al. [2]	3.119	3.794	1.205	1.546	1.187	1.439
Yang et al. [11]	4.066	4.056	1.615	1.701	1.069	1.386
He et al. [5]	2.934	3.788	1.162	1.572	1.095	1.434
Park et al. [7]	2.833	3.498	1.195	1.495	1.064	1.349
Ferstl et al. [3]	3.032	3.785	1.290	1.603	1.129	1.459
Ours	**2.144**	**3.457**	**1.025**	**1.230**	**0.993**	**1.334**

3.2   Upsample the Noised LR Depth Map

In reality, captured depth maps always have plenty of noise. For fair comparison, we employ the **noisy input** dataset used in Frestl et al. [3].

Our method has a great advantage both in quantitative and qualitative evaluations compared with previous methods. Experiments show that our method can remove lots of noise and preserve sharp edges.

The quantitative results in terms of the Root Mean Square Error (RMSE) are shown in Table 2. Beside the bilinear interpolation, the proposed method

Table 2. RMSE comparisons on Middlebury 2007 datasets **with noise**(upscaled ×2, ×4).

	Art		Books		Moebius	
	×2	×4	×2	×4	×2	×4
Bilinear	4.580	5.621	3.948	4.309	4.200	4.565
Diebel et al. [2]	3.489	4.514	2.064	3.002	2.127	3.105
Yang et al. [11]	3.005	4.021	1.874	2.383	1.917	2.418
Chan et al. [1]	3.437	4.464	2.091	2.773	2.076	2.759
He et al. [5]	3.546	4.412	2.375	2.737	2.481	2.831
Park et al. [7]	3.759	4.564	1.946	2.607	1.956	2.508
Ferstl et al. [3]	3.188	4.063	1.522	2.213	1.475	2.030
Ours	**2.305**	**3.747**	**1.495**	**2.098**	**1.455**	**1.821**

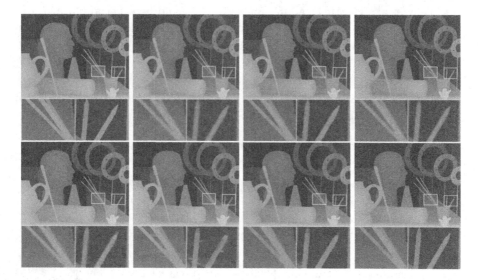

Fig. 5. Visual quality comparisons of ×4 upsampling on the Middlebury *Art* datasets *with noise*. Row 1 Column 1: Ground Truth. Row 1 Column 2: Diebel et al. [2]. Row 1 Column 3: Yang et al. [11]. Row 1 Column 4: Chan et al. [1]. Row 2 Column 1: He et al. [5]. Row 2 Column 2: Park et al. [7]. Row 2 Column 3: Ferstl et al. [3]. Row 2 Column 4: Ours. (Zoom in for better view.)

is compared with six recent methods: Diebel et al. [2], Yang et al. [11], Chan et al. [1], He et al. [5], Park et al. [7], and Ferstl et al. [3]. Our results always rank first in terms of RMSE.

A visual comparison for the different methods is given in Fig. 5. For noised inputs, Diebel et al. [2], Yang et al. [11], Chan et al. [1], He et al. [5] methods generate the noised results with blurrings. Although Park et al. [7], and Ferstl et al. [3] methods can remove some noise, they still introduce some jaggy artifacts along edges because they depend too much on the guided color image. As can be seen, our proposed method can produce high resolution depth maps with sharper edges, clearer structures and fewer artifacts.

4 Conclusions

In this paper, we propose a novel method for depth map upsampling via progressive manner based on probability maximization. The formulated model is able to reveal and employ the intrinsic properties of the depth map. To preserve the sharp edge of the upsampled depth map and remove the noise, we exploit the progressive framework through accumulating the influence of the initial input and remove noise progressively. Experiments show that our method outperforms the state-of-art methods in terms of quantitative and qualitative comparisons. Our method can produce the HR depth map with sharp edges, more accurate values and less noise.

As future work, we would like to improve our work to meet the need of real-time reconstructions and extend our method to more applications.

References

1. Chan, D., Buisman, H., Theobalt, C., Thrun, S., et al.: A noise-aware filter for real-time depth upsampling. In: Workshop on Multi-camera and Multi-modal Sensor Fusion Algorithms and Applications-M2SFA2 2008 (2008)
2. Diebel, J., Thrun, S.: An application of markov random fields to range sensing. In: NIPS (2005)
3. Ferstl, D., Reinbacher, C., Ranftl, R.: Image guided depth upsampling using anisotropic total generalized variation. In: ICCV (2013)
4. Guomundsson, S.A., Larsen, R., Aanæs, H., Pardas, M., Casas, J.R.: Tof imaging in smart room environments towards improved people tracking. In: IEEE Computer Society Conference on Computer Vision and Pattern Recognition Workshops, CVPRW 2008, pp. 1–6. IEEE (2008)
5. He, K., Sun, J., Tang, X.: Guided image filtering. In: Daniilidis, K., Maragos, P., Paragios, N. (eds.) ECCV 2010, Part I. LNCS, vol. 6311, pp. 1–14. Springer, Heidelberg (2010)
6. Hirschmuller, H., Scharstein, D.: Evaluation of cost functions for stereo matching. In: CVPR (2007)
7. Park, J., Kim, H., Tai, Y., Brown, M., Kweon, I.: High quality depth map upsampling for 3D-tof cameras. In: ICCV (2011)
8. Prasad, T., Hartmann, K., Weihs, W., Ghobadi, S.E., Sluiter, A.: First steps in enhancing 3D vision technique using 2D/3D sensors. In: Computer Vision Winter Workshop, pp. 82–86 (2006)
9. Scharstein, D., Pal, C.: Learning conditional random fields for stereo. In: IEEE Conference onComputer Vision and Pattern Recognition, CVPR 2007, pp. 1–8. IEEE (2007)
10. Xu, L., Yan, Q., Jia, J.: A sparse control model for image and video editing. ACM Trans. Graph. (TOG) **32**(6), 197 (2013)
11. Yang, Q., Yang, R., Davis, J., Nister, D.: Spatial-depth super resolution for range images. In: CVPR (2007)

Perceptual Quality Improvement for Synthesis Imaging of Chinese Spectral Radioheliograph

Long Xu[1(✉)], Lin Ma[2], Zhuo Chen[1], Yihua Yan[1], and Jinjian Wu[3]

[1] Key Laboratory of Solar Activity, National Astronomical Observatories,
Chinese Academy of Sciences, Beijing, China
lxu@nao.cas.cn
[2] Huawei Noah's Ark Lab, Hong Kong, China
[3] School of Electronic Engineering, Xidian University, Xi'an, China

Abstract. Chinese Spectral Radioheliography can generate the images of the Sun with good spatial resolutions. It employs the Aperture Synthesis principle to image the Sun with plentiful solar radio activities. However, due to the limitation of the hardware, specifically the limited number of antennas, the recorded signal is extremely sparse in practice, which results in unsatisfied solar radio image quality. In this paper, we study the image reconstruction of Chinese Spectral RadioHeliograph (CSRH) by the aid of compressed sensing (CS) technique. In our proposed method, we adopt dictionary technique to represent solar radio images sparsely. The experimental results indicate that the proposed algorithm contributes both PSNR and subjective image quality improvements of synthesis imaging of CSRH markedly.

Keywords: Compressed sensing · Solar radio astronomy · Image reconstruction · Aperture syntheis

1 Introduction

Chinese Spectral Radioheliography (CSRH) employs the aperture synthesis (AS) principle for imaging the Sun to generate the images of the Sun. AS principle synthesizes a number of small antennas to produce a larger antenna so that a better resolution can be achieved. For AS imaging system, the image resolution is determined by the maximum baseline length termed by the largest distance of two antennas rather than the diameter of a single antenna. The maximum baseline length of CSRH is 3 km, so it can achieve a good resolution determined by $r = \lambda/D$, where λ represents wavelength and D is the diameter of objective lens. AS devices record the Fourier components of observed objects instead of spatial images, where each two antennas compose of an interferometer to capture one Fourier component each time. Given n antennas, there would be $n \times (n\text{-}1)/2$ interferometers, which can record $n \times (n\text{-}1)/2$ Fourier components for each time of observation. In addition, by making use of the Earth's rotation, more Fourier components can be obtained. Nevertheless, only a small part of Fourier components are recorded, which results in blur synthesized images usually.

The synthesized images of solar radio observation can deliver the viewers the plentiful information about solar radio activities more directly and clearly. However,

© Springer International Publishing Switzerland 2015
Y.-S. Ho et al. (Eds.): PCM 2015, Part II, LNCS 9315, pp. 94–105, 2015.
DOI: 10.1007/978-3-319-24078-7_10

due to the extremely sparse sampling of CSRH in practice, the synthesized images usually appear to be blurring. In this paper, we study the image reconstruction of CSRH by the aid of compressed sensing (CS) technique. In the proposed method, instead of using fixed basis functions, an adaptive dictionary is learned from input images to represent solar radio images sparsely.

Studying imaging process of AS, the image degradation comes from sparse sampling of images in Fourier domain, which is formulated by the Fourier transform of a spatial image multiplied by a sampling matrix. According to Fourier theory, this degradation is equivalent to an original image convolved by a point spread function (PSF), which is characterized by a main lobe surrounded by sidelobes. Applying this PSF to images would result in blurring in images. To eliminate blurring caused by the convolution of PSF, the researchers have proposed the opposite processing, i.e., deconvolution [1–3] to recover images from their degradations. This kind of method is called "clean" algorithm. In this paper, we propose a CS-based algorithm to recover images from their sparse samplings. This method is established on the fact that image degradation of CSRH comes from sparse sampling in Fourier domain. In addition, we discuss how to meaningfully measure image quality of synthesized images of AS and how to improve subjective image quality of AS during the reconstruction process.

The remaining content of this paper is arranged as follows. Section 2 gives the introduction of synthesis imaging of CSRH. Section 3 presents the proposed image reconstruction framework for the recorded data of CSRH. Experimental results are provided in Sect. 4. Finally, the conclusion is given in Sect. 5.

2 Synthesis Imaging Principle of CSRH

As we know, the resolution of a telescope is decided by the diameter of objective lens regardless of optical or radio telescopes. Assume the wavelength of received signal λ, the telescope diameter D, the resolution is computed by

$$R = 1.22 \times \lambda/D, \tag{1}$$

which indicates the larger the diameter of a telescope, the better the resolution. Here, the resolution is given by the farthest two bright points in observed object which can be distinguished by the telescope. This is because electromagnetic waves could result in diffraction through small pinhole (such as the lens of telescope). Two electromagnetic waves with small distance would overlap after diffraction. The unit of resolution is arc of second (") which is 1/3600 deg (°). It represents the angle which is formed by two bright points in the observed object with respect to the lens of telescope. Assume the typical wave length of visible light 555 nm, (1) can be rewritten into

$$\alpha'' = 140''/D, \tag{2}$$

where the unit of D is mm, α is measured by arc of second. From (2), to get the resolution of 0.1", the diameter of objective lens would reach 1.4 m for visible light. Although α is measured by an angle, it implies the distance between two nearest points

that a telescope can distinguish. Assume the Sun is 14960 wkm far from the Earth, 0.1"
represents 72.528 km in the Sun. By the same principle, there would be over 2500 m
for receiving radio wave with the minimum wavelength 1 mm with the same resolution.
It is impossible to build such a huge telescope in practice.

In 1950 s, the scientists proposed the aperture synthesis technique to construct an
aperture synthesis telescope array consisting of a number of small telescopes. This
telescope array can achieve the same resolution of a single big telescope with the
diameter equaling to the largest distance between two small telescopes. This distance is
named the maximum baseline length. In aperture synthesis array, the resolution is
dependent on the maximum baseline length instead of the diameter of a single telescope. This technique is a breakthrough to radio wave observation, and also a milestone
in the history of radio astronomy.

The aperture synthesis array employs interferometry technique to image the
brightness function of the observed object. Each two antennas form an interferometer,
which records Fourier coefficients instead of spatial pixel values of the observed object.
Here, the two dimension Fourier space is involved, and it is also called UV space. An
interferometer records one Fourier coefficient each time. Thus, an aperture synthesis
array records a set of Fourier coefficients each time. In addition, it can record more
Fourier coefficients by taking advantage of the earth rotation. The spatial image of an
observed object is usually termed as brightness distribution/function, while its Fourier
transform is named as visibility distribution/function. Assume the brightness function I
(l, m), the visibility function V(u, v), I and V are Fourier transform pairs, i.e.,

$$I(l,m) = \iint V(u,v)e^{-2\pi\tau(ul+vm)}dudv. \tag{3}$$

In practice, $V(u, v)$ is not known everywhere but is sampled at particular positions on
the u-v plane. The sampling can be described by a sampling function $S(u,v)$, which is
zero where no data have been taken. Assume the original visibility function $V(u,v)$, the
actual visibility function recorded by an AS is:

$$V^D(u,v) = V(u,v) \times S(u,v). \tag{4}$$

Applying inverse Fourier transform to (4), one can calculate the reconstructed
spatial image from $V^D(u,v)$ as:

$$I^D(l,m) = \iint V(u,v)S(u,v)e^{-2\pi\tau(ul+vm)}dudv, \tag{5}$$

which is usually referred to dirty image $I^D(l,m)$ instead of the desired intensity distribution, namely clear image $I(l,m)$. According to the convolution theorem for Fourier
transform, $I^D(l,m)$ is related to $I(l,m)$ by

$$I^D(l,m) = I(l,m) \otimes B^D(l,m), \tag{6}$$

where the symbol '$\otimes$' denotes convolution, and $B^D(l, m) = \int\int S(u, v)e^{-2\pi\tau(ul+vm)}dudv$ is the synthesized beam. Dirty beam or point spread function (PSF) corresponds to the sampling function $S(u,v)$.

Based on above analysis, AS has a different imaging principle from the one of a telescope with a single antenna. However, the image quality is still dependent on the PSF which terms the imaging efficiency of a general telescope. From (6), the dirty image is related with clean one by convolving a dirty beam. To recover $I(l,m)$ from $I^D(l, m)$ is an ill-posed problem. With respect to convolution operator, the straightforward method to recover $I(l,m)$ from $I^D(l,m)$ is named "deconvolution", which tries to eliminate the convolution of $B^D(l,m)$ from right side of (6). In astronomy, this kind of method is termed "Clean" algorithm, which was proposed firstly by Högbom in 1974 [1]. The clean algorithm finds the strength and position of the peak (i.e., the greatest absolute intensity) in the dirty image one-by-one, and subtracts it from the dirty image after multiplying it by the PSF. Then, an idealized "clean" beam is applied to these accumulated point sources to output a clean image. It was claimed that "clean" algorithm is good at point source which was usually taken as the model of radio sources, including solar radio source. The point radio source can be assumed to consist of pulse signals. For handling more complicated situations, such as extended source relative to point source, the variants of Högbom clean algorithm were proposed in [2] and [3].

3 Image Reconstruction for CSRH Image System

In computer science, the regularization methods with certain prior image constraints are widely used to solve ill-posed problems. The widely used prior image constraints include smoothness, local similarity, non-local similarity and sparsity. Accordingly, these corresponding regularization methods have been proposed in the literatures. In our case, we have a very sparse sampling of visibility function $V(u, v)$ in Fourier domain, so the sparse constraint is taken into consideration to formulate optimization problem, which is regarded as the decoding part of CS methods.

CS depicts a theory that an original signal can be recovered from its degradation if the signal is subject to a certain property, namely restricted isometry property (RIP) [7–9]. Assume that the degradation process of an image is depicted by:

$$y = Hx + n, \tag{7}$$

where y represents the observed image, x represents the original one, H is the degradation model and n is the random noise usually assumed to be the Gaussian white noise. Since H is irreversible, it is impossible to resolve x from y by multiplying H-1 at the both sides of (7). By imposing constraints of prior image models, x can be derived from

$$\text{argmin}_x \frac{1}{2}\|Hx - y\|_2^2 + \lambda \cdot \Psi(x), \tag{8}$$

where λ is the regularization parameter, $\Psi(x)$ represents a prior image constraint, e.g., sparse constraint of natural images. Based on the sparse constraint, the researchers have proposed corresponding regularization methods to address image denoising, super-resolution, deblur and etc. [10–12]. In our work, the sparse sampling is closely associated with the CSRH imaging system, so sparsity model is exploited in image reconstruction of CSRH imaging system.

3.1 Image Reconstruction with Sparse Constraint

Given an input vector x has N coefficients, it is sparse if there are only K ($K < <N$) non-zero coefficients after a certain transform. In compressed sensing, the sparse signal needs to be further measured by the measurement matrix for compression. The perfect reconstruction can be achieved if and only if RIP is allowed [7–9]. Assume the sparse transform matrix Ψ, the sparse constraint means $\alpha = \Psi^T x$ has a small number of nonzero coefficients. Assume the measurement matrix Φ, we have the measurement on α by $y = \Phi\alpha$. Combining these two operators, we have the following equation:

$$y = \Phi\Psi^T x = \theta x, \tag{9}$$

where θ is a $N \times M$ ($M < <N$) matrix, which means that the linear equation array has more unknown parameters than the number of equations. At this situation, there are some free variables, which mean that we cannot get a determined solution for this equation array. Actually, there are many solutions for this equation array. Such a problem is named an underdetermined system of equations. It is impossible to have the determined answer for an underdetermined system. So there must be other constraints for solving an underdetermined problem. Usually, we have prior models about images to be the constraints of underdetermined problem. In this work, the sparsity constraint is considered, so Eq. (8) is rewritten as

$$\text{argmin}_x \frac{1}{2}\|Hx - y\|_2^2 + \lambda \cdot \|x\|_0, \tag{10}$$

where $\|x\|_0$ represents l_0 norm which indicates the number of nonzero coefficients.

3.2 Sparse Representation of Image by Dictionary

It is widely accepted that natural images are sparse in Fourier, wavelet, and DCT domain. It means that natural images can be represented by a small number of coefficients after one kind of transforms mentioned above, so they can be dramatically compressed in transform domain. That's why compressed sensing was extensively exploited in image compression during the past decades. Besides compression, this prior model of natural image was exploited to establish optimization being a constraint to solve ill-posed problem, which was used to image denoising, debluring and super-resolution.

Apart from fixed basis function for representing images, the user-defined dictionary is more efficient since it is adaptive to input signal. For our concerning, the images for processing are specific instead of general. User's dictionary can be expected to be more competitive than fixed basis functions. For establishing dictionary, there are also a plenty of methods. A general method is to cluster input image patches (blocks) into several groups. The centroids of these clusters compose the dictionary, where each centroid is named as a code word.

In this work, we employ group-based sparse representation [13, 14], which establishes dictionary over a group of patches instead of an individual patch. These patches are from both local and non-local similar patches. This group-based sparse representation processes a group of patches simultaneously in a unified framework for exploring both local and non-local similarity of natural images. An effective self-adaptive dictionary learning method for each group was designed in [13], where an adaptive dictionary D_{G_k} $(Bs \times c)$ was learnt for each group x_{G_k} directly from its estimate r_{G_k} since the original image x does not exist. In practice, r_{G_k} is initialized by the observation y. Then, it is updated in each iteration by the recovered image. After obtaining r_{G_k}, we then apply SVD to it as:

$$r_{G_k} = U_{G_k} \sum G_k V_{G_k}^T = \sum_{i=1}^{m} \gamma_{r_{G_k}(i)} (u_{G_k(i)} v_{G_k(i)}^T), \qquad (11)$$

where $\sum G_k$ is a diagonal matrix with the elements on its main diagonal, and $u_{G_{k(i)}}$ and $v_{G_{k(i)}}$ are the columns of U_{G_k} and V_{G_k}. Each atom in D_{G_k} for group x_{G_k} is defined as

$$d_{G_k(i)} = u_{G_k(i)} v_{G_k(i)}^T, \ i = 1, 2, \ldots, m, \qquad (12)$$

where $d_{G_{k(i)}} \in R^{Bs \times c}$, so the dictionary $D_{G_k} = \left[d_{G_{k(1)}}, d_{G_{k(2)}}, \ldots, d_{G_{k(m)}} \right]$.

Each basis image, $u_i v_i$ specifies a layer of image geometry, and the sum of these layers denotes the complete image structure. The first few singular vector pairs account for the major image structure, whereas the subsequent u_i and v_i account for the finer details in an image. We illustrate this point through an example shown in Fig. 1, where the image size is 512×512. We can see that the first 20 pairs ($z = 20$) of u_i and v_i capture the major image structure, and the subsequent pairs of u_i and v_i signify the finer details in image structure. As an increasing number of u_i and v_i pairs are used, the finer image structural details appear. u_i and v_i can therefore represent the structural elements in images well.

3.3 Optimization Formulation for CSRH Imaging System

After obtaining dictionary, an image can be represented sparely by the codewords of this dictionary. In compressed sensing, this sparse representation will further undergo random measurement of the measurement matrix. In our work, a sample pattern given by the antennas configuration of CSRH [15–17] is imposed on the Fourier transform of the brightness function of the Sun (spatial image) to provide a sparse measurements of the brightness function. Relative to brightness function, the Fourier transform of brightness function is named visibility function. Given a dictionary D_G, the sparse representation of a brightness image can be written

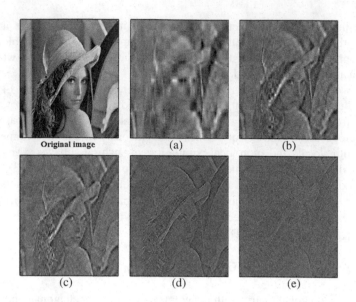

Fig. 1. The basis images $(u_i v_i)$ coming from SVD ($z = 10, 20, 30, 100$ and 512 for (a), (b), (c), (d) and (e) respectively).

$$x = D_G \alpha_G, \tag{13}$$

which further undergoes the sampling of CSRH antenna configuration to have the recorded signal of visibility function as

$$y = HD_G \alpha_G. \tag{14}$$

Since (14) is an underdetermined problem, we have to impose sparse constraint to solve α_G from (14). With sparse constraint, we have a constrained optimization

$$\min \|\alpha_G\|_0 \; s.t. \; y = HD_G \alpha_G, \tag{15}$$

which is not convex due to l_0 norm, so it is usually to replace l_0 by l_1. Thus, the optimization formulation is finalized by

$$\hat{\alpha}_G = \mathrm{argmin}_{\alpha_G} \|HD_G \alpha_G - y\|_2^2 + \lambda \|\alpha_G\|_1, \tag{16}$$

which can be solved efficiently by some recent convex optimization algorithms, such as iterative shrinkage/thresholding, split Bregman algorithms. Generally, there are three kinds of methods: greedy algorithm, convex optimization and combination of them for resolving (16).

4   Experimental Results

CSRH consists of a high frequency array (CSRH-II) with 60 antennas and a low frequency array (CSRH-I) with 40 antennas. These antennas are located in three spiral arms as shown in Fig. 2 (a) to form an AS system. For AS observation, the interferometry technique is employed to record input signal in frequency domain instead of spatial domain. Specifically, each two antennas form an interferometer which can capture one frequency component in Fourier/UV domain at each time of observation. There are 40 antennas, so we have $40 \times 39/2$ frequency components for one time of observation of CSRH-I. By applying inverse Fourier transform, we can obtain spatial images of observed objects.

In our simulation, we use the sample pattern of CSRH-I to generate degraded image from original ones. Here, we use the photos taken by Atmospheric Imaging Assembly (AIA) of Solar Dynamics Observatory (SDO) [18]. The sample pattern of CSRH-I is shown in Fig. 2(b), which is named UV coverage, and represented by a matrix in UV domain or frequency domain. The corresponding spatial form of Fig. 2(b) is called PSF or "dirty beam" which is the inverse Fourier transform of UV coverage. It can be

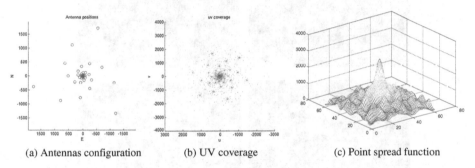

(a) Antennas configuration   (b) UV coverage   (c) Point spread function

Fig. 2. Aperture synthesis of CSRH-I. (a) Antennas configuration (b) UV coverage (c) Point spread function.

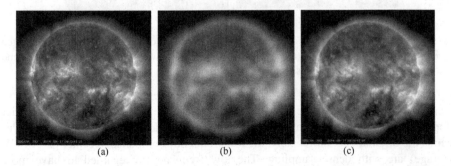

(a)   (b)   (c)

Fig. 3. Image reconstruction of CSRH. (a) An image recorded by SDO/AIA at 193; (b) The imaging result of CSRH ((a) is convolved by the PSF show in Fig. 2 (c)); (c) The recovered image from (b) by using the proposed algorithm.

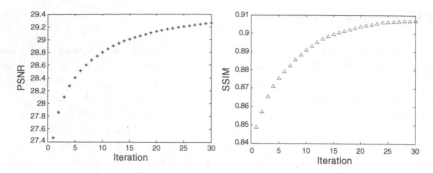

Fig. 4. PSNR/SSIM of reconstructed image over iteration.

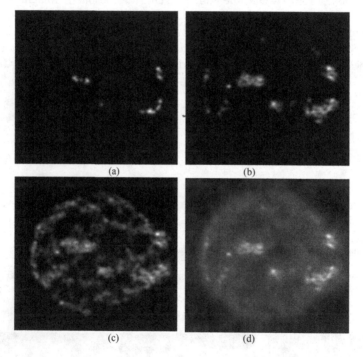

Fig. 5. The reconstructed image by using Högbom clean algorithm ((d) is the reconstructed image with 400 iterations; (a)-(c) only show the peaks without solar disk background (a) 40 iterations; (b) 400 iterations; (c) 4000 iterations).

observed that CSRH-I has a very sparse UV coverage, so only a small part of Fourier components can be recorded by CSRH-I. It is impossible to have a spatial image as clear as the original one. Fortunately, the low frequency components (center part of UV coverage) are with dense sampling. The low frequency is regarded to have more important geometry information of an image.

Simulating the process of AS imaging system, an original image is given in Fig. 3 (a). However, we can only obtain its degraded one shown in Fig. 3(b) by CSRH-I system. This process can be represented that an image is convolved by a dirty beam, e.g., Fig. 2(c). As shown in Fig. 2(c), there are a certain amount of sidelobes apart from main lobe in a PSF, so an original image would be blurred by convolving such a PSF. To improve image quality regarding such a convolution process, "clean" algorithms [1–6] have been proposed to fulfill corresponding deconvolution process, i.e., eliminate convolution effect of the PSF on the original image. These "clean" algorithms are straightforward, and therefore behave more applicable and robust. Associating sparse sampling process of an AS system, CS-based [19–23] algorithms have been explored to recover images from degraded ones. The captured signals were regarded to be sparse with respect to Fourier or wavelet basis in these algorithms.

In this work, we explore user-defined dictionary for more efficient sparse representation of image. This dictionary based CS algorithm is adaptive to input signal, and therefore could acquire better efficiency. The proposed algorithm for CSRH imaging is formulated in (16). To solve (16), the Split Bregman Iteration (SBI) is employed. It concerns a series of interactions. In each round of iteration, the reconstructed image is improved with respect to image quality. The iteration is terminated after 30 iterations or beyond a given threshold which is given by the PSNR difference between two successive iterations. We give an example of the iteration in Fig. 4 to illustrate the procedure of iteration. It can be observed the image quality of reconstructed image is improved step by step. From Fig. 4, the improvement of image quality is significant. There is 5 dB PSNR gain of the final reconstructed image over the worst one. For better measuring subjective image quality, we also employed SSIM [24] to evaluate image quality. The SSIM is regarded to be better for representing image structures. The SSIM for each iteration is also shown in Fig. 4. The final reconstructed image is shown in Fig. 3(c), it can be obviously observed that the better subjective image quality is obtained. Comparing with Fig. 3(b), the reconstructed image presents clear edges and structures can be identified in Fig. 3(c).

For comparing with the state-of-the-art algorithms, we implement Högbom clean algorithm on the blurred image of CSRH. The experimental results are shown in Fig. 5. From Fig. 5, we can observe that more and more peaks which represent point sources in the blurred image are identified with the increase of iterations. In theory, all these point sources can be figured out. Then, convolving these point sources with an idea beam instead of the dirty beam, one can obtain a cleaned image. However, these point sources interfere with each other if these point source are closely related. That's why the results of Högbom clean algorithm is not as satisfied as that it claimed in our simulation.

5 Conclusion

This paper presented a CS-based image reconstruction algorithm for CSRH imaging system. By using user-defined dictionary, the images of CSRH can be better represented sparsely. The better image quality can be obtained by the proposed CS-based image reconstruction algorithm. Comparing with clean algorithms, the proposed algorithm can better recover the high frequency components which represent fine image geometry.

Acknowledgment. This work was partially supported by a grant from the National Natural Science Foundation of China under Grant 61202242, 100-Talents Program of Chinese Academy of Sciences (No. Y434061V01).

References

1. Högbom, J.A.: Aperture synthesis with a non-regular distribution of interferometer baselines. Astron. Astrophys. Suppl. **15**, 417 (1974)
2. Thompson, A.R., Moran, J.M., Swenson, G.W., Wakker, B.P., Schwarz, U.J.: The Multi-Resolution CLEAN and its application to the short-spacing problem in interferometry. Astron. Astrophys. **200**, 312–322 (1988)
3. Cornwell, T.J.: Multi-Scale Clean Deconvolution of Radio Synthesis Images. arXiv: 0806.2228
4. Weir, N.: A multi-channel method of maximum entropy image restoration. In: ASP Conference Series 25: Astronomical Data Analysis Software and Systems I, p. 186 (1992)
5. Cornwell, T.J., Evans, K.F.: A simple maximum entropy deconvolution algorithm. Astron. Astrophys. **143**, 77–83 (1985)
6. Starck, J.L., Pantin, E., Murtagh, F.: Deconvolution in astronomy: a review. Publ. Astr. Soc. Pacific **114**, 1051–1069 (2002)
7. Cand`es, E., Romberg, J., Tao, T.: Robust uncertainty principles: exact signal reconstruction from highly incomplete frequency information. IEEE Trans. Inform. Theory **52**, 489–509 (2006)
8. Donoho, D.L.: Compressed sensing. IEEE Trans. Inf. Theory **52**(4), 1289–1306 (2006)
9. Donoho, D.L., Huo, X.: Uncertainty principles and ideal atomic decompositions. IEEE Trans. Inform. Theory **47**, 2845–2862 (2011)
10. Elad, M., Aharon, M.: Image denoising via sparse and redundant representations over learned dictionaries. IEEE Trans. Image Process. **15**(12), 3736–3745 (2006)
11. Buades, A., Coll, B., Morel, J.M.: A review of image denoising algorithms, with a new one. Multiscale Model. Simul. **4**(2), 490–530 (2005)
12. Yang, J., Wright, J., Huang, T.S., et al.: Image super-resolution via sparse representation. IEEE Trans. Image Process. **19**(11), 2861–2873 (2010)
13. Zhang, J., Zhao, D.B., Gao, W.: Group-based sparse representation for image restoration. IEEE Trans. Image Process. (TIP) **23**(8), 3336–3351 (2014)
14. Zhang, J., Zhao, D.B., Xiong, R.Q., Ma, S.W., Gao, W.: Image restoration using joint statistical modeling in a space-transform domain. IEEE Trans. Circuits Syst. Video Technol. (TCSVT) **24**(6), 915–928 (2014)
15. Yan, Y., Wang, W., Liu,F., Geng, L., Zhang, J.: Radio imaging-spectroscopy observation of the sun in decimeteric and centimetric wavelengths. In: Solar and Astrophysical Dynamos and Magnetic Activity, Proceeding of IAU Symposium, no. 294, pp. 489–494 (2012)
16. Yan, Y., Zhang, J., Wang, W., Liu, F., Chen, Z., Ji, G.: The Chinese spectral Radioheliograph-CSRH. Earth Moon Planet. **104**, 97–100 (2009)
17. Du, J., Yan, Y.H., Wang, W.: A simulation of imaging capabilities for the Chinese Spectral Radioheliograph. In: IAU Symposium, pp. 501–502 (2013)
18. http://sdo.gsfc.nasa.gov/

19. Li, F., Cornwell, T., de Hoog, F.: The application of compressive sampling to radio astronomy I: deconvolution. Astron. Astrophys. **528**(A31), 1–10 (2011)
20. Li, F., Brown, S., Cornwell, T., de Hoog, F.: The application of compressive sampling to radio astronomy II: faraday rotation measure synthesis. Accepted Astron. Astrophys. **531**, A126 (2011)
21. Wenger, S., Magnor, M., Pihlström, Y., et al.: SparseRI: A compressed sensing framework for aperture synthesis imaging in radio astronomy. Publ. Astron. Soc. Pac. **122**(897), 1367–1374 (2010)
22. Bobin, J., Starck, J.L., Ottensamer, R.: Compressed sensing in astronomy. IEEE J. Sel. Top. Sign. Process. **2**(5), 718–726 (2008)
23. Wiaux, Y., Jacques, L., Puy, G., et al.: Compressed sensing imaging techniques for radio interferometry. Mon. Not. Roy. Astron. Soc. **395**(3), 1733–1742 (2009)
24. Wang, Z., Bovik, A.C., Sheikh, H.R., Simoncelli, E.P.: Image quality assessment: from error visibility to structural similarity. IEEE Trans. Image Process. **13**(4), 600–612 (2004)

Social Media Computing

Real-Life Voice Activity Detection Based on Audio-Visual Alignment

Jin Wang[1], Chao Liang[1,2]($\boxtimes$), Xiaochen Wang[1,3], and Zhongyuan Wang[1]

[1] National Engineering Research Center for Multimedia Software,
School of Computer, Wuhan University, Wuhan, China
cliang@whu.edu.cn
[2] Collaborative Innovation Center of Geospatial Technology, Wuhan, China
[3] Research Institute of Wuhan University in Shenzhen, Shenzhen, China

Abstract. Voice activity detection (VAD) is a technology to identify whether the persons in multimedia are speaking. Most of the research efforts focused on utilizing audio and visual information to implement voice activity detection, which outperform audio or visual approach alone proposed earlier. However, current methods explore a supervised classifiers using new feature consist of audio and visual information. In the paper, we propose a novel method to detect voice activity by audio-visual alignment. Since the temporal order relationship of voice activity detection over the whole audio and visual information, we use Needleman-Wunsch algorithm to align two different sequences. Compared to existing VAD algorithms,our experimental results indicate that the proposed approach presents better results, and the accuracy rate reaches about 85 % in real-life environment.

Keywords: Voice activity detection · Sequence alignment · Needleman-Wunsch algorithm

1 Introduction

Automatic voice activity detection (VAD) has been an area of extensive research in recent years, such as speech recognition systems, advanced teleconferencing, and video analysis [1]. Speech is a bimodal signal, involving audio and visual information [2].Traditional methods are mainly based on audio information or visual information individually, which would make poor performance and robustness to undesired environmental changes. To solve the issue, Multi-model VAD (M-VAD) algorithm has been proposed in recent years, which combines audio and visual information.

Early approaches of audio voice activity detection (A-VAD) are based on simple energy thresholds or correlation coefficients and zero-crossing rate rules [3]. More complex algorithms exploit more than one feature to detect speech [4,5]. None of the above-mentioned features are shown to perform uniformly well under low SNR conditions. More recent approaches consider statistical

© Springer International Publishing Switzerland 2015
Y.-S. Ho et al. (Eds.): PCM 2015, Part II, LNCS 9315, pp. 109–117, 2015.
DOI: 10.1007/978-3-319-24078-7_11

model can ensure the performance of the A-VAD. Sohn et al. [6] propose an A-VAD method that uses frequency bands of the speech signal as input features for the likelihood ratio test, which is then followed by an Hidden Markov Model(HMM) hang-over scheme to impose temporal coherence to the detector. J. Ramirez et al. [7] consider contextual information in a multiple observation LRT (MO-LRT), which benefits for speech/pause discrimination in high noise environments. This paper shows a revised MO-LRT VAD that extends the number of hypotheses on the individual multiple observation window that are tested. Youngjoo Suh et al. [8] employ probabilistically derived multiple acoustic models to effectively optimize the weights on frequency domain likelihood ratios with the discriminative training approach for more accurate voice activity detection.

Visual cues can be used for both speech recognition and speech detection, which can help in the voice activity recognition task, since the lips move more than 80 % of the time in human speech [9]. However, not all lip movements correspond to visual speech, yawning and chewing being two such examples. Liu and Wang [10] present an approach, using PCA on the intensities and the first order intensity differences of the pixels in the mouth region. The outcome of PCA is used as the feature vector modeled by Gaussians distributions. Libal et al. [11] propose an approach calculate intensity difference images over a certain time constant. Once the black bar of the mouth has been found on motion history image, lip opening/closing are determined by counting all pixels, recording a value for the 90th percentile, and comparing it to a running average. Siatras et al. [12] present a method based on the significant variation of the intensity values of the mouth region as demonstrated by a speaking person. The number of pixels of the mouth region, in each video frame in which the grayscale value is below an intensity threshold, is employed as visual cue. The speaking and non-speaking intervals are determined by applying an energy detector and an averager to a sliding window that moves frame by frame. P. Tiawongsombat et al. [13] focus on lip movement and speaking assumptions, embracing two essential procedures into a single model. A bi-level Hidden Markov Model (HMM) is an HMM with two state variables in different levels, where state occurrence in a lower level conditionally depends on the state in an upper level.

In [14] an Multi-model VAD algorithm has been proposed, where A-VAD and V-VAD (Visual voice activity detection)are performed separately using Gaussian Mixture. Models are fused by a SNR-based weighting approach. They use mono audio recordings with simulated white noise, showing good VAD accuracy, which is however affected when testing the algorithm on speakers different from those used for training it. Takeuchi et al. [15] use Mel-Frequency Cepstral Coefficients (MFCC) as audio features and optical flow of the speakers mouth as visual features through an multi-stream HMM system. Their algorithm show good performance in a relatively controlled environment, and decreases proportionally to noise levels in the audio modality. Minotto, V.P. et al. [16] propose a multimodal VAD (M-VAD) approach based on a single monocular camera and an array of microphones, focusing on a videoconferencing or human-computer interaction setup where the participants are facing the camera at a roughly known distance.

In this paper, we propose a novel method to accomplish VAD with the audio and visual sequence alignment. It can be assumed that the temporal distribution of A-VAD is approximately similar to the temporal distribution of their V-VAD along the time line. We firstly cluster label set in audio and visual respectively, then describe the M-VAD by the Needleman-Wunsch algorithm to align the audio and video sequence. To our best knowledge, so far no one has used the orders between audio and video information in VAD.

2 Visual Voice Activity Detection

Before applying the proposed algorithm, we detect the face in the video and assign at each frame a rectangle involving the lip region. The faces are detected using the method proposed by [17], based on Haar Cascade Classifier, which is widely used systems. Additionally, we find 68 landmarks on face image by a Coarse-to-Fine Auto-encoder Networks approach [18], which cascades a few successive Stacked Auto-encoder Networks (SANs). Then the lip region is extracted by the landmarks. Figure 1 illustrates sample images of the automatically located face region and landmarks.

(a)   (b)

Fig. 1. (a) Detected face (b) 68 landmarks

The traditional appearance-based approaches require proper image quality to accurately obtain the positions of features, which are impractical caused by natural head motion, uncontrolled lighting conditions, and the dynamic variations during speaking. In order to overcome these limitations, instead of relying on mouth appearances, we calculate the image energy of the lip region [13] as the key of V-VAD. The image energy of the single image frame is denoted by:

$$E[n] = \sum_{i=1}^{M} \sum_{j=1}^{N} v_{i,j}^2(t) \qquad (1)$$

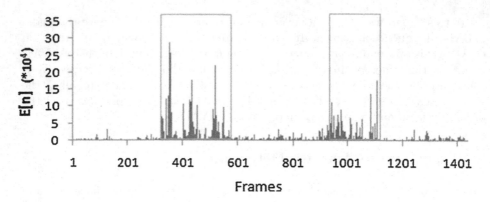

Fig. 2. The image energy of the lip region.The rectangle encompasses the frames where the person is speaking.

where pixels velocity in y-direction, computed by Pyramidal Lucas- Kanade method [19], at image coordinate (i, j) in the mouth image of the size. As Fig. 2, it is obvious that E[n] obtains much higher and large deviations values when the person is speaking. And when the person is non-speaking, E[n] is much lower at average and a smaller deviation. So we calculate conditional probability by HMM [20] as the feature vector of voice activity detection. At last, it uses the k-means algorithm to cluster the new feature vectors. However, V-VAD approach tends to fail when the mouth is covered or when the lips move without speech (e.g., smile or yawn). Next we use audio-visual method to detect voice activity, which produce better results than using audio and video cues individually.

3   Audio Voice Activity Detection

For existing LRT-based VADs [6], the decision rule is formulated by taking the geometric mean of the LRs of all K DFT bins to get the current LR, $\Lambda(t)$ of frame t.

$$log\Lambda(t) = 1/K \sum_{n=0}^{K} log\Lambda_k(t) \qquad (2)$$

where $\Lambda_k(t)$ is the LR of the kth Discrete Fourier Transform(DFT) bin. But When speech is present, it does not always manifest energy in all DFT bins [21]. This is especially true for high-pitched voiced utterance, where most of the energy is located in the harmonic frequency bins. To boost the LR score in such situations, a new method for evaluating the LR for voiced frames is proposed. We calculate the DFT bins LR which just choose in the harmonic peaks, because harmonic spectral peaks are stronger and more flexible than the other noise component in speech spectrum. Finally, the log LR for voiced frame is computed as follows:

$$logΛ(t) = 1/K \sum_{n=0}^{K} logΛ_{Hu,v}(t) \tag{3}$$

As for unvoiced frames, the $logΛ_{u,v}(t)$ is still calculated using (2). Finally, A-VAD is decided based on a time interval of samples, similarly to the V-VAD approach.

4   Audio-Visual Voice Activity Detection

Each of the approaches described above (V-VAD and A-VAD) may be used independently to detect voice activity, as previously mentioned. However, there are situations that are vulnerable to noisy observation and environmental changes, and leads to poor performance and robustness. For instance, when the mouth is covered with speech or when the lips move without speech (e.g., smile or yawn), the V-VAD approach tends to decrease the detection rate. In A-VAD, if there are background noise (e.g. air conditioner or music), the approach based on audio information also tends to decrease the detection rate. Here we use a novel method to achieve M-VAD,which is considered that the temporal order relationship of voice activity detection over the whole audio and visual information, then align two sequences by the Needleman-Wunsch algorithm. The Needleman-Wunsch algorithm [22] is an algorithm used in bioinformatics to align protein or nucleotide sequences. It was one of the first applications of dynamic programming to compare biological sequences which is also sometimes referred to as the optimal matching algorithm and the global alignment technique. Here we align the A-VAD with V-VAD by the Needleman-Wunsch algorithm. Given a A-VAD sequence $X = x_1, x_2, \cdots, x_n$ with each $x_i = A\{1,2\}$ where A is A-VAD cluster label set, and a V-VAD sequence $Y = y_1, y_2, \cdots, y_n$ with each $y_i = B\{1,2\}$ where B is V-VAD cluster label set. Since the lengths of the two sequences are not the same, the delete and insert actions will be taken to certain elements in the sequence. Scores for aligned characters are specified by a similarity matrix. Here, S(A,B) is the similarity of characters A and B. and d is gap penalty. (S(A,B) is calucated by default and d=−5). The entry in row i and column j are denoted here by A two-dimensional matrix $W_{i,j}$. The goal is to find the optimal surjection from A to B with the highest score $W_{i,j}$. The principle of optimality is then applied as follows:

$$W_{i,j} = max(W_{i-1,j-1} + S(A_i, B_j), W_{i,j-1} + d, W_{i-1,j} + d) \tag{4}$$

Once the $W_{i,j}$ matrix is computed, the entry gives the maximum score among all possible alignments. To compute an alignment that actually gives this score, you start from the bottom right cell, and compare the value with the three possible sources (Match, Insert, and Delete above) to see which it came from. If Match, then A and B are aligned, if Delete, then A is aligned with a gap, and if Insert, then B is aligned with a gap.

5 Experimental Results

In this section, we will calculate the accuracy of different VAD algorithm. We use individually our A-VAD approach, our V-VAD approach, the M-VAD approach, and other VAD methods. The videos consist of one-speaker in real-life environment which generate by door slams, speech in background, air condition etc. The experimental results show our M-VAD algorithm better performance than other VAD methods. The algorithm has been implemented in a PC equipped with a Intel Core i5-3210M CPU 2.5 GHz, 4 GB RAM. The ground truth has been determined by manually marking frames of the image sequences, as speaking or non-speaking.

Our approaches are tested on two videos A, B and C (see Fig. 3)(about 1 min). The speakers alternate between 15 s of non-speech and 15 s of speech, starting with non-speech. For the speaking interval, the speakers are directed to naturally read the provided material. Sequences A and B are recorded by the same speaker in different environment. In video A, there are soft music (without singer) in non-speanking periods. The speaker in B moves her mouth during non-speaking moments, and play music (with singer) at the same time. The video C is from the Chinese Television programs - News broadcast. Here, we consider the male as our experimental subject. And the people in C are keeping speaking across the entire video. The males speaking periods account for about 46 % in all video.

(a) video A

(b) video B

(c) video C

Fig. 3. Experimental images.

Table 1. The accuracy of different VAD.

Video	A	B	C
V-VAD(Libal et al[11])	85.14 %	61.67 %	50.98 %
V-VAD(Minotto et al[16])	84.12 %	73.33 %	60.78 %
V-VAD(our approach)	88.89 %	80.02 %	68.63 %
A-VAD(G 729 B[23])	78.62 %	59.68 %	43.14 %
A-VAD (our approach)	85.48 %	67.74 %	45.10 %
M-VAD(our approach)	93.28 %	86.67 %	80.39 %

The results are shown in Table 1, which indicate that the proposed approach presents better results when compared to existing VAD algorithms. The A-VAD method proposed by Minotto et al. [16] can not satisfy the real environment, which requires a microphone array, so we will not compare this with our approach. In V-VAD, the method [11,16] can not conclude the movement of the lips, when there are no talks, resulting in poor performance. And we can see the pixel intensity performance worse than the height in equipped with jamming environment. But there are actually small during the speaking state in our approach, due to the small lip movements, resulting in the speakingness interval mistaken as non-speaking. In video C, the male is away from the camera. As a result, the detection of lip region is worse than video A and B, and the vad result is also poor. And the people in C are keeping speaking across the entire video. This means the male is non-speaking, but the female is speaking. In A-VAD, the feature in our method is more resilient to noise environment others(energy or correlation coefficients). As a result, our method outperforms than G729 B [23] in non-stationary noise. But video B are created to purposely impose challenges to the audio-based and video-based VAD, respectively. When another person nearby kept speaking, the VAD is mainly based on V-VAD. And our M-VAD method increased the accuracy rates in most cases, which the correct detection rate reaches about 85 %.In conclusion, the experimental results confirm our M-VAD presents better classification accuracy and robustness compared to the A-VAD or V-VAD approach.

6   Conclusion

A novel multi-model voice activity detection based on audio-visual alignment is proposed in this paper. The experiments indicate our approach outperform other existing VAD methods, and our features are robust in real-life environments. We consider the temporal order relationship of voice activity detection over the whole audio and visual information, then align two different sequences by Needleman-Wunsch algorithm.This is different from other M-VAD, which explore a supervised classifiers using new feature consist of audio and visual information.

Acknowledgement. The research was supported by the National Nature Science Foundation of China (61231015,61201169, 61303114), the Specialized Research Fund for the Doctoral Program of Higher Education (20130141120024), the Technology Research Project of Ministry of Public Security (2014JSYJA016), Nature Science Foundation of Hubei Province (2014CFB712), the Fundamental Research Funds for the Central Universities(2042014kf0250,2042014kf0025),Science and Technology Plan projects of Shenzhen(ZDSYS2014050916575763),The Project Sponsored by the Scientific Research Foundation for the Returned Overseas Chinese Scholars, State Education Ministry ([2014]1685)

References

1. Liang, C., Xu, C., Cheng, J., et al.: TVparser: An automatic TV video parsing method. In: 2011 IEEE Conference on Computer Vision and Pattern Recognition (CVPR). IEEE (2011)
2. Sodoyer, D., Rivet, B., Girin, L., Savariaux, C., Schwartz, J.-L.: A study of lip movements during spontaneous dialog and its application to voice activity detection. J. Acoust. Soc. Am. **125**(2), 1184–1196 (2009)
3. Woo, K., Yang, T., Park, K., Lee, C.: Robust voice activity detection algorithm for estimating noise spectrum. IET Electron. Lett. **36**(2), 180–181 (2000)
4. Soleimani, S.A., Ahadi, S.M.: Voice activity detection based on combination of multiple features using linear/kernel discriminant analyses. In: Proceedings of the 3rd International Conference on Information and Communication Technologies (2008)
5. Lee, B., Muhkerjee, D.: Spectral entropy-based voice activity detector for videoconferencing systems. In: Kobayashi, T., Hirose, K., Nakamura, S., (eds.) Proceedings of INTERSPEECH (2010)
6. Sohn, J., Kim, N.S., Sung, W.: A statistical model-based voice activity detection. IEEE Signal Process. Lett. **6**(1), 1–3 (1999)
7. Ramirez, J., Segura, J., Benitez, C., Garcia, L., Rubio, A.: Statistical voice activity detection using a multiple observation likelihood ratio test. IEEE Signal Process. Lett. **12**(10), 689–692 (2005)
8. Suh, Y., Kim, H.: Multiple acoustic model-based discriminative likelihood ratio weighting for voice activity detection. IEEE Signal Process. Lett. **19**(8), 507–510 (2012)
9. Wang, L., Wang, X., Xu, J.: Lip detection and tracking using variance based Haarlike features and Kalman filter. In: Proceedings of the 5th International Conference on Frontier of Computer Science and Technology (FCST) (2010)
10. Liu, P., Wang, Z.: Voice activity detection using visual information. In: Proceedings of the IEEE International Conference on Acoustics, Speech, and Signal Processing (ICASSP), Montreal, Canada (2004)
11. Libal, V., Connell, J., Potamianos, G.: An embedded system of in-vehicle visual speech activity detection. In: International Workshop on Multimedia Signal Process (MMSP) (2007)
12. Siatras, S., Nikolaidis, N., Krinidis, M., et al.: Visual lip activity detection and speaker detection using mouth region intensities. IEEE Trans. Circuits Syst. Video Technol. **19**(1), 133–137 (2009)
13. Tiawongsombata, P., Jeongb, M.-H., Yun, J.-S.: Robust visual speakingness detection using bi-level HMM. Pattern Recogn. **45**(2), 783–793 (2012)

14. Almajai, I., Milner, B.: Using audio-visual features for robust voice activity detection in clean and noisy speech. In: Proceedings of the 16th European Signal Processing Conference (EUSIPCO 2008) (2008)
15. Hashiba, T., Tamura, S., Takeuchi, S., Hayamizu, S.: Voice activity detectionbased on fusion of audio and visual information. In: Proceedings of the International Conference on Auditory-Visual Speech Processing (2009)
16. Minotto, V.P., Lopes, C.B.O., Scharcanski, J., et al.: Audiovisual voice activity detection based on microphone arrays and color information. IEEE J. Sel. Top. Sign. Process. 7(1), 147–156 (2013)
17. Viola, P., Jones, M.: Robust real-time face detection. International Journal of Computer Vision(IJCV) 57(2), 137–154 (2004)
18. Zhang, J., Shan, S., Kan, M., Chen, X.: Coarse-to-fine auto-encoder networks (CFAN) for real-time face alignment. In: Fleet, D., Pajdla, T., Schiele, B., Tuytelaars, T. (eds.) ECCV 2014, Part II. LNCS, vol. 8690, pp. 1–16. Springer, Heidelberg (2014)
19. Lucas, B.D., Kanade, T.: An iterative image registration technique with an application to stereo vision. In: IJCAI (1981)
20. Rabiner, L.R., et al.: A tutorial on hidden markov models and selected applications in speech recognition. Proc. IEEE 77(2), 257–286 (1989). AT&T Bell Lab, Murray Hill
21. Tan, L.N., Borgstrom, B.J., Alwan, A.: Voice activity detection using harmonic frequency components in likelihood ratio test. In: Acoustics Speech and Signal Processing (ICASSP) (2010)
22. Needleman, S.B., Wunsch, C.D.: A general method applicable to the search for similarities in the amino acid sequence of two proteins. J. Mol. Biol. 48(3), 443–453 (1970)
23. ITU-T Rec. G.729, Annex B (2007)

Emotion Recognition from EEG Signals by Leveraging Stimulus Videos

Zhen Gao and Shangfei Wang[✉]

Key Lab of Computing and Communication Software of Anhui Province,
School of Computer Science and Technology,
University of Science and Technology of China,
Hefei 230027, Anhui, People's Republic of China
gzgqllxh@mail.ustc.edu.cn, sfwang@ustc.edu.cn

Abstract. This paper proposes a new emotion recognition method from electroencephalogram (EEG) signals by leveraging video stimulus as privileged information, which is only required during training. A Restricted Boltzmann Machine (RBM) is adopted to model the intrinsic relations between stimulus videos and users' EEG response, and to generate new EEG features. Then, the support vector machine is used to recognize users' emotion states from the generated EEG features. Experiments on two benchmark databases demonstrate that stimulus videos as the privileged information can help EEG signals construct better feature space, and RBM can model the high-order dependencies between stimulus videos and users' EEG response successfully. Our proposed emotion recognition method leveraging video stimulus as privileged information outperforms the recognition method only from EEG signals.

Keywords: Emotion recognition · Privileged information · RBM · EEG

1 Introduction

Automatically analysis of users' emotions is essential to understand users' behavior. Users' emotions can be detected from facial expressions, speeches, and physiological signals. Facial expressions and speeches are considered as outer presentation of emotions, and not always the same as the inner emotions, since people can conceal their emotions in facial expressions and speeches. While, physiological signals reflect unconscious changes in bodily functions, which are controlled by the Sympathetic Nervous System (SNS). Compared with facial expression and speech, it is more reliable to recognize emotions from physiological signals, since physiological signals are infeasible to be changed consciously. Recently, considerable progresses have been made on emotion recognition from philological signals, especially from EEG signals, due to its various real-world applications of brain-computer interface for normal people.

Emotion recognition from EEG signals includes two steps: feature extraction and emotion classification. Frequency features, such as power spectrum density

© Springer International Publishing Switzerland 2015
Y.-S. Ho et al. (Eds.): PCM 2015, Part II, LNCS 9315, pp. 118–127, 2015.
DOI: 10.1007/978-3-319-24078-7_12

(PSD) [3,5], are often employed to describe EEG signals, and classifiers, such as hidden Markov model [6] and Support Vector Machine (SVM) [7], are widely used to recognize emotions. Although many advances have been made in feature extraction and machine learning recently, accurate emotion recognition from EEG signals still remains a very challenging problem. Apart from users' spontaneous response, the context, in which the emotional behavior is displayed, is very important for emotion recognition. Specifically, for analyzing the emotions from video-induced EEG signals, the stimulus–video, has intrinsic relations with the induced EEG response, and direct influence on the users' emotions. Therefore, stimulus videos, as external data, could provide important assistance in building the video-based context modal for emotion recognition.

To the best of our knowledge, only two works explore context related factors, such as stimulus, for emotion recognition from EEG signals till now. Wang et al. [10] proposed three Bayesian Networks to annotate videos by combining the video and EEG features at Independent Feature-Level fusion (IFL), Decision-Level fusion (DL) and Dependent Feature-Level fusion (DFL). In this work, videos and EEG signals are required for both training and testing. Zhu et al. [11] proposed to use Canonical Correlation Analysis (CCA) to construct a new EEG feature space for emotion recognition with the help of video content. Stimulus videos are used as privileged information [8], which is only required during training, and is exploited during training to construct better EEG features for emotion recognition. The complexity of emotion recognition method increases if both stimulus videos and EEG signals are required during testing. Therefore, we prefer to incorporate stimulus videos as privileged information, which is only required during training, in this paper. Other than CCA, which models the linear relations among EEG features and video features, a Restricted Boltzmann Machine (RBM) can model high-order dependencies among visible variables by introducing hidden nodes [2]. Furthermore, as a generative model, RBM is naturally good at learning with privileged information, since it allows to capture the joint probability distribution function of the target variable (i.e. emotion labels), available information (i.e. EEG signals), and privileged information (i.e. stimulus videos) during training. Given the joint probability distribution function, the target variable can be estimated from the available information during testing by marginalized over the privileged information. Therefore, in this paper, we propose RBM to model the high-order dependencies between EEG features and video features. We divide the visible units of RBM into two parts, one presents the EEG feature and the other is on behalf of the video features. During training, both EEG features and video features are used to learn the parameters of our RBM model. After that, new EEG feature space can be obtained from EEG features only by sampling over video features. Then, SVM is used to recognize emotions from the generated new EEG features.

Our paper is organized as follows. We presented the framework of our method and explain our method in details in Sect. 2. The experimental results and detailed analysis are shown in Sect. 3. In Sect. 4, we draw a conclusion.

2 Emotion Recognition from EEG Signals with the Help of Stimulus Videos

The framework of our method is shown in Fig. 1. First, EEG features and video features are extracted from EEG signals and stimulus videos respectively. During training, a RBM is proposed to model the high-order dependencies between EEG features and video features by introducing hidden nodes. After learning the parameters of RBM, new EEG feature space can be obtained from EEG features only by sampling over video features. Then, a SVM is trained for emotion recognition from the generated EEG features. The video features are only available during training to help construct a better EEG feature space.

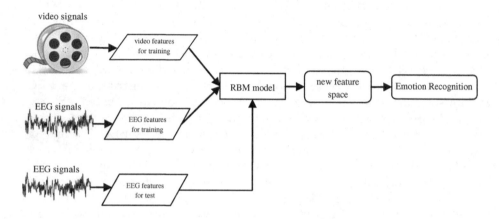

Fig. 1. The framework of our method

2.1 Feature Extraction

We extract both visual and audio features from stimulus videos as video features. For visual features, we extract lighting key, color energy and visual excitement, which are powerful to reflect the mood of the scenes [9]. For audio features, we extract 31 commonly used audio features, including average energy, average energy intensity, spectrum flux, Zero Crossing Rate(ZCR), standard of deviation of ZCR, 12 Mel-frequency Cepstral Coefficients (MFCCs), log energy of MFCC, the standard deviations of 13 MFCCs.

Power spectrum (PS) features are extracted from EEG signals. First, noise attenuation is carried out. Then, the power spectrum of five frequency sub-bands (i.e. delta(0–4 Hz), theta(4–8 Hz), alpha(8–13 Hz), beta(13–30 Hz) and gamma(30–45 Hz)) from 32 electrodes are extracted. Furthermore, the power spectrum asymmetry between 14 pairs of electrodes from alpha, beta, theta and gamma sub-bands, and the ratio of the power in each sub-band to the overall power are obtained.

2.2 Modeling Relations Between EEG Signals and Stimulus Videos by RBM

RBM can capture high-order dependencies among the visible variables by connecting all the visible units through the latent units, therefore, we propose to use RBM to model the natural connection between stimulus videos and users' EEG signals.

The structure of our RBM is shown in Fig. 2(a). We divide the visible nodes into two parts, one part represents the EEG features and the other represents the video features. EEG features and video features are continuous, hence we use the Gaussian units for the visible nodes. We consider EEG features as visible real-valued units $V^{(E)} \in R^{D^E}$ and video features as $V^{(V)} \in R^{D^V}$ where D^E, D^V represent the dimensions of EEG and video features respectively, and make $H \in \{0,1\}^{nhidden}$ be binary stochastic hidden units, where $nhidden$ means the number of hidden nodes. The energy of the state V^E, V^V, h of our Gaussian-Bernoulli RBM [2] is shown as follows:

$$E(V^E, V^V, h|\theta) = \sum_{i=1}^{D^E} \frac{(v_i^{(E)} - b_i^{(E)})}{2\sigma_i^{(E)2}} + \sum_{i=1}^{D^V} \frac{(v_i^{(V)} - b_i^{(V)})}{2\sigma_i^{(V)2}} - \sum_{i=1}^{D^E} \sum_{j=1}^{nhidden} \frac{v_i^{(E)}}{\sigma_i^{(E)}} W_{ij}^E h_j$$

$$\tag{1}$$

$$- \sum_{i=1}^{D^V} \sum_{j=1}^{nhidden} \frac{v_i^{(V)}}{\sigma_i^{(V)}} W_{ij}^V h_j - \sum_{j=1}^{nhidden} b_j^h h_j$$

where $\theta = \{\mathbf{b}, \boldsymbol{\sigma}, \mathbf{W^E}, \mathbf{W^V}\}$ are the parameters of our model. The joint distribution over visible units is defined by

$$P(V^E, V^V|\theta) = \frac{1}{Z(\theta)} \sum_H exp(-E(V^E, V^V, h|\theta)), \tag{2}$$

where $Z(\theta) = \int_{v^E} \int_{v^V} \sum_H exp(-E(V^E, V^V, h|\theta)) dv^V dv^E$

The conditional distributions are factorized as shown in Eqs. 3, 4 and 5.

$$P(V^E|H;\theta) = \prod_{i=1}^{D^E} p(v_i^E|H), \text{ with } v_i^{(E)}|h \sim N(\sigma_i^{(E)} \sum_{j=1}^{nhidden} W_{ij}^{(E)} \cdot h_j + b_i^{(E)}, \sigma_i^{(E)2})$$

$$\tag{3}$$

$$P(V^V|H;\theta) = \prod_{i=1}^{D^V} p(v_i^V|H), \text{ with } v_i^{(V)}|h \sim N(\sigma_i^{(V)} \sum_{j=1}^{nhidden} W_{ij}^{(V)} \cdot h_j + b_i^{(V)}, \sigma_i^{(V)2})$$

$$\tag{4}$$

$$P(H|V^E, V^V; \theta) = \prod_{i=1}^{nhidden} p(h_i|V^E, V^V)$$

$$\textit{where } p(h_j|V^{(E)}, V^{(V)}) = g(\sum_{i=1}^{D^E} W_{ij}^{(E)} \frac{v_i^{(E)}}{\sigma_i^{(E)}} + \sum_{i=1}^{D^V} W_{ij}^{(V)} \frac{v_i^{(V)}}{\sigma_i^{(V)}} + b_j^h)$$

(5)

where $N(\mu, \sigma^2)$ represents a Gaussian with the mean value μ and variance σ^2, $g(x)$ is the sigmoid function $1/(1 + exp(-x))$. Given a set of observations $\{V_n^E, V_n^V\}_{n=1}^N$, the derivative of the log-likelihood respect to $\mathbf{W^E}$ is shown in Eq. 6. It is similar to parameters $\{\mathbf{W^V}, \mathbf{b}\}$.

$$\frac{1}{N} \sum_{n=1}^{N} \frac{\partial log P(V_n^E, V_n^V; \theta)}{\partial W_{ij}^E} = E_{P_{data}} \left[\frac{v_i^E}{\sigma_i^E} h_j \right] - E_{P_{model}} \left[\frac{v_i^E}{\sigma_i^E} h_j \right]$$

(6)

We use the Contrastive Divergence (CD) learning algorithm to learn the parameters θ of our model. For details, please refer to [1].

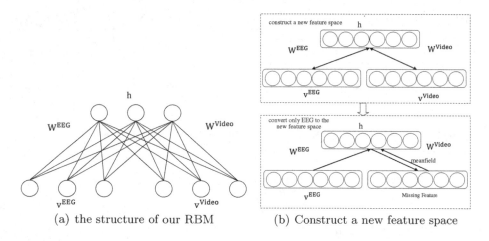

(a) the structure of our RBM   (b) Construct a new feature space

Fig. 2. Our RBM model

Connecting all the visible units through the latent units, RBM is able to capture higher-order dependencies among the visible units (i.e. EEG nodes and video nodes). The hidden units represent a better EEG feature space inferring from EEG signals with the help of stimulus videos.

2.3   New EEG Features Generated Through the Learned RBM

After learning the model parameters, we clamp the EEG modality at the inputs, and sample the hidden layer and the video modality. We update the hidden layer using Eq. 7 and the video modality using Eq. 8 with Gibbs sampling. After $u_i^{(V)}$ and u_j^h are converged, we use the u_j^h as the new features for emotion recognition.

Algorithm 1. convert the EEG feature to the new feature space

---

Input: the parameters $\{W^E, W^V, b\}$ and the EEG features V^E of one sample
Initialize: U^h, U^V
Output: the new feature U^h.
while not converge **do**

1. Update the states of each hidden unit.

$$u_j^h \leftarrow g(\sum_{i=1}^{D^E} W_{ij}^{(E)} \frac{v_i^{(E)}}{\sigma_i^{(E)}} + \sum_{i=1}^{D^V} W_{ij}^{(V)} \frac{u_i^{(V)}}{\sigma_i^{(V)}} + b_j^h)$$

2. Update the states of video visible units.

$$u_i^{(V)} \leftarrow N(\sigma_i^{(V)} \sum_{j=1}^{nhidden} W_{ij}^{(V)} \cdot u_j^h + b_i^{(V)}, \sigma_i^{(V)2})$$

    end while

The video features are only available for training the RBM parameters for a better feature space and they are the missing modality in this phase. The procedure is shown in Algorithm 1.

$$u_i^{(V)} \leftarrow N(\sigma_i^{(V)} \sum_{j=1}^{nhidden} W_{ij}^{(V)} \cdot u_j^h + b_i^{(V)}, \sigma_i^{(V)2}) \tag{7}$$

$$u_j^h \leftarrow g(\sum_{i=1}^{D^E} W_{ij}^{(E)} \frac{v_i^{(E)}}{\sigma_i^{(E)}} + \sum_{i=1}^{D^V} W_{ij}^{(V)} \frac{u_i^{(V)}}{\sigma_i^{(V)}} + b_j^h) \tag{8}$$

3 Experimental Results and Analysis

3.1 Experimental Conditions

We conduct the experiments on two databases to evaluate the performance of our method, the MAHNOB-HCI database [4] and the database in [10].

The MAHNOB-HCI database is a multi-modal database for affect recognition and implicit tagging. There are totally 533 EEG segments recorded from 27 participants watching 20 emotional videos. Each segment is labeled self-reported feelings using 9-scale ratings (1–9). We divide them into two classes, positive (rating 6–9) and negative (rating 1–5). Totally, in the valence space, 289 samples are positive and the rest are negative. In the arousal space, there are 268 positive and 265 negative samples.

The database [10] collected 197 EEG segments recorded from the users responding to 92 video stimulus. Each EEG sample is labeled by the user using 5-scale evaluations (i.e. -2,-1,0,1,2). These samples are divided into two classes based on the value zero. Finally, there are 149 positive and 48 negative samples in the arousal space and, 77 positive and 120 negative samples in valence space.

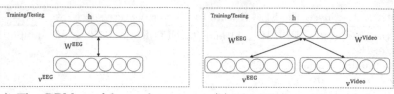

(a) The RBM model merely using EEG

(b) The RBM models for fusion

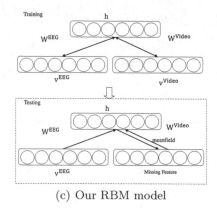

(c) Our RBM model

Fig. 3. The RBM model for three experiments

We conduct three experiments on each database: emotion recognition based on the features generated by RBM using merely EEG features, emotion recognition based on the features generated by RBM using EEG features with video features as the privileged information and emotion recognition based on the features generated by RBM using both EEG features and video features, as shown in Figs. 3(a), 3(c) and 3(b) respectively. We adopted leave-one-video out cross-validations in our experiments and used three commonly-used metrics accuracy, F1-score and kappa to verify the effectiveness of our method.

3.2   Results and Analysis

Table 1 shows the results on the database [10] in valence space and arousal space. From this table, we can find that the fusion method has the best results and our method using video features as privileged information is better than merely using EEG features in both spaces.

Compared with the method merely using EEG features, our method improves the accuracy by 2.6 %, F1-score by 0.033 and kappa by 0.054 in valence space. In arousal space, the accuracy, F1-score and kappa are improved by 0.5 %, 0.006, and 0.010, respectively. Our method constructs a better space for EEG features with help of video features. The fusion method has a better performance than our method. The fusion method exploits EEG and video features in both the training and test phases to obtain more information to achieve better performance. However, our method only uses the video features in the training phase.

Table 1. Emotion recognition results on the database [10]

	Valence						Arousal					
	RBM EEG		RBM Fusion		Ours		RBM EEG		RBM Fusion		Ours	
	Pos	Neg	Pos	Neg	Pos	Neg	Pos	Neg	Pos	Neg	Pos	Neg
Positive	45	32	58	19	47	30	132	17	126	23	133	16
Negative	17	103	18	102	14	106	23	25	16	32	22	26
Accuracy	75.1%		81.2%		77.7%		79.7%		80.2%		80.2%	
F1 score	0.648		0.758		0.681		0.556		0.621		0.562	
Kappa	0.459		0.605		0.513		0.425		0.488		0.435	

Table 2. Emotion recognition results on the MAHNOB-HCI database

	Valence						Arousal					
	RBM EEG		RBM Fusion		Ours		RBM EEG		RBM Fusion		Ours	
	Pos	Neg	Pos	Neg	Pos	Neg	Pos	Neg	Pos	Neg	Pos	Neg
Positive	182	107	185	104	186	103	179	89	199	69	185	83
Negative	115	129	106	138	107	137	96	169	109	156	96	169
Accuracy	58.3%		60.6%		60.6%		65.3%		66.6%		66.4%	
F1 score	0.538		0.568		0.566		0.646		0.637		0.654	
Kappa	0.159		0.206		0.206		0.306		0.331		0.328	

Table 2 displays the results on the MAHNOB-HCI database in both the valence space and arousal space. The results demonstrate that our method not only outperform the method merely using EEG features but also is comparable with the fusion method.

In valence space, our method increases accuracy, F1-score and kappa by 2.3%, 0.028 and 0.047 respectively compared with the model merely using EEG features. Compared with the fusion method, our method has the same accuracy and kappa, and only the F1-score is slightly worse by 0.002. In arousal space, our method outperforms the method merely using the EEG features. Accuracy, F1-score and kappa are increased by 1.1%, 0.008 and 0.022 respectively. Our method is even better than the fusion method by 0.017 in F1-score and the accuracy and kappa is slightly decreased by 0.2% and 0.003.

From the results in Tables 1 and 2, we can find that the fusion method has the best performance. Fusing EEG and video features can obtain more information to achieve better performance. The RBM model can model the high-order dependence between EEG and video features. Our method using video features as privileged information is not as good as the explicit fusion method but better than merely using EEG features. It demonstrates that our method constructs a better feature space with help of video features. Compared with the fusion method, our method is a much more practical approach than explicit fusion models since the video features are not needed during the actual recognition while fusion method increases computation load. From the results, we also conclude that it is easier to recognize emotions in the arousal space than in the valence space.

3.3   Compared with Related Works

Since the used features and experimental conditions of this work are similar with those in Zhu et al. [11] and Wang et al. [10], we compare our work with theirs [11] [10]. The comparison results are shown in Table 3.

Table 3. Compared with related works

Database Models	NVIE						MAHNOB-HCI					
	Valence			Arousal			Valence			Arousal		
	Acc	F1	Kap	Acc	F1	Kap	Acc	F1	Kap	Acc	F1	Kap.
RBM EEG	75.1 %	0.648	0.459	79.7 %	0.556	0.425	58.3 %	0.538	0.159	65.3 %	0.646	0.306
Ours	77.7 %	0.681	0.513	80.2 %	0.562	0.435	60.6 %	0.566	0.206	66.4 %	0.654	0.328
RBM Fusion	81.2 %	0.758	0.605	80.2 %	0.621	0.488	60.6 %	0.568	0.206	66.6 %	0.637	0.331
SVM [11]	73.6 %	0.600	0.413	73.6 %	0.422	0.252	55.7 %	0.514	0.108	60.2 %	0.578	0.204
Zhu et al's [11]	76.6 %	0.697	0.507	68.5 %	0.456	0.242	58.2 %	0.564	0.163	61.4 %	0.631	0.227
IFL Fusion [10]	78.7 %	0.720	0.548	73.1 %	0.531	0.348	—	—	—	—	—	—
DFL Fusion [10]	76.6 %	0.689	0.502	76.1 %	0.373	0.265	—	—	—	—	—	—
DL Fusion [10]	77.7 %	0.722	0.535	73.1 %	0.442	0.240	—	—	—	—	—	—

Comparing the first line and the fourth line of Table 3, we can find that for emotion recognition from EEG signals only, the method using the features extracted by RBM (the 1st line) outperforms that using the power spectrum features directly (the fourth line), demonstrating the effectiveness of the hidden nodes of RBM for new feature representation. For emotion recognition by combining both videos and EEG signals, the measurements on the third line are higher than those on the sixth, seventh, and eighth line, further proving the potential of RBM to generate better features. Compared our method and Zhu et al's, our method is superior to Zhu et al's, since almost all the measurements of ours are higher than theirs. It means the new EEG features generated by RBM under the help of videos are better than those constructed by CCA.

4   Conclusions

In this work, we propose to use RBM to model the high-order dependencies between EEG signals and stimulus videos, and construct a better feature space for emotion recognition from EEG signal with the help of videos. Experimental results on two benchmark databases show that RBM model can model high-order dependencies between EEG and video features effectively, and generated better EEG features with the help of video features. They both make contributions to improving the performance of emotion recognition.

Acknowledgement. This work has been supported by the National Natural Science Foundation of China (61175037, 61228304, 61473270), and Project from Anhui Science and Technology Agency(1508085SMF223).

References

1. Hinton, G.: A practical guide to training restricted boltzmann machines. Momentum **9**(1), 926 (2010)
2. Hinton, G.E., Salakhutdinov, R.R.: Reducing the dimensionality of data with neural networks. Science **313**(5786), 504–507 (2006)
3. Soleymani, M., Koelstra, S., Patras, I., Pun, T.: Continuous emotion detection in response to music videos. In: 2011 IEEE International Conference on Automatic Face & Gesture Recognition and Workshops (FG 2011), pp. 803–808. IEEE (2011)
4. Soleymani, M., Lichtenauer, J., Pun, T., Pantic, M.: A multimodal database for affect recognition and implicit tagging. IEEE Trans. Affect. Comput. **3**(1), 42–55 (2012)
5. Subasi, A.: Eeg signal classification using wavelet feature extraction and a mixture of expert model. Expert Syst. Appl. **32**(4), 1084–1093 (2007)
6. Torres-Valencia, C.A., Garcia-Arias, H.F., López, M.A.A., Orozco-Gutiérrez, A.A.: Comparative analysis of physiological signals and electroencephalogram (eeg) for multimodal emotion recognition using generative models. In: 2014 XIX Symposium on Image, Signal Processing and Artificial Vision (STSIVA), pp. 1–5. IEEE (2014)
7. Tseng, K.C., Lin, B.S., Han, C.M., Wang, P.S.: Emotion recognition of eeg underlying favourite music by support vector machine. In: 2013 International Conference on Orange Technologies (ICOT), pp. 155–158. IEEE (2013)
8. Vapnik, V., Vashist, A.: A new learning paradigm: learning using privileged information. Neural Netw. **22**(5), 544–557 (2009)
9. Wang, H.L., Cheong, L.F.: Affective understanding in film. IEEE Trans. Circuits Syst. Video Technol. **16**(6), 689–704 (2006)
10. Wang, S., Zhu, Y., Wu, G., Ji, Q.: Hybrid video emotional tagging using users eeg and video content. Multimedia Tools Appl. **72**(2), 1257–1283 (2014)
11. Zhu, Y., Wang, S., Ji, Q.: Emotion recognition from users' eeg signals with the help of stimulus videos. In: 2014 IEEE International Conference on Multimedia and Expo (ICME), pp. 1–6. IEEE (2014)

Twitter Event Photo Detection Using both Geotagged Tweets and Non-geotagged Photo Tweets

Kaneko Takamu, Nga Do Hang, and Keiji Yanai[✉]

Department of Informatics, The University of Electro-Communications,
1-5-1 Chofugaoka, Chofu-shi, Tokyo 182-8585, Japan
{kaneko-t,dohang,yanai}@mm.inf.uec.ac.jp

Abstract. In this paper, we propose a system to detect event photos using geotagged tweets and non-geotagged photo tweets. In our previous work, only "geotagged photo tweets" was used for event photo detection the ratio of which to the total tweets was very limited. In the proposed system, we use geotagged tweets without photos for event detection, and non-geotagged photo tweets for event photo detection in addition to geotagged photo tweets. As results, we have detected about ten times of the photo events with higher accuracy compared to the previous work.

Keywords: Event photo detection · Microblog · Twitter

1 Introduction

Because microblogs such as Twitter and Weibo has unique characteristics which are different from other social media in terms of timeliness and on-the-spot-ness, they include much information on various events in the real world. By mining photos related to events, we can get to know and understand what happens in the world visually and intuitively. Previously, we have proposed a system to discover events and related photos from the Twitter stream automatically [5,6], which especially helps us to know about regional events such as local festival, sport game, special natural phenomena including heavy snow, rainbow and earthquake.

In our previous work, however, only "geotagged photo tweets" were used for detecting events and event photos. Since the ratio of geotagged photo tweets to the total tweets is very limited, they detected only limited number of events and event photos.

Then, in this paper, we extend and improve our previous Twitter event photo mining system so that the system uses geotagged tweets without photos for event detection and non-geotagged photo tweets for event photo detection in addition to geotagged photo tweets.

By the experiments, we confirmed the proposed system detected about ten times of the photo events with higher accuracy compared to the existing work.

© Springer International Publishing Switzerland 2015
Y.-S. Ho et al. (Eds.): PCM 2015, Part II, LNCS 9315, pp. 128–138, 2015.
DOI: 10.1007/978-3-319-24078-7_13

2   Related Work

Many works on event detection have been proposed in the multimedia community so far. Most of the works used Flickr photos and tags as a target data from which events were detected including the MediaEval SED task [10–12], while the number of the works on Twitter photo data is limited.

Although there exist many works related to Twitter mining using only text analysis such as the work by Sakaki et al. [13], only a limited number of works exist on Twitter mining using image analysis currently.

As the early works on microblog photos, Yanai have proposed "World Seer" [15] which can visualize geotagged photo tweets on the online map in real-time by monitoring the Twitter stream. This system can store geo-photo tweets to a database as well. They have been gathering geo-photo tweets from the Twitter stream since January 2011 with this system. On the average, they gather about half million geo-photo tweets a day, about one third of which are hosted at Instagram. Thus, Twitter can be regarded as more promising data source of geotagged photos than Flickr, because the number of uploaded photos to Flickr a day in 2014 was officially announced as 1.5 million and only 10 to 20 percent of them are estimated to have geotags.

To utilize their Twitter image database, Nakaji et al. [9] proposed a system to mine representative photos related to the given keyword or term from a large number of geo-tweet photos. They extracted representative photos related to events such as "typhoon" and "New Year's Day", and successfully compared them in terms of the difference on places and time. However, their system needs to be given event keywords or event term by hand. Kaneko et al. [5] extended it by adding event keyword detection to the visual Tweet mining system. As results, they detected many photos related to seasonal events such as festivals and Christmas as well as natural phenomena such as snow and Typhoon including extraordinary beautiful sunset photos taken around Seattle. All of these works focused on only geotagged tweet photos.

Chen et al. [2] treated photo tweets regardless of geo-information. They analyzed relation between tweet images and messages, and defined the photo tweet which has strong relation between its text message and its photo content as a "visual" tweet. In the paper, they proposed the method which is based on the LDA topic model to classify "visual" and "non-visual" tweets. However, because their method was generic and assumed no specific targets, the classification rate was only 70.5 % in spite of two-class classification.

Recently, Yanai et al. proposed Twitter Food Photo Mining [16] which takes advantage of the characteristics of Twitter that many meal photos are uploaded in the time of meals everyday. They used a real-time food recognition engine of the mobile food photo recognition application, FoodCam [7], to detect one hundred kinds of foods from the Twitter stream. They claimed they had already collected more than half million ramen noodle photos, which will be helpful for research on large-scale fine-grained food image classification.

Cao et al. [3] proposed a method to mine brand product photos from Weibo which employs supervised image recognition in the same ways as [16]. They integrated and used visual features and social factors (users, relations, and locations) as well as textual features. The same authors proposed to use hypergraph construction and segmentation for event detection [4].

In this work, we focus to detect event photos from the Twitter stream data. By extending and improving our previous work by Kaneko et al. [5,6], we will propose a new Twitter event photo detection system.

3 Previous System

In this section, we describe the existing Twitter event photo mining system proposed by Kaneko et al. [5,6], and pointed out its drawbacks.

In the previous system, firstly, we detected events by textual analysis, and secondly selected relevant photos and a representative photo to each of the detected event.

In the first step for detecting event words, we divided tweet messages of geo-photo tweets into words by a Japanese morphological analyzer, and detected the burst of keywords in the tweets posted from specific areas in specific days. We detected keyword burst by examining the difference on the word frequency to the previous day. In the second step for selecting relevant photos to the detected events, we selected geo-tweet photos and representative photos corresponding to the events based on image clustering.

The biggest problem of the previous system was that the number of detected events were limited, since they used only geo-photo tweets for event detection as well as photo detection. To increase the number of events and event photos, in this paper, (1) we use geotagged non-photo tweets (geotagged tweets having no links to photos) as well for event burst detection, and (2) we use non-geotagged photo tweets (photo tweets having no geotags) for event photo selection by estimating their locations with the newly proposed method which is a hybrid method of text-based Naive Bayes (NB) classifier and image-based Naive Bayes Nearest Neighbor (NBNN) [1].

In addition, (3) we change the way to extract words from usage of a morphological analyzer to N-gram, and (4) the way to detect keyword burst from the difference to the previous days to the difference to the average over the month.

For photo selection, (5) we use DCNN (Deep Convolutional Neural Network) activation features which is pre-trained with ImageNet 1000 categories instead of conventional SIFT-based bag-of-feature representation.

Note that the current system assumes the tweet messages written by Japanese language, since keyword extraction needs to be taken into account of the characteristics of target language. However, it is not so difficult to extend the proposed system to other languages, since in the proposed system we use N-gram instead of using a morphological analyzer which alway needs to assume a specific language.

4 Proposed System

4.1 Overview

We overview the proposed system in this subsection, which has been greatly enhanced regarding the five points described in the previous section.

The input data of the system are the tweets having geotags or photos (geo-tweets or photo tweets) gathered via the Twitter streaming API. The output of the system are event sets consisting of event words, geo-locations, event date, representative photos, and event photo sets. The system has GUI which shows detected events on the online maps as shown in Figs. 1 and 2.

The processing flow of the new system is as follows:

(1) Calculate area weights and "commonness score" of words in advance.
(2) Detect event word bursts using N-gram
(3) Estimate locations of non-geotagged photos
(4) Select photos and representative photos corresponding to the detected events
(5) Show the detected events with their representative photos on the map (See Figs. 1 and 2)

4.2   Target Data

Before describing the detail, we explain the target data of the proposed system. Basically we mine events and corresponding photos from tweets containing geotags and/or photos gathered from the Twitter stream. In our system, we use the following four kinds of information contained in tweets: (1) date/time information, (2) text messages, (3) photos and (4) geotags representing the pair values of latitude and longitude.

Note that tweet photos used in the system include the photos posted to other image hosting services than the Twitter official photo hosting service such as Instagram, ImageShack and Twitpic as well. One third of all the gathered photos are from Twitter official photo hosting services, one third are from Instagram, and the others are from other photo hosting sites.

4.3   Preparation

To detect events, we search for bursting keywords by examining difference between the daily frequency and the average daily frequency over a month within each unit area. The area which is a location unit to detect events is defined with a grid of 0.5 degree latitude height and 0.5 degree longitude width. In case that the daily frequency of the specific keyword within one grid area increases greatly compared to the average frequency, we consider that an event related to the specific keyword happened within the area in that day.

To detect bursting keywords, we calculate an adjusting weight, $W_{i,j}$, regarding the number of Twitter unique users in a grid, and a "commonness score", $Com(w)$, of a word over all the target area in advance.

Area Weight. In general, the extent of activity within each grid area depends on the location of the area greatly. The activity of the Twitter users in big cities such as New York and Tokyo is very high, while the activity in countryside such as Idaho and Fukushima is relatively low. Therefore, to boost the areas with low activity and handle all the areas equally in the burst keyword detection,

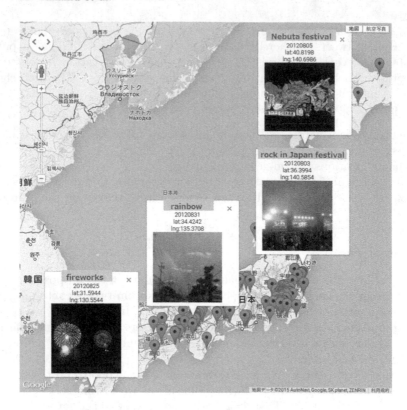

Fig. 1. Example of detected events shown on the online map.

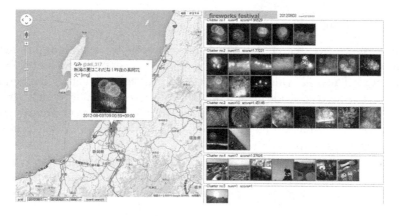

Fig. 2. "Fireworks festival" photos automatically detected by the proposed system.

we introduce $W_{i,j}$ representing a weight to adjust the scale of the number of daily tweet users, which is defined in the following equation:

$$W_{i,j} = \frac{\#users_{max} + s}{\#users_{i,j} + s},$$ (1)

where i, j, $\#users_{i,j}$, $\#users_{max}$ and s represents the index of grids, the number of unique users in the given grid, the maximum number of unique users among all the grids (which is equivalent to the number of the user in downtown Tokyo area in case of Japan), and the standard deviation of user number over all the grids, respectively.

Commonness Score of Words. Next, we prepare a "commonness score" of each of the word appearing in Tweet messages by the following equation:

$$Com(w) = \sum_{i,j} \frac{E(\#users_{w,i,j})^2}{V(\#users_{w,i,j}) + 1},$$ (2)

where i, j, $E(\#users_{w,i,j})$ and $V(\#users_{w,i,j})$ represents the index of grids, and the average number and the variance value of unique users who tweeted messages containing the given word w in the given grid in a day, respectively. The score becomes larger in case that the given word frequently and constantly is tweeted. On the other hand, it becomes smaller in case that the given word does not appear frequently or daily change is large. The "commonness score" is used as a standard value for word burst detection.

4.4   Detect Event Word Burst Using N-Gram

In the previous work, we used only geo-photo tweets, while we detect event keywords from geotagged tweets regardless of attachment of photos. Moreover, the way to detect keyword burst is changed from the difference to the previous days to the difference to the average over the month.

To detect event keywords, in the previous work, we used a morphological analyzer which can extract only words listed in its dictionary. Instead, in this paper, we use N-gram to detect burst words which does not need word dictionaries.

As a unit of N-Gram, we use a character in Japanese texts and a word in English texts. First we count the number of unique users who posted Twitter messages including each unit within each location grid. We merge adjacent units both of which are contained in the messages tweeted by more than five unique users one after another.

We calculate a word burst score, $S_{w,i,j}$, in the following equation:

$$S_{w,i,j} = \frac{\#users_{w,i,j}}{Com(w)} W_{i,j},$$ (3)

where $\#users_{w,i,j}$ is the number of the unique users who tweeted messages containing w in the location grid (i, j). A word burst score, S, represents the extent of burst of the given word taking account of an area weight of the given location grid, $W_{i,j}$, and a "commonness score" of the given word, $Com(w)$. We regard the word the burst score of which exceeds the pre-defined threshold. In the experiments for Japan tweets, we set the threshold as 200. Note that when multiple words which overlap with each other are detected as events, we merge them into one event word.

4.5   Estimate Locations of Non-geotagged Photos

In the previous work, we used only photos embedded in geotagged tweets, the number of which was very limited. Then, in this paper, we extend event photo sets by detecting photos corresponding to the given event from the non-geotagged tweet photos.

The photos embedded in the geotagged tweets from the messages of which the event words were detected in the given day and the given area can be regarded as event photos corresponding to the detected event. In this step, by using them as training data, we detect additional event photos from the non-geotagged photo tweets posted in the same time period as the detected event words. As a method, we adopt two-class classification to judge if each tweet photo corresponds to the given event or not.

To classify non-geotagged tweet photos into event photos or non-event photos, we propose a hybrid method of text-based Naive Bayes (NB) classifier and image-based Naive Bayes Nearest Neighbor (NBNN) [1]. We use Naive Bayes which is a well-known method for text classification to classify tweet messages, and NBNN which is local-feature-based method for image classification to classify tweet photos.

We use message texts and photos of geotagged tweets where the given event word are extracted as positive samples, and message texts and photos of geo-tagged tweets which include the given event words but were posted from the other areas as negative samples. For NB, we count the word frequency in positive and negative samples, while for NBNN, we extract SIFT features from sample images. To classify photos in the same way as NB, we use a cosine similarity between L2-normalized SIFT features instead of Euclid distance used in the normal NBNN.

The equation to judge if the given non-geotagged tweet photo corresponds to the given event or not is as follows:

$$\hat{c} = \arg\max_{c} P(c) \prod_{i=1}^{n} P(x_i|c) \sum_{j=1}^{v} \frac{d_j \cdot NN_c(d_j)}{\|d_j\| \|NN_c(d_j)\|}, \qquad (4)$$

where n, x_i, v, d_j, and $NN_c(d_j)$ represents the number of words in the given tweet, the i-th words, the number of extracted local features from the photo of the given tweet, local feature vectors of SIFT, and the nearest local feature vectors of d_j in the training sample of class c which corresponds to "positive" or "negative", respectively.

Note that we assign the average location of the corresponding event to all the detected non-geotagged event photos for mapping the photos on the online map.

4.6   Select Event Photos and Representative Photos

In the last step, we select suitable photos to represent the given detected event visually and intuitively. In the same way as the previous work [5,6], we carry out event photo selection and representative photo selection based on a modified Ward method which is a kind of hierarchical clustering. The difference to the previous work in this step is that we use an activation feature extracted from Deep Convolutional Neural Network (DCNN) pre-trained with ImageNet 1000 categories [8] instead of standard bag-of-feature representation. We extract 4096-dim L2-normalized DCNN features using Overfeat [14] as a feature extractor.

According to [5,6], we define a cluster score, V_C, to evaluate visual coherence of a cluster so that the score of the cluster the member photos of which are similar to each other becomes larger in the following equation:

$$V_C = \frac{\#images_C}{\sum_{x \in C} \|x - \overline{x}\| + 1},$$ (5)

where $\#images_C$, x and $\overline{x}$ represent the number of images in the cluster C, the DCNN feature of an image, and the average vector of the DCNN features of all the images in the cluster C, respectively.

The cluster is carried out according to the following procedure:

1. Initially regard each of all the elements as an independent cluster.
2. Calculate clustering score, V_C, in case of merging two clusters.
3. Find the cluster pair bringing the maximum cluster score among the possible pairs, and perform merging the cluster pair.
4. Repeat 2 and 3 until the maximum score becomes below the pre-defined threshold.

As a result of clustering, the cluster having the maximum clustering score is regarded as a representative cluster, and the closest photo to the center of the representative cluster in terms of DCNN features is selected as a representative photo to the detected event.

5   Experimental Results

To compare the proposed system with the previous system [5], we used the same tweet data which was collected in August 2012. The number of geotagged photo tweets, geotagged non-photo tweets and non-geotagged photo tweets we collected in August 2012 were 255,455, 2,102,151 and 3,367,169, restpectively. In advance, we calculated area weights and commonness score of words using all the geotagged tweets.

Table 1 shows the statistics of the detected events, the precision of the detected events and the precision of the selected representative photos. The proposed system detected 310 events, while the previous system detected only 35 events which were about one ninth times as many as the proposed system.

Table 2 shows parts of detected events including event names, location, date and event scores. 8 events shown in the table were detected by the proposed

Table 1. Results of detected events

	proposed system	previous system [5]
# detected events	310	35
Precision of detected events(%)	81.3	77.1
Precision of representative event photos(%)	88.7	65.5

Table 2. Part of the detected events.

event name	date	lat,lng	Event Score	# photos	# photos (old)
fireworks	2012/08/01	33,129.5	297.7	38 (10,20)	22
rainbow	2012/08/01	34,134.5	229.1	21 (18,3)	36
ROCK IN JAPAN	2012/08/03	36,140	430.3	51 (32,19)	not detected
Ayu Festival	2012/08/04	34.5,138.5	265.1	28 (10,18)	not detected
Nebuta Festival	2012/08/06	40.5,140	255.7	37 (14,23)	not detected
Awa-odori	2012/08/14	34,134	589.8	31 (16,15)	19
lightning	2012/08/18	34,135	367.5	106 (37,69)†	102
blue moon	2012/08/31	34.5,136	269.7	69 (59,10)	70

Fig. 3. "Nebuta festival" photos. The photos with red bounding boxes come from geotagged photo tweets, while the photos with yellow bounding boxes come from non-geotagged photo tweets (Color figure online).

system, while the previous one detected only 5 out of 8. Regarding the number of detected photos, basically it was increased. However, in some cases, the number of photos was reduced. This is because some events detected by the previous system were decomposed into smaller events by the proposed system. Since the proposed adopted N-gram-based word detection, sometimes multiple event words were extracted from one event. For example, "lightning" shown in Table 2 was detected as six event words, "lightning flash", "lightning and heavy rain", "lightning and power cut" and so on independently by the propose system. Note that the value with † in Table 2 shows the total number of the detected photos of six event words related to the "lightning" event. For future work, we need to improve event word unification as post-processing of event word detection.

Figure 3 shows example photos of the detected events, "Nebuta festival". The representative of this event is shown in Fig. 1. Representative photos are used for mapping detected events on the online map.

6 Conclusions

In this paper, we proposed a system to discover event photos from the Twitter stream. We improved the following five points: (1) use geotagged non-photo tweets for event detection, (2) use non-geotagged photo tweets for event photo detection by the proposed method integrating NB and NBNN, (3) use N-gram and (4) the deference to the average frequency for event word detection, and (5) use the state-of-the-art DCNN features to photo clustering. Compared to the previous system, we have successfully discovered much more regional events and unknown events which cannot be found out by keyword search, and mined their photos which enables us understand the events visually and intuitively.

Currently, we use the temporal unit as one day, and the spatial unit as 0.5 degrees. As future work, we make units variable to discover event photos. In addition, we will introduce spatial-temporal information to unify event keywords. We will also improve usability of the GUI of the system to enable users to understand the detected events more intuitively and visually.

References

1. Boiman, O., Shechtman, E., Irani, M.: In defense of nearest-neighbor based image classification. In: Proceedings of IEEE Computer Vision and Pattern Recognition (2008)
2. Chen, T., Lu, D., Kan, M.-Y., Cui, P.: Understanding and classifying image tweets. In: Proceedings of ACM International Conference Multimedia, pp. 781–784 (2013)
3. Gao, Y., Wang, F., Luan, H., Chua, T.-S.: Brand data gathering from live social media streams. In: Proceedings of ACM International Conference on Multimedia Retrieval (2014)
4. Gao, Y., Zhao, S., Yang, Y., Chua, T.-S.: Multimedia social event detection in microblog. In: He, X., Luo, S., Tao, D., Xu, C., Yang, J., Hasan, M.A. (eds.) MMM 2015, Part I. LNCS, vol. 8935, pp. 269–281. Springer, Heidelberg (2015)
5. Kaneko, T., Yanai, K.: Visual event mining from geo-tweet photos. In: Proceedings of IEEE ICME Workshop on Social Multimedia Research (2013)

6. Kaneko, T., Yanai, K.: Event photo mining from twitter using keyword bursts and image clustering. Neurocomputing (2015) (in press)
7. Kawano, Y., Yanai, K.: FoodCam: a real-time food recognition system on a smartphone. Multimedia Tools Appl. **74**, 5263–5287 (2015)
8. Krizhevsky, A., Sutskever, I., Hinton, G.E.: Imagenet classication with deep convolutional neural networks. In: Proceedings of Neural Information Processing Systems (2012)
9. Nakaji, Y., Yanai, K.: Visualization of real world events with geotagged tweet photos. In: Proceedings of IEEE ICME Workshop on Social Media Computing (SMC) (2012)
10. Petkos, G., Papadopoulos, S., Kompatsiaris, Y.: Social event detection using multimodal clustering and integrating supervisory signals. In: Proceedings of ACM International Conference on Multimedia Retrieval (2012)
11. Reuter, T., Cimiano, P.: Event-based classification of social media streams. In: Proceedings of ACM International Conference on Multimedia Retrieval (2012)
12. Reuter, T., Papadopoulos, S., Petkos, G., Mezaris, V., Kompatsiaris, Y., Cimiano, P., de Vries, C., Geva, S.: Social event detection at MediaEval 2013: Challenges, datasets, and evaluation. In: Proceedings of MediaEval 2013 Multimedia Benchmark Workshop (2013)
13. Sakaki, T., Okazaki, M., Matsuo, Y.: Earthquake shakes Twitter users: real-time event detection by social sensors. In: Proceedings of the International World Wide Web Conference, pp. 851–860 (2010)
14. Sermanet, P., Eigen, D., Zhang, X., Mathieu, M., Fergus, R., LeCun, Y.: Overfeat: integrated recognition, localization and detection using convolutional networks. In: Proceedings of International Conference on Learning Representations (2014)
15. Yanai, K.: World seer: a realtime geo-tweet photo mapping system. In: Proceedings of ACM International Conference on Multimedia Retrieval (2012)
16. Yanai, K., Kawano, Y.: Twitter food photo mining and analysis for one hundred kinds of foods. In: Ooi, W.T., Snoek, C.G.M., Tan, H.K., Ho, C.-K., Huet, B., Ngo, C.-W. (eds.) PCM 2014. LNCS, vol. 8879, pp. 22–32. Springer, Heidelberg (2014)

Weather-Adaptive Distance Metric
for Landmark Image Classification

Ding-Shiuan Ding and Wei-Ta Chu[✉]

National Chung Cheng University, Chiayi County, Taiwan
lzi94u@gmail.com, wtchu@cs.ccu.edu.tw

Abstract. Visual appearance of landmark photos changes significantly in different weather conditions. In this work, we obtain weather information from a weather forecast website based on a landmark photo's geotag and taken time information. With weather information, we adaptively adjust weightings for combining distances obtained based on different features and thus propose a weather-adaptive distance measure for landmark photo classification. We verify the effectiveness of this idea, and accomplish one of the early attempts to develop a landmark photo classification system that resists to weather changes.

Keywords: Weather-adaptive distance metric · Landmark classification

1 Introduction

Landmark image classification has emerged as an important research topic due to its potential usage of location-based service and large-scale image retrieval. Famous landmarks such as Eiffel Tower and Statute of Liberty attract millions of visitors every year, who took pictures of the landmarks from unlimited viewpoints in various conditions, and then shared them on social media platforms. Large amounts of landmark photos thus urge the need of efficient retrieval/access as well as effective recognition/classification.

Many studies of landmark classification and its extended variants, i.e., location prediction/recognition, have been widely proposed in recent years. They mainly focus on integrating multimodal features such as geographical information and visual information, or developing classification models based on large-scale datasets. However, the problem of high intra-class variations caused by drastically different visual conditions still remains.

In this paper, we investigate one factor that largely affects visual appearance of landmark images: *weather types*. Through the whole year many people visit Notre Dame, for example, and take photos under various weather conditions. Figure 1 shows sample photos taken at Notre Dame and Sacre Coeur on sunny and cloudy days, respectively. From this figure we see visual appearances are significantly different in different weathers due to the sky and the intensity of lighting on the building. Such intra-class variations impede accurate image classification. However, the influence of weather types on measuring image similarity was overlooked before. In this work we propose a weather-adaptive distance metric so that better similarity measurement between images can be achieved, and thus better landmark image classification is expected.

© Springer International Publishing Switzerland 2015
Y.-S. Ho et al. (eds.): PCM 2015, Part II, LNCS 9315, pp. 139–148, 2015.
DOI: 10.1007/978-3-319-24078-7_14

When comparing two landmark images, we could calculate their distance from many perspectives, such as texture and local feature points, and then linearly combine distances respectively calculated based on each feature. With weather properties obtained from a weather forecast website, we propose to adjust weightings by formulating this task as an optimization problem. As the first contribution of this work, we consider its analogy to single neuron training and determine the optimal weightings by the gradient method. As the second contribution, more effective features can be discovered and prioritized through the learnt weightings, and more accurate landmark image classification can be achieved.

The rest of this paper is organized as follows. In Sect. 2 literature of landmark image classification will be surveyed. Details of the weather-adaptive distance metric with weight learning are described in Sect. 3. Section 4 provides discussion of the proposed metric and performance of landmark image classification, followed by concluding remarks in Sect. 5.

2   Related Works

Landmark image classification has been widely studied in the past decade. We briefly review some of them in the following. Zheng et al. [14] built an internet-scale landmark dataset by mining true landmark images from GPS-tagged photos and tour guide web pages. Unsupervised clustering techniques and visual models based on feature points were adopted to build a landmark recognition engine. Yi et al. [11] also built a large-scale dataset and adopted the bag of feature approach associated with multiclass SVM to achieve landmark image classification. They also showed that using textual tags and temporal constraints leads to significant performance improvement over the visual only method. Li et al. [13] combined 2D appearance and 3D constraints to discover iconic views of a landmark, which were later used in landmark recognition. Chen et al. [9] proposed a soft bag-of-visual phrase approach for mobile landmark recognition. Visual phrases were learnt in a category-dependent manner to achieve promising recognition performance. Min et al. [12] proposed an efficient mobile landmark search system where the client uploads compressed images to the server, and the server recognizes landmark by matching the uploaded image with landmark texture projected from pre-constructed landmark 3D models.

Since the IMG2GPS system proposed in [6], studies of geographical location estimation emerge in recent years. Hays and Efros [6] estimated the geographical location of a query photo based on a data-driven scene-matching approach. Li et al. [8] improved the scene-matching approach by jointly considering visual similarity and geographical proximity to build a ranking method. Lin et al. [7] greatly extended the scene-matching approach by further considering overhead appearance and land cover survey data. A query photo can be localized even if it has no corresponding ground-level images in the database. Fang et al. [5] adopted latent SVM to discover geo-informative attributes from regions in order to facilitate better location recognition and exploration.

Although there have been many works targeting at landmark or location recognition, few of them specially tackled visual variations caused by lighting, editing, or

Fig. 1. Left to right: sample photos of Notre Dame on sunny days, Notre Dame on cloudy days, Sacre Coeur on sunny days, and Sacre Coeur on cloudy days.

weather change. Shen and Cheng [10] proposed gestalt rule feature points to find visual correspondence between images of different styles (painting vs. photograph, or photographs in different colors) but containing the same semantic meaning. However, methodology or features especially designed to consider visual variations caused by weather conditions are still missing. In this work, we focus on developing a distance metric considering weather conditions.

3   Weather-Adaptive Distance Metric

3.1   Common Distance Metric

Given two images I_p and I_q, assuming that each image can be represented by N types of features, i.e., $I_p = \{p_1,\ldots,p_N\}$ and $I_q = \{q_1,\ldots,q_N\}$, the conventional way to integrate distances derived from features is:

$$D(I_p, I_q) = \sum_{i=1}^{N} w_i d_i(p_i, q_i), \tag{1}$$

where $d_i(p_i, q_i)$ is the normalized distance calculated based on the ith feature. Weightings w_i's are often empirically set or simply follows a uniform distribution, i.e., $w_i = 1/N$. However, the integrated distance $D(I_p, I_q)$ often cannot reflect impacts of different features, yielding limited landmark classification performance.

To show the shortage of this simple metric, from Flickr we collect photos of famous landmarks that were captured on sunny days or cloudy days. We then calculate integrated distance $D(I_p, I_q)$ between photos that are randomly selected following four schemes: (1) I_p and I_q are from the same landmark under the same weather type (sunny or cloudy); (2) I_p and I_q are from the same landmark under different weather types (one is sunny and another is cloudy); (3) I_p and I_q are from different landmarks under the same weather type; (4) I_p and I_q are from different landmarks under different weather types. The integrated distance $D(I_p, I_q)$ is obtained by combining individual features respectively derived from Gabor texture features, haze features, bag of visual words, and CNN features. The individual distance $d_i(p_i, q_i)$ is measured by Euclidean distance. Details of the evaluation dataset and features will be described in Sect. 4.

Figure 2 shows distributions of integrated distances $D(I_p, I_q)$ between photos selected based on four different schemes. Comparing the distributions obtained based on the first and the third schemes, under the same weather condition, distances between photos from the same landmark are similar to that from different landmarks. This shows weather properties dominate calculation of distance measure. The first two distributions (obtained based on the first two schemes) are similar to the last two distributions (obtained based on last two schemes). This means the common distance metric cannot reliably describe that photos from the same landmark are similar, while photos from different landmarks are relatively distinct even when they were captured under the same weather condition.

3.2   Weather-Adaptive Distance Metric

The characteristics shown in Fig. 2 motivate us to propose a weather-adaptive distance metric for measuring landmark photos. The idea is to adjust weightings for combining individual distances in a systematic manner. Let us model whether two photos I_p and I_q belong to the same landmark based on the integrated distances like this:

$$y = \sum_{i=1}^{N} w_i d_i(\boldsymbol{p}_i, \boldsymbol{q}_i) = \boldsymbol{d}^T \boldsymbol{w}, \tag{2}$$

where $\boldsymbol{w} = [w_1, \ldots, w_N]^T$ and $\boldsymbol{d} = [d_1(\boldsymbol{p}_1, \boldsymbol{q}_1), \ldots, d_N(\boldsymbol{p}_N, \boldsymbol{q}_N)]$. The indicator $y = 0$ if I_p and I_q are from the same landmark (no matter whether they were under the same weather condition or not), and $y = 1$ otherwise.

We wish to find the values of weights $w_1, \ldots, w_N$ such that the estimated indication value is as close as the ground truth. Given a training dataset $\{(\boldsymbol{d}_1, y_1), \ldots, (\boldsymbol{d}_M, y_M)\}$ constituted by randomly selecting M photo pairs from the collected landmark photo collection, the training problem is formulated as the following optimization problem:

$$\text{minimize} \quad \sum_{j=1}^{M} (y_j - \boldsymbol{d}_j^T \boldsymbol{w})^2, \tag{3}$$

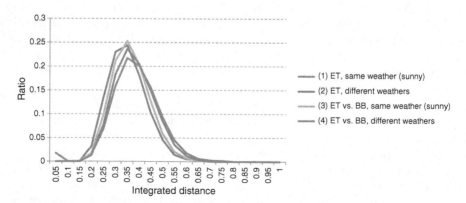

Fig. 2. Distributions of integrated distances calculated based on four settings. ET stands for Eiffel Tower, and BB stands for Big Ben.

where the minimization is taken over all $w = [w_1, \ldots, w_N]^T \in R^N$. The term $d_j^T w$ is the estimated indication value, and the objective function represents the sum of squared errors between the desired output y_j and the estimated result $d_j^T w$. We can write the optimization problem in matrix form:

$$\text{minimize} \quad \left\| y - D^T w \right\|^2, \tag{4}$$

where $D = [d_1 \cdots d_M]$ and $y = [y_1, \ldots, y_M]^T$.

We have more training points than the number of weights. Assuming that rank of D^T is N, the objective function is simply a strictly convex quadratic function of w. In this work, we utilize a fixed-step-size gradient algorithm [3] that iteratively updates the weighting vector w in the following form:

$$w^{(k+1)} = w^{(k)} + \alpha D e^{(k)}, \tag{5}$$

where α is the predefined step size, and $e^{(k)} = y - D^T w^{(k)}$ is the estimation error at the kth iteration.

Through the process mentioned above, we learn the optimal weighting vector $w = [w_1, \ldots, w_N]^T$ that causes the minimum estimation error.

4   Experiments

4.1   Experimental Settings

The collected dataset consists of sunny and cloudy photos of five famous landmarks, including Big Ben, Eiffel Tower, Notre Dame, Sacre Coeur and Winsor Castle. Table 1 shows information of the collected dataset. The numbers of sunny and cloudy photos are roughly balanced, and there are totally 1,210 photos in the dataset.

Table 1.   Information of the evaluation dataset.

Landmark	Sunny	Cloudy
Big Ben	100	100
Eiffel Tower	138	143
Notre Dame	105	104
Sacre Coeur	168	101
Winsor Castle	141	110
Sum	652	558

Features.   We use Gabor texture features [2], haze features [1], bag of feature points [11], and CNN features [15] to describe an image. For Gabor texture features, image pixels' intensity are transformed into the frequency domain, which is then decomposed into 16 ranges by the Gabor Wavelet functions with four scales and four orientations. Mean and standard deviation of the magnitude of the transform coefficients in each range are used to represent each frequency band, and are then concatenated to form a 32-D texture feature vector.

For haze features, dark channel prior [4] is first calculated for each pixel. An image is partitioned by a spatial pyramid scheme, i.e., uniformly partitioned into 2^2, 4^2, and 8^2 non-overlapping regions to obtain 84 sub-regions. The median values of dark channel intensities in these sub-regions are concatenated as an 84-D haze feature vector [1].

Following the single image classification process proposed in [11], we describe an image by a bag of visual words (BoW) model. We utilize the visual vocabulary (with 10,000 visual words) built in Top-SURF [16] to construct an image's 10,000-D BoW representation.

Currently using convolutional neural network (CNN) features largely surpasses hand-crafted features. To extract CNN features, we utilize the MatConvNet package [17] with the pre-trained model obtained based on ImageNet ILSVRC-2012. There are five convolutional layers and three fully-connected layers in the CNN model. The first convolutional layer filters the input image with 64 kernels of size $11 \times 11 \times 3$ with a stride of 4 pixels. The second convolutional layer makes filtering with 256 kernels of size $5 \times 5 \times 64$. The third, fourth, and fifth convolutional layers are connected to one another without any intervening pooling or normalization layers. The third and fourth convolutional layer have 256 kernels of size $3 \times 3 \times 256$, respectively, and the fifth convolutional layer has 4096 kernels of size $6 \times 6 \times 256$. The fully-connected layers have 4096 neurons each. We try to take output of the fifth, sixth, and seventh layers to be CNN features, and found that features from the sixth layer yield the best performance through our preliminary experiments.

Experimental Settings. Based on the dataset, we adaptively adjust weights for measuring distances between photos captured in the same weather condition or in different weather conditions. Particularly, we randomly select pairs of sunny photos to form the training pool $SS = \{I_1^{(s)}, \ldots, I_M^{(s)}\}$. For each pair in SS, if the two photos I_p and I_q belong to the same landmark, the indicator y is set as 0, and set as 1 otherwise. Initial weighted distances between selected pairs, as defined in Eq. (1), and the associated indicators, are treated as the training data, and the updated procedure described in Eq. (5) is used to adjust weightings specifically for measuring distance between sunny photos. We denote the adjusted weightings as $w_{SS} = \{w_1^{(s)}, \ldots, w_4^{(s)}\}$. Similarly, we randomly select pairs of cloudy photos to form the set $CC = \{I_1^{(c)}, \ldots, I_M^{(c)}\}$, and determine the adjusted weightings $w_{CC} = \{w_1^{(c)}, \ldots, w_4^{(c)}\}$. To appropriately measure distance between two photos that were captured in different weather conditions, we also randomly select M photo pairs, where for each pair one photo is sunny and another is cloudy. Based on the corresponding initial weighted distances and associated indicators, the adjusted weightings $w_{SC} = \{w_1^{(t)}, \ldots, w_4^{(t)}\}$ are determined.

Because the captured time and geographical information are available for each photo in our database, we can use this information to obtain weather type through the API provided by the Weather Underground website[1]. Overall, given a pair of photos

---

[1] Weather Underground, http://www.wunderground.com/.

(one may be query, and another may be from the landmark database), we first select appropriate weights from w_{SS}, w_{CC}, or w_{SC}, according to their weather types, and then calculate the weighted distance between them as the foundation for landmark classification or other applications.

4.2   Distributions of Distances

Figure 2 shows that integrated distance distributions are similar no matter photos in the same landmark or in different landmarks are compared. Through the proposed adjustment, we verify that through the adjusted weightings distances between photos can be more appropriately captured.

Figure 3 shows distributions of integrated distances between (a) sunny photos, (b) cloudy photos, and (c) one sunny photo and one cloudy photo, in the same landmark (blue curves) or in different landmarks (red curves). From all these three subfigures, we see that before weighting adjustment (solid curves), distance distributions between photos in the same or different landmarks are similar. After adjustment, distance distributions coming from photos at the same landmark move apart from that for different landmarks.

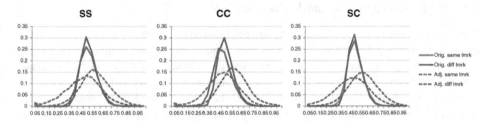

Fig. 3. Distributions of integrated distances before and after weighting adjustment. From left to right: distributions of sunny vs. sunny photos; distributions of cloudy vs. cloudy photos; distributions of sunny vs. cloudy photos.

Table 2. KL divergences of distance distributions.

Type	Before adjustment	After adjustment
Sunny-Sunny	0.0830	0.4034
Cloudy-Cloudy	0.2251	0.5134
Sunny-Cloudy	0.0383	0.3717

To quantitatively show the effect of weighting adjustment, we calculate the symmetric KL divergence between distance distributions respectively derived for same landmark and different landmarks. Table 2 shows detailed information. We can quantitatively observe that the KL divergence between distance distributions largely increases after weighting adjustment.

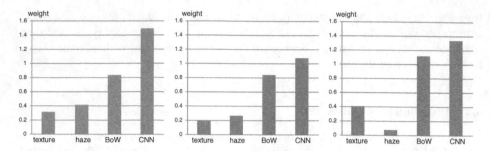

Fig. 4. Absolute weights of different features. Left to right: weights for sunny vs. sunny photos; weights for cloudy vs. cloudy photos; weights for sunny vs. cloudy photos.

Figure 4 shows absolute values of learnt weights for the three different schemes. We especially notice the relative values of these weights, and observe that BoW and CNN features are consistently more important than the other two features. This conforms to recent studies on image classification, and also shows that the proposed method can effectively learn weights.

4.3   Performance of Landmark Classification

We adopt a simple classification method, i.e., K-nearest neighbor classifier, to more clearly show the effectiveness of weighting adjustment in landmark classification. Given a query photo, we find its K-nearest neighbors based on integrated distance, and classify the query photo as one of the landmark according to majority voting. We compare classification performance obtained based on initial integrated distances (Eq. (1)) with that obtained based on adjusted integrated distances (according to weather types). Figure 5 shows accuracy of landmark classification with different settings of the number of nearest neighbors (K). From this figure we clearly see the significant improvement given by appropriately adjusting weightings.

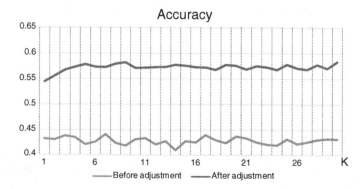

Fig. 5. Accuracy of landmark classification with varied nearest neighbor settings.

5   Conclusion

We have presented a weather-adaptive distance metric that is verified to yield better landmark photo classification based on a pilot database. By considering multiple features, distance between photos is usually calculated by combining individual distance derived from each feature. In this work we advocate that, by further considering weather type of the two compared photos, weightings that can better combine individual features can be learnt. We formulate it as an optimization problem and find the best weighting setting by a gradient algorithm. The reported evaluation verifies that the learnt weightings yield more effective distances between photos and thus improve performance of landmark photo classification increases with adjusted weightings. In the future, a larger-scale evaluation will be conducted, and more elegant methods to combine individual features and the corresponding learning problems will be studied.

Acknowledgements. The work was partially supported by the Ministry of Science and Technology in Taiwan under the grant MOST103-2221-E-194-027-MY3.

References

1. Lu, C., Lin, D., Jia, J., Tang, C.-K.: Two-class weather classification. In: Proceedings of IEEE Conference on Computer Vision and Pattern Recognition, pp. 3718–3725 (2014)
2. Manjunath, B.S., Ma, W.Y.: Texture features for browsing and retrieval of image data. IEEE Trans. Pattern Anal. Mach. Intell. **18**(8), 837–842 (1996)
3. Chong, E.K.P., Zak, S.H.: An Introduction to Optimization, 4th edn. Wiley, Hoboken (2013)
4. He, K., Sun, J., Tang, X.: Single image haze removal using dark channel prior. In: Proceedings of IEEE Conference on Computer Vision and Pattern Recognition, pp. 1956–1963 (2009)
5. Fang, Q., Sang, J., Xu, C.: Discovering geo-informative attributes for location recognition and exploration. ACM Trans. Multimedia Comput. Commun. Appl. **11**(1) (2014). Article No. 19
6. Hays, J., Efros, A.A.: IM2GPS: estimating geographic information from a single image. In: Proceedings of IEEE Computer Vision and Pattern Recognition Conference (2008)
7. Lin, T.-Y., Belongie, S., Hays, J.: Cross-view image geolocalization. In: Proceedings of IEEE Computer Vision and Pattern Recognition Conference, pp. 891–898 (2013)
8. Li, X., Larson, M., Hanjalic, A.: Global-scale location prediction for social images using geo-visual ranking. IEEE Trans. Multimedia **17**(5), 674–686 (2015)
9. Chen, T., Yap, K.-H., Zhang, D.: Discriminative soft bag-of-visual phrase for mobile landmark recognition. IEEE Trans. Multimedia **16**(3), 612–622 (2014)
10. Shen, I.-C., Cheng, W.-H.: Gestalt rule feature points. IEEE Trans. Multimedia **17**(4), 526–537 (2015)
11. Li, Y., Crandall, D.J., Huttenlocher, D.P.: Landmark classification in large-scale image collections. In: Proceedings of IEEE International Conference on Computer Vision, pp. 1957–1964 (2009)
12. Min, W., Xu, C., Xu, M., Xiao, X., Bao, B.-K.: Mobile landmark search with 3D models. IEEE Trans. Multimedia **16**(3), 623–636 (2014)
13. Li, X., Wu, C., Zach, C., Lazebnik, S., Frahm, J. M.: Modeling and recognition of landmark image collections using iconic scene images. Int. J. Comput. Vision **95**(3), 213–239 (2011)

14. Zheng, Y.-T., Zhao, M., Song, Y., Adam, H., Buddemeier, U., Bissacco, A., Brucher, F., Chua, T.-S., Neven, H.: Tour the world: building a web-scale landmark recognition engine. In: Proceedings of IEEE Computer Vision and Pattern Recognition Conference, pp. 1085–1092 (2009)
15. Krizhevsky, A., Sutskever, I., Hinton, G.E.: Imagenet classification with deep convolutional neural network. In: Proceedings of Advances in Neural Information Processing System (2012)
16. Thomee, B., Bakker, E.M., Lew, M.S.: TOP-SURF: a visual words toolkit. In: Proceedings of ACM International Conference on Image and Video Retrieval, pp. 1473–1476 (2010)
17. Vedaldi, A., Lenc, K.: MatConvNet – Convolutional Neural Networks for Matlab (2014). arXiv:1412.4564

Power of Tags: Predicting Popularity of Social Media in Geo-Spatial and Temporal Contexts

Toshihiko Yamasaki[1]([✉]), Jiani Hu[1], Kiyoharu Aizawa[1], and Tao Mei[2]

[1] Department Information and Communication Engineering,
The University of Tokyo, Bunkyō, Japan
`yamasaki@hal.t.u-tokyo.ac.jp`
[2] Microsoft Research, Beijing, China

Abstract. Generating multimedia content and sharing them in social networks has become one of our daily-life activities. Although a lot of people care about the quality of the content itself, much less attention is paid to the text annotations. In our previous work, we have shown that the popularity of the content in social media is strongly affected by its annotated tags, and we have proposed a TF-IDF-like algorithm to analyze which tags are more potentially important to earn more popularity. In this paper, we extend the idea to show how the important tags are geo-spatially varied and how the importance ranking of the tags evolves over time.

Keywords: DF-W · Social media · Social network · Social popularity · Spatial and temporal analysis

1 Introduction

The explosive increase in size of online social network services (SNS) and the availability of large amounts of shared data make it a really difficult task to rank and classify attractiveness of uploaded contents in SNS. In recent years, therefore, significant effort has been expended in evaluating and predicting so-called "social popularity," using social connections [1], textual information [2] and visual information [2,3]. To extract features from millions of shared images or videos and predict social popularity, we focus on text annotations instead of complicated social connections or costly visual features.

In [4], social popularity (which is defined as the number of favorites, views, or comments) is evaluated by combining the tag frequency and the tag weights obtained from a support vector regression (SVR) model. And it is confirmed to have better performance than simply using the tag frequency, the tag weights, and mutual information [2]. Furthermore, differences of such influential tags in SNSs such as Flickr, Instagram, and Vine is analyzed in [5]. Our method can not only be used to predict social popularity, but also to recommend tags that will help to increase attractiveness.

Different from other text information in ordinary documents and web pages, tags usually have the temporal dimension. By analyzing the behavior of users in

© Springer International Publishing Switzerland 2015
Y.-S. Ho et al. (Eds.): PCM 2015, Part II, LNCS 9315, pp. 149–158, 2015.
DOI: 10.1007/978-3-319-24078-7_15

the social tagging over time, we can explore the evolution of community focus, which will help us to increase the effectiveness of tag recommendation as well. Similarly, influential tags may vary spatially because the cultural backgrounds are different. People living in different regions would have different interest. In this paper, we study the spatial and temporal differences of the influential tags in social media and try to reveal cultural differences and the trend of the times.

2   Related Work

Social tagging has been a popular way to allow users to contribute metadata and has gained a lot of attention after explosive increase of the shared content in social network. Reference [6] broadly discussed properties of tag streams, tagging models, tag semantics, tag recommendation, visualizations of tags, and applications of tags. Social popularity analysis and recommendation based on the text tags is one of the attractive applications. Reference [2] combined text annotations with visual features for constructing vector representations of photos and demonstrated high accuracy in image attractiveness classification. Reference [7] considered the problem of visualizing the evolution of tags within Flickr. An animation provided via Flash in a web browser allows the user to observe and interact with the interesting tags as they evolve over time. Reference [8] proposed a power law distribution which can describe the distribution of the frequency of use of tags for popular sites in del.icio.us. with many tags and many users. And it showed tag co-occurrence networks for a sample domain of tags.

In these approaches above, social tags are used for evaluating the social attractiveness and it was shown that tags can help predicting the attractiveness of the images/videos. People paid little attention, however, to tags which make difference to social popularity. The behavior of users in the social networks changes incessantly and dynamically, which brings a great challenge to understand the users' behavior in spatial and temporal context in social networks and predict social popularity effectively. In [9], they proposed a seasonal and temporal information based model which can improve the recommendation accuracy better than previous approaches. In this paper, we focus on the influential tags in social media and study their change over time and space.

2.1   DF-W Algorithm

Although our DF-W algorithm is already presented in [4], we would like to summarize it because it is the core for analyzing the spatial and temporal difference of influential tags. The DF-W algorithm is inspired by the term frequency - inverse document frequency (TF-IDF) algorithm, which is frequently used for calculating the importance of words in documents [10]. TF reflects how often the term (word) is used and IDF corresponds to the rareness of the word across all documents. TF-IDF is very powerful, but cannot be directly applied to tag analysis. Firstly, a tag appears only once at most in each content, thus TF is always one. Secondly, tags are sparse, making IDF meaningless.

In [4], therefore, document frequency - weights from regression (DF-W) is proposed. Assume that we have T kinds of tags, $\{tag_1, tag_2, \cdots, tag_T\}$. First, document frequency (DF) is counted, which represent how popular each tag is:

$$DF_i = \text{The number of counts of the } i\text{th tag in dataset,} \tag{1}$$

Then, a linear SVR model is trained by using the feature vector defined as follows:

$$FV_i^{tag} = \{f_{i1}, f_{i2}, \cdots, f_{ij}, \cdots, f_{iT}\}, \tag{2}$$

where FV_i^{tag} is the feature vector for the ith image and f_{ij} represents whether ith image has the jth most frequently appeared tag. The target value in training the SVR is the social popularity scores such as the number of views, comments, of favorites. After the training, we can obtain the weight vector W, which represents the normal vector of the hyper-plain of the T dimension:

$$W = (w_1, w_2, \cdots, w_T). \tag{3}$$

It is natural to think that W can be used as a measure as importance score of tags and we can sort the tags in the decreasing order of weights. However, we find that a notable number of tags in such a top list appear only in a few images. This means that top tags obtained by this method are not appropriate for predicting the social popularity, because they are only useful for a limited number of images. Therefore, we define the importance score of tags as follows:

$$DF - W_i = DF_i \times |w_i| \tag{4}$$

We sort out all the tags by the importance score, and obtain the top N influential tags to predict social popularity. In Eq. 4, the absolute value of w_i is employed in order to include both positively influential tags and negatively influential ones. If users are interested only in positively influential tags, the raw value of w_i can be used.

One may think that such tag scoring and recommended tags would become useless once certain tags are more influential than the others for obtaining social popularity score. The advantage of the proposed algorithm is its simplicity and the tag scores and tag recommendations can be updated with negligibly small cost.

3   Experimental Results

3.1   Data Collection

We evaluated spatial and temporal differences of influential tags using the dataset of Yahoo Flickr Creative Commons 100M (YFCC100M) [11], which contains 100 million public Flickr photos and videos with creative commons license. Because the dataset do not include the social popularity score such as the numbers of views, comments, and favorites, we collected the data from Flickr during a period from December 1st, 2014 to December 19th, 2014. 68,152,028 photos and videos which had at least one tag and were still available in the public domain during that time were considered. Regarding the spatial difference, only-geotagged photos, which were about 48 million, were used.

In the following subsections, we focus only on the number of views due to the limited number of pages.

3.2 Spatial Analysis

First, we divided geo-tagged photos or videos, the number of which were 40, 389, 114, into nine clusters by using mean-shift clustering on the geographic locations. The nine clusters are approximately in these regions: Europe, North America, Asia, South America, Australia, Africa, North Pacific, South Pacific, and Indian Ocean. And the number of photos and videos taken in each region is summarized in Fig. 1. When we look at the most frequently used tags as shown in Table 1, most of them are names of countries and cities (e.g. *france* and *london*). Such tags are so common that neither are they closely related to social popularity nor do they contribute to increasing the social popularity score.

In Table 2, the top 10 tags that are the most influential to the number of views in each region are summarized. Some regional difference can be observed

Table 1. Top 10 most frequently used tags in each region.

	Europe	North America	Asia	South America	Australia
1	france	california	japan	brazil	australia
2	london	usa	taiwan	brasil	new+zealand
3	europe	united+states	china	argentina	sydney
4	england	canada	tokyo	chile	nsw
5	germany	new+york	travel	peru	melbourne
6	italy	square	india	de	victoria
7	uk	iphoneography	asia	south+america	queensland
8	spain	san+francisco	thailand	buenos+aires	western+australia
9	paris	square+format	(Taiwan)	square	new+south+wales
10	nikon	instagram+app	square	iphoneography	beach

	Africa	North Pacific	South Pacific	Indian Ocean
1	africa	hawaii	french+polynesia	antarctica
2	south+africa	alaska	tahiti	bandai
3	tanzania	oahu	easter+island	gunpla
4	kenya	honolulu	travel	ice
5	safari	maui	island	perfect+grade
6	travel	vacation	chile	strike+freedom
7	cape+town	kauai	south+pacific gundam+seed	
8	dubai	usa	honeymoon	fragments
9	namibia	big+island	moorea	snow
10	uae	waikiki	isla+de+pascua	gundam

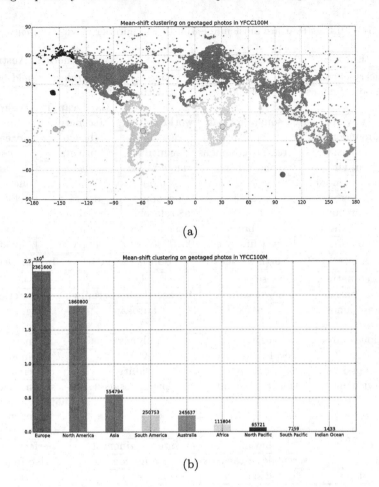

Fig. 1. (a) Region clusters and (b) the number of photos and videos in each cluster.

in Table 2. For instance, photos related to art are popular in Europe area (*abode of chaos, dadaisme*). In North America and Asia area, people care more about famous photographer (*mlhradio, 061028choshi*), photography technology (*hdr, square format*) and camera types (*nikon, range finder*). Photos related to travel are popular in Asia (*world travels, beat*) and South America area (*landscape, nature, explore*, etc.)

Further, Fig. 2 shows the accuracy of social popularity prediction in different areas, using top 1000 influential tags, measured by Pearson's correlation, Spearman's rank correlation, and Kendall's correlation. We can observe that the prediction accuracy of divided areas are generally higher than that of global dataset. It shows that it is more effective to predict social popularity according to different areas.

Table 2. Top 10 tags that are most influential to the number of views in each region.

	Europe	North America	Asia	South America	Australia
1	thierry ehrmann	mlhradio	free	naturaleza	field hockey
2	abode of chaos	90095	attractive	(landscape)	australia
3	raw art	square format	world travels	(scenery)	new
4	salamander spirit	nikon	iso 200	(beach)	oceania
5	dadaisme	creative commons	beautiful	private	nsw
6	paul virilio	phenomenal	range finder	(planet)	nz
7	sculpture moderne	bild	canon	de	queensland
8	groupe	2013	canon	america	beach
8	serveur		a35 datelux	meridionale	
9	stockphoto	hdr	flickr meetup	nature	viaje
10	picture	wet lifestyles	061028choshi	explore	panties
	Africa	North Pacific	South Pacific		Indian Ocean
1	united nations	sexy	kap		ice
2	campaign	women	kite aerial photography		antarctica
3	nasa ames research center	boobs	autokap		fragments
4	naciones unidas	gorgeous	polynésie		nasa
5	nasa	royal caribbean	2013		snow
6	square	booty	tahiti		sanae
7	viewing	midway island	pierre lesage		south african national antarctic programme- arcticantarctic
8	conflict	hawaii	intercontinental		perfect grade
9	ecosystem	serenade of the seas	mururoa		strike freedom
10	plains	disney	blue		dumpr

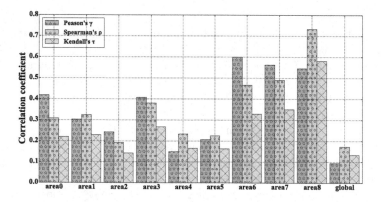

Fig. 2. Accuracy of social popularity prediction using the top 1000 influential tags in spatial dimension.

3.3   Temporal Analysis

To understand the change of influential tags over time, the dataset was divided into groups by the uploaded time, and we use the dataset during the period from January 2009 to December 2013 for our experiments, which contains $43,692,766$ images/videos. Then tags are sorted by the importance score. The total number of unique text tags of every month is shown in Fig. 3. The number changes periodically, with an obvious peak around June and a little one in December of every year, which results from increase of activities in summer and winter holidays.

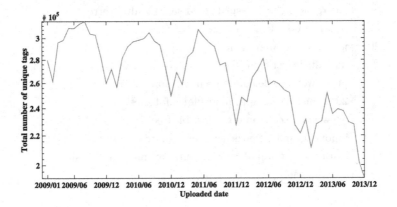

Fig. 3. Total number of unique tags for every month.

Table 3 shows the top 10 influential tags from 2009 to 2013. The most important tags change greatly over time, in the aspects of arts (*abode of chaos, borderline biennale, post-apocalyptique*), artists (*hierry ehrmann, randomok, jmmg*), politics and philosophy (*zionist, fatah, salamander spirit*), photography technology (*hi-res, zoom lens, square format*), and famous places (*90095, southeast asia, atlantic avenue*). And we noticed that the tags of the time of the year (e.g. *2013*) always appeared in the list as well. We see from the table how the community focus and popularity tendency changes over time.

The ranking changes of six typical influential tags are plotted in Fig. 4. For tags such as *instagram app* and *square format*, they started to be popular from 2010 and quickly became one of the most important tags, along with appearance of the online mobile social media Instagram. On the other hand, tags such as *high resolution*, were in one of the most important tags but gradually lost its position after 2011, which results from the fact that high resolution cameras become common and it is not the key point of good photos any more. Strong temporal changes are seen for seasonal tags such as *winter* and *snow*. Peaks are observed around June and December because of the opposite seasons in the northern and southern hemispheres. In addition, tags related to season cycle also show temporal changes. For example, the tag *explore* varied periodically, roughly

Table 3. Top 10 influential tags of the year from 2009 to 2013.

Year	Top 10 influential tags
2009	raw art, zionist, modern sculpture, 2009, maison d'artiste, catalano, tulkarem, alchimie, fatah, hijra
2010	thierry ehrmann, abode of chaos, portland state university, 90095, post-apocalyptique, hi-res, rodents and rabbits, 2010, raw art, demeure du chaos
2011	borderline biennale, randomok, salamander spirit, emergence, 2011, l'esprit de la salamandre, retro, square, zoom lens, vintage
2012	jose maria moreno garcia, jmmg, www.flickriver.com/josemariamorenogarcia, picasaweb.google.com/josemariamorenogarcia, 2012, instagram app, madridejos.fotos.es, www.vimeo.com/madridejos, bi, free
2013	iphoneography, southeast asia, square format, asia, instagram app, malaysia, atlantic avenue, southeast, visit, house

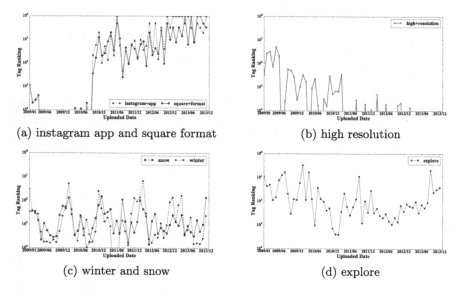

(a) instagram app and square format

(b) high resolution

(c) winter and snow

(d) explore

Fig. 4. The ranking change of typical influential tags.

with two peaks in a year, which corresponds to, as pointed above, the increase of outdoor activities during summer and winter holidays.

Figure 5 shows the accuracy of social popularity prediction in the temporal dimension, using top 1000 influential tags, measured by the three correlation coefficients: Pearson's, Spearman's, and Kendall's. Although the accuracy changed greatly over time, we can still observe that in most cases, the prediction accuracy of each month is better than that of the whole period from 2009 to 2013. Thus we can conclude that it will increase the prediction accuracy to analyze the popularity tendency in different time periods.

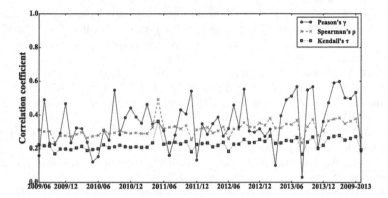

Fig. 5. Accuracy of social popularity prediction using the top 1000 influential tags in temporal dimension.

4   Conclusions and Future Work

In this paper, we studied the spatial difference and temporal evolution of influential tags on social popularity. Basically, the influential tags that affect social popularity change greatly over time and locations. The difference of the top tag lists in different areas showed the difference of culture and community focus among regions. In temporal dimension, we found several typical temporal evolution modes of tags by analyzing the tags in the top 1,000 list. Further, our experiments results also showed that the prediction accuracy would increase if we take spatial and temporal difference into consideration.

In the future work, we would like to use some visualization methods to analyze the spatial difference and temporal evolution of influential tags. Besides, we would like to find a way to describe the evolution, and use it to improve the accuracy and efficiency of social tag recommendation system.

Acknowledgments. This work was partially supported by the Grant-in-Aid for Scientific Research (No. 26700008) from the Japan Society for the Promotion of Science (JSPS), the Microsoft IJARC Core10 project, and Hoso Bunka Foundation. The author also would like to thank Mr. Shumpei Sano for his contribution to this work.

References

1. Nwana, A., Avestimehr, S., Chen, T.: A latent social approach to youtube popularity prediction. In: IEEE Globecom (2013)
2. Pedro, J.S., Siersdorfer, S.: Ranking and classifying attractiveness of photos in folksonomies. In: WWW, pp. 771–780 (2009)
3. Irie, G., Hidaka, K., Satou, T., Yamasaki, T., Aizawa, K.: A degree-of-edit ranking for consumer generated video retrieval. In: IEEE ICME, pp. 1242–1245 (2009)
4. Yamasaki, T., Sano, S., Aizawa, K.: Social popularity score: Predicting numbers of views, comments, and favorites of social photos using only annotations. In: WISMM, pp. 3–8 (2014)
5. Yamasaki, T., Sano, S., Mei, T.: Revealing relationships between folksonomy and social popularity score in image/video sharing services. In: ICCE-TW (2015)
6. Gupta, M., Li, R., Yin, Z., Han, J.: Survey on social tagging techniques. ACM SIGKDD Explor. Newsl. 12(1), 58–72 (2010)
7. Dubinko, M., Kumar, R., Magnani, J., Novak, J., Raghavan, P., Tomkins, A.: Visualizing tags over time. ACM Trans. Web (TWEB), 1(2) (2007)
8. Halpin, H., Robu, V., Shepherd, H.: The complex dynamics of collaborative tagging. In: WWW, pp. 211–220 (2007)
9. Yamasaki, T., Gallagher, A., Chen, T.: Personalized intra-and inter-city travel recommendation using large-scale geotags. In: GeoMM (2013)
10. Rajaraman, A., Ullman, A.D.: "Data Mining" in Mining of Massive Datasets. Cambridge University Press, New York (1978)
11. Thomee, B., Shamma, D.A., Friedland, G., Elizalde, B., Ni, K., Poland, D., Borth, D., Li, L.J.: The new data and new challenges in multimedia research. arXiv preprint arXiv:1503.01817

Human Action Recognition in Social Robotics and Video Surveillance

Recognition of Human Group Activity
for Video Analytics

Jaeyong Ju[1], Cheoljong Yang[1], Sebastian Scherer[2],
and Hanseok Ko[1(✉)]

[1] School of Electrical Engineering, Korea University, Seoul, Korea
{jyju, cjyang}@ispl.korea.ac.kr, hsko@korea.ac.kr
[2] Robotics Institute, Carnegie Mellon University, Pittsburgh, PA, USA
basti@andrew.cmu.edu

Abstract. Human activity recognition is an important and challenging task for video content analysis and understanding. Individual activity recognition has been well studied recently. However, recognizing the activities of human group with more than three people having complex interactions is still a formidable challenge. In this paper, a novel human group activity recognition method is proposed to deal with complex situation where there are multiple sub-groups. To characterize the inherent interactions of intra-subgroups and inter-subgroups with the varying number of participants, this paper proposes three types of group-activity descriptor using motion trajectory and appearance information of people. Experimental results on a public human group activity dataset demonstrate effectiveness of the proposed method.

Keywords: Human group activity · Activity recognition · Video analytics

1 Introduction

Human activity recognition is an important and challenging task for video content analysis and understanding and many notable works have been done in recent years. While the majority of previous works mainly focus on an individual activity [1] or pair activities between two humans [2], and only one group activity [3], recognizing group activity with multiple groups is still a formidable challenge. In complex situations, group activity is composed of activities of one or more sub-groups such as *approach* and *split* between sub-groups as shown in Fig. 1(a). In this case, the interaction information between sub-groups is important to analyze the group activity as a whole. Previous works on group activity recognition [2, 3] focus on analyzing human activities in terms of interactions between individuals using their motion trajectories. Y. Zhou et al. [2] analyzed interactions between two individuals using the Granger Causality Test (GCT) [4]. However, they only focused on the pair-activity recognition problem because of the limitations of the GCT. In consideration of the aforementioned limitations, Ni et al. [3] analyzed individual, pair, and group causalities using motion trajectory. However, these methods address the activities of only one group and cannot be generalized for more complex situations with the activities of one or more subgroups. In order to cope with this situation, some generative-model-based methods [5]

© Springer International Publishing Switzerland 2015
Y.-S. Ho et al. (Eds.): PCM 2015, Part II, LNCS 9315, pp. 161–169, 2015.
DOI: 10.1007/978-3-319-24078-7_16

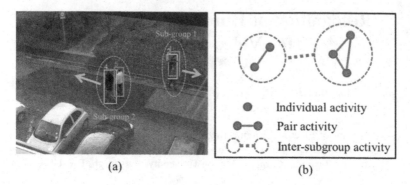

Fig. 1. (a) Example of group activity (*split*) with multiple sub-groups, (b) three types of group activities.

have been proposed. However, they require large-scale training samples, and have difficulties caused by varying number of participants.

In this paper, a novel human group activity recognition method is proposed to deal with more complex situations where there are activities of multiple subgroups. Compared with previous works [2, 3] considering only the interactions of individuals and pairs, this paper proposes three types of group-activity descriptor, i.e., "inter-subgroup" as well as "individual" and "pair" to characterize the inherent interactions of intra-subgroups and inter-subgroups with the varying number of participants. These heterogeneous descriptors are represented in histogram form using the Bag-of-Words (BoW) approach [6] and combined by a Multiple Kernel Learning (MKL) method [7].

The rest of this paper is organized as follows. Section 2 describes the proposed human group activity recognition algorithm. Experimental results are presented in Sect. 3. Finally, conclusion is provided in Sect. 4.

2   The Proposed Method

2.1   Overview

The proposed framework consists of several steps as shown in Fig. 2. In the video pre-processing step, multiple people are detected and tracked using the state-of-art object detection [8] and multi-object tracking algorithms [9, 10]. Individual humans are clustered into non-overlapping subgroups by the mean-shift clustering algorithm based on their positions. In the second step, three types of group activity descriptors are constructed to represent the group activity using motion trajectory and appearance information. Then, these heterogeneous descriptors are represented in histogram form respectively using the BoW approach. A MKL method is used to combine these features and feed the mixture of the kernels to Support Vector Machine (SVM) for generating the fused classifier. Finally, a multi-class SVM classifier makes the recognition decision.

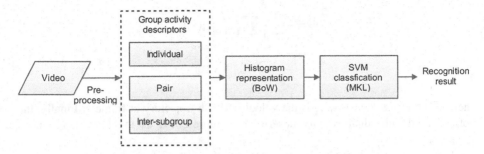

Fig. 2. Framework of the proposed method.

2.2 Problem Definition

Denote the individual trajectories within a video clip which contains M people with Q groups as $S = \{s_1, s_2, \ldots, s_M\}$, where $s_i(t) = \left[s_i^x(t), s_i^y(t)\right]^T$ is the center position of the i-th person at time t. The group trajectories are denoted as $G = \{g_1, g_2, \ldots, g_Q\}$, where $g_j(t) = \left[g_j^x(t), g_j^y(t)\right]^T$ is the center position of the j-th group at time t. As a specific machine learning problem, a set of training samples are given as $\Theta = \{\Theta_1, \Theta_2, \ldots, \Theta_N\}$, where N is the number of training samples and $\Theta_k = (S_k, G_k)$. Each group activity label for Θ_k is denoted as $c_k \in C$, where C is the number of human group activity class. The goal of human group activity recognition is to learn a prediction function of activity class based on the labeled training samples, so as to predict the group activity class of the new samples.

2.3 Group Activity Descriptors

This paper regards group activity as a combination of subgroups and characterizes three types of activity descriptors, i.e., "individual", "pair", and "inter-subgroup" to represent group activity as shown in Fig. 1(b). They are complementary for analyzing group activity. Details of each activity descriptor are introduced in the rest of this section.

2.3.1 Individual Activity Descriptor

Individual activity descriptor characterizes the activity of a single person. By depicting the individual movements within a group, a general knowledge of group activity can be obtained. Individual motion trajectories are useful to represent group activity. Based on the individual trajectory, the location change of human along the trajectory is used to express the movement scale as follows:

$$\zeta_i = \|s_i(1) - s_i(t_e)\| \tag{1}$$

where t_e is the end time of the trajectory. Velocity is also an important property of movement. Hence, the average and variance of velocity are adopted to represent the intensity and variation of the movement respectively as follows:

$$\bar{v}_i = \frac{1}{t_e - 1} \sum_{\tau=1}^{t_e-1} \|v_i(\tau)\| \tag{2}$$

$$\rho_i = \frac{1}{t_e - 1} \sum_{\tau=1}^{t_e-1} \left| \|v_i(\tau)\| - \bar{v}_i \right|^2 \tag{3}$$

where $v_i(\tau) = s_i(\tau + 1) - s_i(\tau)$ is the velocity of i-th trajectory at frame τ. Finally, the motion-based individual activity descriptor is represented as the feature vector of $[\zeta_i, \bar{v}_i, \rho_i]$.

Besides motion information, appearance information such as shape and local motion is also an important clue for recognizing group activity such as fight. It can provide additional discriminative information which is complementary to motion information. To represent the shape and local motion information of individuals, the Histograms of Oriented Gradients (HOG) [11] and the Histogram of Oriented Optical Flow (HOOF) [12] are employed. For each individual trajectory, the sequences of HOG and HOOF features within the bounding box are extracted. And the Principal Components Analysis (PCA) is applied to the HOG features to reduce the computation cost and improve the discriminative power. Thus, the appearance-based activity descriptor is represented by the sequences of PCA-HOG and HOOF features.

2.3.2   Pair Activity Descriptor

Pair activity descriptor characterizes the interaction between an individual pair. In order to consider only meaningful interaction of pairs, only pairs within the same sub-group is used to construct descriptor. First, the relative distance δ_{ij} between two humans i and j is calculated as follows.

$$\delta_{ij}(t) = \left\| s_i(t) - s_j(t) \right\| \tag{4}$$

Based on these relative distances, the variance and maximum change of δ_{ij} are used to provide discriminative activity information such as *walk-in-group* and *approach/split* as follows:

$$\sigma_{ij} = \frac{1}{t_e - 1} \sum_{\tau=1}^{t_e-1} \left| \delta_{ij}(\tau) - \bar{\delta}_{ij} \right|^2 \tag{5}$$

$$\varphi_{ij} = \max_t \delta_{ij}(t) - \min_t \delta_{ij}(t) \tag{6}$$

where $\bar{\delta}_{ij}$ is the average of the relative distance. Additionally, to reflect the tendency of people to get close to each other or the opposite case, the average of relative distance changes is included in a descriptor as follows.

$$\lambda_{ij} = \frac{1}{t_e - 1} \sum_{\tau=1}^{t_e-1} \left\{ \delta_{ij}(\tau + 1) - \delta_{ij}(\tau) \right\} \tag{7}$$

Here, when people get close to each other, λ_{ij} may be a large negative value, or a large positive value for the opposite case.

In addition to the proposed features, the causality and feedback ratios using the GCT [2, 4] are also added to represent causality between humans based on their concurrent motion trajectories. Given two trajectories of human i and j, the GCT models $P(s_i(t)|s_i(1:t-l_i))$ and $P\big(s_i(t)|s_i(1:t-l_i), s_j(1:t-l_j)\big)$ using k-th order linear predictor are as follows:

$$s_i(t) = \sum\nolimits_{\tau=1}^{k} \alpha(\tau)s_i(t-l_i-\tau) + \in_p (t) \tag{8}$$

$$s_i(t) = \sum\nolimits_{\tau=1}^{k} \beta(\tau)s_i(t-l_i-\tau) + \sum\nolimits_{\tau=1}^{k} \gamma(\tau)s_j(t-l_j-\tau) + \in_q (t) \tag{9}$$

where $\alpha(\tau)$'s, $\beta(\tau)$'s, and $\gamma(\tau)$'s are the regression coefficients, and l_i and l_j are the time lag, and $\in_p (t)$ and $\in_q (t)$ are the Gaussian noise with standard deviation $\sigma(s_i(t)|s_i(1:t-l_i))$ and $\sigma\big(s_i(t)|s_i(1:t-l_i), s_j(1:t-l_j)\big)$, respectively. These model parameters can be derived based on the concurrent motion trajectory segment pair of s_i and s_j. Then, the causality ratio r_i and feedback ratio r_j are defined as

$$r_i = \frac{\sigma(s_i(t)|s_i(1:t-l_i))}{\sigma\big(s_i(t)|s_i(1:t-l_i), s_j(1:t-l_j)\big)} \tag{10}$$

$$r_j = \frac{\sigma\big(s_j(t)|s_j(1:t-l_j)\big)}{\sigma\big(s_j(t)|s_i(1:t-l_i), s_j(1:t-l_j)\big)} \tag{11}$$

More details are provided in Y. Zhou et al. [2]. Finally, the pair activity descriptor between two person i and j within the same sub-group is represented as the feature vector of $[\sigma_{ij}, \varphi_{ij}, \lambda_{ij}, r_i, r_j]$.

2.3.3 Inter-Subgroup Activity Descriptor

Compared with previous works [2, 3] only considering the interactions among the individuals and pairs, we further propose the "inter-subgroup" descriptor in the consideration of the interactions between different sub-groups. Inter-subgroup activity descriptor can be useful with the aforementioned activity descriptors for recognizing group activity in complex situations with multiple group behavior such as *approach* or *split* as shown in Fig. 1(a). To cluster individual humans into non-overlapping sub-groups, the mean-shift clustering algorithm is used based on individual position. The reason for using the mean-shift clustering instead of the typical K-means is the advantage of being able to cluster humans adaptively without determining the number of groups in prior. To represent inter-subgroup activity, first, we define the group trajectory of sub-group i which contains u people at frame t as follows.

$$g_i(t) = \frac{1}{u} \sum\nolimits_{k=1}^{u} s_k(t) \tag{12}$$

where u may also be a one. For example, if people are clustered into one group including more than two people and a person, that person is regarded as one sub-group.

Then, the inter-subgroup activity descriptor is constructed in the same way as pair activity descriptor. Thus, the inter-subgroup activity descriptor between two sub-groups i and j is represented as the feature vector of $\left[\tilde{\sigma}_{ij}, \tilde{\varphi}_{ij}, \tilde{\lambda}_{ij}, \tilde{r}_i, \tilde{r}_j\right]$ where each component are represented as

$$\tilde{\sigma}_{ij} = \frac{1}{t_e - 1}\sum\nolimits_{\tau=1}^{t_e-1} \left|\Omega_{ij}(\tau) - \bar{\Omega}_{ij}\right|^2 \tag{13}$$

$$\tilde{\varphi}_{ij} = \max_t \Omega_{ij}(t) - \min_t \Omega_{ij}(t) \tag{14}$$

$$\tilde{\lambda}_{ij} = \frac{1}{t_e - 1}\sum\nolimits_{\tau=1}^{t_e-1} \left\{\Omega_{ij}(\tau + 1) - \Omega_{ij}(\tau)\right\} \tag{15}$$

$$\tilde{r}_i = \frac{\sigma(g_i(t)|g_i(1:t-l_i))}{\sigma(g_i(t)|g_i(1:t-l_i), g_j(1:t-l_j))} \tag{16}$$

$$\tilde{r}_j = \frac{\sigma(g_j(t)|g_j(1:t-l_j))}{\sigma(g_j(t)|g_i(1:t-l_i), g_j(1:t-l_j))} \tag{17}$$

where $\Omega_{ij} = \|g_i - g_j\|$ is the relative distance between two sub-groups i and j.

2.4   Activity Classification with Group Activity Descriptors

When all activity description features are extracted and accumulated from the training video samples, the visual words are generated using K-means clustering for every type of features. These features are then represented in a normalized histogram form by assigning their nearest visual words for each group activity class. Subsequently, the classifiers based on a multi-class SVM [14] with χ^2 kernel are trained using a MKL method [7] which combines various types of these features and feed the mixture of the kernels to SVM. Thus, these classifiers predict the group activity class for new test sample.

3   Experimental Results

In this section, we demonstrate the performance of the proposed method on the public BEHAVE dataset [13] where there are some of the collective behaviors which include multiple groups, and the varying number of participants as well as groups. The BEHAVE dataset was captured by frame rate of 25 fps frame rate and frame size of 640×480 pixels. The dataset consists of ten group activity classes with 2 to 5 people– *In-Group, Approach, Walk-together, Meet, Split, Ignore, Chase, Fight, Run-together,* and *Following*. We consider only *Approach* (A), *Split* (S), *Walk-Together* (W), *Fight*

(F), and *In-Group* (I) to demonstrate the performance of the proposed method because the rest of activity classes do not include group activity or only have few sequences. In this experiment, we use the trajectory and bounding box information provided by the dataset. To evaluate the recognition performance, the leave-one-out cross-validation strategy is used.

The performance comparison of classification accuracy between the proposed method and state-of-art methods [3, 13, 15] are represented in Table 1. In this comparison, four group activity classes that were used in the experiments of the previous work [15] are considered– *Approach*, *Split*, *Walk-Together*, and *In-Group*. As can be seen in Table 1, the performance of the proposed method outperforms other conventional methods in every class.

Table 1. Performance comparison with other methods (%).

Activity/Method	[13]	[3]	[15]	Proposed
Approach	65	57	60	**83.3**
Split	68	66	70	**76.9**
Walk-Together	82	82	45	**96.0**
In-Group	89	82	90	**96.2**
Avearge	76	71.8	66.3	**88.1**

Addtionally, to demonstrate the effectiveness of the proposed inter-subgroup activity descriptor, we compared the performance of group activity classification with and without the proposed inter-subgroup descriptor. Figure 3 shows the performance comparison results by the confusion matrices. In this comparison, the proposed method with the inter-subgroup descriptor performs better than the method without it.

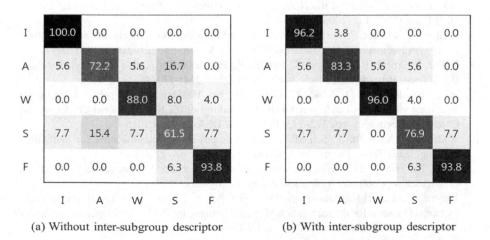

(a) Without inter-subgroup descriptor   (b) With inter-subgroup descriptor

Fig. 3. Comparison of confusion matrices representing the effectiveness of the proposed inter-subgroup descriptor.

4 Conclusion

In this paper, a novel human group activity recognition method is proposed to deal with more complex situation where there are the activities of multiple groups. To characterize the inherent interactions of intra-subgroups and inter-subgroups, we proposed three types of group-activity descriptor, i.e., "inter-subgroup" as well as "individual" and "pair" using motion trajectory and appearance information. Experimental results on the public BEHAVE dataset demonstrated effectiveness of the proposed method in terms of confusion matrix and classification accuracy. For future work, we plan to consider Deep-net based learning method for comparison with MKL to apply the proposed descriptors.

Acknowledgment. This research was supported by BK21 PLUS Program.

References

1. Niebles, J.C., Wang, H., Li, F.: Unsupervised learning of human action categories using spatial-temporal words. Int. J. Comput. Vis. (IJCV) **79**(3), 299–318 (2008)
2. Zhou, Y., Huang, T.S., Ni, B., Yan, S.: Recognizing pair-activities by causality analysis. ACM Trans. Intell. Syst. Technol. **2**(5), 1–20 (2011)
3. Ni, B., Yan, S., Kassim, A.: Recognizing human group activities with localized causalities. In: Proceedings of IEEE Conference on Computer Vision and Pattern Recognition Workshops, pp. 1470–1477, Miami, FL, USA (2009)
4. Granger, C.W.J.: Investigating causal relations by econometric models and cross-spectral methods. Econometrica **37**, 424–438 (1969)
5. Zhang, D., Gatica-Perez, D., Bengio, S., McCowan, I.: Modeling individual and group actions in meetings with layered HMMs. IEEE Trans. Multimed. **8**, 509–520 (2006)
6. Sivic, J., Zisserman, A.: Video Google: a text retrieval approach to object matching in videos. In: Proceedings of the Ninth IEEE International Conference on Computer Vision, pp. 1470–1477. IEEE (2003)
7. Gehler, P., Nowozin, S.: On feature combination for multiclass object classification. In: 2009 IEEE 12th International Conference on Computer Vision, pp. 221–228. IEEE (2009)
8. Dollar, P., Appel, R., Belongie, S., Perona, P.: Fast feature pyramids for object detection. IEEE Trans. Pattern Anal. Mach. Intell. **36**(8), 1532–1545 (2014)
9. Ju, J., Ku, B., Kim, D., Song, T., Han, D.K., Ko, H.: Online multi-person tracking for intelligent video surveillance systems. In: Proceedings of IEEE International Conference on Consumer Electronics, pp. 372–373, Las Vegas, USA, January 2015
10. Andriyenko, A., Roth, S., Schindler, K.: Continuous energy minimization for multitarget tracking. IEEE Trans. Pattern Anal. Mach. Intell. **36**(1), 58–72 (2014)
11. Dalal, N., Triggs, B.: . Histograms of oriented gradients for human detection. In: IEEE Conference on Computer Vision and Pattern Recognition (2005)
12. Rizwan, C., et al.: Histograms of oriented optical flow and binet-cauchy kernels on nonlinear dynamical systems for the recognition of human actions. In: IEEE Conference on Computer Vision and Recognition, CVPR 2009. IEEE (2009)

13. Blunsden, S., Fisher, R.: The behave video dataset: ground trothed video for multi-person behavior classification. In: Proceddings of The British Machine Vision Conference, pp. 1–12, vol. 2010, no. 4, Aberystwyth, UK, August 2010
14. Lin, C.-C., Chang, C.-J.: LIBSVM: a library for support vector machines. ACM Trans. Intell. Syst. Technol 2(1), 27:1–27:27 (2011)
15. Münch, D., Michaelsen, E., Arens, M.: Supporting Fuzzy Metric Temporal Logic Based Situation Recognition by Mean Shift Clustering. In: Glimm, B., Krüger, A. (eds.) KI 2012. LNCS, vol. 7526, pp. 233–236. Springer, Heidelberg (2012)

An Incremental SRC Method
for Face Recognition

Junjian Ye and Ruoyu Yang[(✉)]

State Key Laboratory for Novel Software Technology,
Nanjing University, Nanjing, China
mg1333073@smail.nju.edu.cn, yangry@nju.edu.cn

Abstract. Face recognition has been studied for decades and been used widely in our daily life. However, when the practical application is concerned, not only the occlusion, pose and expression variations, but also the increasing training cost caused by the increasing number of training samples are problems we need to solve. In the paper we present a novel incremental SRC method aimed at solving the practical face recognition problems. On one hand, we divide the face into several components, select out the components affected greatly by face variations and abandon these components, the rest parts are used to rebuild the global face which contributes to the final result. On the other hand, inspired by the strategy of "Divide and Rule", we divide the training samples into multiple groups and train in each group respectively. Therefore, when new training sample is added, we only need to update the model of the group to which the new sample is added, which can greatly decrease the retraining cost.

Numerous experiments are made on the AR and ORL face databases. Experimental results show that the performances of our method outperform the state-of-art linear representation algorithms. In the practical situation of single training sample, our method shows greater advantage than other methods.

Keywords: Sparse representation classification · Incremental · Single training sample

1 Introduction

Face recognition has been studied extensively and achieved great progress in recent years. It has been applied to many fields like attendance checking system and access control system. Various methods have been proposed to address different face recognition problems.

Currently, the face recognitions problems can be classified into two categories. The first one is caused by the face variations. As we know, occlusion of sunglasses, pose variations such as raising one's head, expression variations such as laughing and crying, they all belong to the first category. The other one consists of the problems arisen from the practical systems. Since the training samples of the practical systems are not fixed, it is necessary to consider how to add the new classes and new samples. When new samples are added into the system, the previous trained models are not suitable any more. Therefore, updating the training model quickly is extremely

© Springer International Publishing Switzerland 2015
Y.-S. Ho et al. (Eds.): PCM 2015, Part II, LNCS 9315, pp. 170–180, 2015.
DOI: 10.1007/978-3-319-24078-7_17

important for the face recognition applications. However, as the number of training samples increases, the training cost will increases rapidly.

In the previous decades, many attentions have been focused on the problems of the first category. For example, component-based methods are proposed to extract the face features from several components of the face instead of the holistic face. Since pose, expression variations and occlusions affect only part of the whole face, the rest part is still very helpful to provide discriminative information. For example, [1] extracted features from several components and constructed pose-specific classifiers, which achieved great robustness on pose variations.

Besides the component-based methods, in 2009, Wright et al. [2] applied sparse coding to face recognition and proposed the sparse representation based classification (SRC) scheme, which is proved to have great robustness against occlusions. Its main idea is finding a sparse representation of the test image in terms of the whole training images as the dictionary. In the same year, Naseem et al. [3] proposed LRC method used in face recognition. Different from SRC, LRC represents a testing sample as a linear expression of class-specific training samples. The sum of coefficient (SoC) [4] was proposed to improve the decision rule of SRC. In addition, ICFR [5] proposed an increment coefficient method which simultaneously used the testing sample and training samples of each class to represent a virtual sample for face recognition. In LRSRC [6], the low-rank approximation (LRA) and sparse representation (SR) techniques were combined together. In [7], the authors proposed a prototype plus variation (P + V) model and a corresponding sparsity based classification algorithm which they called superposed SRC (SSRC). In [8], a facial image model based on random projection and sparse representation was used to develop facial variation modeling systems.

Although linear representation-based methods are robust to occlusions, if they build the dictionary using the whole face as [2] did, their robustness against face variations like expression changing can't be high. Therefore, in the current work we propose a face recognition scheme combining SRC with components together. Since face components affected greatly by face variations will greatly influence the face recognition, it is natural to think of the idea to select out these face components and abandon them. The rest components can actively contribute to the face recognition.

Many attentions have been focused on the face recognition problems of first category, however, when the practical face recognition works, it is necessary to take problems of the second category into account. When dealing with continuously increased training samples, we need to update the training model, which means the dictionary. For example, in [9] the dictionary was built by PCA, and the incremental PCA method was applied to update the dictionary. But, it realizes the incremental process of SRC by incremental PCA, which means if the dictionary is not built by PCA, this incremental way will fail. [10] used a subspace learning method to update the dictionary, which is the same way.

In the current work we propose a general framework for incremental SRC. We find that in the dictionary of SRC, each training sample owns its unique code, which means when a new training sample appears, it can be directly added into the dictionary. So we just need to abstract the feature of the new sample and add it directly to the dictionary. However, it meets problems when number of samples is huge. Inspired by the strategy of "Divide and Rule", we break down the whole training samples into multiple groups. Under this framework, when a new sample adds, we firstly choose a group which the new sample joins, then update the dictionary of the group. As to other groups, we can just do nothing.

This paper is organized as follows. In Sect. 2 we introduce the pipeline of our method. In Sect. 3 we show the building process of our dictionary for the incremental SRC method. In Sect. 4 we display our incremental framework. In Sect. 5 we numerically demonstrate the effectiveness of our proposal by extensive experiments on the public databases. In Sect. 6 we conclude our work with discussion of promising future directions.

2   Pipeline Overview

Our scheme is shown in Fig. 1: first, all the training samples are divided into several groups. Then in each group, the training samples need to experience the feature extraction. For a testing image, each group has a result after the prediction process. At last, we synthesize all the group results and get a final prediction among them.

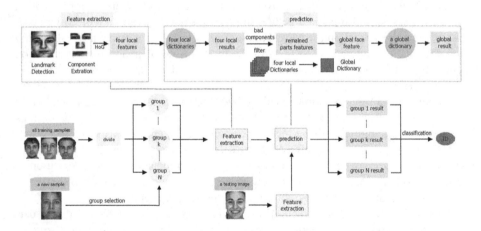

Fig. 1. Pipeline of our method.

In the feature extraction process, a standard landmark detector [13] is used to obtain face landmarks. Then the four components (two eyes, nose, mouth) are localized separately based on detected landmarks and regularized to the same size. HoG descriptor is applied to the four components respectively.

In the prediction process, we will select out the bad face components influenced heavily by face variations according to the results of the four components, and concatenate features of the remained components to get a global face feature. Then we can get a global face result.

When a new sample is coming, we select a proper group for it to add in and the group will update its dictionary.

3   Dictionaries of the Incremental SRC Method

In the previous section, we introduce the pipeline of our method. In this section, we show the generation of the local dictionaries and global dictionary. In each group, we first build four local dictionaries. When a testing image comes, we dynamically build the global dictionary according the results of the four local dictionaries.

3.1   Building Dictionaries

Compared with aligning the whole face, align each face component seems to be much wiser. Therefore, we extract four components from the face, regularize them to the same size, and apply HoG descriptor to obtain the component features.

However, the discriminative ability of single component is not very strong. The idea of combining these components occurs to us. But there is a problem. As we know, when one people smiles, or wears a pair of glasses, some part of his face will have great deformation.

As we can see in Fig. 2, the three rows demonstrate the expression, light variation and occlusion respectively. In Fig. 2(a), we first compare the two faces with all components. We can see the mouth changes greatly. Then we abandon the mouth and compare again using the rest components. Now we can see it's much better compared with before. In Fig. 2(b), the noses are most dissimilar because of the light variation. After we abandon the nose, the two faces look more similar. In Fig. 2(c), the eyes are unable to contribute to identify because of sunglasses, therefore, using the nose and mouth to identify is a better way.

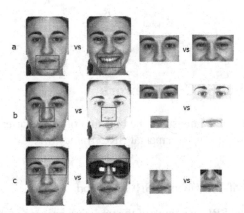

Fig. 2. variations of one people and parts filtering.

Therefore, we use the four local results to select out these bad components. As shown in Fig. 2, the rest components are used to rebuild the global face, which means we concatenate the features of the rest component to build the global face feature.

3.2   Selecting Out the Bad Components

We propose a strategy to select out the bad components from the four local results. For a testing image y, we can get four predicted labels through the four local dictionaries. Instead of using these labels, we focus on the residuals. We use R to indicate the residuals of the person's four components. R1 and R2 denote the smallest and the second smallest residual of R. Good is a vector with four elements indicating whether the corresponding component is good.

$$R = \left\{ R_{lefteye}, \; R_{righteye}, \; R_{nose}, \; R_{mouth} \right\} \tag{1}$$

$$Good = \begin{cases} 1, & R(j) < R1 + R2 \\ 0, & else \end{cases} \tag{2}$$

We choose the corresponding components with 1 in Good to rebuild the global face of the testing person, as to the rest components, they are abandoned.

3.3   Global Result

After we choose the good components, we concatenate features of the good components to build the global face feature. The global dictionary, from which we can get a residual R_g and a prediction label, is generated in the same way. Since the global face can be built by two or three components, we calculate the global residual R_{global} as follows Eq. (3), where $|Good_i|$ represents the value of vector $Good_i$.

$$R_{global} = \frac{R_g}{|Good_i|} \tag{3}$$

4   Our Incremental SRC Framework

In the previous section, we demonstrate the building of dictionaries. In this section, we describe the framework of our incremental SRC framework.

4.1   Basic Problem of Incremental SRC Method

Speaking of incremental SRC, we think of the updating process of the sparse coefficient x when a new training sample is added to the dictionary. Suppose D denotes the dictionary of T training samples, x denotes the sparse coefficient of a test sample y. When a new training sample is added, the new dictionary is updated to $D' = [D, f_{T+1}]$, where f_{T+1} is the feature of the new training sample. The new coefficient x' is computed as:

$$\widehat{x'}_1 = \arg\min ||x'||_1 \quad \text{subject to} \quad ||D'x' - y|| < \varepsilon \tag{4}$$

Inspired by the idea of "Divide and Rule", our incremental SRC framework depends on the dividing process. Since a new sample only belongs to one class, if we can divide different classes into different groups and rule each group independently, we only need to update the result of the group containing the class of the new sample. Except this group, no changes need to be done in the other groups.

4.2   Divide Samples into Multiple Groups

The majority of the SRC-based methods use the way of building one dictionary from the whole training samples. Different from this way, our method divides all the training samples into several groups and then build dictionaries for each group.

Suppose we have K classes. The process of dividing is as follows.

1. Partition the K classes into N groups, a certain class will only appear in one group;
2. For each group, we will build the local dictionaries as Shown in Fig. 1;
3. For a testing image, each group builds the global dictionary according to the four local results and gets a global residual and a predicted label. We synthesize all the group results and decide the final prediction.
4. When a new training sample is added, we first find a group which the sample joins. If it belongs to one class known, we add it to the group this class belongs to.
5. Otherwise, the new sample is put into a random group.

4.3   Decision Among the Groups

Suppose the number of groups is N. For a testing image y, we denote $R_{global}(j)$ as the global residual of group j by Eq. (3) and P(j) as its prediction. All groups will compete to get the final prediction as Eq. (5):

$$\text{Prediction} = P\left(\min_j R_{global}(j)\right), 1 \le j \le N \tag{5}$$

5   Experimental Results

First, we test our method under situations with no occlusions in both ORL [11] and AR [12] database. Our testing samples contain expression, pose variations in ORL database and expression, lighting variations in AR database.

(a)   (b)

Fig. 3. One of the classes from AR database.

- We design experiments under different number of groups and implement the SRC [2] method to compare with our method.
- Since single training sample problem is an important issue in practical face recognition, we design experiments under only single training sample. Both in ORL and AR databases, the first image with neutral expression is used as the training sample, the rest images are used for testing.
- We design experiments under different scaling sizes in ORL database.
- An incremental experiment is designed on AR database.

Second, we test our method under occlusions in AR database. We use a training set of 7 non-occluded images and the first sunglasses image from session 1 for each person and a testing set of the rest five sunglasses images. Similar to the previous case, scarf occlusion is set the same way.

We compare the recognition rates of our method with other linear representation-based methods. On the ORL database, five linear representation-based methods (i.e. SRC [2], LRC [3], SoC [4], ICFR [5]) are compared. On the AR database, SRC [2], LRSRC [6], SSRC [7] and method [8] are compared.

The AR Database. Experiments were done on a subcollection of the AR database. The subcollection includes 100 persons's (50 men and 50 women) face images, 26 face images for each person. Figure 4 shows one of the classes from the subcollection of the AR database. The images of the first row were taken in the first date, and the images of the second row were taken two weeks later on the same condition. The variations of the same person's images in Fig. 4(a) include different illuminations, poses, expressions. The same person's images in Fig. 4(b) include occlusions such as sunglasses and scarf.

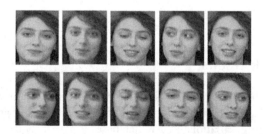

Fig. 4. One of the classes from ORL database.

The ORL Database. The ORL database includes 40 classes with 10 images each class. The variations include facial gestures, and siding face with no more than 20 degree. One of the classes from ORL database is shown in Fig. 3.

5.1 Experiments Under No Occlusions

Experiments Under Different Number of Groups. Table 1 displays our method's recognition rates on ORL and AR database.We randomly choose half samples as training samples for each class.

As shown in Table 1, we can find that our method outperforms the original SRC method. When N is equal to 2, the performance of our method is almost the same with the no-partition SRC method. When N is equal to 4 and 8, the performance of our method is still acceptable.

There is a tradeoff between recognition rate and incremental cost. What we need to do is to find a good tradeoff between them. From the results, the situation of 2 groups seems to be a good tradeoff.

Experiments Under Single Training Sample. As shown in Table 1, we test the single training sample situation under different number of groups. Compared with the original SRC, our method shows better recognition rates and still maintains great performance when group number N is increasing.

Table 1. The recognition rates (%) under different number of groups.

Group number	1	2	4	8	1	2	4	8
Training samples	half	half	half	half	1	1	1	1
Ours(ORL)	98.00	97.50	97.00	95.00	79.17	76.94	73.33	70.56
SRC[2](ORL)	92.00	91.00	89.00	85.00	63.33	34.44	19.17	10.28
Ours(AR)	97.14	95.00	93.75	92.75	75.23	75.69	73.38	67.31
SRC[2](AR)	89.71	75.71	73.86	72.43	63.31	61.54	57.69	56.00

Table 2. The recognition rates under single training sample on ORL database.

Recognition rate(%)	SRC [2]	LRC [3]	SoC [4]	ICFR [5]	Our method
	63.31	66.64	54.03	66.83	**76.94**

Table 3. The recognition rates under single training sample on AR database.

Recognition rate(%)	SRC [2]	SSRC [7]	Method [8]	Our method
	34.92	55.99	61.18	**75.69**

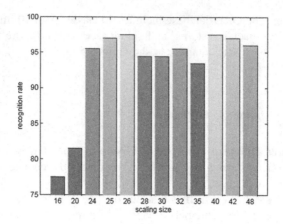

Fig. 5. The recognition rates of different scaling sizes on ORL database.

Table 4. Recognition rates (%) of each segment on AR database.

test \ train	1	2	3	4	5
1	99.23	99.23	99.23	99.23	99.23
2	--	98.46	98.46	97.49	97.49
3	--	--	96.92	94.61	93.07
4	--	--	--	97.69	96.15
5	--	--	--	--	95.38

Tables 2 and 3 display our method's recognition rates on ORL database and AR database under single training sample. The results show that our method can perform better when there is only one training sample.

Experiments Under Different Scaling Sizes. In Fig. 5, we test our method with different scaling sizes in ORL database. The group number is set to be 2. Except when the size is too small, our performance is stable in most cases.

Incremental Experiments. In order to test the incremental situation, we partition the 50 men and 50 women of AR database into 5 segments, each segment consists of 10 men and 10 women. The images from first session are training samples, the rest images are testing samples. The group number is set to 2. We set the training samples in segment 1 as the initial training sets of the two groups. Then segment 2, 3, 4 and 5 are added in turn.

As shown in Table 4, the first row demonstrates the recognition rates of the segment 1 when we add the training samples of segment 1, 2, 3, 4 and 5 in turn into the groups. From the results, we find that the addition of new training samples only reduce the recognition rates little. Therefore, we draw the conclusion that our incremental SRC algorithm can recognize the existing classes well when new classes add.

5.2 Experiments Under Occlusions

Experiments Under Scarf and Sunglass. Results are shown in Table 5, our method achieves the highest recognition rate of 94.44 % and 91.60 % for scarf occlusion and sunglass occlusion respectively. We also find that when N is equal to 2 and 5, the recognition rates under occlusion are better than that when N is equal to 1.

Table 5. Recognition-rates-(%)under-occlusions-on-AR-database.

	SRC [2]	LRSRC [6]	Ours (N = 1)	Ours (N = 2)	Ours (N = 5)	Ours (N = 10)
Sunglasses	87.32	90.87	89.00	**91.60**	90.00	87.40
Scarf	76.56	83.07	90.00	**94.44**	93.20	90.60

6 Conclusions

Another important direction for future investigation is to better find out the discriminative components of the face.We have demonstrated that misalignment can be naturally handled within the sparse representation framework through component extraction. But, if the position of the component isn't located well, the performance will be influenced greatly. It is still an open and important issue to automatically and accurately find the location of the components.

Acknowledgments. Supported by the National Natural Science Foundation of China under Grant Nos. 61321491, 61272218

References

1. Cao, Z., Tang, X., Yin, Q., Sun, J.: Face recognition with learning based descriptor. In: CVPR, pp. 2707–2714. IEEE (2010)
2. Wright, J., Yang, A.Y., Ganesh, A., Sastry, S.S., Ma, Y.: Robust face recognition via sparse representation. Pattern Anal. Mach. Intell. **31**(2), 210–227 (2009)
3. Naseem, A.I., Togneri, B.R., Bennamoun, C.M.: Face identification using linear regression. In: Image Processing (ICIP), pp. 4161–4164. IEEE (2009)
4. Li, J., Lu, C.: A new decision rule for sparse representation based classification for face recognition. Neurocomputing **116**, 256–271 (2013)
5. Li, C., Miao, X., Xiao, L., Li, M., Hu, Z., Pan, Z.: An increment coefficient method for face recognition. In: Image and Signal Processing (CISP), pp. 665–669. IEEE (2014)
6. Gia, Q.K., Nhan, D.C., Tien, D.B.: Sparse representation and low-rank approximation for robust face recognition. In: ICPR. IEEE (2014)
7. Deng, W., Hu, J., Guo, J., Extended, S.R.C.: Undersampled face recognition via intraclass variant dictionary. IEEE Trans. Pattern Anal. Mach. Intell. **34**, 1864–1870 (2012)

8. Cai, J., Chen, J., Liang, X.: Single-sample face recognition based on intra-class differences in a variation model. Sensors **15**(1), 1071–1087 (2015)
9. Qiu, H., Pham, D., Venkatesh, S., Liu, W., Lai, J.: A fast extension for sparse representation on robust face recognition. In: ICPR. IEEE (2010)
10. Wang, C., Wang, Y., Zhang, Z., Wang, Y.: Face tracking and recognition via incremental local sparse representation. In: Image and Graphics (ICIG). IEEE (2013)
11. http://www.faceplusplus.com.cn
12. Samaria, F., Harter, A.: Parameterisation of a stochastic model for human face identification. In: Workshop on Applications of Computer Vision. IEEE (1994)
13. Martinez, A.M., Benavente, R.: The AR face database. In: CVC Technical report (1998)

A Survey on Media Interaction in Social Robotics

Lu Yang, Hong Cheng$^{(\boxtimes)}$, Jiasheng Hao, Yanli Ji, and Yiqun Kuang

Center for Robotics, School of Automation Engineering,
University of Electronic Science and Technology of China, Chengdu 611731, China
hcheng@uestc.edu.cn

Abstract. Social robots have attracted increasing research interests in academic and industry communities. The emerging media technologies greatly inspired human-robot interaction approaches, which aimed to tackle important challenges in practical applications. This paper presents a survey of recent works on media interaction in social robotics. We first introduce the state-of-the-art social robots and the related concepts. Then, we review the visual interaction approaches through various human actions such as facial expression, hand gesture and body motion, which have been widely considered as effective media interaction ways with robots. Furthermore, we summarize the event detection approaches which are crucial for robots to understand the environment and human intentions. While the emphasis is on vision-based interaction approaches, the multimodal interaction works are also briefly summarized for practitioners.

Keywords: Social robot · Hand gesture · Action recognition · Event detection · Human-robot interaction

1 Introduction

Robots play growing number of roles in our world from industry manufacturing to daily services and entertainment. While robots were initially used in repetitive tasks given instructions, recent social robots have been autonomous robots which can interact and communicate with humans. Due to its huge potential to assist people in our daily lives, social robotics has drawn increasing research attention especially in how robots can understand human behaviors by media interactions [57].

It is generally agreed that Human-Robot Interaction (HRI) capability is the heart of a social robot [20,57]. The fundamental goal of HRI is to develop the principles and algorithms for robot systems that make them capable of natural, safe and effective interaction with humans [17]. From the cognitive viewpoint, the emphasis of social HRI raises the issue of cognitive architecture in which robotic agents and memory structures are implemented to accurately model the behavior of interacting humans [5]. Following the taxonomy proposed in

© Springer International Publishing Switzerland 2015
Y.-S. Ho et al. (Eds.): PCM 2015, Part II, LNCS 9315, pp. 181–190, 2015.
DOI: 10.1007/978-3-319-24078-7_18

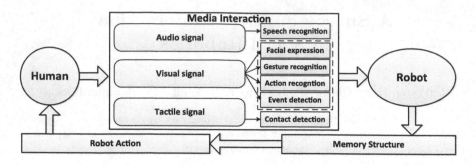

Fig. 1. The architecture of human-robot interaction

[21,57], we can divide the cognitive HRI into three parts: perception(sense), brain(memory) and action(control). As illustrated in Fig. 1, there are mainly three media for interactions in the perception part: video, audio and touch. These three senses are fulfilled by the recognition of human behaviors. Among them, the visual interaction is the most important and popular interaction category for HRI. In this paper, we will survey several visual interaction methodologies: face expression recognition, hand gesture recognition and HRI event detection. Besides, we will also give an overview of audio and tactile interaction approaches which have been applied in social robotics. Although the memory structure and robot action are also important fields for HRI, they are out of the scope of this survey and we will mainly focus on the media interaction for the perception.

There have been a few social robots developed by academic institutes from early MIT's Kismet [8] to CMU's HERB [47], to recent Hobbit [19] of Vienna University of Technology. Meanwhile, commercial social robot products have also appeared in last decade. Representative examples include Sony's AIBO [27], Honda's ASIMO [24], KAIST's HUBO [44] *etc.* Typical state-of-the art social robots and their HRI capabilities are summarized in Table 1. We can see most existed social robots can recognize audio speech and simple interactive events. The face recognition and facial expression are also widely adopted for human-robot interaction. However, as two important media interaction categories, hand gesture and human body action are still less investigated for social robotics. There have been more recent commercial humanoid robots such as SoftBank's Pepper, Toshiba's ChihiraAico and Hitachi's EMIEW, for which facial expression and voice recognition are the two main interaction ways. Although the mechanical design and robot action have been well studied, the related media based perception approaches need to be further explored for natural human-robot interaction.

The rest of the paper is organized as follows. The visual media interaction approaches are summarized in Sect. 2, in which facial expression (Sect. 2.1), hand gesture recognition (Sect. 2.2), body action recognition (Sect. 2.3) and event detection (Sect. 2.4) are reviewed as main visual HRI techniques. Other media interaction approaches in audio and tactile modalities are summarized in Sect. 3. Finally, Sect. 4 concludes the paper.

Table 1. The representative social robots

Social Robot	Facial	Gesture	Action	Event	Audio	Tactile
Kismet [8](2002)	✓			✓	✓	
HERB [47](2010)		✓		✓	✓	
Hobbit [19](2014)	✓	✓	✓	✓	✓	
AIBO [27](1999)	✓			✓	✓	✓
ASIMO [24](2000)	✓	✓	✓	✓	✓	✓
HUBO [44](2004)			✓	✓	✓	

2 Visual Media Interaction

2.1 Facial Expression

Facial expression is the very first HRI capability towards social robots [32]. The fundamental face to face interaction requires real-time face detection and recognition. While rapid face detection by AdaBoost [53] has been widely employed by mobile visual devices, reliable face recognition in wild [54] is still a challenging problem. On the other hand, facial expressions convey emotions such as happiness, sadness and surprise, which are more crucial for the social communication.

Generally, the face expression in HRI involves representation, detection and synthesis. The representation of face expression can leverage the study of human perception with facial emotions [12]. For appearance based representation, Liu et al. [33] used locations of the crucial parts of a face as facial landmarks to encode facial expressions which can be recognized. For model based representation, Trovato et al. [50] employed the 24 DoF head model to present facial expressions in a multi-cultural environment. The facial expression detection has been successfully applied in assistive robotic systems. Jabon et al. [26] developed a driver-assistance system which can predict minor and major accidents by analyzing driver face expressions. To get a greater satisfaction degree during the human-robot dialog, Alonso-Martin et al. [2] proposed a face expression based user-emotion detection system for the social robot to adapt its decision rule. More literatures have been devoted to face expression synthesis [31] owing to the extensive requirements for highly realistic animation in movies and video games. Actually, the rendered face expressions were also displayed on several state-of-the-art social robots as one important interaction medium. Examples include CMU's early roboceptionist 'Tank', UCSD's social robot 'RUBI-5' [35] and MIT's recent family robot 'Jibo'.

Other facial interaction media include eye gaze and blinking. In the natural human-robot interaction, the attention of the robot should match the attention of the user to perform the socially acceptable behaviors [18]. The key for natural eye-contact in HRI is reliable eye gaze estimation under free head motion [13]. Eye gaze tracking modules have been implemented on humanoid social robots serving in group meeting [60] and in the therapy of children with Autism [9].

2.2   Hand Gesture Recognition

Hand gesture recognition is an efficient and robust way for humans to issue instructions to a robot. Hand gesture recognition is the initial effort towards the goal of assistive robot in which the response of the robot converged the user's desired response. Hand gesture recognition approaches which have been applied to social robotics fall into three categories: static or posture based approaches, trajectory or motion based approaches and hybrid approaches.

Yin *et al.* [59] implemented a posture recognition system on a humanoid service robot HARO-1 to enable untrained users to instruct service robots easily. Hand postures can be extracted from the silhouette of the segmented hand region by skin color or hand depth. Among various static hand gestures, pointing gesture is perhaps the most intuitive command or message to a robot even in noisy environments where speech is not effective [43]. Nickel *et al.* [40] presented an approach for recognizing pointing gestures in the context of human-robot interaction using stereo camera. The system aims at run-on gesture recognition in real-time and allow for the robot ego-motion. Bergh *et al.* [6] built a robot which can detect a pointing gesture and 3D pointing direction using a Kinect sensor (Fig. 2 Left). Pointing gesture shows the direction in which the robot should move, or even helps a robot to find a target which humans might have seen.

To handle the challenge of hand tracking in crowded and dynamic HRI environments, McKeague *et al.* [37] proposed a sensor fusion based hand tracking algorithm for crowded environments. Based on Kinect's gesture recognition, they defined four gestures for robot control: A wave for attracting a robot's attention, a subtle push to indicate that interaction has finished, a subtle follow me motion, and a subtle raised hand for stopping the robot's movement. Yanik *et al.* [58] also used Kinect sensor for gesture based robot control where three basic gestures (come closer, go away and stop) from American Sign Language Dictionary were recognized. The main challenge is the hand detection and tracking in cluttered environment [34]. However, the progress of low-cost depth sensors (e.g. Kinect) has significantly promoted the research of 3D hand gesture recognition [10].

Hybrid hand gestures consist of both static and dynamic hand configurations thus can express more complex human intentions to social robots. The hybrid hand gesture interface [39] was developed as part of the Hobbit robot to interpret physical arm/hand gestures to robot commands [19] (Fig. 2 Middle). The intuitively hand gestures can facilitate the communication and even the teleoperation [16] for humans with robots (Fig. 2 Right).

Fig. 2. Hand gestures for social robots [6, 16, 39]

2.3   Body Action Recognition

In addition to facial expressions and hand gestures, body actions are also considered as crucial media to express emotions between human and robots. Similar to hand gestures, body actions are natural and have intuitive meanings. They are related to the human's emotional state and behavior intention [56].

Towards artificial emotions to assist social coordination in HRI, Novikova et al. [42] found that team members routinely monitored their collaborators's attitudes to their individual and joint activity, as well as expressing their own attitudes to progress, through the presentation and interpretation of emotional signals. Especially, upper body gestures [36,56] have been interpreted in HRI as bodily emotional expressions. Based on socially assistive robot Brian of Toronto University (Fig. 3 Left), McColl et al. [36] presented a real-time methodology capable of interpreting the 3D body pose of a person during natural one-on-one human-robot interaction. The final objective is to develop a social robot which can take into account the emotions, moods or affect of the person and show vivid demeanors to humans.

The body gesture language may simultaneously involve head, arm and hand. Xiao et al. [56] applied upper body gesture recognition system to a social robot Nadine (Fig. 3 Middle), where the user's upper body joints were measured by Kinect sensor for the action understanding. Similarly, Keizer et al. [30] developed a Nao-based robot bartender using Kinect to track body actions of all customers in the view. For the joint action in human-robot interaction, handover is a typical task which requires accurate body action recognition to avoid the potential harm. Strabala et al. [49] proposed a coordination structure which carefully considered giver orientation and reaching actions for human-robot handovers. Meanwhile, Grigore et al. [22] used the VICON motion capture (MoCap) system for a humanoid robot BERT2 to track the motion of human body parts and hand over a drink (Fig. 3 Right). However, cognitive aspects of joint action between human and robot still need to be investigated to achieve the continuous joint activities in HRI.

2.4   Event Detection

The daily HRI environment is usually uncontrolled with interactive motion and complex background. Social robots require reliable and early detectors of visual events with humans or contextual objects to induce a cognitive decision on what should be done next [38]. The HRI event detection is highly related to facial expression recognition, gesture recognition and action recognition but there are two major differences: (1) Not only human signals, the environmental cues such as objects, texts and scenario categories also need to be recognized to understand an event [11]; (2) Due to the temporal nature of an event, it requires online time-varying analysis in a continuous way. This means that the social robot should detect spatial-temporal events before they finish so that appropriate responses can be made in a timely manner [25].

Karg et al. [29] used human motion tracking data and semantic maps of the environment to detect unexpected events and failures for table setting tasks in a

Fig. 3. Body actions in human-robot interaction [22,36,56]

typical kitchen environment. By event detection, the social robot can understand about their human partners and the tasks that are to be performed. The human location, action and furniture objects are related to the event representation. For robots operating in real world environments, detecting entering event of dynamic object (e.g. human) is an important capability. Wang *et al.* [55] proposed an Hidden Markov Model (HMM) based approach to enable online prediction of entering event occurrence. For the detection of obstacle event during HRI, Aigner *et al.* [1] utilized a discrete event framework to avoid an obstacle for the robotic system. Berghofer *et al.* [7] further combined visual and laser scan data to let the robot distinguish persons from non-interactive obstacles by incremental online learning. Although the state-of-the-art social robots (Table 1) have implemented some emergency event detection modules such as fall detection, the general HRI event detection is still an open problem due to the absence of explicit begin/end hints for potential events.

3   Multimodal Media Interaction

3.1   Audio Interaction

Speech is a common communication way in HRI. Most social robots have been endowed with microphones to receive the audio signal of the people they interact with [2]. The key is to recognize the verbal voice in natural interaction environment.

Speech recognition has been applied to healthcare robots to help people with motor disabilities. Alves *et al.* [3] employed Google Speech API in a mobile robot system where the user can control the robot by giving speech commands. Similarly, Tsui *et al.* [51] designed speech-based interfaces for telepresence robot for people with disabilities. Although the verbal speech recognition has got significant progress, the emotional states of human voices still need to be further inferred to drive the social behavior of the robot. To reach the goal, Devillers *et al.* [14] used six emotional models to investigate the emotional behavior of human subjects interacting with a robot using audio input. Niculescu *et al.* [41] explored the effects of voice pitch, humor and empathy for the interaction with a social robot receptionist. They found that the right voice pitch is particularly important for the entire interaction.

Voice can be fused with other modalities to achieve the multimodal HRI. Viciana-Abad *et al.* [52] developed an audio-visual perception system for a humanoid robotic head to localize the user. The acoustic and visual modules were fused based on Bayes inference to exploit benefits from both modalities. Alonso-Martin *et al.* [52] proposed a multimodal user-emotion detection system which analyzed the voice and face expression of the user. They further fused multichannel input of audio, vision, radio frequency and touch into communicative acts to represent the transmitted message during HRI. Although audio based HRI has generated efficient interaction and communication for conventional robot [23] and shopping robot [15], the semantic understanding of the dialog in specific context is still challenging for social robots.

3.2 Tactile Interaction

It is obvious that a tangible robot will be more attractive than the one with just stiff surface at home. For human-human social interaction, physical contacts such as hand shaking and embracing are natural ways to express our emotions. For social robots, the capability to sense the touch not only enable the safe operation but also the comfortable interaction with humans.

A typical tactile social robot is MIT's Huggable robot, which has over 1000 force, 400 temperature and 9 electric field sensors [48]. Huggable has a soft multimodal sensory skin with fur covering its entire teddy-bear-shaped body, and classifies the human touching to perform tactile interactive behaviors [4]. Silvera-Tawil *et al.* [45] implemented Electrical Impedance Tomography (EIT) based sensitive skin on an artificial arm. It can classify six emotions and six social messages transmitted by humans when touching. Especially, to make the vivid and acceptable comedian robot, Kaefer *et al.* [28] developed a responsive humanoid robot using tactile feedback for computational awareness. For more tactile HRI works in social robotics, we refer to the survey papers [4,46].

4   Conclusions

In this paper we have given a comprehensive survey of the emerging progress on media interaction in social robotics. The survey reviewed state-of-the-art human-robot interaction works in facial expression, hand gesture recognition, body action recognition, event detection, audio interaction and tactile interaction for social robots. Although visual, acoustic and tactile sensing technologies have been applied to various robots, the semantic and emotion understanding of humans is still the open problem for the future research. We envision the booming of emotional social robots with more human-like cognitions in our daily lives.

Acknowledgment. This work was partially supported by NSFC (No.61305033, 61273256), Fundamental Research Funds for the Central Universities (ZYGX2013J088, ZYGX2014Z009) and SRF for ROCS, SEM

References

1. Aigner, P., McCarragher, B.J.: Modeling and constraining human interactions in shared control utilizing a discrete event framework. IEEE Trans. Syst. Man Cybern. Part A: Syst. Hum. **30**(3), 369–379 (2000)

2. Alonso-Martin, F., Malfaz, M., Sequeira, J., Gorostiza, J.F., Salichs, M.A.: A multimodal emotion detection system during human-robot interaction. Sens. **13**(11), 15549–15581 (2013)
3. Alves, S., Silva, I., Ranieri, C., Ferasoli Filho, H.: Assisted robot navigation based on speech recognition and synthesis. In: ISSNIP-IEEE Biosignals and Biorobotics Conference (2014)
4. Argall, B.D., Billard, A.G.: A survey of tactile human-robot interactions. Rob. Auton. Syst. **58**(10), 1159–1176 (2010)
5. Baxter, P.E., de Greeff, J., Belpaeme, T.: Cognitive architecture for humancrobot interaction: towards behavioural alignment. Biol. Inspired Cogn. Archit. **6**, 30–39 (2013)
6. Van den Bergh, M., Carton, D., de Nijs, R., Mitsou, N., Landsiedel, C., Kuehnlenz, K., Wollherr, D., Van Gool, L., Buss, M.: Real-time 3D hand gesture interaction with a robot for understanding directions from humans. In: IEEE RO-MAN (2011)
7. Berghofer, E., Schulze, D., Rauch, C., Tscherepanow, M., Kohler, T., Wachsmuth, S.: ART-based fusion of multi-modal perception for robots. Neurocomput. **107**, 11–22 (2013)
8. Breazeal, C.: Designing Sociable Robots. MIT Press, Cambridge (2002)
9. Cabibihan, J.J., Javed, H., Ang, M.J., Aljunied, S.M.: Why robots? a survey on the roles and benefits of social robots in the therapy of children with Autism. Int. J. Soc. Rob. **5**(4), 593–618 (2013)
10. Cheng, H., Luo, J., Chen, X.: A windowed dynamic time warping approach for 3D continuous hand gesture recognition. In: IEEE ICME (2014)
11. Cheng, H., Yu, R., Liu, Z., Yang, L., Chen, X.: Kernelized pyramid nearest-neighbor search for object categorization. Mach. Vis. Appl. **25**(4), 931–941 (2014)
12. Costa, S., Soares, F., Santos, C.: Facial expressions and gestures to convey emotions with a humanoid robot. In: Herrmann, G., Pearson, M.J., Lenz, A., Bremner, P., Spiers, A., Leonards, U. (eds.) ICSR 2013. LNCS, vol. 8239, pp. 542–551. Springer, Heidelberg (2013)
13. Coutinho, F.L., Morimoto, C.H.: Improving head movement tolerance of cross-ratio based eye trackers. IJCV **101**(3), 459–481 (2013)
14. Devillers, L., Tahon, M., Sehili, M., Delaborde, A.: Inference of human beings' emotional states from speech in human-robot interactions. Int. J. Soc. Robot. **7**, 1–13 (2015)
15. Doering, N., Poeschl, S., Gross, H.M., Bley, A., Martin, C., Boehme, H.J.: User-centered design and evaluation of a mobile shopping robot. Int. J. Soc. Robot. **7**(2), 203–225 (2015)
16. Dragan, A., Srinivasa, S.: Formalizing assistive teleoperation. In: Robotics: Science and Systems (2012)
17. Feil-Seifer, D., Mataric, M.J.: Human robot interaction. In: Encyclopedia of Complexity and Systems Science (2009)
18. Ferreira, J., Dias, J.: Attentional mechanisms for socially interactive robots - a survey. IEEE Trans. Auton. Ment. Dev. **6**(2), 110–125 (2014)
19. Fischinger, D., Einramhof, P., Papoutsakis, K., Wohlkinger, W., Mayer, P., Panek, P., Hofmann, S., Koertner, T., Weiss, A., Argyros, A., Vincze, M.: Hobbit, a care robot supporting independent living at home: first prototype and lessons learned. In: Robotics and Autonomous Systems (2014). In Press
20. Fong, T., Nourbakhsh, I., Dautenhahn, K.: A survey of socially interactive robots. Robot. Auton. Syst. **42**(3–4), 143–166 (2003)
21. Goodrich, M.A., Schultz, A.C.: Human-robot interaction: a survey. Found. Trends Hum. Comput. Inter. **1**(3), 203–275 (2007)

22. Grigore, E., Eder, K., Pipe, A., Melhuish, C., Leonards, U.: Joint action understanding improves robot-to-human object handover. In: IEEE/RSJ IROS (2013)
23. Han, J., Gilmartin, E., Campbell, N.: Herme, yet another interactive conversational robot. In: Humaine Association Conference on Affective Computing and Intelligent Interaction (2013)
24. Hirose, M., Ogawa, K.: Honda humanoid robots development. Philos. Trans. Royal Soc. Lond. A: Math. Phys. Eng. Sci. **365**(1850), 11–19 (2007)
25. Hoai, M., De la Torre, F.: Max-margin early event detectors. IJCV **107**(2), 191–202 (2014)
26. Jabon, M., Bailenson, J., Pontikakis, E., Takayama, L., Nass, C.: Facial expression analysis for predicting unsafe driving behavior. IEEE Pervasive Comput. **10**(4), 84–95 (2011)
27. Jones, C., Deeming, A.: Affective human-robotic interaction. In: Peter, C., Beale, R. (eds.) Affect and Emotion in HCI. LNCS, vol. 4868, pp. 175–185. Springer, Heidelberg (2008)
28. Kaefer, P., Germino, K., Venske, D., Williams, A.: Computational awareness in a tactile-responsive humanoid robot comedian. In: IEEE International Conference on Systems, Man, and Cybernetics (2013)
29. Karg, M., Kirsch, A.: Acquisition and use of transferable, spatio-temporal plan representations for human-robot interaction. In: IEEE/RSJ IROS (2012)
30. Keizer, S., Kastoris, P., Foster, M., Deshmukh, A., Lemon, O.: Evaluating a social multi-user interaction model using a Nao robot. In: IEEE RO-MAN (2014)
31. Li, K., Xu, F., Wang, J., Dai, Q., Liu, Y.: A data-driven approach for facial expression synthesis in video. In: IEEE CVPR (2012)
32. Littlewort, G.C., Bartlett, M.S., Fasel, I.R., Chenu, J., Kanda, T., Ishiguro, H., Movellan, J.R.: Towards social robots: automatic evaluation of human-robot interaction by face detection and expression classification. In: NIPS (2004)
33. Liu, C., Ham, J., Postma, E., Midden, C., Joosten, B., Goudbeek, M.: Representing affective facial expressions for robots and embodied conversational agents by facial landmarks. Int. J. Soc. Robot. **5**(4), 619–626 (2013)
34. Liu, H., Sun, F.: Semi-supervised particle filter for visual tracking. In: IEEE ICRA (2009)
35. Malmir, M., Forster, D., Youngstrom, K., Morrison, L., Movellan, J.: Home alone: Social robots for digital ethnography of toddler behavior. In: IEEE ICCVW (2013)
36. McColl, D., Zhang, Z., Nejat, G.: Human body pose interpretation and classification for social human-robot interaction. Int. J. Soc. Robot. **3**(3), 313–332 (2011)
37. McKeague, S., Liu, J., Yang, G.-Z.: An asynchronous RGB-D sensor fusion framework using monte-carlo methods for hand tracking on a mobile robot in crowded environments. In: Herrmann, G., Pearson, M.J., Lenz, A., Bremner, P., Spiers, A., Leonards, U. (eds.) ICSR 2013. LNCS, vol. 8239, pp. 491–500. Springer, Heidelberg (2013)
38. Menna, M., Gianni, M., Pirri, F.: Learning the dynamic process of inhibition and task switching in robotics cognitive control. In: ICMLA (2013)
39. Michel, D., Papoutsakis, K., Argyros, A.A.: Gesture recognition supporting the interaction of humans with socially assistive robots. In: Bebis, G., Boyle, R., Parvin, B., Koracin, D., McMahan, R., Jerald, J., Zhang, H., Drucker, S.M., Kambhamettu, C., El Choubassi, M., Deng, Z., Carlson, M. (eds.) ISVC 2014, Part I. LNCS, vol. 8887, pp. 793–804. Springer, Heidelberg (2014)
40. Nickel, K., Stiefelhagen, R.: Visual recognition of pointing gestures for human-robot interaction. Image Vis. Comput. **25**(12), 1875–1884 (2007)

41. Niculescu, A., van Dijk, B., Nijholt, A., Li, H., See, S.: Making social robots more attractive: the effects of voice pitch, humor and empathy. Int. J. Soc. Robot. **5**(2), 171–191 (2013)
42. Novikova, J., Watts, L.: Towards artificial emotions to assist social coordination in HRI. Int. J. Soc. Robot. **7**(1), 77–88 (2015)
43. Park, C., Lee, S.: Real-time 3D pointing gesture recognition for mobile robots with cascade HMM and particle filter. Image Vis. Comput. **29**(1), 51–63 (2011)
44. Park, I., Kim, J., Lee, J., Oh, J.: Mechanical design of humanoid robot platform KHR-3 (KAIST Humanoid Robot 3: HUBO). In: HUMANOIDS (2005)
45. Silvera-Tawil, D., Rye, D., Velonaki, M.: Interpretation of social touch on an artificial arm covered with an EIT-based sensitive skin. Int. J. Soc. Robot. **6**(4), 489–505 (2014)
46. Silvera-Tawil, D., Rye, D., Velonaki, M.: Artificial skin and tactile sensing for socially interactive robots: a review. Robot. Auton. Syst. **63**(3), 230–243 (2015)
47. Srinivasa, S., Ferguson, D., Helfrich, C., Berenson, D., Collet, A., Diankov, R., Gallagher, G., Hollinger, G., Kuffner, J., Weghe, M.V.: HERB: a home exploring robotic butler. Auton. Robot. **28**(1), 5–20 (2010)
48. Stiehl, W., Lieberman, J., Breazeal, C., Basel, L., Lalla, L., Wolf, M.: Design of a therapeutic robotic companion for relational, affective touch. In: IEEE RO-MAN (2005)
49. Strabala, K., Lee, M.K., Dragan, A., Forlizzi, J., Srinivasa, S.S.: Towards seamless human-robot handovers. J. Hum. Robot. Inter. **1**(1), 1–23 (2013)
50. Trovato, G., Kishi, T., Endo, N., Zecca, M., Hashimoto, K., Takanishi, A.: Cross-cultural perspectives on emotion expressive humanoid robotic head: recognition of facial expressions and symbols. Int. J. Soc. Robot. **5**(4), 515–527 (2013)
51. Tsui, K., Flynn, K., McHugh, A., Yanco, H., Kontak, D.: Designing speech-based interfaces for telepresence robots for people with disabilities. In: ICORR (2013)
52. Viciana-Abad, R., Marfil, R., Perez-Lorenzo, J.M., Bandera, J.P., Romero-Garces, A., Reche-Lopez, P.: Audio-visual perception system for a humanoid robotic head. Sens. **14**(6), 9522–9545 (2014)
53. Viola, P., Jones, M.: Rapid object detection using a boosted cascade of simple features. In: IEEE CVPR (2001)
54. Wang, Y., Cheng, H., Zheng, Y., Yang, L.: Face recognition in the wild by mining frequent feature itemset. In: Li, S., Liu, C., Wang, Y. (eds.) CCPR 2014, Part II. CCIS, vol. 484, pp. 331–340. Springer, Heidelberg (2014)
55. Wang, Z., Ambrus, R., Jensfelt, P., Folkesson, J.: Modeling motion patterns of dynamic objects by IOHMM. In: IEEE/RSJ IROS (2014)
56. Xiao, Y., Zhang, Z., Beck, A., Yuan, J., Thalmann, D.: Human-robot interaction by understanding upper body gestures. Presence **23**(2), 133–154 (2014)
57. Yan, H., Ang, M.H.J., Poo, A.N.: A survey on perception methods for human-robot interaction in social robots. Int. J. Soc. Robot. **6**(1), 85–119 (2014)
58. Yanik, P., Manganelli, J., Merino, J., Threatt, A., Brooks, J., Green, K., Walker, I.: Use of kinect depth data and growing neural gas for gesture based robot control. In: PervasiveHealth (2012)
59. Yin, X., Xie, M.: Finger identification and hand posture recognition for human-robot interaction. Image Vis. Comput. **25**(8), 1291–1300 (2007)
60. Zaraki, A., Mazzei, D., Giuliani, M., De Rossi, D.: Designing and evaluating a social gaze-control system for a humanoid robot. IEEE Trans. Hum. Mach. Syst. **44**(2), 157–168 (2014)

Recognizing 3D Continuous Letter Trajectory Gesture Using Dynamic Time Warping

Jingren Tang, Hong Cheng[✉], and Lu Yang

Center for Robotics, School of Automation Engineering,
University of Electronic Science and Technology of China, Chengdu, China
hcheng@uestc.edu.cn

Abstract. Letter trajectory gesture recognition is widely used in Human Computer Interaction. Many approaches for letter trajectory gesture recognition have been proposed in the past several years. Most of the traditional approaches detect letters based on the beginning/end points provided by the user. It causes low writing speed and uncomfortable writing experience. Moreover, traditional Dynamic Time Warping cannot classify the letters which have the familiar trajectory. In this paper, we combine Dynamic Time Warping with structured points of letters to overcome those problems. The main contribution of this paper is that we introduce the structured points information of letters in Time Warping process to detect letters from hand trajectories. Based on this, we can successfully recognize the letter from the weak inter-class feature and the continuous trajectory without beginning point and end point given by the user. Furthermore, we can handle the self-contained trajectory based on the complexity of letters. We evaluate this system in our gesture dataset, and it shows that the proposed approach can significantly outperform the traditional begin-end gesture approach.

Keywords: Letter trajectory gestures · Dynamic time warping · Human computer interaction

1 Introduction

Vision-Based 3D gesture interaction approach has drawn much attention in recent years thanks to the emerging techniques of 3D sensors [1,2,13]. It is a natural and efficient way of human computer interaction (HCI). Moreover, it provides an attractive, user-friendly alternative that using an interface device (keyboard, mouse and other controller)[8] without physical contact.

The trajectory gesture is one kind of most important gestures. In this work, we use this information to build a system and detect letter from hand motion trajectory. The Microsoft Kinect is a 3D sensor which is wildly used now. Also we use it to get RGB data and depth data. The information of beginning point and end point in trajectory is very important to detect letter. In traditional approach, those points are marked by user, which cause writing speed slow and uncomfortable experience. Bhuyan *et al.* [8] proposed a novel continuous hand

© Springer International Publishing Switzerland 2015
Y.-S. Ho et al. (Eds.): PCM 2015, Part II, LNCS 9315, pp. 191–200, 2015.
DOI: 10.1007/978-3-319-24078-7_19

gesture recognition approach by using new features including writing speed. They assumed that the writing speed will slow down at beginning point and end point. According to this assumption, their system will not work if the writing speed of user remains constant. So, the same issue occurred as bad experience and poor efficiency. Furthermore, how to classify the letter from continuous trajectory is a difficult issue. We use normalized vector from different frame as the feature to classify letter at first. However, lots of vector features are similar. For example the vector feature of letter b and letter p are similar and it is hard to distinguish them with this approach.

In this paper, we propose a novel approach for 3D continuous letter trajectory gesture recognition without writing speed restriction and marked points. Series approaches have been proposed for gesture recognition such as Dynamic Time Warping (DTW) [9], Hidden Markov Model (HMM) [3], Finite State Machines (FSM) [10], *etc.* We use the improved Dynamic Time Warping to recognize letters from continuous trajectory for it's high accuracy and easy to be trained with few samples. Also, lots of improved DTW have been proposed such as Multidimensional dynamic time warping (MD-DTW) [4], memory efficient Dynamic Time Warping (MES-DTW) [12] *etc.* The main contribution of this paper is that we recognize letter from continuous trajectory by combining structured points with Dynamic Time Warping algorithm, which uses a natural way to recognize letter without low speed restriction. Furthermore, we handle the self-contained issue between letters which is based on the complexity of letters. This approach is evaluated on new data set and the experiment results show good performance.

The rest of the paper is organized as follows: Section 2 introduces the state of art of trajectory gesture recognition. Section 3 presents the detail of DTW with structured points approach and the solution of self-contained issue. Section 4 designs the experiment and shows the results. Section 5 gives the conclusions of this paper.

2 Related Work

The phases of continuous trajectory gesture recognition include hand location, hand tracking and extracting, classification. In this paper, we mainly concentrate on classification.

The approaches to improve the performance of classification can be carried out in two ways. The first is to improve the classifier. Kristensson *et al.* proposed a approach by using probabilistic algorithm to incrementally predict users intended gestures [11]. Though it has high accuracy, they use a zoning technique which means they detect the distance between user and Kinect. Once the distance below a threshold they define the input zone. The beginning point can be detected easily by the input zone. And they use two hands to select the gesture from some similar results. The approach mentioned above restrict the users hands in the input zone and user has to select gesture from results, for those reasons the writing speed is limited and the user gets uncomfortable experience. Cheng *et al.* proposed Windowed Dynamic Time Warping (WDTW) to classify trajectory gesture [6]. They clustered general gestures into a set of strokes then use the parameterized searching window to recognize the gesture. However, the length of the window cannot be find by a certain process or formula. Lichtenauer *et al.* propose a approach to recognize sign language by combining statistical DTW and independent classification [7], they separated the time warping and classification to satisfy conflicting similar modeling demands, by doing so, the features which without distinction can be abandoned to simplify calculation and enhance robustness.

The second way is to improve features. Bhuyan *et al.* proposed a novel set of features for continuous hand gesture recognition [8], they use the velocity of hand motion trajectory as the new feature and use it to detect begin-end point. This feature also works for distinguishing intentional movements from unintentional movements. Though it is an effective feature to classify trajectory, it assumed that the velocity of hand motion would be decrease when the user beginning and finish writing. So the restriction of velocity slows down the writing speed.

3 The Proposed Continuous Letter Trajectory Recognition System

We use the 3D camera (Microsoft Kinect) to locate the hand center point, then we get the hand motion trajectory. The system support user write in air and then give the output. Once the trajectory have been gained, we use the motion vector from different frame as feature. The scale of letter is unfixed for the variant distance between user and camera, we use motion vector and then normalize it, the normalized vector is calculated as

$$n = \frac{(x_t - x_{t-1}, y_t - y_{t-1})}{\|(x_t - x_{t-1}, y_t - y_{t-1})\|_2}. \tag{1}$$

where x_t, y_t are the points in current frame and x_{t-1}, y_{t-1} are the points in last frame. Subtract x_{t-1}, y_{t-1} from x_t, y_t we can get the motion vector, and then n has been calculated by normalize the vector. In fact, n can be shown as

$$n = (\cos \theta, \sin \theta). \tag{2}$$

So, the feature reflects angle between new hand point and last hand point as shown in Fig. 1(a) and the hand trajectory as shown in Fig. 1(b).

3.1 Traditional Dynamic Time Warping Algorithm

Dynamic Time Warping algorithm is wildly used as a matching algorithm for it is easy to be trained and high accurate. With those advantages, lots of improved Dynamic Time Warping approaches have been proposed, the traditional DTW and improved DTW will be introduced. And we will give more details about DTW with structured points in this section.

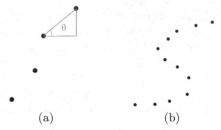

(a)   (b)

Fig. 1. An illustration of a letter trajectory gesture: (a) The reflection of a point angle; (b) Letter 's' trajectory gesture

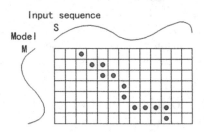

Fig. 2. The DTW algorithm

Assume that the trajectory model $M = \{m_1, m_2...m_n\}$ and the trajectory segment $S = \{s_1, s_2...s_m\}$ in long input stream, as shown in Fig. 2. The similarity of model vector and input vector should be calculated by similarity measure. We use the Euclidean distance to measure the similarity. Then we can get the similarity matrix $G_{n\times\infty}$. To find the optimized path the restrictions in DTW algorithm should be followed which means the next point $G(i, j)$ in path should be selected from neighbour points $G(i - 1, j), G(i, j - 1)$ or $G(i - 1, j - 1)$. This restriction simplify the algorithm and make it more reasonable, the final similarity is calculated by

$$\omega(P_{(i,j)}) = d(P_{(i,j)}) + \min(\omega(P_{(i-1,j-1)}), \omega(P_{(i-1,j)}), \omega(P_{(i,j-1)})). \qquad (3)$$

Where $P_{(i,j)}$ is the location in similarity matrix $G_{n\times\infty}$, and $d(P_{(i,j)})$ is the Euclidean distance at (i, j) , and $\omega(P_{(i,j)})$ is accumulated Euclidean distance.

Once we detect the last row value which is smaller than threshold in the similarity matrix $G_{n\times\infty}$, the gesture segment in input stream match with the model trajectory. Thats means we can detect M from S while

$$\omega(P_{(n,j)}) < \alpha, \qquad j \in [0, \infty), \qquad (4)$$

α is the threshold which is subject to different gestures, and it can be learned by using the leave-one-out cross validation strategy [9]. Now, we detect the same segment between model trajectory and input sequence, and this is the typical dynamic time warping algorithm.

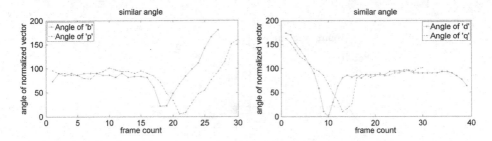

Fig. 3. Similar vectors

3.2 Dynamic Time Warping with Structured Points

In this subsection, we will give specific detail about the novel approach. As for features of trajectory, motion vector is the simple, visualized and efficient one. So, lots of approaches use this feature for it's advantages. To overcome the influence of different velocity, the motion vectors should be normalized. However, the issues occured while we use the normalized motion vectors. The motion vectors are similar of some letters which are hard to classify such as letter b and letter p, shown as in Fig. 3. To improve the performance of classification, we need find more information in DTW process. The DTW with structured points framework is shown as Fig. 4. The red points are beginning point and end point, the green points are turn points. All of them are points on the optimal path. The number of turn points is unconstant, it's depends on the structure of letter.

Next, the detail of how to find the structured points which include the beginning point, end point and turn point will be provided. We detect the similar trajectory while the final cost of optimal path below the threshold value, we find the end point which is the last point of path at the same time. To find the beginning point, we must record the direction of every point which means that we need record the next point of (i, j) is $(i + 1, j)$, $(i, j + 1)$ or $(i + 1, j + 1)$. After we detected the end point, we can find the beginning point by backtracking which use the direction data.

To find the turn point, we use the formula

$$\theta_t = \arccos(\delta_{x(t)}) \pm \arccos(\delta_{x(t-1)}) \tag{5}$$

to calculate the θ_t which is the angle between two vector. $\delta_{x(t)}$ is the normalized motion vector in x axis. To get the index of turn point in the cost matrix, every point of it should be calculated. θs of different letters are shown as Fig. 5. Then, we can detect the turn point only if

$$(\theta_i - \theta_{i-1}) * (\theta_i - \theta_{i+1}) \leq \tau, \tag{6}$$

where τ is threshold to detect the turn point. Now, we find the structured points of letters, the sample of them is shown as Fig. 6.

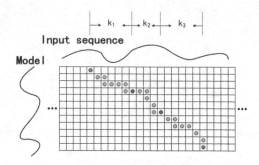

Fig. 4. The framework of DTW with structured points approach

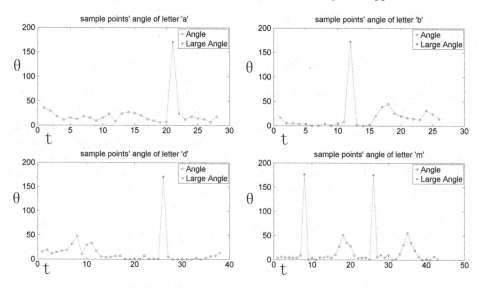

Fig. 5. Θs of different letters

As mentioned above, we can use the structured points to classify the trajectories which have the similar vector. The unit of cost matrix should include the distance data which can express as $(\delta_{x(t)}, \delta_{y(t)}; l(t))$. Define k_i as the distance structured points which is shown in Fig. 4. Once we detect the structured point, we calculate k_i using

$$k_i = (\sum_{t=l}^{m} \delta_{(x(t) \times l(t))}, \sum_{t=l}^{m} \delta_{(y(t) \times l(t))}). \tag{7}$$

l is the index of latest calculated structured point, m is the index of latest structured point, in this way we can reduce the calculation. Assume that the distant data of model is k_i', then we can calculate the similarity of them $||k_i' - k_i||_2$. After getting the similarity, we change the cost value of optimal path according the similarity. Now we can classify the letters which have similar vectors. Actually,

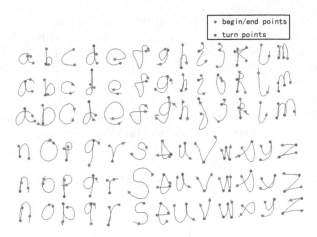

Fig. 6. Structured points of 26 letters

this approach use the relative location of points in trajectories to improve the performance of classifier.

3.3   Determine the Output Letter

Though we have handled the weak inter-class feature issue, there is another problem exist. Think about all letters, we will raise the question: the output is letter d or c while the input trajectory is d. Obviously, letter c contain with letter d, as shown in Fig. 7. The red is the common part between letter d and c.

m b d h w

Fig. 7. Self-Contained issue

One approach to handle this problem is that using the speed of movements to locate the beginning/end point and extract intentional movements [8]. However, this approach restrict writing speed and it does not work if all states of trajectory are same which including speed, location, acceleration, depth and so on. In this case we propose a rule-based approach which determine the output letter by letters' complexity for isolated letter detection. Actually, we find that the complex letters always contain with the simple letters, so, we will choose the more complex letter in output buffer as the system output. Moreover, we should discriminate whether two letters in the input sequence is contained or not. For example, the input buffer is c, d, c while user write d and c, the first c in the buffer is the contain part and another c is isolate letter which should be output.So, we record the location of each point in the frame, then, the sample points location of letter in the buffer should be compared with each other, if one letter contain with another, they share the same location data in common part.

Finally, we find out whether the letter in the output buffer contain with each other, then we will choose the more complex letter as the output if one letter contain with another. We can know that the segment 1 have multi-outputs. Also, we can confirm that the optimal path of contained letters in similarity matrix are similar. So, whole process in Dynamic Time Warping as shown in Fig. 8.

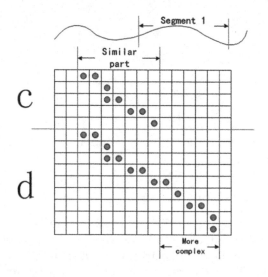

Fig. 8. An illustration of determining the output letter

4   Experimental Results and Analysis

Dataset: We have designed a new data set of 26 lowercase letters trajectory using Kinect devices. We record both RGB and depth clips, and get the hand trajectory points by NITE at the same time. Each letter is performed 10 times continuously by 5 volunteers. There are 1300 samples in total. Setup: we used 26 samples for training and the rest data for testing. The results are obtained by averaging 10 different trials to evaluate the performance of our approach. Assume that each letter is performed times and then we use correct detection rate as follow:

$$cRate = \frac{\sum_{i=1}^{m} C_i^2}{\sum_{i=1}^{m}(O_i + NC_i)}. \tag{8}$$

Where C_i means correct match and O_i is the number of all outputs, NC_i refer to the outputs without correct letter, in addition, the value of C_i and NC_i can

be only 1 or 0. Then we average of 10 different trials to obtain the final results.
Results:

Table 1. Using DTW with motion vector only

Char	cRate	Char	cRate	Char	cRate	Char	cRate	Char	cRate
a	40.4 %	g	28.5 %	m	24.8 %	s	53.4 %	y	88.4 %
b	32.5 %	h	32.4 %	n	51.9 %	t	51.0 %	z	67.2 %
c	81.0 %	i	72.1 %	o	92.7 %	u	77.6 %		
d	45.5 %	j	50.4 %	p	30.7 %	v	83.4 %		
e	50.4 %	k	15.2 %	q	54.7 %	w	78.1 %		
f	64.8 %	l	93.6 %	r	75.3 %	x	69.5 %		

Table 2. Our approach

Char	cRate	Char	cRate	Char	cRate	Char	cRate	Char	cRate
a	83.5 %	g	61.4 %	m	53.8 %	s	83.9 %	y	90.9 %
b	81.3 %	h	87.4 %	n	90.4 %	t	78.5 %	z	66.5 %
c	81.6 %	i	84.5 %	o	93.4 %	u	78.5 %		
d	52.9 %	j	56.9 %	p	66.7 %	v	92.2 %		
e	86.8 %	k	53.8 %	q	63.8 %	w	91.7 %		
f	63.7 %	l	94.0 %	r	91.3 %	x	81.8 %		

We can conclude from the results, the performance of system which combining DTW with structured points of letters and have the self-contained solution
is better than the system which using DTW and motion vector only. Note that
the recognition rate of some letters in Table 1 is extremely low, because the letters contain with lots of other letters and the system cannot separate them. In
addition, writing habits differ from person to person, thus causing some letters
are hard to be recognized (Table 2).

5 Conclusions

One critical issue in continuous gesture recognition research is that how to find
the effective approach to get the correct classification. And another issue is that
the letter usually contain with each other. In this paper, we combine Dynamic
Time Warping with structured points of letters to get the correct classification
for 3D continuous hand trajectory gesture recognition. Moreover, we propose the
novel approach to overcome the self-contained problem between letters which

use the complexity of letters. The evaluation shows that the approach improves performance compares with classical DTW.

Acknowledgment. This work was partially supported by NSFC (No.61305033, 61273256), Fundamental Research Funds for the Central Universities (ZYGX2013J088, ZYGX2014 Z009) and SRF for ROCS, SEM.

References

1. Chaudhary, A., Raheja, J.L., Das, K.: Intelligent approaches to interact with machines using hand gesture recognition in natural way: a survey. arXiv:1303.2292 (2013)
2. Kurakin, A., Zhang, Z., Liu, Z.: A real time system for dynamic hand gesture recognition with a depth sensor. In: EUSIPCO. IEEE (2012)
3. Gehrig, D., Kuehne, H., Woerner, A.: Hmm-based human motion recognition with optical flow data. In: HR. IEEE (2009)
4. Ten Holt, G.A., Reinders, M.J.T., Hendriks, E.A.: Multi-dimensional dynamic time warping for gesture recognition (2007)
5. Cheng, H., Zhongjun, D., Liu, Z.: Image-to-class dynamic time warping for 3D hand gesture recognition. In: ICME. IEEE (2013)
6. Cheng, H., Luo, J., Chen, X.: A windowed dynamic time warping approach for 3D continuous hand gesture recognition. In: ICME. IEEE (2014)
7. Lichtenauer, J.F., Hendriks, E., Reinders, M.J.T.: Sign language recognition by combining statistical DTW and independent classification. Pattern Anal. Mach. Intell. **30**(11), 2040–2046 (2008)
8. Bhuyan, M.K., Kumar, D.A., MacDorman, K.F.: A novel set of features for continuous hand gesture recognition. J. Multimodal User Interfaces **8**(4), 333–343 (2014)
9. Reyes, M., Dominguez, G., Escalera, S.: Featureweighting in dynamic timewarping for gesture recognition in depth data. In: ICCV Workshops. IEEE (2011)
10. Hong, P., Turk, M., Huang, T.S.: Gesture modeling and recognition using finite state machines. In: AFGR. IEEE (2000)
11. Kristensson, P.O., Nicholson, T., Quigley, A.: Continuous recognition of one-handed and two-handed gestures using 3D full-body motion tracking sensors. In: IUI. ACM (2012)
12. Anguera, X., Ferrarons, M.: Memory efficient subsequence DTW for query-by-example spoken term detection. In: ICME. IEEE (2013)
13. Ren, Z., Yuan, J., Meng, J.: Robust part-based hand gesture recognition using kinect sensor. Multimedia **15**(5), 1110–1120 (2013)

Rapid 3D Face Modeling from Video

Hong Song[1(✉)], Jie Lv[2], and Yanming Wang[1]

[1] School of Software, Beijing Institute of Technology, Beijing 100081, China
anniesun@bit.edu.cn
[2] School of Computer Science, Beijing Institute of Technology,
Beijing 100081, China

Abstract. In this paper, an efficient technique is developed to construct textured 3D face model from video containing a face rotating from frontal to profile. After two manual clicks on a profile to tell the system where the eye corner and bottom of the chin are, the system automatically generates a realistic looking 3D face model. The proposed method consists of three components. Firstly, based on the facial feature points extracted from frontal and profile images, an individual 3D geometric face model is generated by deforming the generic model with improved Radial basis function. Then the model is refined by using improved $\sqrt{3}$-Subdivision. Secondly, the multi-resolution technique and weighted smoothing algorithm are combined to synthesize individual facial texture image. Finally, a realistic 3D face model is built by mapping the individual texture to the individual 3D geometric model. The accuracy and robustness of the method are demonstrated with a set of experiments.

Keywords: Face modeling · Generic model · Multi-resolution · Texture synthesis · Subdivision

1 Introduction

3D face modeling has extensive application prospects in teleconference, human computer interaction (HCI), multimedia entertainment, and face recognition etc. It is a challenging task to generate realistic 3D face model. Since the pioneering work of Parke in 1972 [1], many researchers have obtained many important results in the next forty years. According to the input source data, face modeling technique can be roughly divided into the methods based on 3D scanning and those based on images.

The method based on 3D scanning is that the 3D information is directly captured from 3D digital scanners, cameras and other hardware devices. Guenter et al. developed techniques to clean up and register data generated from laser scanners [2]. This method can accurately generate 3D face models, but it requires special high-cost hardware and a powerful workstation.

The image-based modeling is roughly divided into methods those are based on a single image, two images and video.

Park et al. proposed a method to reconstruct 3D face model and estimate 3D pose using Active Appearance Model (AAM), expectation maximization and 3D Morphable Model (3DMM) [3], which is time-consuming. The method proposed by Fan et al. is

© Springer International Publishing Switzerland 2015
Y.-S. Ho et al. (Eds.): PCM 2015, Part II, LNCS 9315, pp. 201–211, 2015.
DOI: 10.1007/978-3-319-24078-7_20

suitable for real-time and mobile applications [4]. But the texture mapping method is not accurate and the face model is not smooth.

Lin et al. also proposed a modeling method based on orthogonal images [5]. The feature points on the profile are detected automatically by using LMCT (Local Maximum-Curvature Tracking). It selects feature points with local maximum absolute curvature value. However, LMCT is designed to calculate concave and convex points, which works very well only for Mongoloid looking people. Beeler et al. proposed a passive stereo vision system that computes the 3D geometry of the face [6]. But the intrinsic camera parameters need to be calibrated and pixel matching need to be done pair wise very accurately.

Xin et al. proposed a method for automatic 3D face modeling from video [7]. Based on image sequence segmentation, they estimated 2D feature points, head poses and underlying 3D face shape with a morphable model.

In this paper, a method is proposed for rapid 3D face modeling from video containing a face rotating from frontal to profile. It is assumed that the neck is regarded as the axis of rotation. There are three key contributions in this paper. Firstly, improved frame difference method and Canny algorithm are combined to extract profile facial feature points. Secondly, Sub-regional Multi-step RBF method is applied to deform the generic face model. Finally, improved $\sqrt{3}$-Subdivision algorithm is used to make the model smoother.

2 System Overview

In this paper, the proposed individual 3D face modeling system framework mainly consists of three parts, as shown in Fig. 1, including generating individual 3D geometric face model, synthesizing individual facial texture image, and texture mapping.

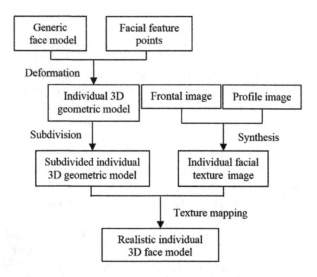

Fig. 1. The block diagram of the proposed algorithm

3   Generating Individual 3D Geometric Face Model

3.1   Extracting 2D Facial Feature Points

Extracting Frontal Facial Feature Points. In this paper, Active Shape Model (ASM) is used to extract 68 facial feature points (eyes, nose, mouth, and facial silhouette, etc.) on the frontal image [5]. The training data sets are provided by Song [8]. The extracted feature points are shown in Fig. 2(a).

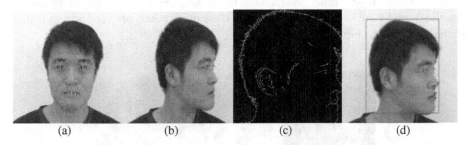

$\quad$ (a) $\qquad\qquad$ (b) $\qquad\qquad$ (c) $\qquad\qquad$ (d)

Fig. 2. (a) Extracted 68 frontal facial feature points (b) Profile face (c) Extracted profile facial silhouette (d) Profile facial feature points (two manually labeled points indicated by green dots, five automatically detected points indicated by red dots, face region indicated by red rectangle) (Color figure online)

Extracting Profile Facial Feature Points. The depth information of the face is got by extracting facial feature points on the profile image (see Fig. 2(b)). The detail of the process to extract feature points is described in the following.

$\quad$ Firstly, improved frame difference method that fuses the last few frame difference results is used to detect profile face region [9]. Then, Canny algorithm is applied to extract a facial silhouette on the profile face region, as shown in Fig. 2(c). Finally, the other 5 points on the profile face are extracted automatically based on two manually labeled points and profile facial silhouette. Since height of the frontal and profile face is not necessarily same, normalization based on two manually labeled points is a necessary step. In this paper, the profile face is aligned to frontal face by scaling transformation. The normalized coordinates of eye corner on the frontal and profile face are denoted by $P_1(x_1, y_1)$ and $P_2(x_2, y_2)$ respectively. The normalized coordinates of nose tip on the frontal and profile face are denoted by $F_1(x_3, y_3)$ and $F_2(x_4, y_4)$ respectively. P_1, P_2 and F_1 are known, so

$$y_4 = y_2 + |y_3 - y_1| \tag{1}$$

where $x_i(i = 1, 2, 3, 4)$ are horizontal coordinates, and $y_i(i = 1, 2, 3, 4)$ are vertical coordinates.

$\quad$ It is assumed that the face turns to the left. After scanning the profile silhouette (see Fig. 2(c)) horizontally on the height of y_4, then x_4 is horizontal coordinate of leftmost intersection. Other points on the profile face silhouette can be got in the same way. The result is shown in Fig. 2 (d).

3.2 Deforming Generic Face Model

Generic Face Model. To decrease the complexity of modeling, in this paper, Candide-3 model is used as a generic model [10]. While Candide-3 model only includes 113 points and 168 surfaces, subdivision algorithm (see Sect. 3.2.3) is applied to increase the number of points and surfaces, which makes the 3D face model more realistic.

The points on Candide-3 model are compared with frontal facial feature points shown in Fig. 2(a) by using orthogonal projection method, there are 44 points as their correspondences (eyes, nose, mouth and facial silhouette etc.). The 44 points on Candide-3 model are used as feature points, while other points are used as non-feature points, as shown in Fig. 3(a).

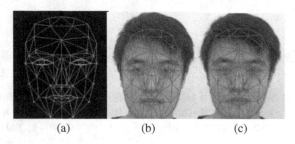

      (a)   (b)   (c)

Fig. 3. (a) Candide-3 face model (feature points on Candide-3 model indicated by red dots) (b) Result of deformation on the plane XOY with traditional RBF (c) Result of deformation on the plane XOY with improved RBF (Color figure online)

Deformation. Radial Basis Function (RBF) (see Eq. (2)) is a multivariate interpolation function of scattered data [11]. In the process of deformation, solving c, M and t is a crucial step based on the 3D location of feature points. Then 3D location of non-feature points is determined by Eq. (2). In this paper, the Gaussian $g(r) = \exp(-\frac{r^2}{\sigma^2})$ is used as the kernel function.

$$f(x) = \sum_{i=1}^{n} c_i g(\|x - x_i\|) + Mx + t, x \in R^d \tag{2}$$

where c_i is weight, $g(\|.\|)$ is a kernel function, x is a scattered data set, x_i is center of kernel function, M is a coefficient matrix, and t is a constant matrix.

The points on the face model are distributed unevenly. Some parts (eyes, nose, and mouth) are dense, while other parts are sparse. Furthermore, the number of points on the frontal face is more than that on the profile face. In this paper, Sub-regional Multi-stepRBFmethod is applied to deform the generic model [12]. Specific algorithm of deformation is as follows:

Step 1 Deformation on the plane XOY:

Firstly, the unit between Candide-3 model and face images is normalized. Secondly, 2D frontal information of Candide-3 model is obtained by using orthogonal projection method. Then the bottom of the chin on Candide-3 model and frontal face is selected as a reference point. The difference $(\varDelta x, \varDelta y)$ of feature points between Candide-3 model and frontal face are calculated on the plane XOY. Finally, the whole face is divided into different areas (eyes, nose, and mouth etc.). The RBF method is applied to these areas respectively. The results of deformation by using traditional and improved RBF methods are shown in Fig. 3(b) (c) respectively. As shown in Fig. 3(b) (c), The latter covers the face more completely than the former, especially on both sides of skull.

Step 2 Deformation on the axis Z.

Deformation method on the axis Z is similar with that on the plane XOY. The RBF method is applied to the whole face on the axis Z.

Figure 4 shows an individual 3D geometric face model in different views. Material, color and lighting are also added on the models, as shown in Fig. 5.

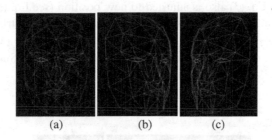

(a)   (b)   (c)

Fig. 4. Individual 3D geometric face model in different views

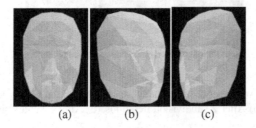

(a)   (b)   (c)

Fig. 5. Individual 3D geometric face model in different views with material color and lighting

Subdivision. Due to sparse facial structure, the slower increase of the mesh complexity and the suitability for adaptive refinement, $\sqrt{3}$-Subdivision is applied to refine the individual 3D geometric face model (see Fig. 4) in this paper [13]. $\sqrt{3}$-Subdivision consists of two smoothing rules shown in Eqs. (3) and (4), one for the placement of the newly inserted vertices and the other for the relaxation of the old ones.

$$V_F = \frac{1}{3}(V_0 + V_1 + V_2) \tag{3}$$

where V_0, V_1 and V_2 are three vertices of a triangle, and the new vertex V_F is simply inserted at the center of the triangle.

$$V_v = (1 - \alpha_n)V + \frac{\alpha_n}{n}\sum_{i=0}^{i-1}V_i \tag{4}$$

where $\alpha_n = \frac{1}{9}(4 - 2\cos\frac{2\pi}{n})$, $V_0, V_1, \ldots, V_{n-1}$ are directly adjacent neighbors of vertex V in the unrefined mesh.

The result of subdivision is shown in Fig. 6. After 3 subdivision iterations, although the model has 2505 vertices and 4968 smooth surfaces, its volume becomes smaller than the original model (see Fig. 4), especially in nose and mouth etc. The improved algorithm for $\sqrt{3}$-Subdivision proposed in this paper is inspired by [14]. The vertex V_v shown in Eq. (4) after several subdivision iterations is adjusted to original position. The vertex V_F shown in Eq. (3) also is adjusted to new position (see Eq. (5)), which makes the mesh keep original shape. It is assumed that three vertices of a triangle for original model are denoted by (V_0, V_1, V_2) respectively, they are denoted by (V_0', V_1', V_2') after 1 subdivision iteration respectively. The result of using improved $\sqrt{3}$-Subdivision method is shown in Fig. 7.

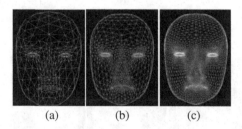

(a)   (b)   (c)

Fig. 6. Individual 3D geometric face model using $\sqrt{3}$ -Subdivision method (a) After 1 subdivision iteration (b) After 2 subdivision iterations (c) After 3 subdivision iterations

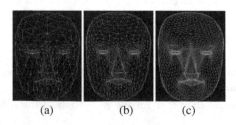

(a)   (b)   (c)

Fig. 7. Individual 3D geometric face model using improved $\sqrt{3}$-Subdivision method (a) After 1 subdivision iteration (b) After 2 subdivision iterations (c) After 3 subdivision iterations

$$V'_F = V_F + \frac{d_1}{d_1 + d_2 + d_3}S_1 + \frac{d_2}{d_1 + d_2 + d_3}S_2 + \frac{d_3}{d_1 + d_2 + d_3}S_3 \qquad (5)$$

where V'_F is new position of V_F, d_1 is the distance between V_F and V_1, d_2 is the distance between V_F and V_2, d_3 is the distance between V_F and V_3, S_1 is the offset between V_1 and V'_1, S_2 is the offset between V_2 and V'_2, and S_3 is the offset between $V3$ and V'_3.

4   Synthesizing Individual Facial Texture Image

To improve the speed of modeling, only the frontal and the profile face images are used to synthesize the individual facial texture image. The column of eye corner is selected as a seam line. Stitching the images directly will form the cracks. So the multi-resolution technique and the weighted smoothing algorithm are combined to eliminate the cracks [15].

Firstly, Laplacian pyramids of frontal and profile face images are generated respectively.

Then, the Laplacian pyramids of frontal and profile face images at each level are fused by using weighted smoothing algorithm shown in Eq. (6).

$$imageC = (1 - m) * imageB + m * imageA(0 \leq m \leq 1) \qquad (6)$$

where m is set to j/W, j represents the column number of overlapping region, W is width of overlapping region.

Finally, an individual facial texture image is synthesized by summing all the levels of the Laplacian pyramids, as shown in Fig. 8.

Fig. 8. Synthesized individual facial texture image

5   Texture Mapping

It is crucial for texture mapping to build the correspondences between the 3D face model and 2D texture image, as shown in Eq. (7).

$$(u,v) = F(x,y,z) \qquad (7)$$

In Sect. 3.2, some correspondences (eyes, nose, mouth and facial silhouette) shown in Fig. 3(a)are established,other correspondences are established by using linear interpolation method. It is assumed that two known 2D texture coordinates are denoted by $a(u_1, v_1)$ and $b(u_2, v_2)$ respectively, their corresponding 3D face model coordinates are denoted by $A(x_1, y_1, z_1)$ and $B(x_2, y_2, z_2)$, then a 3D face model coordinate is denoted by $C(x, y, z)$, its corresponding 2D texture coordinate (u, v) is expressed by Eq. (8).

$$\begin{cases} u = \frac{(x-x_1)*(u_2-u_1)}{(x_2-x_1)} + u_1 \\ v = \frac{(y-y_1)*(v_2-v_1)}{(y_2-y_1)} + v_1 \end{cases} \tag{8}$$

To establish the correspondences more accurately, the whole face is divided into different areas (eyes, nose and mouse etc.). Multi-step linear interpolation which is similar with Multi-step RBF is applied in the process of texture mapping.

6   Experiments and Evaluation

The system was written in C++ with OpenCV and OpenGL. Videos with 1024 * 768 pixels were captured as our input data. All the experiments were performed on a quad 3.40 GHz core i7-2600 CPU and 2 GB RAM machine, some experimental results are shown in Fig. 9.

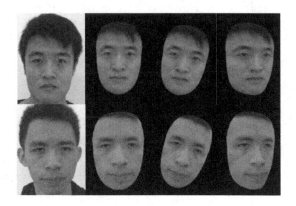

Fig. 9. Frontal face image and 3D face model in different views

The "Step 1", "Step 2" and "Step 3" in Table 1 represent the process of generating individual 3D face model, synthesizing individual facial texture image and texture mapping respectively. We can see that generating an individual face model cost about 15.085s totally, in which -Subdivision after 3 subdivision iterations cost half the time nearly. Because many parameters (head pose, 2D feature points etc.) need to be estimated in [3, 7], the total running time is about several minutes.

Table 1. 3D face modeling time consumption

Experiment	Step 1 / s	Step 2 /s	Step 3 / s	Total / s
Experiment 1	11.289	2.928	0.249	14.466
Experiment 2	12.312	3.377	0.250	15.939
Experiment 3	11.359	3.216	0.276	14.851
Average (s)	11.653	3.174	0.258	15.085

Because it is a very important step that profile facial feature points are detected accurately for the subsequent steps. But Snake model [16] is very sensitive to initial position and LMCT works well only for Mongoloid looking people [5]. Our method is used to avoid these limitations, which only needs profile facial silhouette and frontal facial points (see Fig. 2(d)).

Owing to space constraints, partial deformation results (right side of skull, nose tip and outer corner of right eye etc.) of different methods (see Fig. 3(b)(c)) are shown in Table 2. Compared with traditional RBF method [11, 12], the result based on improved RBF method is more close to that based on manual method. Equation (9) can be used to evaluate the result of deformation. All the experiments are used to calculate the error, when error is less than or equal to five pixels, it accounts for more than 95 percent.

$$Error = \sqrt{(x - X)^2 + (y - Y)^2} \tag{9}$$

where (x, y) represents the result of deformation with improved RBF (see Fig. 3(c)), (X,Y) represents manually labeled points.

Table 2. Partial results of deformation with different methods

Method	Outer corner of right eye (X, Y) / pixel	Nose tip (X, Y) / pixel	Bottom of the chin (X, Y) / pixel
Manual method	(270, 406)	(508, 534)	(513, 755)
Traditional RBF	(277, 408)	(512, 540)	(510, 758)
Improved RBF	(271, 405)	(506, 535)	(512, 756)

Candide-3 model has only 113 points, which can decrease deformation running time. But the final 3D face model will lose lots of details especially in boundaries [4, 10]. In this paper, improved $\sqrt{3}$-Subdivision is used to make the model become dense (see Fig. 7). Furthermore, only a frontal image is used as texture image [4, 10], which can not reflect the final texture completely. As shown in Fig. 9, 2D texture images are mapped to 3D models correctly by using Multi-step linear interpolation method in this paper. But only the frontal and the profile face images are used to synthesize the individual facial texture image so that some parts (eye corner etc.) of the 3D models at large angles may become distorted.

7 Conclusions and Future Work

In this paper, a method is proposed for face modeling from videos with minimal user intervention. After two manual clicks by the user, our system can quickly generate an individual face model. Furthermore, a set of experiments were implemented to show the accuracy and robustness of the pro-posed method. Our system can be used by an ordinary user at home to generate their own face models, which can be used in virtual conferencing, online chatting, and face recognition etc.

In future study, based on the method in this paper, the algorithm of extracting profile facial feature points automatically will be developed. And we will try to synthesize individual facial texture image by using face image sequences in different views.

References

1. Parke, F.I.: Computer generated animation of faces. In: Proceedings of the ACM Annual Conference, pp. 451–457 (1972)
2. Guenter, B., Grimm, C., Wood, D., et al.: Making faces. In: Proceedings of the 25th Annual Conference on Computer Graphics and Interactive Techniques, pp. 55–66. ACM (1998)
3. Park, S.W., Heo, J., Savvides, M.: 3D face reconstruction from a single 2D face image. In: IEEE Computer Society Conference on Computer Vision and Pattern Recognition Workshops, CVPRW 2008, pp. 1–8 (2008)
4. Fan, X., Peng, Q., Zhong, M.: 3D face reconstruction from single 2D image based on robust facial feature points extraction and generic wire frame model. In: International Conference on Communications and Mobile Computing, pp. 396–400 (2010)
5. Lin, Y., Lin, Q., Tang, F., et al.: Creating 3D realistic head: from two orthogonal photos to multiview face contents. In: International Society for Optics and Photonics (2011)
6. Beeler, T., Bickel, B., Beardsley, P., et al.: High-quality single-shot capture of facial geometry. ACM Trans. Graph. (TOG) **29**, 40 (2010)
7. Xin, L., Wang, Q., Tao, J., Tang, X., et al.: Automatic 3D face modeling from video. In: Tenth IEEE International Conference on Computer Vision, ICCV, vol. 2, pp. 1193–1199 (2005)
8. Song, H., Huang, X., Wang, S.: Automatic generation of portraits with multiple expressions. Acta Electronica Sin. **41**(8), 1494–1499 (2013). Chinese
9. Yin, L., Basu, A.: Integrating active face tracking with model based coding. Pattern Recogn. Lett. **20**(6), 651–657 (1999)
10. Huang, J., Su, Z., Wang, R.: 3D Face Reconstruction based on Improved CANDIDE-3 model. In: 2012 Fourth International Conference on Digital Home (ICDH), pp. 438–442. IEEE (2012)
11. Liu, S., Wang, C.C.L.: Quasi-interpolation for surface reconstruction from scattered data with radial basis function. Comput. Aided Geom. Des. **29**(7), 435–447 (2012)
12. Du, P., Xu, D., Liu, C.: Research of individual 3D face model and its application. J. Shanghai Jiaotong Univ. **37**(3), 435–439 (2003)
13. Huang, J., Schröder, P.: $\sqrt{3}$-Based 1-form subdivision. In: Boissonnat, J.-D., Chenin, P., Cohen, A., Gout, C., Lyche, T., Mazure, M.-L., Schumaker, L. (eds.) Curves and Surfaces 2011. LNCS, vol. 6920, pp. 351–368. Springer, Heidelberg (2012)

14. Zhang, H., Wang, G.: Semi-stationary push-back subdivision schemes. J. Softw. **13**(9), 1830–1839 (2002)
15. Zhang, C., Burt, P.J., van der Wal, G.S.: Multi-scale multi-camera adaptive fusion with constrast normalization: U. S. Patent. 8,411, 938 (2013)
16. Lee, W.-S., Thalmann, N.M.: Head modeling from pictures and morphing in 3d with image metamorphosis based on triangulation. In: Magnenat-Thalmann, N., Thalmann, D. (eds.) CAPTECH 1998. LNCS (LNAI), vol. 1537, pp. 254–267. Springer, Heidelberg (1998)

Recent Advances in Image/Video Processing

Score Level Fusion of Multibiometrics Using Local Phase Array

Luis Rafael Marval Pérez[✉], Shoichiro Aoyama, Koichi Ito, and Takafumi Aoki

Graduate School of Information Science, Tohoku University,
6-6-05, Aramaki Aza Aoba, Aoba-ku, Sendai-shi 980-8579, Japan
{lmarval,aoyama,ito}@aoki.ecei.tohoku.ac.jp

Abstract. Local phase array for biometric recognition have demonstrated efficient performance in face, palmprint and finger knuckle recognition. If the matching score for each trait is calculated by one matcher using local phase array, the size of the system can be reduced and the simple score level fusion can be used to exhibit good performance for person authentication. In this paper, we consider the score level fusion of face, iris, palmprint, and finger knuckle whose matching scores are calculated using local phase array. Through a set of experiments using public databases, we demonstrate effectiveness of local phase array for multibiometric recognition compared with the combination of the state-of-the-art recognition algorithm for each trait.

Keywords: Multibiometrics · Score level fusion · Local phase array

1 Introduction

Person authentication systems that use various biometric traits such as fingerprint, iris, and vein, are now becoming extensively used as the applicability of biometric authentication expands [8]. Biometric systems that utilize a single trait do not always exhibit high quality performance because one trait is no longer universal in applications with a large number of users, and the noise levels in sensed data increase due to the imperfect conditions during acquisition. To overcome these limitations within unibiometric systems, person authentication systems that make use of multiple biometric traits have recently attracted considerable interest [10].

Multibiometric systems improve performance by the complementary use of multiple traits, and exploiting distinctive advantages such as the capacity to: (A) address limited population coverage, (B) hinder spoofing by impostors, and (C) assess noise in sensed data, which are previously unmanageable with unibiometric systems. Fusion levels for multbiometric systems can be classified into five categories: (i) sensor level, (ii) feature level, (iii) score level, (iv) rank level, and (v) decision level. In this paper, we focus on score level fusion of multiple biometric traits, since matching scores are accessible and relatively simple to combine regardless of the algorithms or traits used.

© Springer International Publishing Switzerland 2015
Y.-S. Ho et al. (Eds.): PCM 2015, Part II, LNCS 9315, pp. 215–224, 2015.
DOI: 10.1007/978-3-319-24078-7_21

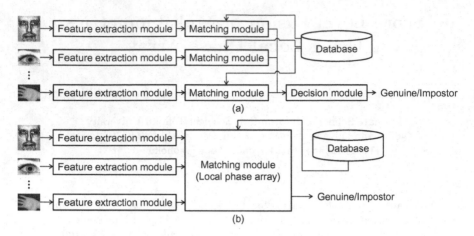

Fig. 1. Multibiometric systems: (a) conventional system using multiple matcher, and (b) proposed system using local phase array.

In general, the matching score for each trait is calculated by using its dedicated recognition algorithm. Therefore, the increase in the number of traits results in a large-scale system as shown in Fig. 1 (a). In contrast, the unified recognition algorithm for multibiometric recognition is expected to realize a compact system as shown in Fig. 1 (b).

In a previous study, we proposed a biometric recognition algorithm using local phase array, and demonstrated its efficiency for face, palmprint and finger knuckle [5]. If the matching score for each trait is calculated by one matcher using local phase array, the size of the system can be reduced and the simple score level fusion can be utilized to exhibit high quality performance for person authentication. In this paper, we consider the score level fusion of face, iris, palmprint and finger knuckle whose matching scores are calculated using local phase array. Through a set of experiments using public databases, we demonstrate the effectiveness of local phase array for multibiometric recognition compared with the combination of the state-of-the-art recognition algorithms for each trait.

2   Biometric Recognition Using Local Phase Array

This section describes the fundamentals of biometric recognition using local phase array [5].

In general, biometric recognition systems normalize the position and illumination of images according to the type of biometric traits. For example, in the case of face recognition, we detect the face region, extract feature points such as eyes, nose, mouth, etc., and then normalize the position of the face according to feature points.

To perform accurate similarity evaluation taking into consideration nonlinear deformation of normalized images, we employ local phase array extracted from

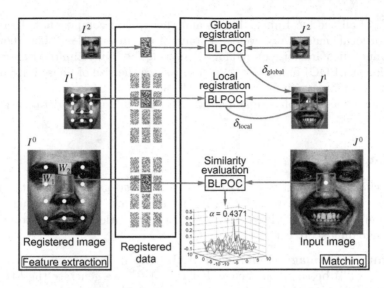

Fig. 2. Flow of biometric recognition using local phase array.

multi-scale image pyramids with 3 layers. Figure 2 shows a flow of biometric recognition using local phase array which consists of 2 steps: (i) feature extraction and (ii) matching. In the following, we describe the detailed procedure of each step.

2.1 Feature Extraction

The hierarchical local phase array consists of the phase feature of the entire image in the top layer and the phase features of local block images in the middle and bottom layers. Phase features in the middle and bottom layers are extracted according to the position of reference points. The feature extraction step consists of (i) reference point placement, (ii) hierarchical image generation and (iii) phase feature extraction.

(i) Reference Point Placement

The reference points are the center coordinates of each local block image. Let a reference point be $p = (p_1, p_2) \ (= p^0)$ and the registered image be $I \ (= I^0)$, respectively.

(ii) Hierarchical Image Generation

For $l = \{1, 2\}$, we generate the l-th layer images $I^l(n_1, n_2)$, which are $1/2^l$ times the size of $I^0(n_1, n_2)$. Also, we calculate coordinate $p^1 = (p_1^1, p_2^1) = \left(\lfloor p_1^0/2 \rfloor, \lfloor p_2^0/2 \rfloor \right)$, corresponding to p^0 on $I^1(n_1, n_2)$.

(iii) Phase Feature Extraction

In the top layer, we calculate 2D DFT of I^2 and its phase components. In the middle and bottom layers, we extract $W_1 \times W_2$-pixel local block images with its center on p^1 and p^0 from I^1 and I^0, respectively. Then, we calculate

2D DFTs of all the local image blocks and their phase components. To reduce the size of local phase array, we can eliminate the meaningless high frequency components which are not required for calculating the Band-Limited Phase-Only Correlation (BLPOC) function [6]. We also reduce the size of phase information based on the symmetry property of DFT. In addition, the amount of registered data can be reduced by phase quantization. Refer to [5] for more details on phase quantization.

2.2   Matching

The matching step consists of (i) hierarchical image generation of the input image, (ii) global image registration in the top layer, (iii) local image block registration in the middle layer, (iv) similarity evaluation in the bottom layer and (v) matching score calculation.

(i) Hierarchical Image Generation of the Input Image
  Let J $(= J^0)$ be the input image. For $l = \{1, 2\}$, we generate the l-th layer images $J^l(n_1, n_2)$, which are $1/2^l$ times the size of $J^0(n_1, n_2)$.
(ii) Global Image Registration in the Top Layer
  In the top layer, we estimate the translational displacement between I^2 and J^2 using BLPOC. We denote the estimated global translations as $\delta_{\text{global}} = (\delta_{\text{global},1}, \delta_{\text{global},2})$.
(iii) Local Image Block Registration in the Middle Layer
  In the middle layer, we estimate the translational displacement between local block images of I^1 and J^1. We extract the $W_1 \times W_2$-pixel image blocks with its center on $q^1 = p^1 + 2\delta_{\text{global}}$ from J^1. Using BLPOC for each local block image pair of I^1 and J^1, we estimate the local translations δ_{local}.
(iv) Similarity Evaluation in the Bottom Layer
  We evaluate the similarity between each local block image pair in the bottom layer. We extract the $W_1 \times W_2$-pixel local block images with its center on $q^0 = 2(q^1 + \delta_{\text{local}})$ from J^0. Then, we calculate the BLPOC function between each local block image pair of I^0 and J^0 and obtain the correlation peak value α.
(v) Matching Score Calculation
  We evaluate the matching score between I and J according to the correlation peak values obtained in the step (iv). In this paper, we employ the matching score S defined by

$$S = \frac{\sum_i \alpha_i}{N_{\text{block}}}, \tag{1}$$

where N_{block} is the number of local image block pairs, and α_i ($i = 1, 2, \cdots, N_{\text{block}}$) is the correlation peak value of i-th local image block pair.

3   Score Fusion Approaches

This section describes score fusion approaches considered in this paper.
  Score fusion approaches are broadly classified into 3 approaches: (i) density-based approach, (ii) classifier-based approach, and (iii) transformation-based

approach [10]. The density-based approach estimates the Probability Density Function (PDF) of the matching scores for both genuine and impostor pairs of each trait, then, this approach calculates the combined matching score according to the relation between genuine and impostor PDFs. Given an accurate estimation of these PDFs, the density-based approach exhibits the best performance in score fusion approaches, however, an accurate estimation is not always possible under practical situations where the amount of training data is limited. Transformation-based approach allows us to approximate easily the relation between the PDFs compared with classifier-based approach. Therefore, we employ the density-based and transformation-based approaches in this paper. From this point, the matching scores of face, iris, palmprint and finger knuckle are denoted by S_x ($x \in T = \{F, E, P, K\}$), and the matching score vector is denoted by $\boldsymbol{S} = [S_F, S_E, S_P, S_K]$, where the high value of S_x indicates the high possibility of genuine match. The set of biometric traits to be fused is indicated by T' as

$$T' \in \mathfrak{P}(T) \smallsetminus \{\phi, \{F\}, \{E\}, \{P\}, \{K\}\}, \tag{2}$$

where $\mathfrak{P}(T)$ is a power set of T.

3.1 Density-Based Approach

This approach uses the PDF of matching score S to combine matching scores calculated from different traits. In the training stage, the PDFs $p_x(S \mid \omega)$ of each $x \in T'$ for $\omega \in \{\text{genuine}, \text{impostor}\}$ are estimated from the training data set, in this paper the PDF $p_x(S \mid \omega)$ is modeled by a Gaussian mixture. In the testing stage, the values of $p_x(S_x \mid \text{genuine})$ and $p_x(S_x \mid \text{impostor})$ for $\boldsymbol{S}$ of the input data. Then, we calculate the combined matching score S_{fusion} as a Likelihood Ratio (LR) between genuine and impostor defined by

$$S_{\text{fusion}} = \frac{p(\boldsymbol{S} \mid \text{genuine})}{p(\boldsymbol{S} \mid \text{impostor})} = \frac{\prod_{x \in T'} p_x(S_x \mid \text{genuine})}{\prod_{x \in T'} p_x(S_x \mid \text{impostor})}. \tag{3}$$

3.2 Transformation-Based Approach

This approach employs simple fusion rules to calculate the combined matching scores by transforming input matching scores of different traits into a common domain. The parameters for score transformation, i.e., score normalization, are calculated from the training data set. In this paper, we employ 3 normalization techniques [7]: (i) Min-max, (ii) Double sigmoid, and (iii) tanh-estimators. Using the parameters, each element S_x of the matching score vector $\boldsymbol{S}$ is transformed into S'_x. We then calculate the combined matching score S_{fusion} from the normalized matching score vector $\boldsymbol{S}' = [S'_F, S'_E, S'_P, S'_K]$ using the simple fusion rules:

•**Average** : $S_{\text{fusion}} = \frac{\sum_{x \in T'} S'_x}{|T'|}$

•**Mean Square (MS)** : $S_{\text{fusion}} = \frac{\sum_{x \in T'} S'^2_x}{|T'|}$

•**Residuals** : $S_{\text{fusion}} = 1 - \prod_{x \in T'} (1 - S'_x)$

4   Experiments and Discussion

This section describes a set of experiments to evaluate performance of the proposed multibiometric recognition system using local phase array.

4.1   Virtual Multibiometric Databases

We make a virtual multibiometric database generated from public biometric databases to evaluate performance of score level fusion of face, iris, palmprint and finger knuckle recognition algorithms.

The virtual multibiometric database consists of chimera subjects created by pairing together face, iris, palmprint and finger knuckle images from unimodal databases. In the experiments, the number of chimera subjects is 100 with 4 images for each biometric trait. As for face images, we use 144 subjects with 4 images from FERET database [9]. As for iris images, we use 175 subject with 4 images from Iris Challenge Evaluation 2005 (ICE 2005) database [2], where we have assumed the left eye and right eye of the same person as different subjects. As for palmprint images, we use 600 subjects with 4 images from CASIA Palmprint database [1], where we have assumed the left and right hand of the same person as different subjects. As for finger knuckle images, we employ PolyU FKP database [3] which consists of 7920 images with 165 subjects and 6 different images for each of the left index finger, the left middle finger, the right index finger and the right middle finger in 2 separate sessions. We assume each finger knuckle of the same person as different subjects, i.e., a total of 660 subjects (= 165 subjects ×4 fingers). We use 660 subjects with 4 images from PolyU FKP database, where 2 images are from the first session and the remaining 2 images are from the second session. Subsequently, for all these subjects, we calculate matching scores of all the possible combinations of genuine pairs using local phase array. Then, we made a database for each trait by selecting first 100 subjects in order of increasing the average of those matching scores. In the following experiments, we combine these 100 subjects to make chimera subjects of virtual multibiometric databases. Figure 3 shows examples of images for each database.

For performance comparison, we employ the following conventional algorithms: Local Phase Quantization [4] for face recognition, Ordinal Code [11] for iris recognition, SIFT [13] for palmprint recognition, and Local-Global Information Combination [12] for finger knuckle recognition. These algorithms are known to belong to the state-of-the-art algorithms for the corresponding biometric trait.

For each database, we evaluate Equal Error Rate (EER) of conventional and local phase array algorithms using $100 \times_4 C_2 = 600$ genuine pairs and

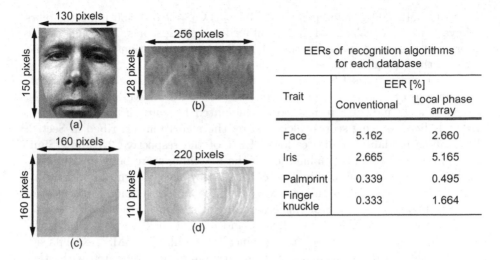

Trait	EER [%]	
	Conventional	Local phase array
Face	5.162	2.660
Iris	2.665	5.165
Palmprint	0.339	0.495
Finger knuckle	0.333	1.664

Fig. 3. Examples of ROI images in each database and EERs [%] of each recognition algorithm: (a) face, (b) iris, (c) palmprint, and (d) finger knuckle.

$_{100}C_2 \times 4 \times 4 = 79,200$ impostor pairs. Figure 3 shows a summary of EERs of conventional algorithms and local phase array for each database. The proposed algorithm exhibits comparable performance with the conventional algorithms specialized for each biometric trait despite selecting worst subjects for the local phase array algorithm.

4.2 Performance Evaluation

We evaluate the error rates statistically by using the bootstrap technique, which is a non-parametric method to estimate the confidence interval by random data sampling.

We create 100 chimera subjects with 4 biometric traits by randomly combining subjects from each database generated in Sect. 4.1. For each trait, the number of all the possible combinations of genuine matching scores is $_4C_2 = 6$, while the number of all the possible combinations of impostor matching scores between different 2 chimeras is $4 \times 4 = 16$, since one chimera subject has 4 images of each trait. In this paper, 2 genuine scores and 6 impostor scores are used in the training step and remaining scores are used in the testing step. We then generate a set of virtual score vectors using the same procedure in [7]. Thus, the total number of genuine combinations of score vectors S for one chimera subject is $4^4 = 256$, since there are 4 genuine matching scores for each trait. On the other hand, the total number of impostor combinations of score vectors S between 2 different chimeras is $10^4 = 10,000$, since there are 10 impostor matching scores for each trait. Among the above score vectors, we randomly select 128 genuine pairs and 512 impostor pairs to generate a set of virtual score vectors. Then, for $|T'| = 2$, we apply the score fusion approaches described in Sect. 3 to

$128 \times 100 = 12,800$ genuine pairs and $512 \times_{100} C_2 = 2,534,400$ impostor pairs
and evaluate EERs calculated from combined matching scores. We repeated the
experiment described above 100 times for different combinations of subjects and
scores in the virtual database.

Table 1 summarizes EERs for each fusion rule when using all the possible
combinations of biometric traits, where EER indicates an average of 100 trials.
"Conventional" indicates fusing scores calculated by conventional algorithms,
"LPA" indicates fusing scores calculated by the algorithm described in Sect. 2.
"Best single modality" indicates lower EER of the respective 2 traits. "Similarity" indicates a simple combination of the matching score between 2 images,
which is given as a value within $[0, 1]$ calculated by each recognition algorithm.

From the EERs of Conventional, we can make 3 observations. First, most of
them are higher than those of single modality cases. In particular, the combination of face, whose EER is the highest, and finger knuckle, whose EER is the
lowest, shows this tendency. Second, specifically "Similarity + MS" exhibits significantly high EER for face-palmprint and iris-palmprint compared with other

Table 1. EERs [%] for each combination rule.

T'			$\{F,E\}$	$\{F,P\}$	$\{F,K\}$	$\{E,P\}$	$\{E,K\}$	$\{P,K\}$
Conventional	Best single modality		2.665	0.339	0.333	0.339	0.333	0.333
	LR		0.749	0.222	0.630	0.138	0.270	0.138
	Similarity	Average	0.973	0.391	0.699	0.240	0.391	0.260
		MS	1.017	1.973	0.706	1.604	0.535	0.791
		Residuals	1.023	0.707	0.645	0.660	0.441	0.332
	Min-max	Average	1.078	0.499	0.677	0.435	0.519	0.194
		MS	0.897	1.722	1.137	1.192	0.790	0.410
		Residuals	0.948	0.804	0.781	0.638	0.605	0.218
	Double sigmoid	Average	1.029	0.321	0.735	0.161	0.448	0.158
		MS	0.820	0.221	0.663	0.082	0.305	0.085
		Residuals	0.826	0.218	0.650	0.077	0.304	0.082
	tanh-estimator	Average	1.024	0.792	0.861	0.739	0.806	0.197
		MS	0.876	0.804	0.854	0.476	0.525	0.196
		Residuals	0.928	0.803	0.856	0.577	0.629	0.195
LPA	Best single modality		2.660	0.495	1.664	0.495	1.664	0.495
	LR		0.777	0.253	0.603	0.330	0.872	0.270
	Similarity	Average	1.071	0.321	0.732	0.683	1.296	0.427
		MS	0.994	0.276	0.686	0.516	1.223	0.334
		Residuals	1.030	0.294	0.703	0.577	1.260	0.366
	Min-max	Average	1.282	0.304	0.742	0.734	1.473	0.379
		MS	1.081	0.302	0.686	0.404	1.316	0.362
		Residuals	1.144	0.300	0.701	0.422	1.370	0.359
	Double sigmoid	Average	1.124	0.330	0.860	0.554	1.340	0.461
		MS	0.891	0.214	0.682	0.341	1.027	0.271
		Residuals	0.891	0.216	0.680	0.340	1.030	0.267
	tanh-estimator	Average	1.155	0.396	0.768	0.476	1.544	0.495
		MS	1.070	0.281	0.734	0.386	1.407	0.370
		Residuals	1.096	0.327	0.742	0.432	1.442	0.421

fusion rules. This is because distribution of matching scores for palmprint is significantly different from other algorithms. Third, EERs of "Double sigmoid + MS" and "Double sigmoid + Residuals" improves the EERs of other fusion rules. These observations indicate that conventional algorithms have to employ normalization to use simple combination rules and to ensure high performance.

For the result of LPA, we can make three observations. First, EERs are lower than those of single modality cases, except for the combination of iris and palmprint for some fusion rules. Second, "Similarity + MS" shows the best EER between simple combinations. Third, in this case also, the EERs of "Double sigmoid + MS" and "Double sigmoid + Residuals" improve the EERs of simple combination and the other normalization methods. These observations indicate that LPA does not always need normalization since it can employ simple combinations for almost all the combination of traits.

In both Conventional and LPA, LR exhibits the lowest EERs for most cases. As mentioned in Sect. 3, LR is expected to show the best performance, if the PDFs for genuine and impostor pairs were estimated accurately. However, the transformation-based approaches, "Double sigmoid + MS" and "Double sigmoid + Residuals," also exhibit efficient performances comparable to the ones of LR. Therefore, these score fusions are robust against limited training data and the diversity of their score distributions compared with LR.

Focusing on "Similarity + Average" and "Double sigmoid + MS/Residuals," EERs of LPA are significantly improved compared with those of Conventional. This result indicates that multiple recognition algorithms require a complex optimization for score normalization and combination approaches to exhibit the efficiency of score fusion observed in LPA, since the optimal score normalization method and optimal fusion rule might be different for each recognition algorithm.

As observed above, successful score fusion with LPA does not depend on normalization methods and fusion rules as it does with the conventional method. Hence, the use of LPA for multi-modal biometric systems makes it possible to improve the performance only with simple combination.

5   Conclusion

This paper proposed a biometric recognition algorithm with score level fusion of multiple matching scores calculated by local phase array. Through a set of experiments, we demonstrate that simple score fusion approach is enough to exhibit good recognition performance for local phase array. Hence, the use of local phase array makes it possible to realize simple and compact multibiometric person authentication systems, since only one matching module and simple score fusion approach are employed. In future, we will consider other types of multibiometric fusion for local phase array. Also, we will develop a fast multibiometric identification system using local phase array.

References

1. CASIA palmprint image database. http://www.cbsr.ia.ac.cn/english/Palmprint Databases.asp
2. Iris Challenge Evaluation (ICE). http://www.nist.gov/itl/iad/ig/ice.cfm
3. PolyU FKP database. http://www4.comp.polyu.edu.hk/~biometrics/FKP.htm
4. Ahonen, T., Rahtu, E., Ojansivu, V., Heikkilä, J.: Recognition of blurred faces using local phase quantization. In: Proceedings of the International Conference on Pattern Recognition, pp. 1–4, December 2008
5. Aoyama, S., Ito, K., Aoki, T.: Similarity measure using local phase feature and its application to biometric recognition. In: Proceedings of IEEE Compututer Society Conference on Compututer Vision and Pattern Recognition Workshops, pp. 180–187, June 2013
6. Ito, K., Nakajima, H., Kobayashi, K., Aoki, T., Higuchi, T.: A fingerprint matching algorithm using phase-only correlation. IEICE Trans. Fundam. **E87–A**(3), 682–691 (2004)
7. Jain, A., Nandakumar, K., Ross, A.: Score normalization in multimodal biometric system. Pattern Recogn. **38**(12), 2270–2285 (2005)
8. Jain, A., Flynn, P., Ross, A.: Handbook of Biometrics. Springer, New York (2008)
9. Phillips, P.J., Moon, H.J., Rizvi, S.A., Rauss, P.J.: The FERET evaluation methodology for face recognition algorithms. IEEE Trans. Pattern Anal. Mach. Intell. **22**(10), 1090–1104 (2000)
10. Ross, A.A., Nandakumar, K., Jain, A.K.: Handbook of Multibiometrics. Springer, New York (2006)
11. Sun, Z., Tan, T.: Ordinal measures for iris recognition. IEEE Trans. Pattern Anal. Mach. Intell. **31**(12), 2211–2226 (2009)
12. Zhang, L., Zhang, L., Zhang, D., Zhu, H.: Ensemble of local and global information for finger-knuckle-print recognition. Pattern Recogn. **44**, 1990–1998 (2011)
13. Zhao, Q., Bu, W., Wu, X.: SIFT-based image alignment for contactless palmprint verification. In: Proceedings of International Conference on Biometrics, pp. 1–6, June 2013

Histogram-Based Near-Lossless Data Hiding and Its Application to Image Compression

Masaaki Fujiyoshi[✉] and Hitoshi Kiya

Department of Information and Commununication Systems,
Tokyo Metropolitan University, 6–6 Asahigaoka, Hino-shi, Tokyo 191–0065, Japan
mfujiyoshi@ieee.org, kiya@tmu.ac.jp

Abstract. This paper proposes a near-lossless data hiding (DH) method for images where the proposed method can improve the image compression efficiency. The proposed method firstly quantizes an image in accordance with a user-given maximum allowed error. This method, then, embeds data to the quantized image based on histogram shifting (HS). Even this method uses HS-based DH which requires to memorize the shifted bins for data extraction, the method, under some conditions, takes data out from the marked image by just applying re-quantization as least significant bitplane (LSB) substitution-based DH. So the proposed method is based on unification of HS- and LSB substitution-based DH. In the method, lossless compression of the marked image can achieve better compression efficiency than lossy compression of the original image. Experimental results show the effectiveness of the proposed method.

Keywords: Reversible watermarking · Visually lossless · Non-uniform quantizer · Lossless compression · Histogram packing

1 Introduction

A data hiding (DH) method for images once distorts an image to hide data to the image [8,10] where the image and data are referred to as the *original* image and the *payload*, respectively. The method takes the payload out from the distorted image called the *stego* image. Most earlier methods and those for security related issues such as copyright protection [24] leave the stego image as distorted, i.e., *lossy*. A simple lossy DH mechanism is based on least significant bitplane (*LSB*) *substitution* [27] which replaces the LSB of an original image with a payload.

Later, in particular, for military and medical imagery, *lossless* DH methods have been proposed [5,17] where a method perfectly restores the original image from a stego image in addition to taking the payload out from the stego image. One major class of lossless DH is based on *histogram shifting* [4,12,20] (HS) where a payload is hidden to an original image based on the tonal distribution of the image. Another class is based on *generalized LSB substitution* [6] where LSB substitution-based DH is reconsidered as the combination of uniform quantization of the original image and payload addition.

© Springer International Publishing Switzerland 2015
Y.-S. Ho et al. (Eds.): PCM 2015, Part II, LNCS 9315, pp. 225–235, 2015.
DOI: 10.1007/978-3-319-24078-7_22

Similar to image compression, *visually-lossless* and *near-lossless* (NLL) DH methods have been also developed. A method removes the hidden payload from a stego image where a pixel in the payload-removed image slightly differs from that in the original image. For this category, two approaches exist; A lossy-based approach controls the distortion of DH [3,16,25], whereas a lossless-based approach applies lossless DH to a pre-distorted image [7,15,28]. This paper focuses on the latter approach.

This paper develops a NLL DH method based on lossless DH, where a payload is hidden to an image by using HS-based DH but the payload can be taken out based on LSB substitution, viz., LSB substitution- and HS-based techniques are unified. The proposed method also include the conventional NLL DH method [15] as its special form. In addition, the proposed method can improve the image compression efficiency by utilizing the sparsity of stego images, whereas distorted images generally deteriorate the compression efficiency.

2   Preliminary

This section mentions LSB substitution-based DH and its generalization, HS-based lossless DH and its generalization, and the conventional NLL DH method.

2.1   LSB Substitution-Based DH

As shown in Fig. 1(a), the essence of the simplest LSB substitution-based method replaces the LSB of $X \times Y$-sized B-bit grayscale original image $\mathbf{f} = \{f(x,y)\}$ with payload $\mathbf{p}_{\mathrm{LSB}} = \{p_{\mathrm{LSB}}(x,y)\}$ to embed $\mathbf{p}_{\mathrm{LSB}}$ to $\mathbf{f}$;

$$\hat{f}_{\mathrm{LSB}}(x,y) = p_{\mathrm{LSB}}(x,y) + \sum_{b=2}^{B} 2^{(b-1)} f_b(x,y), \qquad (1)$$

where $\hat{\mathbf{f}}_{\mathrm{LSB}} = \left\{ \hat{f}_{\mathrm{LSB}}(x,y) \right\}$ is a stego image, $\mathbf{f}_b = \{f_b(x,y)\}$ is the b-th LSB of $\mathbf{f}$, $x = 0,1,\dots,X-1$, $y = 0,1,\dots,Y-1$, $b = 1,2,\dots,B$, $p(x,y) \in \{0,1\}$, $f(x,y) \in \left[0 .. 2^B - 1\right]$, $f_b(x,y) \in \{0,1\}$, and $\hat{f}(x,y) \in \left[0 .. 2^B - 1\right]$. So, the payload size is XY [bits] and the *capacity* which is the maximum conveyable payload size is also XY [bits].

LSB substitution-based DH was generalized [6], c.f., Fig. 1(b), as

$$\hat{f}_{\mathrm{GLSB}}(x,y) = qQ_{\mathrm{F}}\left(f(x,y), q\right) + p_{\mathrm{GLSB}}(x,y) \qquad (2)$$

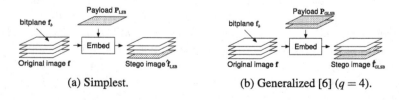

(a) Simplest.   (b) Generalized [6] ($q = 4$).

Fig. 1. LSB substitution-based DH ($B = 4$).

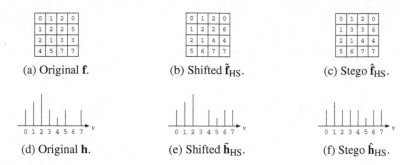

(a) Original **f**.

(b) Shifted $\tilde{\mathbf{f}}_{\text{HS}}$.

(c) Stego $\hat{\mathbf{f}}_{\text{HS}}$.

(d) Original **h**.

(e) Shifted $\tilde{\mathbf{h}}_{\text{HS}}$.

(f) Stego $\hat{\mathbf{h}}_{\text{HS}}$.

Fig. 2. HS-based DH [20] ($X = Y = 4$, $B = 3$, $v_{\max} = 2$, $v_{\min} = 6$, $h\,(v_{\max}) = 4$, $h\,(v_{\min}) = 0$, and $\mathbf{p}_{\text{HS}} = \{0, 1, 1, 0\}$.

by introducing floor function-based q-level uniform scalar quantizer $Q_{\text{F}}\,(f(x,y), q)$ $= \lfloor f(x,y)/q \rfloor$ and payload $\mathbf{p}_{\text{GLSB}} = \{p_{\text{GLSB}}(x,y)\}$ consisting of q-ary symbols, i.e., $p_{\text{GLSB}}(x,y) \in \{0, 1, \ldots, q-1\}$, where $\hat{\mathbf{f}}_{\text{GLSB}} = \left\{\hat{f}_{\text{GLSB}}(x,y)\right\}$ is a stego image and $\lfloor \cdot \rfloor$ returns the integer part of the input, i.e., the floor function. The payload size and capacity are both given as $XY \log_2 q$ [bits]. It becomes the original LSB substitution-based DH when $q = 2$. A lossless DH method was developed based on this generalization [6].

2.2   HS-Based DH

The original HS-based method [20] shown in Fig. 2 firstly derives tonal distribution $\mathbf{h} = \{h(v)\}$ from original image $\mathbf{f}$ where $v = 0, 1, \ldots, 2^B - 1$ and $h(v) = |\{(x,y) \mid f(x,y) = v\}|$. This method then finds two pixel values $v_{\min} = \arg \min h(v)$ and $v_{\max} = \arg \max h(v)$, here it assumes for the simplicity that $h\,(v_{\min}) = 0$ and $v_{\max} < v_{\min}$. The method shifts a part of the *histogram* of $\mathbf{f}$ toward $h\,(v_{\min})$;

$$\tilde{f}_{\text{HS}}(x,y) = \begin{cases} f(x,y) + 1, & v_{\max} < f(x,y) < v_{\min} \\ f(x,y), & \text{otherwise} \end{cases}, \qquad (3)$$

where $\tilde{\mathbf{f}}_{\text{HS}} = \left\{\tilde{f}_{\text{HS}}(x,y)\right\}$ is the histogram shifted image and $\tilde{f}_{\text{HS}}(x,y) \in [0 .. 2^B - 1]$. In tonal distribution $\tilde{\mathbf{h}}_{\text{HS}} = \left\{\tilde{h}_{\text{HS}}(v)\right\}$ of $\tilde{\mathbf{f}}_{\text{HS}}$, the *peak histogram bin* is followed by one *zero histogram bin*, i.e., $\tilde{h}_{\text{HS}}\,(v_{\max}) = \max \tilde{h}_{\text{HS}}(v)$ and $\tilde{h}_{\text{HS}}\,(v_{\max} + 1) = 0$. Finally, stego image $\hat{\mathbf{f}}_{\text{HS}} = \left\{\hat{f}_{\text{HS}}(x,y)\right\}$ conveying payload $\mathbf{p}_{\text{HS}} = \{p_{\text{HS}}(l)\}$ is given by modifying the pixel value of a pixel with $v_{\max}$ in accordance with payload bit $p_{\text{HS}}(l)$ to be hidden;

$$\hat{f}_{\text{HS}}(x,y) = \begin{cases} \tilde{f}_{\text{HS}}(x,y), & (x,y) = m_l \text{ and } p_{\text{HS}}(l) = 0 \\ \tilde{f}_{\text{HS}}(x,y) + 1, & (x,y) = m_l \text{ and } p_{\text{HS}}(l) = 1 \\ \tilde{f}_{\text{HS}}(x,y), & \text{otherwise} \end{cases}, \qquad (4)$$

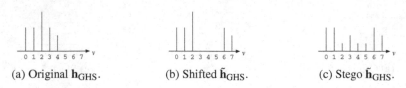

(a) Original $\mathbf{h}_{\mathrm{GHS}}$.   (b) Shifted $\tilde{\mathbf{h}}_{\mathrm{GHS}}$.   (c) Stego $\hat{\mathbf{h}}_{\mathrm{GHS}}$.

Fig. 3. Generalized HS-based DH [2, 13, 14] ($B = 3$, $v_{\max} = 2$, $v_{\min,\min} = 5$, $v_{\min,\max} = 7$, $h(v_{\max}) = 5$, and $q = 4$).

where m_l is the l-th element of set $M = \{(x, y) \mid f(x, y) = v_{\max}\}$, $\hat{f}_{\mathrm{HS}}(x, y) \in [0 .. 2^B - 1]$, $p_{\mathrm{HS}}(l) \in \{0, 1\}$, $l = 0, 1, \ldots, L - 1$, and payload size L should be less than or equal to capacity $|M| = h(v_{\max})$. With pixel values $v_{\max}$ and $v_{\min}$, p_{HS} is extracted by tracing pixels with $v_{\max}$ and $(v_{\max} + 1)$. Decreasing the pixel value of pixels with $(v_{\max} + 1)$ by one gives $\tilde{f}$, and then the inverse shifting is applied to $\tilde{f}$ to restore f. It is noted that this method requires to memorize $v_{\max}$ and $v_{\min}$ to take out p_{HS} from $\hat{f}_{\mathrm{HS}}$ whereas descendent methods overcome this disadvantage [11].

HS-based DH was also generalized [2, 13, 14]. Let $\mathbf{f}_{\mathrm{GHS}} = \{f_{\mathrm{GHS}}(x, y)\}$ be an original image with $(q - 1)$ successive zero histogram bins, i.e., $h_{\mathrm{GHS}}(\omega) = 0$ for $v_{\min,\min} \leq \omega \leq v_{\min,\max}$ and $|v_{\min,\max} - v_{\min,\min} + 1| = (q - 1)$ where $f_{\mathrm{GHS}}(x, y) \in [0 .. 2^B - 1]$ and $\mathbf{h}_{\mathrm{GHS}} = \{h_{\mathrm{GHS}}(v)\}$ is the tonal distribution of $\mathbf{f}_{\mathrm{GHS}}$. It is assumed here that $v_{\max} < v_{\min,\min}$. Generalized HS-based DH shifts a part of $\mathbf{h}_{\mathrm{GHS}}$ as shown in Fig. 3 by

$$\tilde{f}_{\mathrm{GHS}}(x, y) = \begin{cases} f_{\mathrm{GHS}}(x, y) + (q - 1), & v_{\max} < f_{\mathrm{GHS}}(x, y) < v_{\min,\min} \\ f_{\mathrm{GHS}}(x, y), & \text{otherwise} \end{cases}, \quad (5)$$

where $\tilde{\mathbf{f}}_{\mathrm{GHS}} = \{\tilde{f}_{\mathrm{GHS}}(x, y)\}$ is the histogram shifted image and $\tilde{f}_{\mathrm{GHS}}(x, y) \in [0 .. 2^B - 1]$. Tonal distribution $\tilde{\mathbf{h}}_{\mathrm{GHS}} = \{\tilde{h}_{\mathrm{GHS}}(v)\}$ of $\tilde{\mathbf{f}}_{\mathrm{GHS}}$ has $(q - 1)$ successive zero histogram bins preceded by $\tilde{h}_{\mathrm{GHS}}(v_{\max})$. In accordance with q-ary payload symbol $p_{\mathrm{GHS}}(l) \in \{0, 1, \ldots, q - 1\}$ to be inserted, the pixel value of a pixel with $v_{\max}$ is changed to the value between $v_{\max}$ and $(v_{\max} + (q - 1))$;

$$\hat{f}_{\mathrm{GHS}}(x, y) = \begin{cases} \tilde{f}_{\mathrm{GHS}}(x, y) + p_{\mathrm{GHS}}(l), & (x, y) = m_{\mathrm{GHS}, l} \\ \tilde{f}_{\mathrm{GHS}}(x, y), & \text{otherwise} \end{cases}, \quad (6)$$

where $m_{\mathrm{GHS}, l}$ is the l-th element of set $M_{\mathrm{GHS}} = \{(x, y) \mid f_{\mathrm{GHS}}(x, y) = v_{\max}\}$, $\hat{\mathbf{f}}_{\mathrm{GHS}} = \{\hat{f}_{\mathrm{GHS}}(x, y)\}$ is the stego image, and $\hat{f}_{\mathrm{GHS}}(x, y) \in [0 .. 2^B - 1]$. With $v_{\max}$, $v_{\min,\min}$, and $v_{\min,\max}$, p_{GHS} is taken out from $\hat{\mathbf{f}}_{\mathrm{GHS}}$ and $\tilde{\mathbf{f}}_{\mathrm{GHS}}$ are restored. Original image $\mathbf{f}_{\mathrm{GHS}}$ is obtained by applying the inverse shifting to $\tilde{\mathbf{f}}_{\mathrm{GHS}}$. Blind methods which are free from memorizing $v_{\max}$, $v_{\min,\min}$, and $v_{\min,\max}$ were also developed [13, 14].

2.3 NLL DH

HS-based DH mentioned in the preceding section needs one zero histogram bin at least, so when $h(v_{\min}) \neq 0$, it has to change the pixel value of pixels with $v_{\min}$ to another value for ensuring a zero histogram bin, and it simultaneously has to memorize $v_{\min}$ and set $R = \{(x, y) \mid f(x, y) = v_{\min}\}$ for restoring original image $\mathbf{f}$.

The conventional NLL DH method [15] accepts slight distortion less than or equal to user-given maximum allowed error δ in payload-removed images, i.e., NLL, instead of memorizing image-dependent $v_{\min}$ and R. This method firstly change the pixel value of pixels with $(v_{\max} + 1)$ (or $(v_{\max} - 1)$) to $v_{\max}$ by a non-uniform quantizer to get a zero histogram bin and to increase the capacity simultaneously. The method then applies HS-based lossless DH to hide a payload to the pre-distorted image, so the pre-distorted image is obtained from a stego image where a pixel in the pre-distorted image differs from that in the original image by not more than $\delta = 1$.

3 Proposed Method

Figure 4 shows the block diagram of the proposed NLL DH method, where original image $\mathbf{f}$ is fed to a quantizer, for making the image histogram sparse, and payload $\mathbf{p}$ is inserted to quantized image $\tilde{\mathbf{f}}$ by HS-based lossless DH. Even HS-based DH is used, hidden payload $\mathbf{p}$ can be taken out from stego image $\hat{\mathbf{f}}$ by using the same quantizer as that is applied to $\mathbf{f}$. That is, memorizing $v_{\max}$ and so on is not required.

Quantization error is not compensated in this method, so it generally obtains $\tilde{\mathbf{f}}$ from $\hat{\mathbf{f}}$. The method described here accepts any quantizer, whereas scalar quantizers are assumed hereafter for the simplicity. With quantizers guaranteeing that a pixel in $\tilde{\mathbf{f}}$ differs from that in $\mathbf{f}$ by not more than δ, NLL DH can be achieved. Note that the conventional NLL DH method [15] is regarded as the special form of the proposed method.

3.1 Example Algorithms

A simple example of the proposed method is developed here. Quantizer $Q_F(\cdot, q)$ is used here where $q = \delta + 1$. Whereas the prediction error of pixels which give

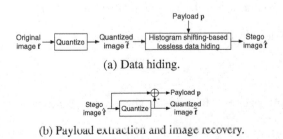

(a) Data hiding.

(b) Payload extraction and image recovery.

Fig. 4. The proposed NLL DH method.

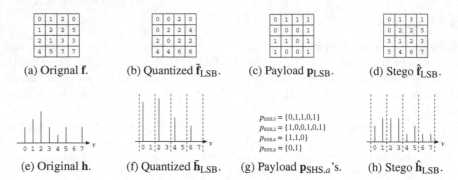

Fig. 5. Unifying LSB substitution- and HS-based techniques ($X = Y = 4$, $B = 3$). Adding $\mathbf{p}_{LSB}$ to quantized image $\tilde{\mathbf{f}}$ in LSB substitution-based DH is interpreted as HS-based DH in split sub histograms.

much larger peak histogram bin is often utilized to increase the capacity [26], this paper uses the original HS-based DH [20] to focus the concept of the proposed method.

Embedding

Step 1. Apply $Q_F(\cdot, q)$ to $\mathbf{f}$ to obtain $\tilde{\mathbf{f}} = \left\{\tilde{f}(x, y)\right\}$ where

$$\tilde{f}(x, y) = qQ_F(f(x, y), q). \tag{7}$$

Step 2. Add q-ary symbols $\mathbf{p}_q = \{p_q(x, y)\}$ to $\tilde{\mathbf{f}}$ to serve $\hat{\mathbf{f}}$ as

$$\hat{f}(x, y) = \tilde{f}(x, y) + p_q(x, y). \tag{8}$$

This algorithm becomes the generalized LSB substitution [6] and the generalized HS [2,13,14] when $\forall(x, y)p_q(x, y)$ satisfies and $\forall(x, y) \in Mp_q(x, y)$, respectively.

Payload Extraction and Image Restoration.

Step 1. Apply $Q_F(\cdot, q)$ to $\hat{\mathbf{f}}$ to reproduce $\tilde{\mathbf{f}}$ where $\tilde{f}(x, y) = qQ_F\left(\hat{f}(x, y), q\right)$.

Step 2. Subtract $\tilde{\mathbf{f}}$ from $\hat{\mathbf{f}}$ to take $\mathbf{p}_q$ out from $\hat{\mathbf{f}}$ as $p_q(x, y) = \hat{f}(x, y) - \tilde{f}(x, y)$.

Payload $\mathbf{p}_q$ is retrieved and quantized image $\tilde{\mathbf{f}}$ is restored.

3.2   Features

This section summarizes five features of the proposed method.

Unification of DH Techniques. LSB substitution- and HS-based DH techniques are preparatorily unified in the proposed method.

LSB substitution-based DH given by Eq. (1) is represented with Eq. (2) as

$$\hat{f}_{\text{LSB}}(x,y) = 2Q_{\text{F}}\left(f(x,y),2\right) + p_{\text{LSB}}(x,y).   \tag{9}$$

Here, tonal distribution $\tilde{\mathbf{h}}_{\text{LSB}} = \left\{\tilde{h}_{\text{LSB}}(v)\right\}$ of quantized image $\tilde{\mathbf{f}}_{\text{LSB}} = \left\{\tilde{f}_{\text{LSB}}(x,y)\right\}$ is focused, where $\tilde{f}_{\text{LSB}}(x,y) = 2Q_{\text{F}}(f(x,y),2)$ and $\tilde{h}_{\text{LSB}}(v) = \left|\left\{(x,y) \mid \tilde{f}_{\text{LSB}}(x,y) = v\right\}\right|$. Similar to the idea which splits the histogram of an image to sub histograms [4], $\tilde{\mathbf{h}}_{\text{LSB}}$ is split to $2^{(B-1)}$ sub histograms where each sub histogram consists of two histogram bins; a peak histogram bin at even pixel value $2a$ and a zero histogram bin at odd pixel value $(2a+1)$ where $a = 0, 1, \ldots, 2^{(B-1)} - 1$. So, adding a payload bit to the quantized image can be reformulated as HS-based DH as shown in Fig. 5:

$$\hat{f}_{\text{LSB}}(x,y) = \tilde{f}_{\text{LSB}}(x,y) + p_{\text{SHS},a}\left(l_a\right),   \quad (x,y) = m_{a,l_a},   \tag{10}$$

where $\mathbf{p}_{\text{SHS},a} = \{p_{\text{SHS},a}\left(l_a\right)\}$ is the sub payload for peak histogram bin $\tilde{h}_{\text{LSB}}(2a)$, m_{a,l_a} is the l_a-th element of set $M_a = \left\{(x,y) \mid \tilde{f}_{\text{LSB}}(x,y) = 2a\right\}$, and $l_a = 0, 1, \ldots, |M_a|$. It is noted that $\sum_{a=0,1,\ldots,2^{(B-1)}-1} |M_a| = XY$.

So, LSB substitution- and HS-based techniques are unified. It is noted that this unification can be extended to any q, viz., generalized LSB substitution- and generalized HS-based techniques are also unified.

NLL DH. By using q-level uniform quantizer $Q_{\text{F}}\left(\cdot, q\right)$ and payload with q-ary symbols $\mathbf{p}_q$ as described in Sect. 3.1, the proposed method serves NLL DH where $\delta = q - 1$. Quantization error $\left|\tilde{f}(x,y) - f(x,y)\right|$ and distortion by DH $\left|\hat{f}(x,y) - \tilde{f}(x,y)\right|$ are up to δ. Consequently, this method achieves NLL DH. It is noteworthy that a histogram-based non-uniform quantizer losslessly quantizes an image in some conditions [18,19], so the method with this quantizer can restore $\mathbf{f}$.

Asymmetric DH. As described in Sect. 3.1, the hidden payload is taken out from a stego image based on LSB substitution, i.e., requantization, even the payload is hidden to the image based on HS. This feature is realized by the unification of two DH techniques and it makes the proposed method blind, viz., no parameter has to be memorized. This feature is important for practical use of DH; When parameters are stored in a parameter database, the stego image is firstly identified among all possible images in the database to retrieve the corresponding parameters. The other possible way is hiding parameters to the stego image by another lossless DH method, but it should introduce multiple lossless DH methods to the system.

Flexible Control of Distortions and Capacity. By controlling δ, the proposed method can flexibly control quantization error and distortion by DH. In addition, the number of sub histograms to be marked can be controlled, the proposed method is able to control the capacity.

Compression Efficiency Improvability. Quantization makes the histogram of an image sparse and images with a sparse histogram can be efficiently compressed by a lossless encoder with histogram packing (HP) [9,21]. This feature is meaningful for distributing stego images to receivers.

4   Experimental Results

Five grayscale images with zero histogram bins [23] shown in Fig. 6 and seven grayscale images without zero histogram bins [15,22] shown in Fig. 7 are used in this section.

Figure 8 shows the averaged peak signal-to-noise ratio (PSNR) between quantized and original images and that between stego and original images of image 'Lena' and '15' by the proposed method with the algorithms described in Sect. 3.1. The number of marked sub histograms are 1, 2, 4, 8, 16, 32, 64, and 128, and ten different payloads consisting of uniformly distributed q-ary random symbols are used for each condition where $q = \delta + 1$. In contrast to the conventional NLL DH method [15] in which the highest embedding rate is 0.26 bits/pixel for image '20' and the blind generalized HS-based method [13,14] in which the highest embedding rate is 0.29 bits/pixel for image 'airplane,' the proposed method is flexible in controlling the achievable embedding rate. It is found that hiding a payload to the quantized image improves the PSNR of stego images (slightly for $\delta = 1$ of 'Lena') because a pixel value once becoming smaller by quantization (Eq. (7)) gets larger by DH (Eq. (8)). It is noted that almost similar results are derived for other images.

(a) Airplane.   (b) Baboon.   (c) Lena.   (d) Peppers.   (e) Sailboat.

Fig. 6. Five images with zero histogram bins [23] ($X = 512$, $Y = 512$, and $B = 8$).

(a) 05.   (b) 08.   (c) 14.   (d) 15.   (e) 17.   (f) 18.   (g) 20.

Fig. 7. Seven images without zero histogram bins [15,22] ($X = 768$, $Y = 512$, and $B = 8$).

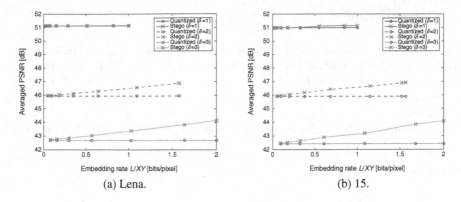

(a) Lena.   (b) 15.

Fig. 8. PSNR versus embedding rate.

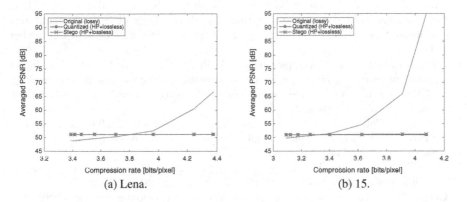

(a) Lena.   (b) 15.

Fig. 9. Compression efficiency Improvement ($\delta = 1$).

Figure 9 shows the compression efficiency of stego images. First, stego images for 'Lena' given by the proposed method with $\delta = 1$ are HPed [9,21] and losslessly compressed by a JPEG 2000 [1] standard encoder (JasPer). Then, the averaged compression rate was derived including the bzip2-compressed table for histogram unpacking. The green curve in Fig. 9 indicated by 'Lossy' shows the PSNR between original and lossily JPEG 2000 compressed original images where the compression rate is the same as the above derived rate. It is found that the proposed method can give images with higher PSNR than those lossily compressed by the standard JPEG 2000 encoder, whereas distorted images generally deteriorates the compression efficiency. For stego images with lower embedding rates or stego images with sparser histograms, the efficiency improvement by HP more than compensates for the efficiency deterioration by quantization and DH. It is noted that almost similar results are confirmed for other images.

5   Conclusions

This paper has proposed a histogram-based NLL data hiding method. By the unification of LSB substitution- and HS-based techniques, data are hidden to and extract from an image based on both techniques, and the method is flexible in capacity and distortion. In addition, stego images can be efficiently lossless compressed.

Performance evaluation with zero skip quantization [18,19] is a further work.

References

1. Information technology - JPEG 2000 image coding system - Part 1: Core coding system. International Standard ISO/IEC IS-15444-1, December 2000
2. Arabzadeh, M., Helfroush, M.S., Danyali, H., Kasiri, K.: Reversible watermarking based on generalized histogram shifting. In: Proceedings of IEEE International Conference on Image Processing, pp. 2797–2800, Brussels, Belgium, September 2011
3. Barni, M., Bartolini, F., Cappellini, V., Magli, E., Olmo, G.: Near-lossless digital watermarking for copyright protection of remote sensing images. In: Proceedings of IEEE International Geoscience and Remote Sensing Symposium, pp. 1447–1449, Toronto, Ontario, Canada, June 2002
4. Caciula, I., Coltuc, D.: Improved control for low bit-rate reversible watermarking. In: Proceedings of IEEE International Conference on Acoustics, Speech and Signal Process, pp. 7425–7429, Florence, Italy, May 2014
5. Caldelli, R., Filippini, F., Becarelli, R.: Reversible watermarking techniques: an overview and a classification. EURASIP J. Inf. Secur. **2010**, 19 (2010)
6. Celik, M.U., Sharma, G., Tekalp, A.M., Saber, E.: Lossless generalized-LSB data embedding. IEEE Trans. Image Process. **14**(2), 253–266 (2005). Article no. 134546
7. Conotter, V., Boato, G., Carli, M., Egiazarian, K.: Near lossless reversible data hiding based on adaptive prediction. In: Proceedings of IEEE International Conference on Image Processing, pp. 2585–2588, Hong Kong, September 2010
8. Cox, I.J., Miller, M.L., Bloom, J.A., Fridrich, J., Kalker, T.: Digital Watermarking and Steganography, 2nd edn. Morgan Kaufmann Publishers, San Francisco (2008)
9. Ferreira, P.J.S.G., Pinho, A.J.: Why does histogram packing improve lossless compression rates? IEEE Sig. Process. Lett. **9**(8), 259–261 (2002)
10. Fridrich, J.: Steganography in Digital Media. Cambridge University Press, Cambridge (2010)
11. Fujiyoshi, M.: A histogram shifting-based blind reversible data hiding method with a histogram peak estimator. In: Proceedings of IEEE International Symposium Communication and Information Technology, pp. 318–323, Gold Coast, QLD, Australia, October 2012
12. Fujiyoshi, M.: A separable lossless data embedding scheme in encrypted images considering hierarchical privilege. In: Proceedings of EURASIP European Signal Process. Conference, Marrakech, Morocco, September 2013
13. Fujiyoshi, M.: A blind lossless information embedding scheme based on generalized histogram shifting. In: Proceedings of APSIPA Annual Summit and Conference, Kaohsiung, Taiwan, R.O.C., October 2013

14. Fujiyoshi, M.: Generalized histogram shifting-based blind reversible data hiding with balanced and guarded double side modification. In: Shi, Y.Q., Kim, H.-J., Pérez-González, F. (eds.) IWDW 2013. LNCS, vol. 8389, pp. 488–502. Springer, Heidelberg (2014)

15. Fujiyoshi, M.: A near-lossless data hiding method with an improved quantizer. In: Proceedings of IEEE International Symposium on Circuits System, pp. 2289–2292, Melbourne, VIC, Australia, June 2014

16. Fujiyoshi, M., Kiya, H.: An image-quality guaranteed method for quantization-based watermarking using a DWT. In: Proceedings of IEEE International Conference on Image Processing, pp. 2629–2632, Singapore, October 2004

17. Fujiyoshi, M., Sato, S., Jin, H.L., Kiya, H.: A location-map free reversible data hiding method using block-based single parameter. In: Proceedings of IEEE International Conference on Image Processing, pp. 257–260, San Antonio, TX, U.S., September 2007

18. Iwahashi, M., Kobayashi, H., Kiya, H.: Lossy compression of sparse histogram images. In: Proceedings of IEEE International Conference on Acoustics, Speech and Signal Process, pp. 1361–1364, Kyoto, Japan, March 2012

19. Iwahashi, M., Yoshida, T., Kiya, H.: Range reduction of HDR images for backward compatibility with LDR image processing. In: Proceedings of APSIPA Annual Sumit and Conference, Siem Reap, Cambodia, December 2014

20. Ni, Z., Shi, Y.Q., Ansari, N., Su, W.: Reversible data hiding. IEEE Trans. Circuits Syst. Video Technol. **16**(3), 354–362 (2006)

21. Pinho, A.J.: An online preprocessing technique for improving the lossless compression of images with sparse histograms. IEEE Sig. Process. Lett. **9**(1), 5–7 (2002)

22. Center for Image Processing, Rensselaer Polytechnic Institute: Kodak photo CD. http://www.cipr.rpi.edu/resource/stills/kodak.html

23. Center for Image Processing, Rensselaer Polytechnic Institute: Misc 1. http://www.cipr.rpi.edu/resource/stills/misc1.html

24. Sae-Tang, W., Fujiyoshi, M., Kiya, H.: Effects of random sign encryption in JPEG 2000-based data hiding. In: Proceedings of IEEE International Conference on Intelligent Information Hiding and Multimedia Signal Processing, pp. 516–519, Kitakyushu, Japan, August 2014

25. Tachibana, T., Fujiyoshi, M., Kiya, H.: A removable watermarking scheme retaining the desired image quality. In: Proceedings of IEEE International Symposium on Intelligent Signal Processing and Communication System, pp. 538–542, Awaji Island, Japan December 2003

26. Tsai, P., Hu, Y.C., Yeh, H.L.: Reversible image hiding scheme using predictive coding and histogram shifting. Sig. Process. **89**(6), 1129–1143 (2009)

27. van Schyndel, R., Tirkel, A.Z., Osborne, C.F.: A digital watermark. In: Proceedings of IEEE International Conference on Image Processing, pp. 86–90. Austin, TX, U.S., November 1994

28. Wu, J.H.K., Chang, R.F., Chen, C.J., Wang, C.L., Kuo, T.H., Moon, W.K., Chen, D.R.: Tamper detection and recovery for medical images using near-lossless information hiding technique. SIIM J. Digit. Imaging **21**(1), 59–76 (2008)

Hierarchical Learning for Large-Scale Image Classification via CNN and Maximum Confidence Path

Chang Lu[1], Yanyun Qu[1(✉)], Cuiting Shi[1], Jianping Fan[2], Yang Wu[3], and Hanzi Wang[1]

[1] Computer Science Department, Xiamen University, Xiamen, China
{luchangxmu, quyanyun, wang.hz}@gmail.com,
shicuitingfei@foxmail.com
[2] Computer Science Department, University of North Carolina at Charlotte,
Charlotte, NC, USA
jfan@uncc.edu
[3] Center for Frontier Science and Technology,
'Nara Institute of Science and Technology, Ikoma, Japan
wuyang0321@gmail.com

Abstract. We propose a framework to integrate the large scale image data visualization with image classification. The Convolution Neural Network is used to learn the feature vector for an image. A fast algorithm is developed for inter-class similarity measurement. The spectral clustering is implemented to construct a hierarchical visual tree. Instead of the flat classification way, a hierarchical classification is designed according to the visual tree, which is transformed to a path search problem. The path with the maximum joint probability is the final solution. Experimental results on the ILSVRC2010 dataset demonstrate that our method achieves the highest top-1 and top-5 classification accuracy in comparison with 6 state-of-the-art methods.

Keywords: Hierarchical learning · Large-scale · Image classification · Convolution neural network

1 Introduction

Nowadays, it is an era of big data. On one hand, the data are rising in an exploded way and are beyond the control of any specific user; on the other hand, users are eager to know the data in a macrograph view and then want to utilize them properly. Images play an important role in the big data era, though they are not as much explored as text contents. How to organize image data and how to understand image data are still challenging tasks. In this paper, we focus on the large-scale image dataset visualization and image classification.

Great progress has been made in large scale image classification [1–5]. However, most of the state-of-the-art methods have the following two limitations: (1) They only target at the classification accuracy and never care about visualizing the inter-class visual hierarchical relationship. However, as the exploding rise of the diverse social media data is beyond the users' administration, proper visualization can make it easier

© Springer International Publishing Switzerland 2015
Y.-S. Ho et al. (Eds.): PCM 2015, Part II, LNCS 9315, pp. 236–245, 2015.
DOI: 10.1007/978-3-319-24078-7_23

for users to grasp the distribution of the image data so that they can effectively manage and organize them; (2) Most of the adopted classifiers treat the classes equally. However, as we know, the real-world object classes have strong hierarchical relationships, and it is usually easier to distinguish coarse-level categories than to differentiate fine-grained subcategories. Furthermore, most flat classification methods are time consuming. In order to avoid the two limitations, we propose a framework to integrate visualization with classification. We first represent each image by the Convolution Neural Network (CNN), and then we organize the large-scale image data in a hierarchical way according to the images' visual appearance attributes. Secondly, we design the hierarchical classifier according to the visual hierarchical structure. Thirdly, we make the classification decision using the hierarchy of classifiers. There are three advantages in our method: (1) A hierarchical structure for organizing and visualizing the image collection; (2) A fast algorithm for measuring the inter-class similarity; (3) Hierarchical classification which saves computational time and resource.

The rest of this paper is organized as follows. In Sect. 2, we introduce the related work. In Sect. 3, we detail the hierarchical visual tree construction. In Sect. 4, we introduce how to make a decision according to the hierarchical structure. Experimental results are shown in Sect. 5. Conclusions are made in Sect. 6.

2   Related Work

There are two types of hierarchical learning approaches: taxonomy related methods and taxonomy unrelated methods. Motivated by the success of taxonomies in the web organization and text document categorization, many researchers in computer vision utilize taxonomies to organize the large scale image collections or improve the performance of the visual systems. Li et al. [1] constructed the ImageNet according to WordNet [6]. After that, the WordNet taxonomy has been widely used for image classification [2–5]. It is worth noting that the WordNet is semantic hierarchy, unrelated to visual effect. The hierarchical structure completely ignores the visual attributes.

It is reasonable to use visual information to learn a hierarchical structure for image classification [7, 8]. Sivic et al. [9] used a hierarchical latent Dirichlet allocation (hLDA) to discover a hierarchical structure from unlabeled images, which is helpful for image classification and image segmentation. Bart et al. [10] utilized an unsupervised Bayesian model to learn a tree structure to organize image collections. Both the two methods are tested on the moderate scale dataset, and their performances are not clear on the large scale image datasets. Moreover, they just aimed at image classification but not giving a visualization result. Some researchers [11–14] built the hierarchical model based on the confusion matrix, which is obtained or computed by the output of image categorization or object classification by using N one-vs-rest SVM classifiers. Griffin et al. [11] constructed a binary branch tree to improve visual categorization. Bengio et al. [13] made a label embedding tree for multi-class classification. Liu et al. [15] made a probabilistic label tree for large-scale classification. There are two limitations of these hierarchical learning methods based on the confusion matrix: (1) the computation of the confusion matrix using one-vs-rest SVM is time consuming; (2) the confusion matrix may not be reliable due to the imbalance of the training data.

Different from the above mentioned methods, we integrate the visualization with the hierarchical learning. Our method can give a distribution of hierarchical memberships for image categories. According to the hierarchical structure, we make the joint decision by levels of classifiers for image classification.

3 Image Presentation and Visual Tree Construction

3.1 Image Representation Based on CNN Features

Motivated by the success of CNN in computer vision, we utilize CNN for feature extraction. We use the pre-trained CNN model in CaffeNet [16] to represent an image on a Tesla C2050 GPU with 3 GB caches. The flowchart of the CNN is shown in Fig. 1. It contains 5 convolution layers and two fully-connected layers. In our scenario, we only extract the output of the second fully-connected layers as the feature vector. In detail, the size of an input image is 224*224*3. The first convolution layer consists of three operations: filter, normalization, and pooling. 96 filters of size 11*11*3 are implemented on a query image with a stride of 4 pixels. The outputs of this layer are 96 maps of the size 55*55. And then the filter results are normalized. In the pooling layer, a sliding window of size 3*3 is implemented with space 2 pixels. The second convolution layer contains 256 filters of size 5*5*48 by which the results of the first layer are filtered. After that, normalization and pooling are implemented followed the second layer whose parameters are the same as those in the first convolution layer. The outputs of the second layer are 256 maps of size 27*27. The third layer contains 384 filters of size 3*3*256, which is fully-connected. The outputs of the third layer are 384 maps of size 13*13. From the third convolution layer to the last convolution layer, no normalization and pooling are implemented. The fourth convolution layer has 384 filters of size 3*3*192 and the fifth convolution layer has 256 filters of size 3*3*192. There are two fully-connected layers, each of which has 4096 neurons. We input a query image in the CNN and extract the output of the second fully-connected layer as the feature vector, denoted by $I \in R^{4096}$.

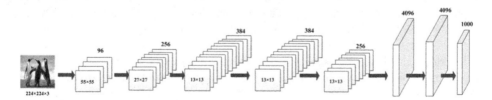

Fig. 1. The flowchart of the employed CNN.

3.2 Visual Tree Construction

In order to visualize the relationship between image categories, we construct a hierarchical visual tree based on clustering results. Suppose that there are M categories of images, $\{C_1, C_2, \ldots, C_M\}$, and the *ith* image category C_i contains N_i images which are represented by the CNN features, $\{I_l^i\}_{l=1}^{N_i}$. Aligning the inter-class semantic similarity context with the inter-class visual similarity context is still an unexplored issue in the multimedia and computer vision communities. Human perceptual factors may play an important role in designing a more suitable cross-modal alignment framework. Some researchers used the inter-class distance to measure the similarity. It is critical that how the inter-class similarity is defined. Here, we propose a measurement which utilizes the sum of the distances of all pairwise instances in the two classes as follow,

$$dis(C_i, C_j) = sqrt(\frac{1}{N_i N_j} \sum_s \sum_t \left\| I_s^i - I_t^j \right\|^2) \tag{2}$$

Equation (2) is time consuming, thus, we further infer Eq. (2) to obtain a fast computing way,

$$dis(C_i, C_j) = \left\| Q_i - Q_j \right\|^2 + \sigma_i^2 + \sigma_j^2 \tag{3}$$

where the mean vector of the category C_i is $Q_i = \frac{1}{N_i} \sum_{l=1}^{N_i} I_l^i$, and the square of the variance

of the category C_i is $\sigma_i^2 = \frac{1}{N_i} * \sum_{l=1}^{N_i} \left\| I_l^i - Q_i \right\|^2$.

Note that the operations of Euclid norm in Eq. (2) is $O(N_i N_j)$ while that in Eq. (3) is about $O(N_i + N_j)$. Thus, the computational complexity of Eq. (3) is much lower than that in Eq. (2). In the following, we implement the spectral clustering to construct a visual tree. We firstly build the similarity matrix. Its element is computed as, $A_{ij} = \exp(-\frac{dis(C_i, C_j)}{\delta_{ij}^2})$, where δ_{ij} is the self-tune parameter according to the method introduced in [17]. We adopt spectral clustering to split the image data. We recursively partition each node and terminate the partition until reaching any of the following conditions: (1) The current node is a leaf node;(2) The number of categories in the current node is less than K;(3) The depth is the maximum depth L. The maximum number of children nodes for each parent node and the values for K and L are pre-defined. Discussions on how they may influence the performance will be given in the experiment section. In Fig. 2, we show parts of a visual tree for the ILSVRC2010 dataset which contain four levels and each level contains no more than 6 branches. We can see that a node contains the classes which are visually similar. For the limit of the space, we just show the nodes containing flowers in detail in Fig. 2. In the first level, there are six groups. From left to right, the first node contains some classes with man-made objects, the second node contains some plant classes such as trees, the third node contains some animal classes, the fourth node contains some flower classes and the fifth node contains some fruit classes and other materials. For more detail, in the second level, the fourth nodes are visualized with its membership classes. From left to right, the flower classes are divided according to their corolla and color. In the bottom level, as shown in Fig. 2, the flower classes are divided according to the shape. Thus, we can see an object tree in which the membership of the object classes is visualized.

4 Hierarchical Learning

After constructing visual tree, the problem becomes how to make the final decision by using this tree for a new query image. Some researchers aimed at designing the pairwise classifier between the parent node and its children nodes, in which the classifiers in the parent node and its children nodes are trained simultaneously. Rather than focusing on the pairwise classifiers between the parent-children pairs, we aim at finding a path with the maximum joint probability. We transform the classification problem into a path search problem.

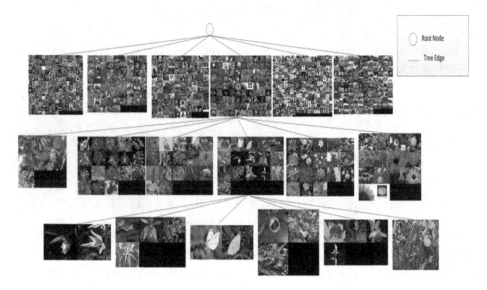

Fig. 2. Part of the visual tree for ILSVRC2010.

4.1 Classifier Training

At each level, we train the classifiers for the nodes with the same parent nodes. For the $(s + 1)th$ layer, we treat each group as a new combination class and take the one-vs-rest SVM to train classifiers. Considering a parent node s and its children node set $\{G_i^s\}_{i=1}^g$, if we train the classifier of the ith group G_i^s, we take the samples from the group G_i^s as the positive data, while we take the samples from the other groups $G_j^s (j \neq i)$ as the negative data. We only train the classifiers for the children nodes under the same parent node, disregarding of the other nodes in the same layer with the other parent nodes.

4.2 Hierarchical Prediction

In this subsection, we discuss how to use the learned classifiers to compute the final decision, which is hardly discussed in the literatures of image classification. In this paper, we transform the classification problem to the path search problem and keep the first N best paths.

Actually, the visual tree is treated as a causal Bayes network. The children node is only decided by its parent node. For the purpose of simplification, we suppose a children node is only related to its parent and is independent to its ancestors. For a query image, we should traverse the whole visual tree and find the path with the maximum joint probability. We formulate the problem as

$$c = \arg\max_{c_i} p(path(c_i))$$
$$= \arg\max_{c_i} \prod_i P(X_i(c_i)|parent(X_i(c_i))) \tag{6}$$

where c_i is the leaf node, $path(c_i)$ denote the path that can arrive at the leaf node c_i, $X_i(c)$ denote the state variable of the nodes which may take the label of the nodes.

Because traversing the whole visual tree is time consuming, we take the maximum confidence path algorithm to find the path with the maximum probability. Concretely, if we want to achieve the path with the maximum joint probability at the Lth layer, we should achieve the maximum probability at the $(L-1)th$ layer, and so on. In general, we formulate the problem as follows,

$$\max p(X_1, X_2, \cdots, X_{t+1}) = \max_{X_{t+1}} p(X_{t+1}|X_t) \max_{X_1,\cdots,X_t} p(X_1, X_2, \cdots, X_t) \tag{(9)}$$

Furthermore, considering the tree structure, the computational complexity is high if we use the maximum confidence path algorithm. Thus, we use an approximation method to find the path with the largest probability. In the ith layer, we just keep the first N best paths. The flowchart of the approximation N-best path algorithm is shown in Fig. 3.

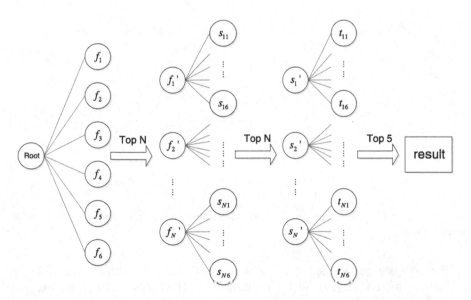

Fig. 3. The flowchart of the approximation N-best path algorithm.

5 Experimental Results

In this section, we implement our method on the ILSVRC2010 dataset which is one of the most popular large scale image datasets for image classification. The ILSVRC2010 dataset contains three parts: (1) The training set which contains 1.2 million training images for 1000 image categories (the number of images per category varying from 668 to 3047); (2) The verification set which contains 50 k images with 50 images per category; (3) The test set which contains 150 K images with 150 images per category. We use the top-1 and top-5 classification accuracy as the criteria. The LIBLINEAR toolbox [18] is used to train the SVM classifier. When training the second layer classifiers, whose parent node is the root node, we just use 600 images from each class training dataset as the new training samples. In order to train a classifier of a group, the positive samples are from all the classes contained in the group, and the negative samples are from the other group with the same parent node. When training other layer classifiers, we take all the samples from each class training set. The positive and negative set is similar to the one in the second layer. In the following, we firstly analyze how the parameters affect the performance and show the classification performance.

We analyze the influence of the number of layers and the number of branches in the visual tree. We construct 3 visual trees with different number of layers and different number of branches: (1) the visual tree contains 3 layers with 10 branches per node, denoted by T_10_3; (2) the visual tree contains 2 layers with 32 branches per node, denoted by T_32_2; (3) the visual tree contains 4 layers with 6 branches per node, denoted by T_6_4. The classification results are shown in Fig. 4. It demonstrates that the visual tree with 2 layers and 32 branches achieves the highest top-one and top-five accuracy. The classification performance are also influenced by the penalty parameter c in SVM, when c is equal to 0.05, the classification accuracy arrive the highest value.

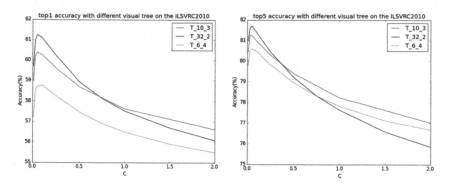

Fig. 4. The performance comparison with different visual trees with different penalty parameters c.

We compare our method (CNN features + visual tree) with 6 state-of-the art methods: sparse coding [19], SIFT + FV [20], JDL [21], Fisher vector [22], NEC [23], visual forest [24]. The results are shown in Table 1. All the results obtained by the competing methods are the original ones published by their authors. In [21, 24], the authors didn't give the top-5 results, so here the corresponding space is denoted as "N/A" (not available). Our method achieves significantly higher top-1 and top-5 classification accuracy than any others. The performance difference between our method and the secondly ranked method is more than 10 %.

Table 1. Comparison with the state-of-the-art methods on the ILSVRC-2010 test set.

Model	Top-1 accuracy	Top-5 accuracy
Sparse coding [19]	52.9 %	71.8 %
SIFT + FV [20]	54.3 %	74.3 %
JDL + AP Clustering [21]	38.9 %	N/A
Fisher Vector [22]	45.7 %	65.9 %
NEC [23]	52.9 %	71.8 %
Visual forest [24]	41.1 %	N/A
CNNs features + visual tree	**61.2 %**	**81.7 %**

6 Conclusions

In this paper, we propose a novel framework which integrates data visualization with image classification. We introduce the flexible CNN features to represent an image and design a fast inter-class similarity computational algorithm. Spectral clustering is adopted to construct a visual hierarchical tree recursively. According to the visual tree, we train one-vs-rest SVM classifiers for each children node (belonging to a parent node). After that, we design a decision rule which use the maximum joint probability to find the best path leading to the leaf node. Experimental results demonstrate that our method is superior to the other state-of-the-art methods in terms of both top-1 and top-5 classification accuracy.

Acknowledgements. This work was supported by the National Natural Science Foundations of China under Grants 61373077, 61472334 and 61170179,the Natural Science Foundation of Fujian Province of China Under Grant 2013J01257,the Fundamental Research Funds for the Central Universities under Grant 20720130720,the 2014 national college students' innovative and entrepreneurial training project, and the Scientific Research Foundation for the Introduction of Talent at Xiamen University of Technology YKJ12023R.

References

1. http://www.image-net.org/
2. Csurka, G., Dance, C., Fan, L., Willamowski, J., Bray, C.: Visual categorization with bags of keypoints. In: Workshop on Statistical Learning in Computer Vision, ECCV, pp. 1–2, Prague (2004)

3. Winn, J., Criminisi, A., Minka, T.: Object categorization by learned universal visual dictionary. In: Tenth IEEE International Conference on Computer Vision, ICCV 2005, pp. 1800–1807. IEEE (2005)
4. Nowak, E., Jurie, F., Triggs, B.: Sampling strategies for bag-of-features image classification. In: Leonardis, A., Bischof, H., Pinz, A. (eds.) ECCV 2006. LNCS, vol. 3954, pp. 490–503. Springer, Heidelberg (2006)
5. Fan, J., Gao, Y., Luo, H., Jain, R.: Mining multilevel image semantics via hierarchical classification. IEEE Trans. Multimed. 10, 167–187 (2008)
6. Miller, G., Fellbaum, C.: Wordnet: An Electronic Lexical Database. MIT Press, Cambridge (1998)
7. Fan, J., Shen, Y., Yang, C., Zhou, N.: Structured max-margin learning for inter-related classifier training and multilabel image annotation. IEEE Trans. Image Process. 20, 837–854 (2011)
8. Fan, J., He, X., Zhou, N., Peng, J., Jain, R.: Quantitative characterization of semantic gaps for learning complexity estimation and inference model selection. IEEE Trans. Multimed. 14, 1414–1428 (2012)
9. Sivic, J., Russell, B.C., Zisserman, A., Freeman, W.T., Efros, A.A.: Unsupervised discovery of visual object class hierarchies. In: IEEE Conference on Computer Vision and Pattern Recognition, CVPR 2008, pp. 1–8. IEEE (2008)
10. Bart, E., Porteous, I., Perona, P., Welling, M.: Unsupervised learning of visual taxonomies. In: IEEE Conference on Computer Vision and Pattern Recognition, CVPR 2008, pp. 1–8. IEEE (2008)
11. Griffin, G., Perona, P.: Learning and using taxonomies for fast visual categorization. In: IEEE Conference on Computer Vision and Pattern Recognition, CVPR 2008, pp. 1–8. IEEE (2008)
12. Marszałek, M., Schmid, C.: Constructing category hierarchies for visual recognition. In: Forsyth, D., Torr, P., Zisserman, A. (eds.) ECCV 2008, Part IV. LNCS, vol. 5305, pp. 479–491. Springer, Heidelberg (2008)
13. Bengio, S., Weston, J., Grangier, D.: Label embedding trees for large multi-class tasks. In: Advances in Neural Information Processing Systems, pp. 163–171 (2010)
14. Deng, J., Satheesh, S., Berg, A.C., Li, F.: Fast and balanced: efficient label tree learning for large scale object recognition. In: Advances in Neural Information Processing Systems, pp. 567–575 (2011)
15. Moosmann, F., Triggs, B., Jurie, F.: Fast discriminative visual codebooks using randomized clustering forests. In: Twentieth Annual Conference on Neural Information Processing Systems (NIPS 2006), pp. 985–992. MIT Press (2007)
16. http://caffe.berkeleyvision.org/
17. Frey, B.J., Dueck, D.: Clustering by passing messages between data points. Sci. 315(5814), 972–976 (2007)
18. http://www.csie.ntu.edu.tw/∼cjlin/liblinear/
19. Berg, A., Deng, J., Fei-Fei, L.: Large scale visual recognition challenge 2010 (2010). www.image-net.org
20. Sánchez, J., Perronnin, F.: High-dimensional signature compression for large-scale image classification. In: 2010 IEEE Conference on Computer Vision and Pattern Recognition (CVPR), pp. 1665–1672. IEEE (2011)
21. Zhou, N., Fan, J.: Jointly learning visually correlated dictionaries for large-scale visual recognition applications. IEEE Trans. Pattern Anal. Mach. Intell. 36, 715–730 (2014)
22. Perronnin, F., Akata, Z., Harchaoui, Z., Schmid, C.: Towards good practice in large-scale learning for image classification. In: 2012 IEEE Conference on Computer Vision and Pattern Recognition (CVPR), pp. 3482–3489. IEEE (2012)

23. Lin, Y., Lv, F., Zhu, S., Yang, M., Cour, T., Yu, K., Cao, L., Huang, T.: Large-scale image classification: fast feature extraction and SVM training. In: 2011 IEEE Conference on Computer Vision and Pattern Recognition (CVPR), pp. 1689–1696. IEEE (2011)
24. Fan, J., Zhang, J., Mei, K., Peng, J., Gao, L.: Cost-sensitive learning of hierarchical tree classifiers for large-scale image classification and novel category detection. Pattern Recognit. **48**, 1673–1687 (2014)

Single Camera-Based Depth Estimation and Improved Continuously Adaptive Mean Shift Algorithm for Tracking Occluded Objects

Jaehyun Im[1], Jaehoon Jung[2], and Joonki Paik[2(✉)]

[1] CIS Division, Design Group, SK Hynix, Seoul, Gyeonggi, Korea
[2] Department of Image Engineering,
Graduate School of Advanced Imaging Science,
Multimedia, and Film, Chung-Ang University, Seoul, Korea
paikj@cau.ac.kr

Abstract. This paper present a novel object tracking algorithm that can efficiently overcome the object occlusion problem by combining depth and color probability distribution information. The proposed algorithm consists of; (i) the depth estimation step using a color shift model (CSM)-based single camera, and (ii) the combination of depth and color probability distribution step using continuous adaptive mean shift (CAMSHIFT) algorithm, which is an adaptive version of the existing mean shift algorithm. In spite of the optimum object segmentation ability, the CAMSHIFT algorithm may fail in tracking if multiple occluded objects have similar colors. In order to overcome this limitation, the proposed algorithm combines depth and color probability distribution information. The experimental results show that the proposed algorithm is real time for well tracking the occluded object which cannot be tracked by the traditional CAMSHIFT algorithm, and the accuracy of depth estimation of the proposed algorithm is about 97.5 %.

Keywords: Object tracking · Depth estimation · Occlusion handling · CAMSHIFT

1 Introduction

Object tracking is a challenging problem in many computer vision applications such as surveillance, robot vision, user perceptual interface, augmented reality, and object-based image compression, to name a few. For a single stationary camera, object occlusion is a common phenomenon that critically degrades the tracking performance. For solving the occlusion problem, a number of methods have been proposed in the literature.

For a single stationary camera, mean shift or continuous adaptive mean shift (CAMSHIFT) algorithms are used [1, 2]. On the other hand, for multiple cameras, stereo vision-based methods are commonly used for estimating depth information that can differentiate multiple occluded objects [3, 4].

Mean shift algorithm is a robust, non-parametric technique that climbs the gradient of a probability distribution to find the mode (peak) of the distribution. CAMSHIFT is

© Springer International Publishing Switzerland 2015
Y.-S. Ho et al. (Eds.): PCM 2015, Part II, LNCS 9315, pp. 246–252, 2015.
DOI: 10.1007/978-3-319-24078-7_24

a modified version of the mean shift algorithm to deal with dynamically changing color probability distributions derived from video frame sequences. Bradski applied the CAMSHIFT algorithm for face tracking [2]. However, the performance of the CAMSHIFT algorithm is degraded by noise and dynamic lighting conditions. Furthermore, if any object has similar color to that of the target object, the tracking fails.

The proposed depth-based tracking algorithm is based on the following assumption. Even if two objects contain similar colors, the distance of each object is generally different, and we can keep tracking the target object using combined depth and position information. The proposed system employs optically modified apertures to estimate depth information using a single camera, and combines estimated object's depth information and color probability distribution estimated by CAMSHIFT algorithm for solving the occlusion problem in object tracking. Li also used color and depth information to track multiple objects in a crowded scene [5]. This method uses dominant color histogram to determine the order objects based on the distance. On the other hand, the proposed system can estimate the object's distance from the camera with significantly reduced computational load because of the special optical device, and at the same time it can solve the occlusion problem by combining the depth and color distribution information obtained by CAMSHIFT algorithm.

The work organized as follows. Section 2 introduces the theory of color shift model-based depth estimation. In Sect. 3, we present the tracking algorithm for occlusion handling. Section 4 provides the experimental results. Finally, Sect. 5 concludes the work.

2   Color Shift Model-Based Depth Estimation

In this paper, we use a computational camera with multiple color-filtered apertures (MCAs) to estimate object's depth information. The MCA camera generates a color shifts among the RGB color channels depending on the depth of the subject [6]. For this reason, it can calculate the distance to the object from camera by estimating the direction and amount of color shifts among the RGB color channels. Figure 1 shows the MCA system and its optical characteristics.

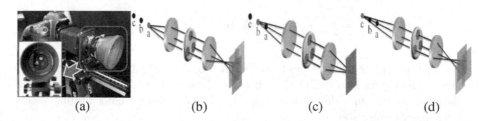

(a)   (b)   (c)   (d)

Fig. 1. The MCA system deployed on a digital single lens reflex (DSLR) camera: (a) the system configuration, (b) an object is closer than the focal length, (c) an object is on the focal length, and (d) an object is farther than the focal length.

Figure 2 shows various patterns of color shifting generated by the MCA camera. When an object is at the in-focusing position, there is no color shift. On the other hand, if the object is farther away or closer from the in-focusing position, color shift occurs. The direction of the color shift for the near-focused object is opposite to that of the far-focused one.

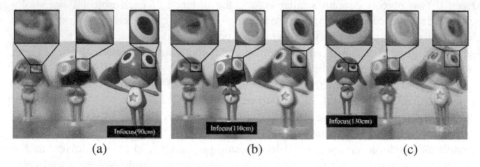

(a)   (b)   (c)

Fig. 2. Various color shifting patterns according to different distances: (a) near-focusing object, (b) in-focusing object, and (c) far-focusing object.

In order to estimate the object's distance, its region is first extracted using a simple background difference in each color channel as:

$$DI_c(x,y) = \begin{cases} 1, & if \ |BG_c(x,y) - I_c(x,y)| > T \\ 0, & otherwise \end{cases}, c \in \{R,G,B\}, \tag{1}$$

where c represents the color channel, DI_c is the extracted object region. BG_c and I_c respectively represent the background, and input image. T is the difference threshold.

Color channel shifting means phase shifting of the RGB channels in the frequency domain. Therefore, in this paper, we use phase correlation matching (PCM) to estimate the amount of color shift as [6]:

$$R_d(k,l) = \frac{F_G F_{R \ or \ B}^*}{|F_G F_{R \ or \ B}^*|}$$
$$= \sum_{m=sx}^{ex} \sum_{n=sy}^{ey} r(m\Delta x, n\Delta y) \exp\left(-j2\pi\left(\frac{mk}{M} + \frac{nl}{N}\right)\right), \tag{2}$$

where $R_d(k,l)$ represents the cross-power spectrum, F_c is the two-dimensional (2D) discrete Fourier transform (DFT) of each color channel, and $(\Delta x, \Delta y)$ represents the displacement of the color shift. The inverse 2D DFT provides the color shifting vector between color channels as:

$$r_d(x,y) = \delta(x - \Delta x, y - \Delta y). \tag{3}$$

In order to infer the relationship between color shifting vectors and object's depth information, we calculate the ratio of change of object's depth information according to color shifting vector. More specifically, we measure distances of an in-focused object and an object at arbitrary distance. Let the in-focusing distance be D_f, and an arbitrary distance D_t, then the rate of change of object's depth information can be expressed as:

$$I_d = \frac{|D_f - D_t|}{r_{df} - r_{dt}},$$

(4)

where I_d represents the ratio of object's depth information according to color shifting, r_{df} and r_{dt} respectively represent color shifting information at the in-focusing position, and an arbitrary position.

Given the ratio of change of object's depth information, we can calculate the depth information of object as:

$$Depth = D_f + I_d r_d.$$

(5)

3 Tracking Algorithm for Occlusion Handling

The entire framework of the proposed algorithm is shown in Fig. 3. In order to use CAMSHIFT to track objects in a video scene, a probability distribution of the desired color in the video scene must be created. In order to perform this, we first create a model of the desired hue using a color histogram in the hue, saturation, and value (HSV) color space. We create the color model for the proposed tracking algorithm by taking one-dimensional (1D) histograms from the hue channel.

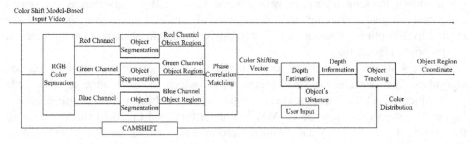

Fig. 3. The proposed object tracking algorithm using single camera-based object's depth estimation and object tracking for occlusion handling.

For discrete 2D image probability distribution, under assumption that $I(x, y)$ represents the probability distribution at position (x, y) in image, the CAMSHIFT algorithm first determines the size of the initial search window, and moves the center of the search window to mean location of data. In order to find the new mean location of the search window, we compute the zero-th moment as:

$$M_{00} = \sum_x \sum_y I(x,y), \tag{6}$$

and the first moments for x and y as:

$$M_{10} = \sum_x \sum_y x I(x,y), \quad M_{01} = \sum_x \sum_y y I(x,y). \tag{7}$$

The new mean location of search window is then defined as:

$$x_c = \frac{M_{10}}{M_{00}}, \quad y_c = \frac{M_{01}}{M_{00}}. \tag{8}$$

For the new mean location of search window, repeat the above steps until convergence.

In order to deal with the occlusion problem, we define the criterion as:

$$Occlusion = \begin{cases} true, & if\ D_{diff} \geq T \\ false, & otherwise \end{cases}, \tag{9}$$

where D_{diff} represents the change of depth, and T represents the corresponding threshold. In other words, the depth of object that exceeds the pre-specified threshold is the sudden change of depth, thus it is considered as occlusion.

4 Experimental Results

For evaluating the proposed occluded object tracking performance, the comparison of the proposed tracking algorithm with general CAMSHIFT algorithm is performed using test images containing multiple objects of similar colors. We also used the ground truth depth information of object to prove the depth estimation performance of the proposed algorithm.

Figure 4 shows experimental results of CAMSHIFT algorithm and the proposed algorithm. The target is the small green object, which moves around another large green object. Occlusion occurs when the target goes behind the large green object. Figure 4(a) shows result of the CAMSHIFT algorithm, and Fig. 4(b) shows results of the proposed algorithm. When the target object goes behind another object, CAM-SHIFT algorithm tracks the one with the largest density. On the other hand, the proposed algorithm can keep tracking the target object even after occlusion by using combined depth information and color probability distribution.

Figure 5 shows another test sequence for depth estimation. The lady is walking toward the camera, and raises her hand at every 4 m in the range from 45 to 21 m.

Figure 6 compares the estimated distances with ground truth data. The blue curve represents the ground truth, and the red curve represents the estimated object's distance using the proposed algorithm. The accuracy of depth estimation of the proposed algorithm is about 97.5 %, and the error range is ± 2 m.

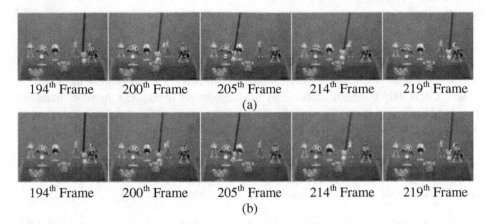

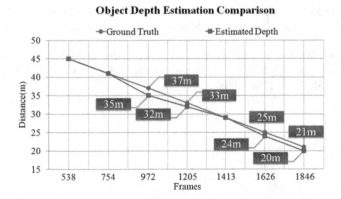

Fig. 4. Comparison of experimental results using; (a) CAMSHIFT algorithm, and (b) the proposed algorithm. Occlusion occurs at the 205th frame. The proposed algorithm can keep tracking the target after the 205th frame whereas the CAMSHIFT algorithm tracks another bigger object.

Fig. 5. Test sequence for evaluating the performance of depth estimation.

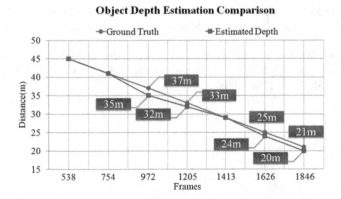

Fig. 6. Comparison of estimated distances using the proposed algorithm with the ground truth data.

5 Conclusion

In this paper, we presented a combined single camera-based depth estimation and robust object tracking algorithm. The existing CAMSHIFT algorithm tracks object, which has the largest color density probability. If two occluded objects have similar colors, CAMSHIFT cannot successfully track the target object. On the other hand the proposed algorithm keeps tracking the target object using additional depth information since two occluded objects have different distances. Major contribution of this paper is a novel configuration of real-time depth estimation using a single camera and its application to improve robustness of tracking occluded object. For this reason the proposed algorithm can be applied to various intelligent vision fields, such as video surveillance, robot vision, and smart space with visual interactions.

Acknowledgment. This work was supported by the ICT R&D program of MSIP/IITP. [14-824-09-002, Development of global multi-target tracking and event prediction techniques based on real-time large-scale video analysis and by the Technology Innovation Program (Development of Smart Video/Audio Surveillance SoC & Core Component for Onsite Decision Security System) under Grant 10047788 and by Ministry of Culture, Sports and Tourism (MCST) and Korea Creative Content Agency(KOCCA) in the Culture Technology (CT) Research & Development Program.

References

1. Li, Z., Gao, J., Tang, Q., Sang, N.L Improved mean shift algorithm for multiple occlusion target tracking. SPIE Opt. Eng. 47, 086402, 1–6 (2008)
2. Bradski, G.: Computer vision face tracking for use in a perceptual user interface. In: IEEE Workshop on Application of Computer Vision, pp. 214–219 (1998)
3. Skulimowski, P., Strumilo, P.: Refinement of depth from stereo camera ego-motion parameters. IET Electron. Lett. **44**, 729–730 (2008)
4. Kwolek, B.: Face tracking system based on color, stereovision and elliptical shape features. In: Proceedings of IEEE Conference Advanced Video, Signal Based Surveillance, pp. 21–26 (2003)
5. Li, L., Huang, W., Gu, I., Luo, R., Tian, Q.: An efficient sequential approach to tracking multiple objects through crowds for real-time intelligent CCTV systems. IEEE Tarns. Syst. Man Cybern. **38**, 1254–1269 (2008)
6. Kim, S., Lee, E., Maik, V., Paik, J.: Real-time image restoration for digital multifocusing in a multiple color-filter aperture camera. SPIE Opt. Eng. 49, 040502, 1–3 (2010)

A Flexible Programmable Camera Control and Data Acquisition Hardware Platform

Fei Cheng[1]($\boxtimes$), Jimin Xiao[1], Tammam Tillo[1], and Yao Zhao[2]

[1] Department of Electrical and Electronic Engineering,
Xian Jiaotong-Liverpool University (XJTLU), 111 Ren Ai Road, SIP,
Suzhou 215123, Jiangsu Province, People's Republic of China
{fei.cheng,jimin.xiao,tammam.tillo}@xjtlu.edu.cn
http://www.mmtlab.com
[2] Beijing Jiaotong University, Institute of Information Science, Beijing, China
yzhao@bjtu.edu.cn

Abstract. There are a number of standard video sequences produced by some organizations and companies, which simulate different scenes and conditions in order to properly test video coding methods. With the rapid development of sensing and imaging technologies, more information can be employed to assist video coding. However, the existed sequences cannot satisfy the recently proposed video coding methods. Moreover, to capture 3D video, multi-cameras, including texture and depth cameras, are used together. Thus, a synchronization mechanism is needed to control different cameras. In addition, some adverse conditions, such as light source flicker, affect the quality of the video. In order to acquire proper video sequences, we proposed a flexible control platform for texture and depth cameras, which could address the above issues. We have applied the platform to test several new methods proposed recently. With the assistance of this platform, we have achieved considerable positive results.

Keywords: Video sequence · Camera control · Synchronization · Flicker · Platform

1 Introduction

Globally, video traffic accounted for 66 percent of all consumer Internet traffic in 2013 [1]. With the development of imaging and display technologies, the video resolution is rapidly increasing. The 4K (3840×2160) resolution has been used in camcorders and TVs. Meanwhile, 3D TVs and films are being accepted in consumer market. Moreover, some sensors, such as motion sensors and GPS, have been integrated on smart devices, which could assist the video coding and enhance the video quality [2]. Figure 1 shows the current development trend of

This work was supported by National Natural Science Foundation of China (No.61210006, No.60972085).

© Springer International Publishing Switzerland 2015
Y.-S. Ho et al. (Eds.): PCM 2015, Part II, LNCS 9315, pp. 253–261, 2015.
DOI: 10.1007/978-3-319-24078-7_25

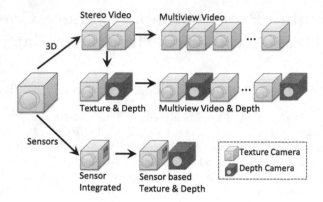

Fig. 1. The development trend of video acquisition

video coding. The compression performance of modern video coding is improving constantly, and new video coding methods are emerging. Therefore, new video sequences with auxiliary information need to be produced to test new methods, and more importantly video acquisition platforms with sensor capability are needed to simulate different emerging scenarios.

In fact, the existing video sequences cannot meet the requirement of some recently proposed video coding methods. For instance, the movement of smart phone while capturing a video results in the global motion between frames, which increases the bit-rate and the computational complexity of the video encoder. The global motion information could be used to assist video coding [3]. The problem is that no existing video sequence includes the camera global motion information. Therefore, it is necessary to develop an open and flexible video capturing platform to obtain different kinds of video and auxiliary information.

In terms of 3D video, complicated camera settings bring a great challenge for 3D video acquisition. Multiple cameras have to work synchronously to capture the 3D scene. In stereo video, two cameras are employed to capture two views, while the multiview video normally deals with cases with more than two cameras. Another 3D video scheme is texture plus depth format, which generates virtual views using the Depth-Image-Based Rendering (DIBR) [4,5]. A depth map, which represents the distance from the objects in the scene to the camera, and its aligned texture, have been exploited to describe 3D scenes. Different approaches to obtain depth exist, however, currently most depth maps are estimated using software-based approaches from several views of the scene, which has many problems. Meanwhile, the hardware depth acquisition technology is rapidly developing. Some depth cameras are now becoming available. Multi-view Video plus Depth (MVD) format is a promising way to represent 3D scene [6,7], where multiple texture and depth cameras are employed. As the alignment of texture and depth is of great importance in 3D video, the position of each camera and the frame rate need to be controlled accurately.

There are some other issues affecting the video sequence acquisition. Firstly, the flicker effect caused by light source makes the brightness fluctuation between frames [8]. This effect would reduce the performance of Motion Estimation (ME) for video encoder [9]. This problem becomes more serious with different types of cameras, which causes luminance difference between different views. Secondly, without a hardware trigger, it is difficult to obtain several video acquisitions for the same movement. Lastly, the existing camera platforms are not flexible and programmable to fully control cameras and their movements.

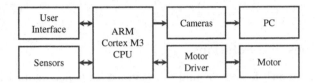

Fig. 2. The diagram of the proposed platform

We proposed and developed a programmable camera control and data acquisition hardware platform as shown in Fig. 2. This platform is designed based on a slider tracker, while a DC motor can move the platform properly. An ARM CPU is exploited to control and manage the motor, sensors and interface. This platform can drive four texture cameras or depth cameras by using Pulse Width Modulation (PWM). To reduce the flicker effect, the frequency and phase of power grid can be sampled. Moreover, in order to fully and conveniently control cameras and related devices, a user-friendly interface is developed. This platform has been used to obtain experiment data for several schemes we proposed recently. We have achieved some positive results based on those video sequences and data. Two works about planar surfaces detection have been published [10]. Figure 3 presents an example of this platform, which is able to record texture plus depth video with global camera motion.

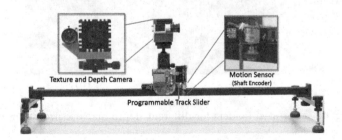

Fig. 3. An example of our texture plus depth record system with global motion information

The rest of this paper is organized as follows. In Sect. 2, the details of our solutions and designs are introduced. Then, tests and related experiment results are presented in Sect. 3. Finally, Sect. 4 concludes this work.

2   Methodology

The proposed platform is developed to satisfy the requirements of several experiments. In this section, the difficulties are specified and the corresponding solutions are explained. Then, the details of developed hardware and software are introduced.

2.1   Difficulties and Solutions

For texture plus depth format or multiview video sequence acquisition, a mechanism should be used to mount two different types of cameras. Their relative positions are pivotal arguments for image alignment and rendering. Although the calibration methods could reduce of position errors, the accurate positions could minimize the errors of image alignment and rendering. As different cameras have different sizes and shapes, it is difficult to mount them properly. In order to make the camera positions relatively accurate, the 3D molding and printing technology is exploited to customize the cameras holder. By using 3D molding, the holder could be designed to adjust to camera size and shape. More importantly, it could guarantee that the center of each camera is aligned horizontally or vertically.

(a)   (b)

Fig. 4. 3D printed holders for different depth and texture cameras

Figure 4 displays two examples of 3D printed holders. In Fig. 4 (a), an industry camera Blaser acA640-90gc is used as texture camera, while the depth camera is a SwissRanger SR4000 [11]. The Fig. 4 (b) is a vertically alignment of a DSLR camera and depth camera.

The flicker effect is caused by unstable light sources. As the power grid is AC power, the lamp luminance would change with the frequency of power grid. A kind of common light source is fluorescent lamp, which would cause serious flicker effect. Even incandescent produced light suffers from flicker effect. As shown in Fig. 5, when the frame rate of video capturing dose not exactly match the frequency of the power grid, the video frame would flicker. Therefore, we sample the frequency and the phase of the power grid in order to adjust the frame rate and trigger the acquisition. Consequently, the video flicker effect can be minimized.

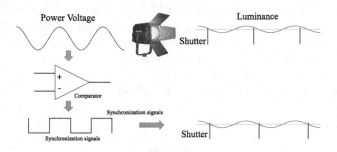

Fig. 5. The flicker effect and the proposed solution

In some video experiments, the video is captured when the camera is moving. If we want to obtain the same video content with different camera settings, the video capturing should start at a fixed position. We use a mechanical switch as shown in Fig. 6 to generate a trigger signal. In Fig. 6, the longer bar represents the slider tracker, while the shorter one is a trigger block. When the platform is moving towards left, the switch on the platform will hit the trigger block and generate a signal. This trigger signal is sent to the CPU, which starts the capturing process. This solution is low cost and reliable.

Fig. 6. The proposed position trigger using a mechanical switch

2.2 Implementation

In order to increase the reliability and flexibility, a Printed Circuit Board (PCB) was developed to integrate all the components and circuits. The PCB schematic consists of a CPU, a power supply, sensors, communication modules, camera drivers, a motor driver and an user interface, etc.

The selected CPU is an ARM 32-bit cortex M3 core based STM32 F103. The supported frequency is up to 72 MHz, which is enough for the developed control strategy. This CPU supports seven timers, which could generate PWM signals to drive motor and cameras. The power is supplied by a 12V DC battery to avoid interference from AC power. To minimize the flicker effect, a comparator is used to convert the AC signal to digital synchronization signal as shown in Fig. 5. A shaft encoder is exploited as motion sensor to adjust the speed of motor and record the position of the platform. In terms of communication, BlueTooth and RS232 are employed to communicate with smart devices and PCs. To drive different kinds of cameras, a inferred (IR) emitter and four PWM signal sockets are used. The IR emitter can control most DSLR cameras, while the PWM signal can drive industry cameras. The motor driver is composed of two half bridge ICs. In order to operate the platform conveniently, the user interface consists of an IR remote controller, buttons and a LCD screen.

Figure 7 shows the platform and each key component on it. The PCB schematic and layout are available for download at http://www.mmtlab.com/platform[1].

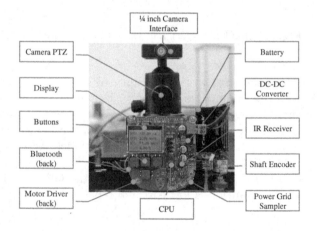

Fig. 7. The hardware design for this platform

The software of the proposed platform is developed based on the standard peripheral library published by STMicroelectronics. By using the library, the development cycle can be reduced.

The system resources need to be allocated reasonably. Four timers are exploited for motor driving, camera driving, IR signal capturing and speed capturing. The interrupt mechanism is used for real-time capturing. In order to debug the software system, an UART based command line interface is designed. By using the PC command line, each parameter can be on-line adjusted and

---

[1] All the hardware schematics and software codes will be available for download after the paper is accepted.

stored. Moreover, with the assistance of script programs, cameras and other related devices can be controlled automatically. The motor close loop control method is based on PID algorithm [12], which can control the speed and distance of the movement.

The current version of software utilizes only half of the CPU resources, which means the software is able to be updated in the future. More functions can be implemented on this platform. The source code of the proposed platform is available for download at http://www.mmtlab.com/platform.

3   Test and Related Experiments

3.1   Tests

After PCB customization and system assembly, the proposed platform was tested. The PID parameters were adjusted to make the platform move smoothly and accurately. The movement of this platform can satisfy the requirement of variety of experiments. It is worth mentioning that the software could be modified based on specific requirement of different experiments. Therefore, this platform is flexible and programmable.

Fig. 8. A related experiment based on the proposed platform

3.2   Related Experiments and Results

This platform has been used to conduct two experiments related to some proposed methods [10]. Moreover, some new methods are being developed based on this platform. Figure 8 shows a related experiment using this platform. In this case, we use one texture camera and one depth camera, while the platform is moving. The video sequences of texture and depth are captured synchronously with the motion information. By using the produced sequence, Fig. 9 reports that a proposed method which exploits the camera global motion information is better than the baseline codec (i.e., JM software [13] without using motion information). The BD-Rate is -11.87 %, while the BD-PSNR is 0.49 dB.

The other experiment is depth map coding using motion sensor information. This experiment verified that the proposed global motion information assisted depth map coding is more efficient than the baseline codec (i.e., JM software). The BD-Rate is -41.12 %, and the BD-PSNR is 2.04 dB. The experiments based on the platform indicate that the platform can be beneficial to various emerging video coding scenarios (Fig. 10).

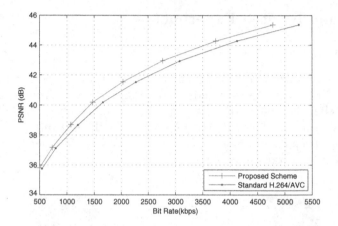

Fig. 9. The RD-performance of one experiment of texture coding based on camera global motion and depth information

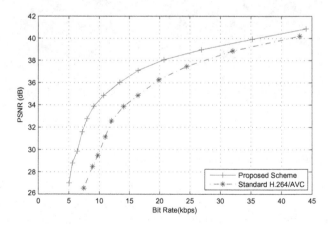

Fig. 10. The RD-performance of one experiment of depth map coding based on camera global motion information

4   Conclusions

This paper introduced a flexible and programmable camera control and video data acquisition platform. This platform could control the movement of cameras and record the motion information. Multiple types of cameras could be driven and captured synchronously. The sampling of power grid is exploited to adjust the frame rate of video capturing in order to reduce the flicker effect. It is worth mentioning that the hardware schematic, layout and the software source codes are published and available for download.

With the assistance of this platform, we completed some video coding experiments and obtained positive results. In the future, this platform will be updated continuously. More functions will be implemented on it, while the performance of the existing functions will be improved.

References

1. CISCO White Paper, Cisco Visual Networking Index: Forecast and Methodology, 20132018. http://www.cisco.com/
2. Angelino, C.V., Cicala, L., Persechino, G., Baccaglini, E., Gavelli, M., Raimondo, N.: A sensor aided H.264 encoder tested on aerial imagery for SFM. In: IEEE International Conference on Image Processing (ICIP), Paris, France, 10/2014 (2014)
3. Chen, X., Zhao, Z., Rahmati, A., Wang, Y., Zhong, L.: Sensor-assisted video encoding for mobile devices in real-world environments. IEEE Trans. Circuits Syst. Video Technol. **21**(3), 335–349 (2011)
4. Merkle, P., Smolic, A., Muller, K., Wiegand, T.: Multi-view video plus depth representation and coding. In: IEEE International Conference on Image Processing, ICIP 2007, vol. 1, pp. I-201. IEEE (2007)
5. Fehn, C.: Depth-image-based rendering (DIBR), compression, and transmission for a new approach on 3D-TV. In: International Society for Opticsand Photonics Electronic Imaging 2004, pp. 93–104 (2004)
6. Hannuksela, M., Chen, Y., Suzuki, T., Ohm, J.-R., Sullivan G., (ed.) AVC draft text 8. In: JCT-3V document JCT3V-F1002, vol. 16 (2013)
7. Chen, Y., Hannuksela, M.M., Suzuki, T., Hattori, S.: Overview of the MVC+ D 3D video coding standard. J. Vis. Commun. Image Represent. **25**(4), 679–688 (2014)
8. Bucci, G., Fiorucci, E., Landi, C.: A digital instrument for light flicker effect evaluation. IEEE Trans. Instrum. Meas. **57**(1), 76–84 (2008)
9. Qiao, S., Zhang, Y., Wang, H.: PI-frames for flickering reduction in H.264/AVC video coding. In: 2012 International Conference on Computer Science Service System (CSSS), pp. 1551–1554, August 2012
10. Jin, Z., Tillo, T., Cheng, F.: Depth-map driven planar surfaces detection. In: IEEE Visual Communications and Image Processing Conference (IEEE VCIP), Valletta, Malta, 12/2014 (2014)
11. MESA IMAGING. SR4000. http://www.mesa-imaging.ch/products/sr4000/
12. Shumway-Cook, A., Woollacott, M.H.: Motor Control: Theory and Practical Applications, vol. 157. Williams & Wilkins Baltimore, Baltimore (1995)
13. HHI Fraunhofer Institute, H.264/AVC Reference Software. http://iphome.hhi.de/suehring/tml/download/

New Media Representation and Transmission Technologies for Emerging UHD Services

Comparison of Real-time Streaming Performance Between UDP and TCP Based Delivery Over LTE

Sookyung Park, Kyeongwon Kim, and Doug Young Suh$^{(\boxtimes)}$

Department of Electronics and Radio Engineering, Kyung Hee University,
Seoul 446-701, South Korea
{thunder3000, beatiful72}@naver.com, suh@khu.ac.kr

Abstract. Video traffic becomes dominant in mobile networks since users use mobile networks for enjoying more instant and handy video service. Currently, TCP and UDP are major transport layer protocols for video streaming service. This paper compares performance of those protocols over LTE in terms of Maximum Available Bitrate (MAB). The results will become the key clue to select one between DASH (Dynamic Adaptive Streaming over HTTP) and MMT (MPEG Media Transport).

Keywords: Mobile network · UDP · TCP · Video streaming service · LTE · Maximum available bitrate · MMT · DASH

1 Introduction

Mobile market is gradually growing in the size, and the ratio of the mobile data traffic occupied by the video is 55 % in 2012. The part occupied by the video traffic is expected that more than 69 % in 2019 [1]. Using mobile is more comfortable than other equipment (laptop or desktop) to use consumer goods. This causes the result of be offered the real-time service. The video streaming service is a type of real-time service and there are two types of video streaming services which are Video on Demand (VOD) Streaming Service and Real-time Streaming Service. In case of VOD streaming, the server has various media files and sends the requested media file to the client. Another case which is the real-time streaming service, the server does not have whole of video file but has fragments of the filming video. Both the VOD service and the real-time streaming service must have buffer for smooth video playback, because of the network delay which is caused by various factors. To create a robust system in delay, the de-jittering buffer must be existed. Presence of the buffer causes the buffering which induce that the video played after a period time as well as the larger the buffer, the longer the delay to be axiomatic. Therefore, we think that it is important to measure the accurate available bitrate to decrease the buffering delay. We measure Maximum Available Bitrate (MAB) of UDP and TCP which one can be the main transport layer protocol to transfer streaming video data as well as analyze which protocol is more suitable for the estimate available bitrate.

© Springer International Publishing Switzerland 2015
Y.-S. Ho et al. (Eds.): PCM 2015, Part II, LNCS 9315, pp. 265–274, 2015.
DOI: 10.1007/978-3-319-24078-7_26

2 Related Work

In this chapter, we present features of UDP and TCP as well as the estimate available bitrate algorithm to improve End-to-End QoS.

2.1 Features of UDP and TCP

UDP is the best-effort protocol which does not guarantee reliable transmitting. It causes the need to add Forward Error Correction (FEC) code. TCP is the connection-oriented protocol which guarantees reliable transmitting with congestion and flow control. For this reason, TCP is a reliable protocol than the UDP. Data rate, on the other hand, is higher UDP than TCP since UDP does not control the rate for reliability. Because of the trade-off in reliability and data rate, the service provider is used to select the protocol suitable for the service that it provides. To transmit video stream to contents consumer, UDP and TCP are both commonly used in recent years.

2.2 Estimate Available Bandwidth

In this section, we briefly describe the existing estimate available bitrate algorithm.

Pathload uses a one-way delay of periodic packet stream and observes that the stream rate is how much increase as compared to the available bitrate [2]. It can apply the amount of bandwidth similar with the estimate available bandwidth to the adaptive application, there is a small disadvantage to large time convergence.

pathChirp is the method using the concept of "self-induced congestion". Exponentially time spaced packets are used and available bandwidth is estimated by using that the amount of delay is decreased or increased [3].

TOPP use the analysis method by segmented regression. The well-separated probe packet pair is used, the available bitrate is measured using a delay, regardless of the network condition [4].

pathQuick2 utilizes the probe slope model, including the concept of the probe rate model [6]. PSM-based method measures the effective UDP throughput quickly, and estimates the available bandwidth at the same time. Each packet in the packet train is transmitted in the same time interval, *pathQuick2* uses a delay of time interval [5].

[7] presumes available bitrate using the one-way delay distribution. The Bayesian mechanism is proposed to model bandwidth variations. The algorithm can estimate the available bandwidth very fast, because it does not use the RTT, but uses the one-way delay which is receiver-based mechanism. However, service provider sets some parameters to estimate the available bandwidth, but those parameters are not shown on.

3 Experiment Method and Mathematical Model

In reality, the change of the MAB is irregular and difficult to predict due to various factors. The irregular characteristics make it difficult to adapt streaming service application to the current channel, which immediately appears to the degradation of

QoS and QoE. In this chapter, we propose some parameters which are the criterion that which one is more appropriate between the UDP and the TCP, also those are the value of applicability on adaptive streaming service.

3.1 Measure Method

To analyze network statement, one PC sender and one laptop receiver were used. Laptop computer connect Internet using LTE network by the tethering using mobile phone. Sender transmits more than 1 Mbytes using 10 kbytes packet over best-effort on UDP. If receiver takes 1 Mbytes size data, it sends a signal which uses flag and measuring RTT to server. Sender receives this signal and sends 1 Mbytes segment immediately. Repeating this simple test for a period of time, MAB is measured. The method is as follows Fig. 1.

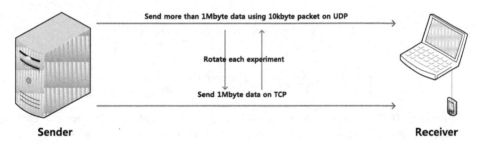

Fig. 1. Experimental method

1. **UDP MAB**
   Receiver measures UDP MAB using the getting started time (T_{1st}) and the finishing received time (T_{last}). The equation is as follows (1).

$$MAB_{UDP}(t) = \frac{1Mbyte \times 8(bit/byte)}{(T_{last\_i} - T_{1st\_i}) \times 10^{-3}s}(Mbps) \qquad (1)$$

   Also, $T_{last\_i} - T_{1st\_i}$ is defined as the inter arrival time J_{i_UDP}.
2. **TCP MAB**
   Receiver measures TCP MAB using the getting started time (T_{T1st}) and the finishing received time (T_{Tlast}). The equation is as follows (2).

$$MAB_{TCP}(t) = \frac{SegmentSize(byte) \times 8(bit/byte)}{(T_{Tlast\_i} - T_{T1st\_i}) \times 10^{-3}s}(Mbps) \qquad (2)$$

   Also, $T_{Tlast\_i} - T_{T1st\_i}$ is defined as the inter arrival time J_{i_TCP}.

3. **Markov model**

   The Markov model is a valuable model that can be used to "predict the future through the past". In this paper, bitrate is how sustainable and changeable between a good environment and a bad environment. The threshold of two probability variable is 14 *Mbps* for UDP and 9 *Mbps* for TCP. Value to be obtained through the model are shown in the following (3) ~ (4).

$$P_{g \to b} = \frac{P_{gb}}{P_{gg} + P_{gb}} \tag{3}$$

$$P_{b \to g} = \frac{P_{bg}}{P_{bg} + P_{bb}} \tag{4}$$

   S.Time is a formula on how much each state is maintained. The equations are shown in the following (5) ~ (6).

$$S.Time_{gb} = \frac{1}{P_{g \to b}} \times PacketTimeInterval(sec) \tag{5}$$

$$S.Time_{bg} = \frac{1}{P_{b \to g}} \times PacketTimeInterval(sec) \tag{6}$$

   UDP and TCP is interpolated every 500 ms to ease the calculations, because each experiments run at random time space. So, *PacketTimeInterval* is 0.5 s in this paper.

4. **Second Order Differential**

   Second-order differential of bitrate means instantaneous rate of change of bitrate, using equations shown in (7) ~ (8).

$$f'_{MAB}(t_k) = \frac{f_{MAB}(t_{i+1}) - f_{MAB}(t_i)}{t_{i+1} - t_i} \quad when, \ t_k = \frac{t_{i+1} + t_i}{2} \tag{7}$$

$$f''_{MAB}(t_l) = \frac{f'_{MAB}(t_{k+1}) - f'_{MAB}(t_k)}{t_{k+1} - t_k} \quad when, \ t_l = \frac{t_{k+1} + t_k}{2} \tag{8}$$

5. **Correlation**

   If the random variable has more than two, the joint PDF can be used to perform various calculations. Among them, covariance and correlation coefficient is used to learn correlation between random variable X and Y. The formula for covariance is the same as (9).

$$\sigma_{XY} = Cov(X, Y) = E[(X - \mu_X)(Y - \mu_Y)] = E(XY) - \mu_X \mu_Y \tag{9}$$

   μ_X and μ_Y is $E(X)$ and $E(Y)$, the correlation coefficient through covariance is defined by the following (10).

$$\rho = \frac{Cov(X,Y)}{\sqrt{Var(X)Var(Y)}} \tag{10}$$

ρ has a value between -1 and 1, if $\rho = 0$, X and Y is independent. In this paper, we analyze correlation using MAB and RTT.

4 Comparison Between UDP and TCP

4.1 Maximum Available Bitrate

This experiment carried out aboard the bus around Gangnam of Seoul, Korea as shown in Fig. 2. It measured during about 1000 s. The result of experiment is represented in Fig. 3. The blue line is the MAB of UDP and the red spotted line is the MAB of TCP. Generally, the MAB of TCP is higher than the MAB of UDP. In addition to, we can confirm that the variation of TCP is larger than the UDP. If the MAB is low, the client cannot receive high quality video. Large variance means that quality of service is often changed. It force to use larger buffer, it leads to longer delay that the time was taken to play video. It is not desirable for real-time streaming service, because real-time streaming service is sensitive for delay. Especially, when offered service is sports, client was spoiled crucial moments, because of delay.

Additionally, the loss packet and the bitrate are closely related. About 20,000 of the UDP packets were sent and only 5 packets were lost. The loss caused by the congestion and the fading is substantially not occur, this part is ignored in this paper.

Fig. 2. Gangnam, Seoul, Korea, the map which is place of experiment

How long network conditions maintain is shown in Table 1. At good statement, UDP are kept during 33.93 s, TCP are maintained just 1.3 s. However, at bad statement, UDP are kept during 9.3 s, TCP are maintained 6.23 s. It is the best that good statements are maintained longer and bad statements are maintained shorter. But, the threshold of UDP is 14 Mbps is higher than the one of TCP 9 Mbps.

Table 1. Sustainment Time of each condition

			Second
UDP	Avg. Bitrate: 17.2 Mbps	$S.Time_{gb}$	33.93 s
	Threshold: 14 Mbps	$S.Time_{bg}$	9.3 s
TCP	Avg. Bitrate: 11.7 Mbps	$S.Time_{gb}$	1.3 s
	Threshold: 9 Mbps	$S.Time_{bg}$	6.23 s

4.2   Distribution of J_i

Measured J_i fits to the Erlang distribution. In general, J_i of UDP is smaller than TCP, because UDP has higher MAB. J_{i_UDP} and J_{i_TCP} are both right-skewed, but J_{i_TCP} has heavy-tail. The heavy-tail means that probabilities were not concentrated on average, and were spread both sides. This characteristic makes hard to estimate MAB, and is represented in Fig. 4. In this graph, inter arrival times of UDP are concentrated from 380 ms to 526 ms, but inter arrival times of TCP don't have appropriately concentrated point. We show that J_{i_UDP} and J_{i_TCP} are fit $k_{UDP} = 3, \mu_{UDP} = 25$ Erlang distribution and $k_{TCP} = 3, \mu_{TCP} = 80$ Erlang distribution. Consequently, the UDP protocol is suitable for the adaptive streaming, because of easier estimation.

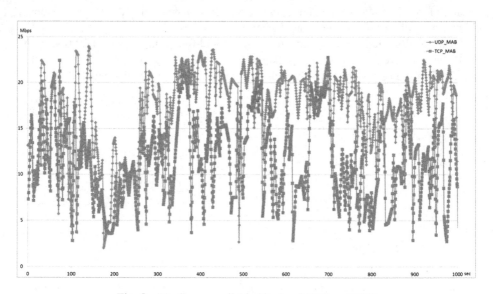

Fig. 3. Maximum available bitrate of UDP and TCP

4.3   Instantaneous Change of Rate

If the MAB maintains steady slope, it means that channel condition is gradually bad or good. Especially the slope is kept zero, it means that channel condition does not change. So when the slope is retained some values, estimating MAB using previous values is not difficult. However, when the slope is changeable, estimating MAB is more difficult. Then estimation error probability is increase.

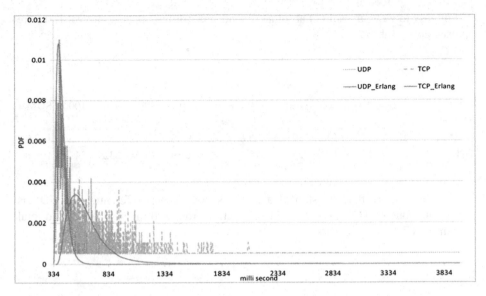

Fig. 4. A PDF of the inter arrival time

In this section, we propose second order differential of MAB for analysis which is easier to estimate MAB TCP or UDP. By using measured MAB, we draw second order differential of MAB graph at Fig. 5. for analysis

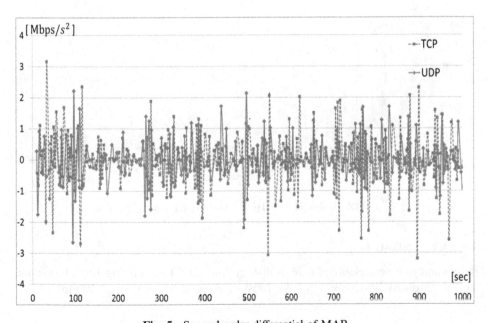

Fig. 5. Second order differential of MAB

For easy to view, the sum of absolute second order differentials at interval 100 s are represented in Table 2.

Table 2. Second order differential of each time section

TIME Section	100	200	300	400	500	600	700	800	900	1000	SUM
UDP	18.75	11.06	13.59	17.07	12.59	10.61	9.385	16.72	15.1	17.91	142.8
TCP	26.24	12.85	11.98	23.72	14.52	20.39	12.39	28.58	23.28	24.28	198.6

We can acquire that the sum of absolute second order differentials of UDP is smaller than one of TCP, except for TIME Section from 300 ms to 400 ms, also total the sum of UDP is smaller too.

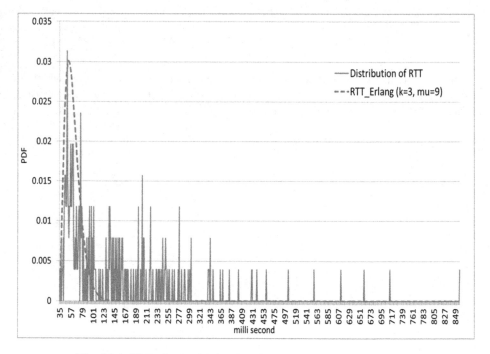

Fig. 6. A PDF of Round Trip Time following the Erlang distribution

4.4   Correlation

Before analyzing correlation, Fig. 6. is distribution of RTT with fitting graph following the Erlang distribution ($k = 3, \mu = 9$). Most of value is not more than 200 ms.

We analyze correlation between the MAB and the RTT. First of all, covariance $\text{Cov}(\text{MAB}_{UDP}, \text{MAB}_{TCP})$ and correlation $\rho_{MAB_{UDP}, MAB_{TCP}}$ are calculated by the equation in correlation part of Sect. 3.1 (9) and (10).

$$\text{Cov}(\text{MAB}_{UDP}, \text{MAB}_{TCP}) = 6.062, \rho_{MAB_{UDP}, MAB_{TCP}} = 0.29$$

And correlation values between RTT and TCP as well as UDP are calculated below.

$$\text{Cov}(MAB_{TCP}, \text{RTT}) = 5.254, \rho_{MAB_{TCP}, \text{RTT}} = 0.009$$

$$\text{Cov}(MAB_{UDP}, \text{RTT}) = -30.026, \rho_{MAB_{UDP}, \text{RTT}} = -0.055$$

We expected that the MAB has correlation with the RTTs, but correlations almost zero. So we can conclude the MAB is scarcely correlated with the RTT.

5   Conclusion

In this paper, we compare end-to-end performance of the UDP and the TCP, by using LTE, which is the most commonly used. The UDP has four advantages in comparison with TCP. First, the MAB of UDP is higher than the one of TCP, so the UDP can transmit higher quality video. Second, the MAB estimation is easy at the UDP, because the 2nd order of differential of the UDP is smaller than the TCP. Third, maintenance of state at the UDP is longer than the TCP. It means that the video quality of UDP is maintained. It is comfortable to watch for user. Last, Distribution of the MAB_{UDP} is right-skewed.

In conclusion, above advantages represent that UDP is suitable protocol for real-time streaming service. In future works, we will design the MAB estimation algorithm, and using the algorithm with Forward Error Correction (FEC), confirm the UDP-based streaming service is better than TCP.

Acknowledgement. This research was funded by the MISP(Ministry of Science, ICT & Future, Planning), Korea in the ICT R&D Program 2015.

References

1. Cisco Systems, Inc. Cisco Visual Networking Index: Global Mobile Data Traffic Forecast Update, 2014–2019, 3 February 2015. http://www.cisco.com/
2. Jain, M., Dovrolis, C.: Pathload: a measurement tool for end-to-end available bandwidth. In: Passive Active Meas (PAM) Workshop, pp. 14–25 (2002)
3. Ribeiro, V.J., Riedi, R.H., Baraiuk, R.G., Navratil, J., Cottrell, L.: pathChirp: efficient available bandwidth estimation for network paths. In: Passive and Active Monitoring Workshop, April 2003

4. Melander, B., Bjorkman, M., Gunningberg, P.: A new end-to-end probing ad analysis method for estimation bandwidth bottlenecks. In: Global Telecommunications Conference, GLO-BECOM 2000, IEEE, December 2000
5. Oshiba, T., Nakajima, K.: Quick and simultaneous estimation of available bandwidth and effective UDP throughput for real-time communication. In: 2011 IEEE Symposium on Computers and Communications (ISCC), pp. 1123–1130. IEEE, July 2011
6. Lao, L., Dovrolis, C., Sanadidi, M.Y.: The probe gap model can underestimate the available bandwidth of multihop paths. ACM SIGCOMMM CCR 36(5), 29–34 (2006)
7. Javadtalab, A., Semsarzadeh, M., Khanchi, A., Shirmohammadi, S.I., Yassine, A.: Continuous one-way detection of available bandwidth changes for video streaming over best-effort network. IEEE Trans. Instrum. Measur. 64, 190–203 (2015). IEEE

Video Streaming for Multi-cloud Game

Yoonseok Heo, Taeseop Kim, and Doug Young Suh[✉]

Electronic and Radio Engineering Department,
Kyung Hee University, Seoul 446-701, South Korea
{hyscokr, terr34, suh}@khu.ac.kr

Abstract. Depending on the development of the game industry, required hardware performance rises steadily. High-end game cannot be enjoyed by outmoded computer or smart mobile device. As a way to solve this problem, studies on the cloud gaming being actively conducted. However, latency of cloud gaming is greater than the latency of regular game and it interfere the smooth game play. Enjoying multi user game is not fair by different latency between each user. Thus we propose new cloud gaming system. It reduce the processing time in the cloud using the distribute processing, makes the same latency of each user through the appropriate distribution. In addition it corresponds to a loss by using the FEC. Accordingly, the proposed system enables a fair and good quality game on the thin client with the hardware of the lower-performance.

Keywords: Cloud gaming · Distributed computing · Cloud offloading · Multi-clouds · Multi- nodes · Video streaming

1 Introduction

Penetration of smart mobile devices with the explosive growth of smart mobile device market has already crossed the penetration of the PC. And smart mobile device has been essential in life of modern. In top application list of App market(Apple Store, Google Play, etc.) for smart mobile device, games are took the majority. In Smart mobile game market, there are many types of game more than PC game market. However, in terms of specifications of game, the level of smart mobile game market is lower than PC game market. The reason for this is that the poor performance of the mobile device relative to the PC. This low performance device, We called thin client even PC. Depending on the performance improvement and various sensors of mobile device, smart mobile game market has grown. Thus smart mobile game genre is becoming more diverse and quality is improved. But the limitation due to low device capability problem is still exist. But introduced in [8], as 3D Rendering technology advances, fighting games and FPS games which has dynamic and colorful images are more popular than Arcade game or others. Those games require high specifications for the terminal. Accordingly, due to the many users who want more quality game on smart mobile devices, cloud gaming has emerged which is harmonized online gaming and cloud concept.

Cloud gaming is processed the process required to play the game in cloud instead of client and just send processed data to client. It enables the work was a limit to the processing in the client. Therefore, unlike the existing online game, client receives the video data as opposed game data and usage of network is increased. The main problem

© Springer International Publishing Switzerland 2015
Y.-S. Ho et al. (Eds.): PCM 2015, Part II, LNCS 9315, pp. 275–284, 2015.
DOI: 10.1007/978-3-319-24078-7_27

to be considered due to the transmission of video data over the network is latency, loss, the bitrate. It uses UDP in the common video stream. Loss occurs in UDP according to the network status. But clear screen is required when game playing. To solve this problem we use the raptor which is one of FEC code. When serving a high-definition video, this can occur a node in cloud unable to afford. In addition, bps of a node in cloud cannot always be a certain. Since we are using a P2P method uses a method of transmitting data to client from multiple nodes in cloud. And for reducing latency, we use distribute Processing. It is for reducing encoding delay to make image. In this paper, we propose cloud gaming system having lower delay and less image quality deterioration compared with existing cloud gaming on Non-guaranteed network. In Sect. 2, we explicate background and Sect. 3 introduce proposed system. Finally, in Sect. 4 gives a conclusion.

2 Background

2.1 Cloud Gaming

Cloud gaming is the novel service which is combined the concept of cloud with on-line game. There are many type of on-line games using computer. Among them, the large size championship is held by using on-line game. In order to survive between a number of game, game producer take best effort to satisfy needs of consumer. In most cases, hardware specification is needed to make better game playing environment. To enjoy game on the better environment, the upper specification of part such as CPU, GPU and RAM also sufficient storage to install game is required to service consumer. This kind of reason demand upgrades their PC periodically. Moreover, PC does not have mobility, which causes not to enjoy game anytime, anywhere.

A lot of people have mobile device, and it is possible to enjoy game using mobile equipment like a smart phone or tablet which called thin client. Those kind of equip-ment does not have high specification which results of impossible to enjoy service demands high specification. So, many games using the mobile device have light volume of content and enjoy the light and instant content anywhere.

Cloud-gaming disencumber from disadvantage of PC and mobile equipment. Cloud-gaming solve this problem not to install game on mobile device and to opera-tion at the Cloud-Server to enjoy service. This work possible due to rapidly develop of the Internet. If you have thin device which have display and can connect the Inter-net, you can enjoy game anytime and anywhere not to download the content [5, 8].

A cloud gaming consumer use one game engine and one encoding server shown as Fig. 1.

2.2 Response Delay

Response delay is overall delay playing the cloud gaming. This is time which begin user's action for command and finished display game frames. Response delay is consist of following delays [2]

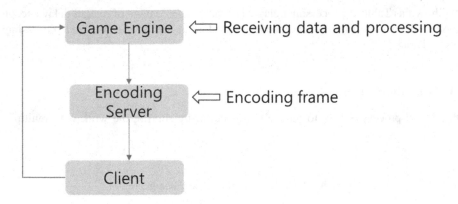

Fig. 1. Structure of general cloud gaming

- Network delay (ND): ND is considered as time which transmit user's command to server and video information back to client. It does not differ much from RTT
- Processing delay (PD): PD is considered time which receive user's command and process it by server. PD also includes time to encode and packetize the corresponding video information.
- Game delay (GD): GD is considered as time which process user's command and render video information of next game to apply user's command.
- Playout delay (OD): OD is considered as time which receive video information and decode and display a frame.

Response delay can be calculated ND + PD + GD + OD. In Table 1, there are delay measurement of cloud gaming at SMG(Stream My Game) and OnLive. [2] Network delay is not in Table 1 because it is highly dependent on distance between cloud server and client.

Table 1. Delay measurement of cloud gaming at SMG and OnLive

Service	Delay type	Game		
		Tomb(ACT)	FEAR(FPS)	DOW(RTS)
OnLive	PD	119ms	110ms	191ms
	GD	50ms	17ms	16ms
	OD	33ms	31ms	21ms
PD rate		**59%**	**70%**	**84%**
SMG	PD	380ms	362ms	365ms
	GD	50ms	17ms	16ms
	OD	23ms	18ms	21ms
PD rate		**84%**	**91%**	**91%**

Shown as Table 1, processing delay takes largest percentage of overall delay except network delay. The time which encode game data at server is considered processing delay (Table 2).

2.3   Distributed Process

Distributed process is way to reduce computation by sharing the work with multiple devices.

Table 2. Delay threshold each game genre

Game type	Perspective	Delay threshold(ms)
First Person Shooter (FPS)	First person	100
Role Playing Game (RPG)	Third person	500
Real-Time Strategy (RTS)	Omnipresent	1000

There are each different threshold people can recognize delay in several genre games. If performance of encoding server is increased, it is possible to reduce processing delay. In this paper, we decrease processing delay using distributed process method which employ low performance instances replacing a high performance instance because price of two low performance instances is lower than price of one high performance instance.

There are experiment result of distributed encoding comparing one 4CIF video and four CIF videos which are made up quadrisected 4CIF. As shown Table 3, Sum of four CIF videos size is larger than that of one 4CIF, but there are not much in it. In this experiment, it can save the encoding time about 26 %. If we use Distributed process method with N nodes, processing delay is decreased about 1/N.

2.4   P2P(Peer to Peer)

It is not satisfied performance which transmit HD(1280 × 720) resolution video in non-guaranteed network environment by existing cloud-gaming services. Higher-quality videos, such as HD resolution at 50 fps (frame-per-second), inherently result in higher bit rates, which render the cloud gaming systems vulnerable to higher network latency, and thus longer response delay [9]. The P2P algorithm is that multiple peers send data to client because of high transmission performance. This structure is useful for cloud which is free for number of nodes. In cloud gaming, it is better UDP than TCP because cloud gaming is delay sensitive. In order to solve the problem appearing loss in each node and the problem there are performance gap in each node because of network condition, we use Raptor FEC method in suggested system.

Table 3. Measurement of distributed encoding experiment using JSVM

	Type	Size	Encoded size(KB)	Processing time	
Soccer	**Entire**	704*576	1352	810.951	
	Quarter#0	352*288	342	210.479	
	Quarter#1	352*288	284	207.749	
	Quarter#2	352*288	440	216.429	
	Quarter#3	352*288	362	211.896	
	Sum of quarter	704*576	1428	Max value	216.43
	Ratio (entire : sum of Quarters)		105.62%	**Ratio (entire : Max Quarter)**	26.69%

	Type	Size	Encoded size(KB)	Processing time	
City	**Entire**	704*576	1153	796.217	
	Quarter#0	352*288	315	212.622	
	Quarter#1	352*288	267	213.537	
	Quarter#2	352*288	295	204.913	
	Quarter#3	352*288	330	204.862	
	sum of quarter	704*576	1207	Max value	213.54
	Ratio (entire : sum of Quarters)		104.68%	**Ratio (entire : Max Quarter)**	26.82%

	Type	Size	Encoded size(KB)	Processing time	
Ice	**Entire**	704*576	512	626.675	
	Quarter#0	352*288	149	162.306	
	Quarter#1	352*288	210	172.602	
	Quarter#2	352*288	57	144.583	
	Quarter#3	352*288	113	155.636	
	sum of quarter	704*576	529	Max value	172.6
	Ratio (entire : sum of Quarters)		103.32%	**Ratio (entire : Max Quarter)**	27.54%

2.5   Raptor

When multiple nodes transmit the encoded data to client, it is possible to occur loss in some nodes. In order to prevention this situation, we use Raptor(Rapid tornado) which is one of the FEC method. We divide the original frame into K symbols and make (N-K) extra symbols which is relative existing symbols, then server send N symbols to client. If client received more than K symbols, receiver can restore original. In our

Table 4. Variable for optimizing number of encoding server

Variable	Description
N_E	The number of encoding server
D_P	Processing(encoding) delay(ms)
D_D	Delivery Delay from game server to encoding server(ms)
D_N	Network delay(ms)
D_O	Play out delay(ms)
D_G	Game delay(ms)
D_{Th}	Delay threshold(ms)
D_{fd}	Fixed delay defined by main game server.

system, the FEC method ensure the stable transmission. General purpose using FEC is not only for error correction, but also reducing delay effectively because it is needed only K symbols to display game screen (Table 4).

2.6 Multi-cloud

Raptor FEC method have disadvantage the amount of transmission increase, then network condition can be congested. The solution of this problem is using Multi-cloud method. Multi-cloud method means cloud nodes at different physical location share the task. If a cloud in position near client take on the role which encode the video instead of distant cloud, it is possible to increase performance.(i.e., PLR or Network load balancing).

3 Proposed System

Overall outline of the proposed system is shown in Fig. 2.

Figure 2 Shows that the two user away physically play the same game. Client I and Client II are connected to the cloud where located each of the near and transmit the control data for play the game to game server in each cloud. Game servers integrate game control data to a game server which has more client. We called as main game server. In Fig. 2, game server in cloud A is main game server. There is just one client for each game server. So main game server is selected randomly. Main game server send the game control data corresponding to the screen in each of the clients to encoding server. Encoding server makes frame by received game control data and encode frame. And encoding server distribute encoded frame and send to sending server for FEC. Sending server do the FEC encoding by received data and send to Client.

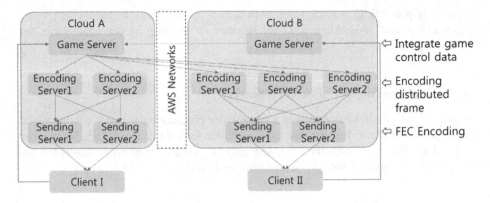

Fig. 2. Structure of proposed system

3.1 Optimizing Number of Encoding Server

In Fig. 2, client II has more network delay than client I. because of physical location, delivery time of client II sent game control data to game server in cloud A is longer than client I. And the time of client II for selected control data in main game server is delivered to encoding server is also longer than client I. Therefore, client I and client II are hard to a fair game. When playing Starcraft or League of Legend, difference of delay between each client is large, having a longer delay client is absolutely disadvantageous. To solve this problem, we propose using multiple encoding server for reducing processing delay occupied the largest percentage of the total delay for long network delay client.

Total delay have to lower than delay threshold.

$$D_P + D_D + D_N + D_O + D_G \leq D_{Th} \tag{1}$$

If this condition is satisfied for client I of cloud A where the main game server is existed, client can use just one encoding server. And D_{Th} is changed as $D_P + D_D + D_N + D_O + D_G$. But this condition is not satisfied for client I, the number of encoding server for client I N_{1E} got as below equation.

$$N_{1E} = ceil\left(\frac{D_P}{D_{Th} - D_D - D_N - D_O - D_G}\right) \tag{2}$$

Getting N_{1E}, the number of encoding server of client II in cloud B N_{2E} have to be got by D_P satisfying below equation.

$$D_P + D_D + D_N + D_O + D_G \cong D_{fd} \tag{3}$$

Therefore,

$$N_{2F} = celi\left(\frac{D_P}{D_{fd} - D_D - D_N - D_O - D_G}\right) \tag{4}$$

So it is possible to service game with a fair overall delay to each client using obtained N_{1E} and N_{2E}. After that, the number of sending server responsible for transmitting be got.

3.2   Optimizing Number of Sending Server

Sending server should be operated as a variable by considering network condition and data to be sent to client. It is because the cost. Since the cost increase when using more server. Therefore, sending server is used only as needed over time. Algorithm for changing the number of Sending server follows Fig. 3.

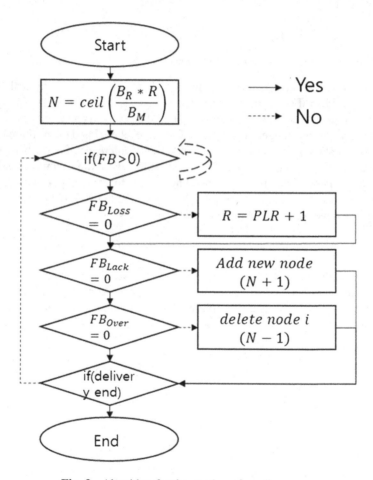

Fig. 3. Algorithm for the number of sending server

N_S is decided considering bitrate at first time and average bitrate to transmit by one cloud node. It is also depend on R which is FEC code rate. After prior work, it is needed to redefine N_S considering variables FB_{Lack}, FB_{Loss} and FB_{RTO}. When amount of data to transmit is larger than amount of available data, FB_{Lack} is occurred. In this situation, the number of node increase and multiplying amount of data to transmit by variable S because of prevention of resolution deterioration with preparing sudden increasing of bitrate. FB_{Lack} is occurred in situation which is impossible to restore original format from a mount of loss, and then it is needed to feedback from client. In this situation, we increase the R after checking PLR of client. FB_{RTO} is the element to check node condition and bitrate at regular time intervals and eliminate unnecessary nodes. Increasing the number of node is changing immediately for prevention of resolution deterioration, but decreasing the number of node regularly for stability. In this system, cloud gaming service producer can provide high-resolution and stable video to client (Table 5).

Table 5. Variable for optimizing number of sending server

Variable	Description
B_R	Required bitrate for game stream
B_M	Mean available bitrate of nodes
R	FEC code rate
S	Extra rate
$M()$	Mean of time $(t - \tau) \sim t$
N_S	The number of Sending server
$B_i(t)$	Bitrate of i_th node at t $(0 \leq i \leq N)$
$B_C(t)$	Bitrate of all node $\sum_{i=0}^{N} B_i$ at t
FB_{Lack}	$\begin{cases} 1 & M(B_C(t)) < (B_R * R) * S \\ 0 & O.W \end{cases}$ from server
FB_{RTO}	$\begin{cases} 1 & rest\ time(\tau)\ over \\ 0 & O.W \end{cases}$ from server
FB_{Over}	$\begin{cases} 1 & M(B_C(t) - B_i(t)) > (B_R * R) * S \\ 0 & O.W \end{cases}$ from server
FB_{Loss}	$\begin{cases} 1 & R > 1 + PLR \\ 0 & O.W \end{cases}$ from client
FB	$FB = FB_{Lack} + FB_{RTO} + FB_{Loss}$

4   Conclusion

Cloud gaming is way to playing games for thin client. Any devices which is available the internet with display equipment can play the games anywhere. In the near future, it is possible to play the games displayed video in windshield while people is in driverless-car. In this paper, we suggest novel cloud gaming system which can decrease delay and is adapt to variable network condition. It is possible to enjoy the high resolution video games fairly at reasonable price for cloud gaming users.

Acknowledgement. This research was funded by the MSIP(Ministry of Science, ICT & Future, Planning), Korea in the ICT R&D Program 2015.

References

1. Wu, X.: Peide Qian – Suzhou University, Suzhou, China, towards the scheduling of access requests in cloud storage. In: 2013 8th International Conference on Computer Science and Education (ICCSE), pp. 37–41. IEEE, April 2013
2. Chen, K.-T., Chang, Y.-C., Hsu, H.-J., Chen, D.-Y., Huang, C.-Y., Hsu, C.-H.: On the quality of service of cloud gaming systems. Multimedia, IEEE Trans. **16**, 480–495 (2013). IEEE
3. Lucas-Simarro, J.L., Moreno-Vozmediano, R., Montero, R.S., Llorente, I.M.: Dynamic Placement of Virtual Machines for Cost Optimization in Multi-Cloud Envireonments, IEEE, July 2011
4. Cai, W., Leung, V.C.M.: Multiplayer Cloud Gaming System with Cooperative Video Sharing, IEEE, December 2012
5. Fullerton, T., Chen, J., Santiago, K., Nelson, E., Diamante, V., Meyers, A.: That cloud game: dreaming (and doing) innovative game design. In: Sandbox 2006 Proceedings of the 2006 ACM SIGGRAPH Symposium on Videogames, ACM, July 2006
6. Shea, R., Liu, J., Ngai, E.C.-H., Cui, Y.: Cloud Gaming: architecture and performance. IEEE Netw. **27**, 16–21 (2013). IEEE
7. Cai, W., Leung, V.C.M., Chen, M.: Next generation mobile cloud gaming. In: SOSE 2013 IEEE 7th International symposium, pp. 551–560. IEEE, March 2013
8. Chen, K.-T., Chang, Y.-C., Tseng, P.-H., Huang, C.-Y., Lei, C.-L.: Measuring the latency of cloud gaming systems. In: MM 2011 Proceedings of the 19th ACM International Conference on Multimedia, pp. 1269–1272. ACM, November 2011
9. Huang, C.-Y., Hsu, C.-H., Chang, Y.-C., Chen, K.-T.: Gaming anywhere; an open cloud gaming system. In: MMSys 2013 Proceedings of the 4th ACM Multimedia Systems Conference, pp. 36–47. ACM, February 2012
10. Jarschel, M., Schlosser, D., Scheuring, S., Hoßfeld, T.: An evaluation of QoE in cloud gaming based on subjective test. In: 2011 Fifth International Conference on Innovative Mobile and Internet Services in Ubiquitous Computing (IMIS), pp. 330–335. IEEE, July 2011
11. Choy, S., Wong, B., Simon, G., Rosenberg, C.: The brewing storm in cloud gaming: a measurement study on cloud to end-user latency. In: NetGames 2012 Proceedings of the 11th Annual Workshop on Network and Systems Support for Games, ACM, November 2012

Performance Analysis of Scaler SoC for 4K Video Signal

Soon-Jin Lee$^{(\boxtimes)}$ and Jong-Ki Han

Sejong University, Seoul, South Korea
sincere0831@naver.com, hjk@sejong.edu

Abstract. This paper considers some issues to implement the scaler SoC. Because the size and power of SoC are constrained, the bit number to represent the coefficients of scaler kernel and LPF should be limited. In addition, the interpolation position should be located at the quantized phases. We analyze the effects of the various constraints in the performance of scaling system. The simulation results provide the guidance to implement the scaler SoC.

Keywords: Cubic convolution scaler · Interpolation · System on chip

1 Introduction

Scaler is one of the most important modules in image and video broadcasting systems to get a better overview, where the supported resolutions arc from QCIF to Ultra High Definition (UHD). In these systems, the original image data is resampled by interpolation schemes according to the resolution supported by the display module.

During last decades, several researchers have proposed various scaling algorithms to increase the quality of the resized images. Among them, the bilinear [1], cubic B-spline [2, 3], and bi-cubic (Cubic Convolution Scaler) [4] are practically used, because they provide the resized image having high quality with simple operations.

When the scaling module is implemented in a System on Chip (SoC), there are several constraints for the size of the internal memory, the number of data bus, the quantized data format and so on. In this paper, we analyze the performances of scaling systems for a variety of constraints about the quantized coefficients of the interpolation filter and the quantized interpolation positions.

This paper is organized as follows. Section 2 explains the scaling algorithm of cubic convolution scaler [4]. The problem to be solved is formulated in Sect. 3, where the constraints about the bit number to represent the coefficients of scaler and LPF are considered to increase the efficiency of implementation of SoC. In addition to the constraints related to the coefficients, the number of phase which is related to the interpolation position is considered to implement the scaler SoC. In Sect. 4, various

This work was supported in part by the Technology Innovation Program, Industrial Strategic Technology Development Program (Development of Super Resolution Image Scaler for 4 K UHD), under Grant K10041900, and in part by the National Research Foundation of Korea (NRF) grant funded by the Korea government (MSIP) (NRF-2015R1A2A2A01006193).

Y.-S. Ho et al. (Eds.): PCM 2015, Part II, LNCS 9315, pp. 285–291, 2015.
DOI: 10.1007/978-3-319-24078-7_28

simulation results show that the parameters optimized to implement the scaler SoC provide the reasonable performances. We conclude this paper briefly in Sect. 5.

2   Scaling Algorithm

Figure 1 shows the scaling process for the original pixel $f(x_k)$ and scaled pixel value $g(y_k)$, where $f(x_k)$'s are the given digital sample values. In this process, the interpolated signal $\hat{f}(x)$ is represented as follows.

$$\hat{f}(x) = \sum f(x_k)\beta(x - x_k) \tag{1}$$

where $\beta(x)$ is the interpolation kernel x and x_k represent continuous and discrete values, respectively. The scaled pixels $g(y_k)$ are obtained by sampling the continuous interpolation function, i.e., $g(y_k) = \hat{f}(x = y_k)$. In the scaling process using bi-cubic [4], the kernel is

$$\beta(x, \alpha) = \begin{cases} (\alpha + 2)|x|^3 - (\alpha + 3)|x|^2 + 1, 0 \le |x| \le 1 \\ \alpha|x|^3 - 5\alpha|x|^2 + 8\alpha|x| - 4\alpha, 1 \le |x| \le 2 \end{cases} \tag{2}$$

The scaler module calculates the interpolated pixel values by using a set of its nearest neighboring pixels. If $f(x_{k-1}), f(x_k), f(x_{k+1})$, and $f(x_{k+2})$ are the neighbor pixels, the interpolated function can be calculated as follows.

$$\hat{f}(x) = f(x_{k-1})A_0(s) + f(x_k)A_1(s) + f(x_{k+1})A_2(s) + f(x_{k+2})A_3(s) \tag{3}$$

where

$$\begin{aligned}
A_0(s) &= \alpha s^3 - 2\alpha s^2 + \alpha s \\
A_1(s) &= (\alpha + 2)s^3 - (3 + \alpha)s^2 + 1 \\
A_2(s) &= -(\alpha + 2)s^3 + (3 + 2\alpha)s^2 - \alpha s \\
A_3(s) &= -\alpha s^3 + \alpha s^2
\end{aligned} \tag{4}$$

In (3) and (4), the spacing of the sampling grid is assumed as '1'. $x_k < x < x_{k+1}$, $s = x - x_k$, $0 < s < 1$. The bi-cubic is known as one of the most efficient schemes, because the complexity of the bi-cubic is low while the performance is comparable to that of cubic B-spline [5].

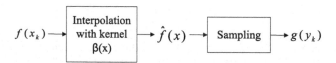

Fig. 1. Scaling process.

3 Optimization of Scaler SoC

3.1 Constrained Coefficients of Scaler in SoC

Because the size of SoC chip is limited, it is important to choose the optimal bit number to represent the coefficients of scaler filter. As the bit number increases, the quality of the resized image is more enhanced, because the accuracy of the interpolation increases. If the coefficients are represented with the constrained bit number, those are altered as follows.

$$I_i(s) = (\text{int})A_i(s), \quad i = 0, 1, 2, 3 \tag{5}$$

$$F_{i,N}(s) = (A_i(s) - (\text{int})A_i(s))_N, \quad i = 0, 1, 2, 3 \tag{6}$$

where $I_i(s)$ is the integer part of $A_i(s)$. $F_{i,N}(s)$. is the number below the decimal point, where $F_{i,N}(s)$ is represented with N bits. Based on the altered coefficients, the interpolated function is represented as follows.

$$\begin{aligned}
\tilde{f}(x) = {} & f(x_{k-1})(I_0(s) + F_{0,N}(s)) + f(x_k)(I_1(s) + F_{1,N}(s)) \\
& + f(x_{k+1})(I_2(s) + F_{2,N}(s)) + f(x_{k+2})(I_3(s) + F_{3,N}(s))
\end{aligned} \tag{7}$$

3.2 Quantized Phase of Scaler in SoC

The interpolation position depends on s in (3)-(7), where $0 < s < 1$ when no constrained is assigned. However, although it is $0 < s < 1$, arbitrary s value is not permitted in implementation of SoC, because of the limitation about the complexity and the size of chip area. Thus, the phase s is permitted on the finite M positions. In this paper, we decide the finite optimal positions to interpolate the new pixel values. It implies that the quantized values of s, $0 < s < 1$, are used in SoC. When M positions are permitted for interpolation, the quantized phase is represented as follows.

$$\tilde{s} = (\text{int})(s \times M)/M \tag{8}$$

3.3 Constrained Coefficients of LPF in SoC

When the resolution of the original image is reduced to be displayed on the specific devices, the Low Pass Filter (LPF) should be used to cutoff the components of high-frequency of the original data, before applying the scaling algorithm. In this procedure, the problem related to the bit number to represent the filter coefficients should be considered also. Because the bit number L to represent the coefficients of LPF affects the quality of the low passed image, the optimal bit number should be used in the implementation of Soc.

4   Simulation Results

4.1   Performance for Constrained Coefficients of Scaler

Figures 2 and 3 show the performances of the scalers when the bit number to represent the coefficients of scaler is constrained from 2 to 6 bits.

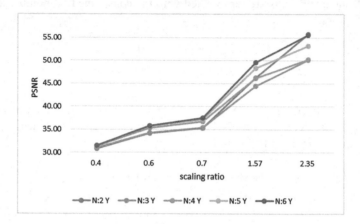

Fig. 2. Image qualities according to the bit number to represent the coefficients of scaler.

Fig. 3. When scaling ratio is 1.57, (a) original image (above), (b) the resized image by bi-cubic without any constraints about the coefficients of scaler (left), (c) the resized image by bi-cubic with N = 5 (right).

4.2 Performance for Quantized Phase of Scaler

Figures 4 and 5 show the performances of the scalers when the number of the permitted phases is from 2 to 32.

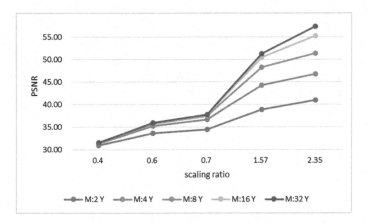

Fig. 4. Image qualities according to the number of the quantized phase in scaling procedure.

Fig. 5. When scaling ratio is 1.57, (a) original image (above), (b) the resized image by bi-cubic without any constraints about the phase (left), (c) the resized image by bi-cubic with M = 32 (right).

4.3 Performance for Constrained Coefficients of LPF

Figures 6 and 7 show the performances of the scalers when the bit number to represent the coefficients of LPF is constrained from 2 to 6 bits.

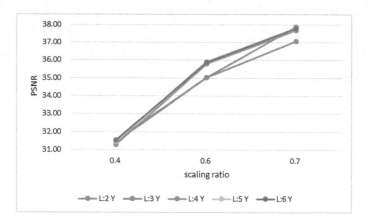

Fig. 6. Image qualities according to the bit number to represent the coefficients of LPF.

Fig. 7. When scaling ratio is 0.7, (a) original image (above), (b) the resized image by bi-cubic without any constraints about the coefficients of LPF (left), (c) the resized image by bi-cubic with L = 6 (right).

5 Conclusions

In this paper, we have considered the problems related to implement the scaler SoC. Due to the constraints about the size and power of SoC, the bit number to represent the coefficients of scaler kernel and LPF, and the phase number have been limited. We have analyzed the effects of the various constraints. Based on the computer simulation results, we are guided to implement the SoC efficiently.

References

1. Dodgson, N.A.: Quadratic interpolation for image resampling. IEEE Trans. Image Process. **6**, 1322–1326 (1997)
2. Hou, H.S., Andrews, H.C.: Cubic splines for image interpolation and digital filtering. IEEE Trans. Acoust. Speech Sig. Process. **26**, 508–517 (1978)
3. Unser, M., Aldroubi, A., Eden, M.: Fast B-Spline transforms for continuous image representation and interpolation. IEEE Trans. Pattern Anal. Mach. Intell. **13**, 277–285 (1991)
4. Keys, R.G.: Cubic convolution interpolation for digital image processing. IEEE Trans. Acoust. Speech Signal Process. **ASSP-29**(6), 1153–1160 (1981)
5. Miklós, P.: Comparison of convolutional based interpolation techniques in digital image processing. In: International Symposium on Intelligent Systems and Informatics, pp. 87–90, August 2007

Deblocking Filter for Depth Videos in 3D Video Coding Extension of HEVC

Yunseok Song and Yo-Sung Ho(⊠)

Gwangju Institute of Science and Technology (GIST), 123 Cheomdangwagi-ro,
Buk-gu, Gwangju 500-712, Republic of Korea
{ysong,hoyo}@gist.ac.kr

Abstract. This paper presents a modified deblocking filter for depth video coding in the 3D video coding extension of High Efficiency Video Coding (3D-HEVC). The conventional 3D video coding extension of HEVC (3D-HEVC) employs a deblocking filter and sample adaptive offset (SAO) in the loop filter in which both tools are applied to color video coding only. Nevertheless, the deblocking filter can smooth out blocking artifacts existing in coded depth videos, resulting in improving the coding efficiency. In this paper, we modify the original deblocking filter of HEVC and apply it to depth video coding. The goal is to enhance the depth video coding efficiency. The modified filter is executed when a set of conditions regarding the boundary strength are satisfied. In addition, the impulse response is altered for more smoothing between block boundaries. Experiment results show 5.2 % BD-rate reduction in depth video coding in comparison to the conventional 3D-HEVC.

Keywords: Depth video coding · Deblocking filter · HEVC

1 Introduction

The latest video coding standard, High Efficiency Video Coding (HEVC), was developed by Joint Collaborative Team on Video Coding (JCT-VC). Experts from Moving Picture Experts Group (MPEG) of ISO/IEC and Video Coding Experts Group (VCEG) of ITU-T contributed to the development. Compared to the previous coding standard, advanced video coding (AVC), HEVC is capable of doubling the compression efficiency. In 2011, the 3D video coding (3DVC) group of the MPEG issued a call for proposals (CfP) on 3D video coding technology [1]. The coding tools were required to be compatible with either AVC or HEVC. Since July 2012, the Joint Collaborative Team on 3D Video Coding Extension (JCT-3 V) has governed the standardization activities of 3D-AVC and 3D-HEVC. The development of 3D-HEVC is expected to be finalized in 2015.

In comparison to AVC, HEVC provides enhanced conventional coding tools including intra/inter prediction, transform/quantization and entropy coding [2, 3]. Sample adaptive offset (SAO) is a newly introduced tool used in the loop filter which also includes a deblocking filter. SAO is used for pixel-wise error compensation. The deblocking filter reduces blocking artifacts caused by block-based coding. In addition

© Springer International Publishing Switzerland 2015
Y.-S. Ho et al. (eds.): PCM 2015, Part II, LNCS 9315, pp. 292–299, 2015.
DOI: 10.1007/978-3-319-24078-7_29

to such techniques, flexible prediction units with varying sizes are used, i.e., from 4×4 to 64×64.

Further, on top of HEVC, 3D-HEVC contains tools designed for inter-view prediction and depth video coding specifically. Redundancy between color video and its corresponding depth video is also taken into consideration. Notable tools include disparity-compensated prediction (DCP), advanced residual prediction (ARP), depth modeling modes (DMM), and depth-based block partitioning (DBBP) [4]. In this paper, we modify the deblocking filter for depth video coding. We briefly describe the procedures of the deblocking filter and present the proposed method.

2   Deblocking Filter

In HEVC, the loop filter consists of a deblocking filter and SAO. The deblocking filter is executed first, followed by SAO. These tools compensate errors to enhance the overall picture quality prior to the outputting process. Specifically, the deblocking filter is designed to reduce blocking artifacts which exhibit sudden variation of pixels at block boundaries which are caused by block-based transform coding followed by quantization [5]. Prediction of adjacent blocks also cause this problem. 3D-HEVC does not use the deblocking filter in depth video coding due to the color video-targeted design and complexity problems.

Depending on boundary strength (Bs) estimation and a number of thresholds, one of three actions is carried out: no action, normal filtering or strong filtering. Filtering is performed only when the block boundary is either a prediction unit (PU) boundary or a transform unit (TU) boundary.

2.1   Boundary Strength

8×8 sized blocks are considered when calculating Bs. Figure 1 shows an example of block boundary for deblocking filter. Figure 2 represents the Bs determining criteria. First, if at least one of the blocks is intra coded, Bs is the highest value, two. In the case that neither is intra coded,

1) One of the blocks has non-zero coded residual coefficients.
2) The blocks have different reference pictures.

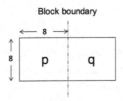

Fig. 1. Block boundary defined in the deblocking filter.

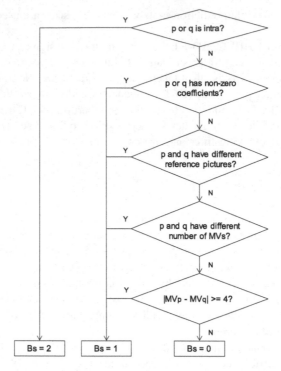

Fig. 2. Boundary strength decision.

3) They have different number of motion vectors.
4) The absolute difference of their motion vectors (MV) is greater than four.

Bs is zero if none of such conditions are satisfied. This means the block boundary is smooth with little variation. Thus, filtering is skipped in this case.

2.2   Strong/Normal Filter

The deblocking filter is executed if Bs is two or one for luma samples. For chroma samples, Bs value of two is required. Deblocking process is performed by either normal filtering or strong filtering. The type of filtering is determined by several conditions based on sample values and two defined thresholds.

3   Proposed Method

We present two modifications to the existing deblocking filter of HEVC. The modified deblocking filter is applied to depth video coding while the original deblocking filter is applied to color video coding.

3.1 Boundary Strength and Filter Type

We have analyzed the used percentages of filter types that are triggered in depth video coding. The test conditions of JCT3 V were used in this simulation [6]. Table 1 represents the results. For simple design, we adopt the most probable scenario. The strong filter is executed if Bs value is one or two. The normal filter is disabled, skipping the filter-selecting process.

Table 1. Filter type statistics.

Filter type	Percentage (%)
No action (Bs = 0)	83.9
Strong filter	14.5
Normal filter	1.6

3.2 Impulse Response

In [5], strong filtering operations is explained extensively. We denote that $p0$ and $q0$ are at the boundary and $p3$ and $q3$ are the farthest samples from the center. In this case, originally, the impulse responses for $p0$, $p1$ and $p2$ are (1 2 2 2 1) / 8, (1 1 1 1) / 4 and (2 3 1 1 1) / 8, respectively. Since depth videos show more homogeneous regions than color videos, we intensify the smoothing operation. Instead of (1 2 2 2 1) / 8 and (2 3 1 1 1) / 8, impulse responses of (1 1 1 2 1 1 1) / 8 and (1 2 1 1 1 1 1) / 8 are employed. (1 1 1 1) / 4 response remains the same.

4 Experiment Results

The proposed method was implemented on 3D-HTM 13.0. Tests were conducted on four sequences that possess three views each: Poznan_Hall2 (1920×1088), Poznan_Street (1920×1088), Kendo (1024×768) and Newspaper (1024×768). These are four of the eight sequences used in JCT-3 V activities [6]. Figure 3 displays color and depth images of such test sequences. Bjontegaard delta rates (BD-rate) are measured for objective evaluation [7].

The number of coded frames is 50. Common test conditions in [5] were used. Quantization parameters (QP) are set to 25, 30, 35 and 40 for color videos. Accordingly, QPs for depth are 34, 39, 42 and 45. QPs for depth video coding are set higher since depth video coding requires less accuracy in comparison to color video coding.

Tables 2 and 3 report depth video coding performances on base view and dependent views, respectively. The base view is generally denoted by View 0 while dependent views are represented by View 1 and so on.

Tables 4 and 5 represent BD-rate results of depth and color video coding, respectively. BD-rate is employed for computing average PSNR differences between

(a) Poznan_Hall2 (1920×1088)

(a) Poznan_Street (1920×1088)

(c) Kendo (1024×768)

(d) Newspaper (1024×768)

Fig. 3. Test sequences used in the evaluation

rate-distortion curves. A negative percentage represents the amount of bit savings that can be achieved. A 5.2 % average gain 5.2 % was achieved in depth video coding. Only 0.1 % BD-rate increase was observed in color video coding. Thus, the impact on color video coding is negligible. High gains were achieved in lower QPs. This is due to the fact that blocking artifacts occur more at lower QPs.

Table 2. Depth video coding results of base view

Test sequence	QP	3D-HTM 13.0		Proposed Method	
		Bitrate (kbps)	PSNR (dB)	Bitrate (kbps)	PSNR (dB)
Poznan_Hall2	25	72.38	46.04	73.00	46.19
	30	30.17	42.55	29.69	42.74
	35	15.53	39.71	15.26	40.01
	40	8.84	37.16	8.74	37.40
Poznan _Street	25	192.16	43.76	200.79	43.59
	30	58.15	40.80	60.10	40.94
	35	23.98	37.43	24.34	37.84
	40	11.53	33.94	11.44	34.32
Kendo	25	107.66	40.23	106.84	40.56
	30	37.44	36.18	37.34	36.45
	35	16.77	33.36	16.55	33.65
	40	8.40	30.97	8.41	31.22
Newspaper	25	171.65	39.56	177.47	39.44
	30	62.43	36.18	63.68	36.39
	35	26.80	33.37	27.18	33.77
	40	12.78	30.10	12.84	30.52

Table 3. Depth video coding results of dependent views

Test sequence	QP	3D-HTM 13.0				Proposed Method			
		View 1		View 2		View 1		View 2	
		Bitrate (kbps)	PSNR (dB)	Bitrate (kbps)	PSNR (dB)	Bitrate (kbps)	PSNR (dB)	Bitrate (kbps)	PSNR (dB)
Poznan_Hall2	25	53.60	44.43	48.91	44.44	54.83	44.60	50.36	44.57
	30	20.63	41.43	18.66	41.59	20.78	41.67	19.06	41.77
	35	10.16	38.21	9.24	38.56	9.99	38.48	8.90	38.94
	40	5.44	35.50	5.14	36.16	5.29	35.67	5.14	36.41
Poznan _Street	25	117.82	42.06	136.04	41.64	124.35	42.02	141.66	41.54
	30	28.40	39.42	32.04	38.92	29.21	39.62	33.19	39.02
	35	11.21	36.39	11.22	36.23	11.10	36.70	11.74	36.57
	40	5.49	33.38	5.18	33.13	5.40	33.78	5.24	33.47
Kendo	25	88.19	37.02	113.66	35.60	89.35	37.44	114.26	35.81
	30	28.75	33.17	36.32	31.93	28.56	33.41	36.24	32.12
	35	12.69	30.65	15.72	29.31	12.54	30.79	15.71	29.62
	40	6.58	28.10	7.53	26.59	6.40	28.29	7.72	26.90
Newspaper	25	119.10	38.00	107.18	37.38	125.27	37.81	112.23	37.32
	30	44.36	34.89	35.62	34.34	45.66	35.00	36.42	34.47
	35	19.66	32.17	15.29	31.98	20.89	32.53	15.45	32.16
	40	9.10	28.94	7.63	28.95	9.49	29.41	8.03	29.64

Table 4. Depth video coding performance (BD-rate, %)

Test sequence	View 0	View 1	View 2
Poznan_Hall2	−6.4	−6.2	−6.5
Poznan_Street	−3.7	−5.9	−2.6
Kendo	−7.9	−7.3	−6.9
Newspaper	−5.2	−1.1	−2.5

Table 5. Color video coding performance (BD-rate, %)

Test sequence	View 0	View 1	View 2
Poznan_Hall2	0.0	0.0	0.3
Poznan_Street	0.0	0.2	0.0
Kendo	0.0	−0.3	0.8
Newspaper	0.0	0.5	0.0

5   Conclusion

In this paper, we presented a modified deblocking filter for depth video coding in 3D-HEVC. The conventional 3D-HEVC uses the deblocking filter for color video coding only; nevertheless, with some modifications the tool can enhance the quality of coded depth videos, increasing the practicality. The proposed method executes the strong filter if the boundary strength value is two or one. The normal filter is disabled for simplicity. In addition, the impulse response is changed to intensify the smoothing operation considering depth video characteristics. The proposed method was implemented on 3D-HTM 13.0, following the test configurations used in common test conditions employed by JCT-3 V. Experiments were conducted on four test sequences. Experiment results exhibited 5.2 % BD-rate reduction in depth video coding while maintaining the performance of color video coding. Thus, the proposed method successfully enhanced the depth video coding performance in 3D-HEVC.

Acknowledgements. This research was supported by Basic Science Research Program through the National Research Foundation of Korea(NRF) funded by the Ministry of Science, ICT & Future Planning(No. 2011-0030079).

References

1. Video and requirement group: call for proposals on 3D video coding technology. In: ISO/IEC MPEG N12036, pp. 1–20 (2011)
2. Bross, B., Han, W.J., Ohm, J.R., Sullivan, G.J., Wang, Y.K., Wiegand, T.: High efficiency video coding (HEVC) text specification draft 10. In: ITU-T/IEC/ISO JCTVC-L1003, pp. 1–88 (2013)

3. Ohm, J.R., Sullivan, J., Schwarz, H., Tan, T.K., Wiegand, T.: Comparison of the coding efficiency of video coding standards—including high efficiency video coding (HEVC). IEEE Trans. Circuits Syst. Video Technol. **22**(12), 1669–1684 (2012)
4. Chen, Y., Tech, G., Wegner, K., Yea, S.: Test model 10 of 3D-HEVC and MV-HEVC. In: ITU-T/ISO/IEC JCT3V-J1003, pp. 1–92 (2014)
5. Norkin, A., Bjontegaard, G., Fuldseth, A., Narroschke, M., Ikeda, M., Andersson, K., Zhou, M., Van der Auwera, G.: HEVC deblocking filter. IEEE Trans. Circuits Syst. Video Technol. **22**(12), 1746–1754 (2012)
6. Muller, K., Vetro, A.: Common test conditions of 3DV core experiments. In: ITU-T/IEC/ISO JCT3V-G1100, pp. 1–10 (2014)
7. Bjontegaard, G.: Calculation of average PSNR differences between RD-curves. In: ITU-T VCEG-M33, pp. 1–4 (2001)

Sparcity-Induced Structured Transform in Intra Video Coding for Screen Contents

Je-Won Kang[(✉)]

Department of Electronics Engineering, Ewha Womans University, Seoul, Korea
jewonk@ewha.ac.kr

Abstract. In this paper, we propose a novel transform method based on learning a sparse model of residue in intra video coding. The proposed method considers generation of transformed coefficients locally grouped in a block as well as the sparsity of the coefficients, which can improve the coding efficiency of the screen content videos such as computer synthetic videos. The proposed method trains the transform on-line with using the residue, applied to the current frame. It is demonstrated with experiments that the proposed method improves the coding gain over HEVC Range extension standard.

Keywords: HEVC · Sparsity-induced structured transform · Computer synthetic videos · Screen contents · HEVC range extension

1 Introduction

Sparse coding [1] has gained significant research interests in fields of image/video processing and coding applications as it provides a systematic learning framework to find a large set of element vectors, contained in a dictionary. The vectors are designed to capture underlying high level features in a signal, and, thus the sparse coding is to efficiently approximate a signal with a smaller number of the element vectors than a dimension of an input signal. In video coding, the sparse coding can be applied to residue after an inter prediction or intra prediction and provide a dictionary including useful features of the residue. For example, the residue along boundary regions show slant characteristics, and the features embedded in a dictionary can be used for sparse representation of the residue [2].

Practical hybrid image/video coding standards such as High Efficiency Video Coding (HEVC) employed non-adaptive and analytic transform functions such as the discrete cosine transform (DCT), which is approximated into Karhunen-Loeve (K-L) transform. However, DCT suffers from undesirable noise such as ringing artifacts along edges, and K-L transform is sub-optimal in non-Gaussian sources [3]. In particular, screen content videos including graphical contents and animations synthesized by a computer are challenged to use the coding tools incorporated in HEVC main profile because the coding tools are developed mainly for natural videos. The screen content videos often show different characteristics e.g. including sharp contrast edges, so the coding efficiency becomes lower. For this reason, transform skipping mode [4] bypassing DCT is frequently selected in the screen contents as there are a lot of high frequency components in a block.

© Springer International Publishing Switzerland 2015
Y.-S. Ho et al. (Eds.): PCM 2015, Part II, LNCS 9315, pp. 300–307, 2015.
DOI: 10.1007/978-3-319-24078-7_30

Recently, learning-based transform exploiting statistical characteristics of residue has been developed. Dictionary-based video coding scheme learned from actual residue was developed in [2, 5]. The elements in the trained dictionary were used for the sparse representation as the training process pertained the sparse constraint, imposing the number of coefficients, i.e., L_0 norm or L_1 norm for the computational complexity reason. Early research works in the dictionary-based video coding considered only the sparse constraint, but it could be also desirable for the representation to keep intrinsic structural information of the data since salient residual samples were often localized and connected in a block. Kang *et al.* categorized residual samples with a classifier in the dictionary training and created different types of dictionaries for two-layered transform [6]. However, the same L_1 optimization problem was considered in the dictionary.

In this paper, we propose a novel transform method based on a sparse model of residue in intra video coding. In previous dictionary-based methods, the number of coefficients becomes sparse, but increasing size of a dictionary should be problematic for an efficient entropy coding, e.g. $\log_2 N$ extra bits for signaling a vector in a dictionary of size N [6]. As compared, in this paper, we consider not only the sparsity of the transformed coefficients but also their structured forms in the transformed domain, leading to the improved coding gain. The idea behind the structured sparse transform is motivated by [7, 8], and applied to the screen content video coding application in this paper. It is demonstrated with experiments that the proposed method improves coding performance for screen content videos.

The rest of the paper is organized as follows. We present the proposed method in Sect. 2. Experimental results are given in Sect. 3. We conclude with remarks and show future works in Sect. 4.

2 Proposed Method

2.1 Structured Sparse Transform Design

Given an input matrix $\mathbf{X} \in R^{N \times p}$ of N residue observations, a data vector $\mathbf{x} \in R^p$ can be represented as a linear combination of a set of column vectors $\mathbf{v}_k \in R^p$ in a matrix $\mathbf{V} = [\mathbf{v}_1, \mathbf{v}_2, \ldots, \mathbf{v}_p]$ and a coefficient vector $\mathbf{u}$. Denote the coefficient matrix by $\mathbf{U}$. In training, we learn $\mathbf{V}$ and $\mathbf{U}$ simultaneously. Note we aim to obtain the transform whose coefficients are sparse while, in previous dictionary-based learning schemes, the number of non-zero coefficients in $\mathbf{u}$ is minimized. Mathematically, we solve the following constraint optimization problem,

$$\min_{\mathbf{U},\mathbf{V}} \left\| \mathbf{X} - \mathbf{U}\mathbf{V}^T \right\|_F^2 + \lambda \sum_{k=1}^{p} \|u_k\|_1 \tag{1}$$

$$s.t. \|\mathbf{v}_k\|_1 \leq 1, \forall k \in \{1, 2, \ldots, p\}$$

where λ is the regularization parameter which controlling the trade-off between the first term fitting the data and the second term penalizing the sparsity term in $\mathbf{u}_k$. We consider

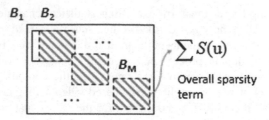

Fig. 1. Group of nonzero pattern augmentation.

only the sparsity of $\mathbf{u}_k$ in (1) but also another constraint to a local pattern of residue, so that they are grouped together in a block boundary. To this aim, we change the second term by masking a logical block patch denoted by $B \in \mathrm{B}$, which is a specific set of the overlapping block patch whose m-by-m nonzero pattern is geometrically connected as shown in Fig. 1. We modify (1) as

$$\min_{\mathbf{U},\mathbf{V}} \left\| \mathbf{X} - \mathbf{U}\mathbf{V}^T \right\|_F^2 + \lambda \sum_{k=1}^p S(\mathbf{u}_k),$$ (2)

and $S(\mathbf{u}_k)$ is defined as

$$S(\mathbf{u}_k) = \sum_{B \in \mathrm{B}} \sum_{j \in B} \left| b^B(j) \mathbf{u}_k(j)^2 \right|^{1/2} = \sum_{B \in \mathrm{B}} \left\| (b^B \cdot \mathbf{u}_k) \right\|_2,$$ (3)

where b^B is a vector where $b^B(j)$ is one if a pixel coordinate j belongs to a group B and 0, otherwise. The operator $\cdot$ is the component-wise multiplication. Thus, $S(\mathbf{u}_k)$ sets several pixels to 0 in a block, leading to the sparsity of the coefficients as well as the constrained distribution in the block due to the penalized term. A large value of λ converts the problem to a mere least square optimization problem while a small value of λ confines a strict boundary of a coefficient distribution in a block. In our test, λ is decided as an empirical value set to 0.0001.

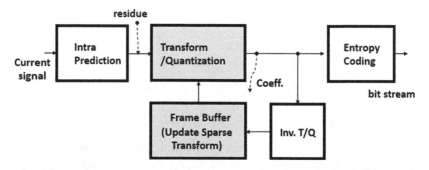

Fig. 2. Block diagram of the proposed codec design.

2.2   Codec Design

The proposed codec design is described in Fig. 2, which includes the frame buffer conducting the update of the sparse transform on top of the conventional intra video coding block diagram. We train residual samples by iteratively solving $\mathbf{U}$ and $\mathbf{V}$ in (2), and the transform is learned on the fly by using previously coded video frames. Specifically, for coding the ith video frame, we use a given sample matrix $\mathbf{X}_{i-1}$ in the i-1th video frame to solve the minimization problem. $\mathbf{V}_{i-1}$ is already given, and the iteration is given as below:

$$\mathbf{U}_i = \underset{\|\mathbf{v}\| \leq 1}{\mathrm{argmin}} \left\| \mathbf{X}_{i-1} - \mathbf{U}\mathbf{V}_{i-1}^T \right\|_F^2 + \lambda \sum_{k=1}^{p} S(\mathbf{u}_k), \tag{4}$$

where $\mathbf{V}_{i-1}$ is given in (4), and we use $\mathbf{U}_i$ to obtain $\mathbf{V}_i$ as follows:

$$\mathbf{V}_i = \underset{\|\mathbf{v}\| \leq 1}{\mathrm{argmin}} \left\| \mathbf{X}_{i-1} - \mathbf{U}_i\mathbf{V}^T \right\|_F^2 + \lambda \sum_{k=1}^{p} S(\mathbf{u}_{k,i}). \tag{5}$$

In the optimization, it is worth to note the second term regarding $\mathbf{u}_k$ in (4) is non-convex with respect to $\mathbf{U}$ while (5) is the quadratic term with the fixed $\mathbf{u}_k$. Thus, in the update, the problem is converted into a variational method using an auxiliary term as in [8].

We show the graphical results of the transformed coefficients in an 8-by-8 block-sized transform in Fig. 3, where a gray color represents a small value while a white or block color represent a large value. We see the distribution is close to the dirac delta distribution while there are some local bumps in the portion of the block, achieved with the grouping constraint. The coefficients are quantized with the dead-zone uniform scalar quantization used for HEVC, and entropy coded. The reconstructed samples are stored in the frame buffer, and iteratively used for learning the transform in coding the i + 1th frame as shown in Fig. 2.

We employ the rate-distortion (R-D) optimized transform selection between the conventional DCT mode in HEVC and the proposed method. We use the best transform to minimize the Lagrangian cost function defined as,

$$J = \left| \mathbf{X} - V^T\mathbf{U}_q \right| + \mu(R_C + R_H), \tag{6}$$

where X is the current block, $\mathbf{u}_q$ is the quantized coefficient. The first term in (6) is the distortion as mean-squared error (MSE), and in the second term R_C is the estimated bit-rates used for coding the coefficients and R_H is the overhead flag used for indicating if the mode is DCT mode or the proposed method. For the estimation we compute the proportion of the non-zero quantized coefficients using ρ domain estimation [9] for the simplicity while the coding gain can be improved with a more accurate estimator. We remain this in the future work. The Lagrangian multiplier μ in (6) is the same as the value used for the mode decision in HEVC. We choose an optimal transform by minimizing (6) as follows:

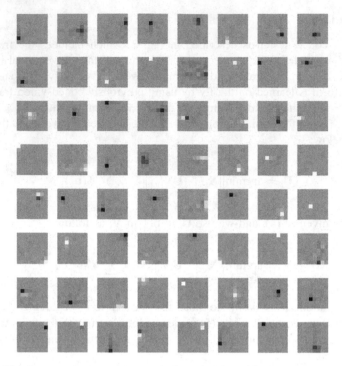

Fig. 3. Examples of transformed coefficients of the proposed method

$$T^* = \underset{\mathbf{V} \in \{T_{DCT}, T_{\text{PROP}}\}}{\arg\min} J \tag{7}$$

where we choose the best transform among DCT and the transform denoted by T_{PROP} derived by the proposed method.

3 Experimental Results

We conduct experiments to show coding efficiency of the proposed method in this section. We use the HEVC Range Extension (HEVC/RExt.) [10] reference software configured for the intra-only coding to obtain the residue. The reference software is available online [11]. In the software, we turn off few coding tools, including "Residual Rotation" and "Residual DPCM" because they change the structure of the residue. Moreover, we use MATLAB to train the residue and obtain the sparse structured transform. For test sequences, we use screen content videos [12] including computer synthetic videos "DeskTop," "Programming," and "Web Browsing", and so on formatted with RGB 4:4:4 sequences, used for HEVC/RExt standardization. We use all Intra-Coding structure. QPs are 20, 24, 28, and 32. We turned off the transform skipping mode to investigate the contribution of the proposed method in the improved coding gain.

Table 1. BD-rate saving (in the unit of %) of the proposed method.

Sequence	BD-rate Saving		
	R	G	B
Web Browsing	−1.7 %	−2.0 %	−2.0 %
Programming	−1.8 %	−1.9 %	−1.9 %
Conference	−2.1 %	−2.2 %	−1.8 %
Map	−1.5 %	−1.9 %	−1.6 %
Desktop	−2.2 %	−2.3 %	−1.9 %
FlyingGraphics	−2.1 %	−2.1 %	−2.2 %
Viking	−0.3 %	0.0 %	0.3 %
Robot	0.1 %	0.0 %	0.1 %
Big Buck	0.0 %	−0.3 %	0.0 %
Average	−1.28 %	−1.41 %	−1.26 %

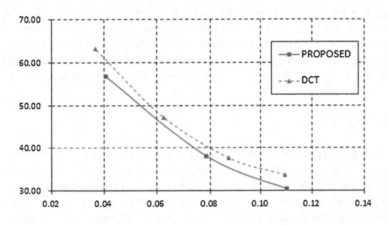

(a) "Programming" test sequence (RGB 444)

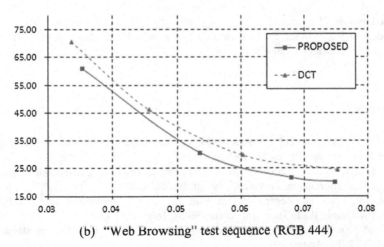

(b) "Web Browsing" test sequence (RGB 444)

Fig. 4. Estimated R-D curves using the proportion of the non-zero coefficients

The proposed method outperforms HEVC quantitatively in terms of BD-rate saving about −1.28 %, −1.41 %, and −1.26 % in R, G, and B components, respectively, as shown in Table 1. The coding gain in G component is slightly better than the other two components because the residues are trained using the G component. As shown in the table, the coding performance of the proposed method relies on the characteristics of the test sequences. For example, the sequences including many of texts and graphical polygons shows the visible coding gain more than 2 % BD-rate saving while the only very small gain or loss are shown in the computer animation typed videos such as "Robot" and "Viking," which are even close to natural videos.

Though the coding gain is not much significant, we observe the proposed method provides significantly better coding efficiency envisioned by the estimated rates from the ρ domain analysis, which computes a ratio of the non-zero coefficients in a block. We show the estimated R-D curves in Fig. 4 where x-axis refers to the proportion of the non-zero coefficients, and y-axis refers the MSE as distortion. As shown, the intrinsic capability of the proposed method may provide more than 9 % BD-rate saving in "Web Browsing" test sequence, however, there might be coding loss from the HEVC entropy coding method, which is not tailored to the proposed method. Though the proposed method groups the coefficients in a local region, the HEVC transform coefficient coding scheme with the diagonal search would be not suitable for the characteristics. The improvements of the entropy coding is remained as the future work.

4   Conclusion

We proposed a novel transform method using a sparse model of residue for intra video coding, which is designed to capture a local nonzero pattern in a block. The transform was incorporated into the state-of-the-art video codec and was shown in experimental results to improve the coding efficiency, particularly, in computer synthetic videos. As future works, we will develop an efficient transform coefficient coding scheme, and extend the proposed method to the inter-coding.

Acknowledgement. This research was supported by Basic Science Research Program through the National Research Foundation of Korea(NRF) funded by the Ministry of Education (NRF-2014R1A1A2056587).

References

1. Chen, S.S., Donoho, D.L., Saunders, M.A.: Atomic decomposition by basis pursuit. SIAM J. Sci. Comp. **20**, 33–61 (1999)
2. Kang, J.W., Cohen, R., Vetro, A., Kuo, C.-C.J.: Efficient dictionary based video coding with reduced side information. In: Proceedings of the ISCAS, May 2011
3. Effros, M., Feng, H., Zeger, K.: Sub-optimality of the Karhunen-Loeve transform for transform coding. IEEE Trans. Inf. Theor. **50**(8), 1605–1619 (2004)
4. Mrak, M., Xu, J.: Improving screen content coding in HEVC by transform skipping. In: EUSIPCO 2012, August 2012

5. Ventura, R., Vandergheynst, P., Frossard, P.: Lowrate and flexible image coding with redundant representations. IEEE Trans. Image Process. **15**(3), 726–739 (2006)
6. Kang, J., Gabbouj, M., Kuo, C.-C.J.: Sparse/DCT (S/DCT) two-layered representation of prediction residuals for video coding. IEEE Trans. Image Process. **22**(7), 2711–2722 (2013)
7. Zou, H., Hastie, T., Tibshirani, R.: Sparse principal component analysis. J. Comput. Graph. Statist. **15**(2), 265–286 (2006)
8. Bach, F., Jenatton, R., Obozinski, G.: Structured sparse principal component analysis. J. Mach. Learn. Res. 366–373
9. He, Z., Mitra, S.: Optimum bit allocation and accurate rate control for video coding via rho-domain source modeling. IEEE Trans. Circuits Syst. Video Tech. **12**(10), 840–849 (2012)
10. Sullivan, G., Boyce, J., Chen, Y., Ohm, J.R., Segall, A., Vetro, A.: Standardized extensions of high efficiency video coding. IEEE J. Sel. Topics Sign. Process. **7**(6), 1001–1016 (2013)
11. ISO/IEC JTC1/SC29/WG11 N14978. PDAM1 Reference software for format range extensions profiles, July 2014
12. ISO/IEC JTC1/SC29/WG11 and ITU-T SG16 Q.6, Joint call for proposals for coding of screen content, January 2014

Special Poster Sessions

High-Speed Periodic Motion Reconstruction Using an Off-the-shelf Camera with Compensation for Rolling Shutter Effect

Jeong-Jik Seo, Wissam J. Baddar, Hyung-Il Kim, and Yong Man Ro[✉]

Image and Video Systems Lab, School of Electrical Engineering,
Korea Advanced Institute of Science and Technology (KAIST),
Daejeon, Republic of Korea
{jj.seo,wisam.baddar,hyungil.kim,ymro}@kaist.ac.kr

Abstract. In recent years, high-speed signal reconstruction with sub-Nyquist sampling have attracted the attention of researchers in the signal processing field. Nonetheless, such methods have been limited either by the need to utilize multiple cameras, or relying on newly designed imaging hardware. In this paper, we propose a high-speed periodic motion reconstruction method, obtained by randomly delaying the camera exposure. This allows it to utilize a conventional off-the-shelf camera. In addition, the proposed method compensates the rolling shutter effect, which is inevitable if the camera's image sensor is made of complementary metal-oxide semiconductor (CMOS), while reconstructing the high-speed periodic motion. Exhaustive and comparative experiments have been conducted to validate the proposed method, which showed promising performance in terms of reconstruction error, and effective compensation of the rolling shutter effect.

Keywords: Computational imaging · Compressive sensing · High speed motion · Sparse reconstruction · Rolling shutter effect

1 Introduction

According to the Nyquist sampling theorem, the sampling rate should be at least twice the bandwidth of a signal in order to avoid information loss (i.e., aliasing). Therefore, high-speed motions which can be observed in real-life (e.g., hand mixers or electric fan motions rotating over 60 rounds per second) cannot be captured by a conventional low-frame rate cameras (e.g., 30 frames per second (fps)). In other words, in order to capture such fast motions, a high-speed camera which can cover high sampling rate (with at least over 120 fps) is required. However, the high-speed cameras have been known to be relatively expensive [7]. In addition, the short aperture time of high-speed cameras induces light-inefficiency [5]. Recently, in image processing area, there have been various research efforts [4,6–8] concerning the reconstruction of high-speed motion from the low-frame rate cameras. In [4,8], the authors proposed reconstructing the high-speed video by

© Springer International Publishing Switzerland 2015
Y.-S. Ho et al. (Eds.): PCM 2015, Part II, LNCS 9315, pp. 311–320, 2015.
DOI: 10.1007/978-3-319-24078-7_31

utilizing multiple low-frame rate cameras. It should be noted that such methods require a large number of cameras in order to achieve reasonable reconstruction. In contrast, Mergell et al. tried to capture a high-speed signal through intentional aliasing induced by band-pass sampling, assuming that the motion in the scene is periodic [4]. However, since the motion was exposed for a short time, light-inefficient problem could not be resolved. More recently, coded strobing photography, which is a method for sampling high-speed periodic signal based on random coded exposure, was proposed [7]. Because the randomly coded exposure technique does not affect the exposure time for a frame, this method was able to overcome the light-inefficient problem with the effective reconstruction capability of high-speed video. Nevertheless, this method requires a ferro-electric shutter in order to control the exposure with high temporal resolution, which is not always available off-the-shelf. Furthermore, in [7], a charge-coupled device (CCD) image sensor was adopted for image acquisition to achieve high image quality, because it is more sensitive to light and robust to noise. However, due to the low-cost and low-power characteristics of the complementary metal-oxide semiconductor (CMOS) image sensor, it has been received a wider adoption in camera systems [3]. Unlike CCD image sensors, each scan-line of the CMOS image sensor has a different exposure time due to the sequential read-out nature of the most CMOS image sensor array [3]. This inevitably results in rolling shutter effect which leads to geometric distortion in the captured image [3]. If the speed of the motion is slow, the rolling shutter effect can be negligible, but it steeply increases when the motion is very fast. Therefore, the rolling shutter effect must be considered when reconstructing high-speed motion from a low-frame rate camera with a CMOS image sensor.

In this paper, we propose a method for reconstructing high-speed periodic signals from a low-frame rate camera with CMOS image sensor based on a randomly introduced exposure delay. In addition, the rolling shutter effect is compensated while reconstructing the high-speed periodic signal. The main contributions of this paper are follows:

- By making use of random delays between the acquisition times of some frames, which are achieved by randomly turning the camera off and on, high-speed periodic motion is reconstructed without the aid of newly designed hardware or multiple cameras. Therefore, high-speed periodic signals can be reconstructed with an off-the-shelf camera only (e.g., a camera built in mobile phone or webcam). Moreover, the long exposure time of the conventional camera resolves the light-inefficiency problem.
- Through time-shifting of the Fourier basis functions in the process of high-speed periodic motion reconstruction, we can compensate the rolling shutter effects caused by the read-out operation of CMOS image sensor. Therefore, the geometric distortion due to the rolling shutter effect can be avoided when reconstructing the high-speed periodic motion.

The remainder of this paper is organized as follows. Section 2 reviews coded strobing photography briefly. In Sect. 3, the proposed high-speed motion reconstruction method is described. In Sect. 4, we present experimental results that demonstrate the effectiveness of the proposed framework. Finally, conclusion is drawn in Sect. 5.

2 Brief Review of Coded Strobing Photography

In this section, we briefly review the coded strobing photography [7], which aims to reconstruct high-speed periodic signals from a low-frame rate video. For that purpose, an imaging system combining a low-frame rate camera and a high-speed shutter is utilized. In the coded strobing photography system, a high-speed signal reconstruction problem is formulated as a structured sparse reconstruction based on a camera observation model and a signal model.

2.1 Camera Observation Model and Signal Model

Given a periodic signal $x(t)$ with the period T_0, i.e., $x(t) = x(t + T_0)$, we can accurately represent and recover the signal if the sampling rate f_h is higher than its Nyquist sampling rate (i.e., temporal resolution δt is $1/f_h$). If the total number of samples is N, the samples can be represented as a N-dimensional vector $\mathbf{x} \in \mathbb{R}$. By using an imaging system that performs amplitude modulation of the incoming radiance value instead of amplitude integration during the entire frame duration [7], the observed intensity values are obtained, which are represented as an M-dimensional vector $\mathbf{y} \in \mathbb{R}^M$. Moreover, the number of measurements is usually much less than that of the samples in $\mathbf{x}$ (i.e., $M \ll N$). The relationship between the samples $\mathbf{x}$ and the observed intensity values y can be represented as following:

$$\mathbf{y} = \mathbf{Cx} + \boldsymbol{\eta}, \tag{1}$$

where $\mathbf{C} \in \mathbb{R}^{M \times N}$ is a code matrix which represents randomly coded exposure and integration of the frame duration [7], and $\boldsymbol{\eta}$ denotes the observation noise. If the periodic signal $x(t)$ is band-limited, the samples x can be represented as a linear combination of Fourier basis:

$$\mathbf{x} = \mathbf{Bs}, \tag{2}$$

where $\mathbf{B}$ and $\mathbf{s}$ denote a Fourier basis matrix and a basis coefficient, respectively. By combining Eqs. (1) and (2), the observed intensity values $\mathbf{y}$ can be modelled by the following equation:

$$\mathbf{y} = \mathbf{Cx} + \boldsymbol{\eta} = \mathbf{CBs} + \boldsymbol{\eta} = \mathbf{As} + \boldsymbol{\eta}. \tag{3}$$

In other words, since the observed intensity values $\mathbf{y}$ and $\mathbf{A}$ are known, we can reconstruct the high-speed periodic signal by obtaining the basis coefficient vector s when solving Eq. (3).

2.2 High-Speed Periodic Motion Reconstruction via Structured Sparse Reconstruction

According to Sect. 2.1, the high-speed motion reconstruction problem is formulated as Eq. (3), which is used to obtain the basis coefficient vector $\mathbf{s}$. Added to that, a prior knowledge about the signal is essential to find robust solutions for $\mathbf{s}$ in Eq.(3). If the signal is periodic, the signal consists of a fundamental frequency components and its harmonics. However, in the real world, the exactly periodic signal is hard to exist because of noise in the periods. Therefore, quasi-periodic signals are usually observed. If the signal is quasi-periodic, the frequency components are concentrated in peripheries of the fundamental frequency and its harmonics [7]. In other words, the quasi-periodic signal is sparsely represented in the frequency domain, i.e., the basis coefficient vector $\mathbf{s}$ is sparse. This knowledge can be used to enforce sparsity on $\mathbf{s}$ for solving Eq. (3).

To represent the set of basis elements that the sparse vector $\mathbf{s}$ can have, the set is defined by S_{f_o} with basis elements $[jf_o - \delta f_o, jf_o + \delta f_o]$ for fundamental frequency f_o, where $j \in \{-Q, \cdots, 0, 1, \cdots, Q\}$ such that all the sparse coefficients $\mathbf{s}$ will lie in this smaller set [7]. In order to estimate the sparse vector $\mathbf{s}$, the problem so called "structured sparse reconstruction" is formulated as follows [7]:

$$\min \|\mathbf{s}\|_0 \ s.t. \ \|\mathbf{y} - \mathbf{As}\| \leq \epsilon, \ \xi(\mathbf{s}) \in S_{f_0}, \tag{4}$$

where $\xi(\mathbf{s})$ is a new vector whose only non-zero entries are the entries in $\mathbf{s}$. The solution $\mathbf{s}$ of Eq. (4) is determined through CoSaMP algorithm [1]. Finally, the signal $\mathbf{s}$ (e.g., high-speed motion) can be simply computed by Eq. (2).

However, even if the method in [7] has shown the effective high-speed motion reconstruction results, the additional ferro-electric shutter with high-frame rate should be equipped, which is not available in a conventional camera. Furthermore, this frame work has been based on CCD image sensor, a rolling shutter effect encountered in the commercialized CMOS image sensor has not been considered.

3 Proposed Method

In this section, we explain the proposed method in detail. In Sect. 3.1, we describe the proposed method to reconstruct a high-speed periodic/quasi-periodic signal without the need for additional hardware, by introducing random exposure delay. We also discuss effects of the introduced random delays. In Sect. 3.2, the method for compensating the rolling shutter effect, caused by the read-out operation of CMOS image sensor, is explained.

3.1 High-Speed Periodic Motion Reconstruction Based on Random Delay

When a conventional camera captures a high-speed periodic motion with a sub-Nyquist sampling rate, information loss is inevitable due to aliasing. To avoid

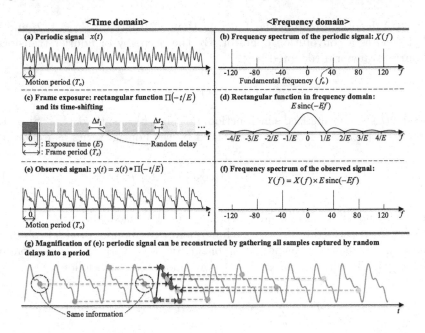

Fig. 1. Visualization of analysis of the proposed high-speed periodic signal reconstruction.

the problem, we propose the high-speed signal reconstruction method based on random delay. From herein, a random delay is defined as the intentionally and randomly generated delay between randomly selected consecutive frames. To apply the random delay into a conventional camera system, the camera should be randomly turned off and on. By delaying exposure time randomly, we can capture different time ranges of a period. This leads to the acquisition of additional information of the periodic signal. Therefore, the high-speed periodic signal can be reconstructed by a sub-Nyquist sampling [2,7]. In terms of light-inefficiency, although the proposed algorithm does not expose light during the random delay, the exposure time for a frame (E) is about twice the average exposure time of the coded strobing approach [7] and multiple times longer than the exposure time in high-speed cameras.

Figure 1 visualizes the analysis of the proposed high-speed periodic signal reconstruction algorithm. The left graphs show the time domain of the signal and the graphs on the right visualize the frequency domain of the signal. The fundamental frequency of the periodic signal $x(t)$ is defined as $f_o = 1/T_o$ (seen in Fig. 1(a)). If a frame is exposed for E seconds (shown in Fig. 1(c)), an observed intensity during the exposure for the time interval $[t - E/2, t + E/2]$ can be computed as follows:

$$y_c(t) = \int_{t-\frac{E}{2}}^{t+\frac{E}{2}} x(\tau)d\tau = \int_{-\infty}^{+\infty} x(\tau)\Pi\left(\frac{\tau - t}{E}\right) d\tau = x(t) * \Pi\left(\frac{t}{E}\right), \quad (5)$$

where $\Pi(k)$ denotes a rectangular function that returns 1 for $|k| < 1/2$, and '*' symbol denotes a convolution operator. By the property of Fourier transform, the Fourier transform (denoted by '$\mathcal{F}$') of Eq. (5) can be represented as:

$$y(t) \xleftrightarrow{\mathcal{F}} Y(f) = X(f) \times E\text{sinc}(-Ef), \tag{6}$$

where $\text{sinc}(k)$ is a normalized sinc function, i.e., $\text{sinc}(k) = \sin(\pi k)/\pi k$. Since the signal $x(t)$ is periodic, it can be represented by the Fourier series as following:

$$x(t) = D_o + \sum_{n=1}^{\infty} D_n \cos(2\pi f_o nt + \theta_n), \tag{7}$$

where D_n denotes the amplitude spectrum, θ_n the phase spectrum, and $n \in \{1, 2, \cdots\}$. As described in Eq. (6), each frequency component of $y(t)$ is obtained by multiplying each frequency component of $x(t)$ and the corresponding $E\text{sinc}(-Ef)$. Therefore, the signal $y(t)$ can be formulated as follows:

$$
\begin{aligned}
y(t) &= ED_o + \sum_{n=1}^{\infty} E\text{sinc}(-Ef_o n)D_n \cos(2\pi f_o nt + \theta_n) \\
&= G_o + \sum_{n=1}^{\infty} G_n \cos(2\pi f_o nt + \theta_n),
\end{aligned}
\tag{8}
$$

where $G_n = E\text{sinc}(-Ef_o n)D_n$. Finally, $y(t)$ can be seen as a periodic signal whose fundamental frequency is the same as $x(t)$ (seen in Fig. 1(e) and (f)). In other words, if we know $y(t)$ we can reconstruct $x(t)$. This way, the objective of the problem is transformed from finding the periodic signal $x(t)$ to finding the sampled observed signal $y(t)$. In order to capture $y(t)$ without loss of information, it should be sampled higher than the Nyquist sampling rate. Nonetheless the signal $y(t)$ is captured by a sub-Nyquist sampling rate, the signal can be reconstructed because of its periodicity and proposed randomly delayed exposure technique. As can be seen in Fig. 1, for example, if the frame rate is $f_s = 1/T_s = 30$fps, the motion frequency is $f_o = 1/T_o = 40$ Hz, and the bandwidth of $y(t)$ is $3f_o$. Since $2 \times 3f_o = 240 \times f_s$, Nyquist sampling theorem is not satisfied. In order to satisfy the theorem, at least a sampling interval shorter than $1/6f_o$ is needed. Therefore, at least 6 different samples for a period of $y(t)$ are necessary. However, since the signal $y(t)$ is periodic, certain points at one period can correspond to the sampled points at the other periods (shown in Fig. 1(g)). In this case, while three frames are being sampled, a period is repeated four times because $T_o \times 4 = T_s \times 3 = 1/10$. Therefore, three values at different points within a motion period of $y(t)$ are obtained (the red points in Fig. 1(e) and (g)). When the frames are captured with a delay δt_i, i.e., captured at $\{\delta t_i, T_s + \delta t_i, 2T_s + \delta t_i, \cdots\}$, additional three values can be observed (the blue points in Fig. 1(e) and(g)). The more random delays, the more additional information is obtained, resulting in reconstructing the high-speed periodic signal with a sub-Nyquist sampling.

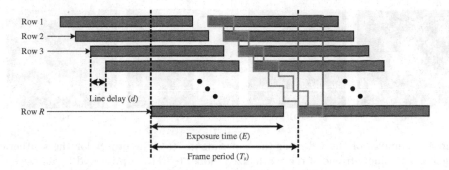

Fig. 2. Timing diagram of the CMOS sensor array.

3.2 Compensation for Rolling Shutter Effect

As aforementioned, CMOS image sensors have intrinsically rolling shutter effects caused by the read-out operation. In particular, this effect occurs due to the line delay that the exposure of the rows within a single frame starts and ends at different times as shown in Fig. 2. For example, if we are about to obtain the information within the red box in Fig. 2 (i.e., the signal in r-th row denoted by $x_r(t)$), the information within the green region is actually obtained (i.e., actually observed signal in r-th row at time t denoted by $x_r^o(t)$). This relationship between $x_r(t)$ and $x_r^o(t)$ is formulated as time-shifting:

$$x_r(t) = x_r^o(t + (R - r)d). \tag{9}$$

Since the Fourier basis matrix $\mathbf{B}$ consists of N orthogonal Fourier bases, we can represent the basis matrix as $\mathbf{B(t)} = [b_0(\mathbf{t}, b_1(\mathbf{t}, \cdots, b_{N-1}(\mathbf{t}]$ with N orthogonal Fourier bases functions, where $\mathbf{t} = [0, \delta t, \cdots, (N-1)\delta t]^T$. By the signal model in Eq. (2), $x_r^o(\mathbf{t})$ is represented as $\mathbf{B(t)}\mathbf{s}_r$, where $\mathbf{s}_r$ denotes a basis coefficient vector at the r-th row. Therefore, by using Eq. (9), $x_r(\mathbf{t})$ can be represented as follows:

$$
\begin{aligned}
x_r(\mathbf{t}) &= x_r^o(\mathbf{t} + (R - r)d) = \mathbf{B(t} + (R - r)d)\mathbf{s}_r \\
&= [b_0(\mathbf{t} + (R - r)d), b_1(\mathbf{t} + (R - r)d), \cdots, b_{N-1}(\mathbf{t} + (R - r)d)]\mathbf{s}_r,
\end{aligned} \tag{10}
$$

where R is the number of rows in a frame. Equation (10) shows that the signal $x_r(\mathbf{t})$ can be reconstructed by using time-shifted Fourier basis matrix. After computing $\mathbf{s}_r$ by the structured sparse reconstruction, $x_r(\mathbf{t})$ can be simply computed with matrix multiplication between $\mathbf{B(t} + (R - r)d)$ and $\mathbf{s}_r$.

4 Experiments

4.1 High-Speed Periodic Motion Reconstruction

In order to validate the effectiveness of the proposed method, synthetic videos comprised of moving objects with texture were generated for the experiments

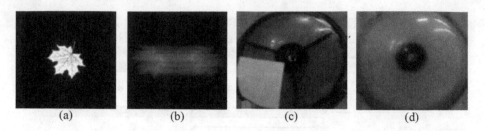

(a)   (b)   (c)   (d)

Fig. 3. Examples of the video for the experiments. (a) The object for the synthetic video. (b) Example frame of the synthesized video. (c) The cap rotated by an electric motor. (d) Example of the observed video when the object in (c) is rotating and capture it with the CMOS camera.

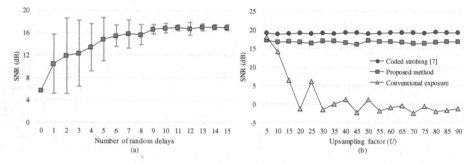

Fig. 4. Examples of the video for the experiments. (a) The object for the synthetic video. (b) Example frame of the synthesized video. (c) The cap rotated by an electric motor. (d) Example of the observed video when the object in (c) is rotating and capture it with the CMOS camera.

as shown in Fig. 3. The videos were captured by a conventional camera with 25 fps (i.e., $f_s = 25$), and a high-speed video was reconstructed with 1,000 fps. The fundamental frequency of the motion captured was 60 Hz (i.e., $f_o = 60$), and the number of random delays was set to 10 delays. The observed frames (M) of the generated video consisted 300 frames. In order to analyze only the effect of random delays, the rolling shutter effect was not considered in the synthetic video.

To analyze the performance of the proposed reconstruction methods under different number of random delays, signal-to-noise ratio (SNR) was utilized. Figure 4(a) shows the SNR values according to the number of random delays. It should be noted that for each random delay, the experiments were repeated 50 times and the average SNR and standard deviation were measured. From Fig. 4(a), it can be observed that the standard deviation as well as the reconstruction error decreases when the number of random delays increase showing more stable reconstruction performance. Compared to the SNR for the coded strobing method [7], which achieved 16.9 dB, the SNR for the proposed method with 10 random delays showed comparable results without any additional hardware.

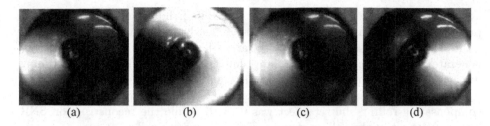

    (a)   (b)   (c)   (d)

Fig. 5. Examples of the video for the experiments. (a) The object for the synthetic video. (b) Example frame of the synthesized video. (c) The cap rotated by an electric motor. (d) Example of the observed video when the object in (c) is rotating and capture it with the CMOS camera.

In particular, the amount of additional information largely depends on the random delay δt_i. When δt_i is the same as the period of a motion, no additional information is obtained. Therefore, when the number of random delays is small, the amount of additional information is largely varying over each trial, thus resulting in large standard deviations. On the other hand, when the number of random delays increases, obtaining additional information is guaranteed because different values of δt_i are randomly generated.

Additionally, we compared the high-speed motion reconstruction performance of the proposed method to the coded strobing and the reconstruction in the conventional exposure in terms of SNRs. Figure 4(b) shows the SNR values for the aforementioned 3 methods under various up-sampling factors U, respectively. When up-sampling factor U increases, the temporal resolution also increases. Therefore, more information must be reconstructed given the limited observed frames. When U increases from 5 to 90, the proposed method and the coded strobing showed stable SNR values. On the other hand, the SNR of the conventional exposure based reconstruction decreases rapidly. Likewise, although our method does not require an additional hardware such as high-speed electric shutter, our proposed method showed comparable SNR values to that of the coded strobing according to varying up-sampling factor.

4.2 Compensation for Rolling Shutter Effect

Further, we analyzed the results of compensation for rolling shutter effect by making use of an off-the-shelf camera, where Logitech HD pro webcam C920 with CMOS image sensor were adopted. The object in Fig. 3(c) was rotating with a frequency of 60 Hz and was captured with a frame rate of 25 fps. An example of the observed frames is shown in Fig. 3(d). We set the up-sampling factor U to 20 and the number of random delays to 10. When the rolling shutter effect was not considered, the reconstruction results are shown in Fig. 5(a) and (b). These results showed that the white area of the object was varying dependent on its location. For example, if the direction of rotation is clock-wise, the white part was larger when it was located at the right side, and smaller when it located at left. However, when we compensated the rolling shutter effect based on the proposed method, the area of white part is nearly the same and independent on its location (please refer to Fig. 5(c) and (d)). For quantitative analysis, we generated a synthetic video with rolling shutter effect by giving different timing in each row. After that, we compared the results when the effect of the rolling shutter was compensated and when it was not. The reconstruction SNR before compensating rolling shutter effect was 0.15 dB, while the reconstruction SNR was 16.9 dB, when the rolling shutter effect is compensated.

5 Conclusion

In this paper, high-speed periodic motion reconstruction with an off-the-shelf conventional camera, by adding randomly delayed exposure, was proposed. In order to mitigate the rolling shutter effect, encountered in the commonly used CMOS image sensor, time-shifted Fourier basis matrix was proposed to compensate it during the reconstruction. The proposed method showed higher reconstruction SNR than reconstruction with a conventional camera. Moreover, the proposed method did not require any additional hardware, yet was able to achieve comparable results with the coded strobing method.

Acknowledgments. This paper is based on a research which has been conducted as part of the KAIST-funded K-Valley RED&B Project for 2015.

References

1. Needell, D., Tropp, J.A.: CoSaMP: iterative signal recovery from incomplete and inaccurate samples. Commun. ACM **53**(12), 93–100 (2010)
2. Baraniuk, R.G., Candes, E., Nowak, R., Vetterli, M.: Compressive sampling. IEEE Signal Process. Mag. **25**(2), 12–13 (2008)
3. Liang, C.-K., Chang, L.-W., Chen, H.H.: Analysis and compensation of rolling shutter effect. IEEE Trans. Image Process. **17**(8), 1323–1330 (2008)
4. Mergell, P., Herzel, H., Titze, I.R.: Irregular vocal-fold vibration - high-speed observation and modeling. J. Acoust. Soc. Am. **108**(6), 2996–3002 (2000)
5. Reddy, D., Veeraraghavan, A., Chellappa, R.: P2C2: programmable pixel compressive camera for high speed imaging. In: IEEE Conference on Computer Vision and Pattern Recognition, pp. 329–336 (2011)
6. Shechtman, E., Caspi, Y., Irani, M.: Increasing space-time resolution in video. In: Heyden, A., Sparr, G., Nielsen, M., Johansen, P. (eds.) ECCV 2002, Part I. LNCS, vol. 2350, pp. 753–768. Springer, Heidelberg (2002)
7. Veeraraghavan, A., Reddy, D., Raskar, R.: Coded strobing photography: compressive sensing of high-speed periodic events. IEEE Trans. Pattern Anal. Mach. Intell. **33**(4), 671–686 (2011)
8. Wilburn, B., Joshi, N., Vaish, V., Talvala, E.-V., Antunez, E., Barth, A., Adams, A., Horowitz, M., Levoy, M.: High performance imaging using large camera arrays. ACM Trans. Graph. **224**(3), 765–776 (2005)

Robust Feature Extraction for Shift and Direction Invariant Action Recognition

Younghan Jeon, Tushar Sandhan, and Jin Young Choi[✉]

Perception and Intelligence Laboratory, Department of Electrical and Computer Engineering, ASRI, Seoul National University, Seoul, Republic of Korea
{yh1992,tushar,jychoi}@snu.ac.kr

Abstract. We propose a novel feature based on optical flow for action recognition. The feature is quite simple and has much lower computational load than the existing features for action recognition algorithms. It has invariance to scale, different time duration and direction of an action. Since raw optical flow is noisy on the background, several methods for noise reduction are presented. Firstly, we bundle up the fixed number of frames as a block and take the median value of optical flow (median flow). Secondly, we take normalization of histogram depending on the total magnitude. Lastly, we do low-pass filtering in frequency domain. Converting the time domain to frequency domain based on Fourier transform makes the feature invariant to shifted time duration of action. In constructing the histogram of optical flow, we align the direction of an action so that we can get direction invariant action representation. Experiments on benchmark action dataset (KTH) and our own dataset for smart class show that the proposed method gives a good performance comparable to the state-of-the-art approaches and has applicability to actual environments with smart class dataset.

Keywords: Action recognition · Activity recognition · Optical flow histogram · Behavior understanding · Video analysis

1 Introduction

Human action recognition (HAR) is an important research area of computer vision. Since figuring out the purpose of human in video is very important and every data is hard to be analyzed by human, automatic action recognition algorithm is required.

In the last decade, various approaches have been proposed for HAR. Previous works are roughly divided into two categories, appearance based [1–5] and motion based approaches [6–18]. Appearance based method captures the visual information in order to construct features. Main drawback of this model is ignoring the inherent temporal information of the action. In other words, it is hard to reflect sequential information when using the feature based on appearance. For this reason, the current researches mostly utilize a motion based algorithm. In particular, [7, 8] have shown outstanding performance in this area, however they require high computational load. In this paper, we aim to develop a motion based feature requiring low computation as well as giving good performance.

© Springer International Publishing Switzerland 2015
Y.-S. Ho et al. (Eds.): PCM 2015, Part II, LNCS 9315, pp. 321–329, 2015.
DOI: 10.1007/978-3-319-24078-7_32

In the motion based approaches, features should include the information on action dynamics which takes an important role in characterizing the actions. The motion features could be categorized into local and global features. Local features such as silhouette, dense trajectories [9], space-time descriptors [10, 11] only use the information of observed region. Computing the saliency is also needed, which requires much computational time. Also they have weakness for background clutter or illumination changes i.e. seriously affected by noise. Moreover, inherent data or constraints are discarded except salient part. Global features such as motion energy images [12] and motion history images [13] use whole sequence but ignore the temporal information. Moreover these features have different representation for same action with different directions and are hard to capture the action of different time duration. Frequencygram [14] discards the steady frames which lead to loss of data and also ignores the direction of action.

In this paper, we propose a new motion based feature resolving above issues with relatively low computational load. The proposed feature has invariance on scale, direction and time duration. It is based on optical flow and several techniques for reducing the noise are applied. To get direction-invariant action representation and recognition, direction is aligned in constructing the histogram of optical flow by considering the total magnitude of each direction (left bins or right bins) in histogram. In addition, frequency domain analysis is done to show the shift invariance property of the proposed feature through a similar framework with Frequencygram [14].

2   Proposed Method

Our feature is based on optical flow which is a sequence of motion patterns of objects, surfaces and edges in a video. Since various methods for computing optical flow has been developed, we use one of these algorithms. Lots of features and algorithms for action recognition use optical flow as it encapsulates information and dynamics of motion. Since its raw form is hard to be used because of noise from background, regulation strategy for the noise is required. Inspired by Dalal et al. [19], we use histogram approach for reducing the noise so as to obtain the representative features of actions.

2.1   Optical Flow Histogram

Each video divided into N_B blocks and each block consists of fixed number of video frames (N_f). Block analysis could help reducing the noise and give other side effects disturbing the representative feature extraction. Optical flow histogram is constructed in each block as follows.

2.1.1   Median Flow

As mentioned above, raw optical flow is hard to be applied intactly, we first reduce the background noise by combining N_f number of frames as a block. For every frames in each block, optical flow is computed. Then in every pixel the median value of optical flow among the included frames in each axis is picked. Median flow vector $\bar{v} = [u, v]^T$ for each pixel of each block is obtained.

Figure 1 shows the alternation between raw optical flow and median flow.

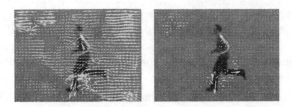

Fig. 1. Comparison between raw optical flow (left) and median flow (right)

2.1.2 Sum of Magnitude in Each Range

Each pixel in the block has median flow vector $\bar{v}$. Initial histograms of blocks are constructed by the following process. Angle from $-\pi$ to π is divided into a number of bins N_{bin}. N_{bin} must be multiple number of 4 in order to be symmetric as shown in Fig. 2. For each range, magnitude of the corresponding median flow vector is summed up to be the weight of each bin as follows:

$$W_i = \sum_{\bar{v} \in \Theta_i} \|\bar{v}\|, \tag{1}$$

where,

$$\Theta_i = \left\{ \bar{v} \mid -\pi + \frac{2\pi(i-1)}{N_{bin}} \le \theta(\bar{v}) \le -\pi + \frac{2\pi i}{N_{bin}} \right\}, \tag{2}$$

$$\theta(\bar{v}) = \frac{(1 - sgn(u))\, sgn(v)}{2} \pi + \tan^{-1} \frac{v}{u}. \tag{3}$$

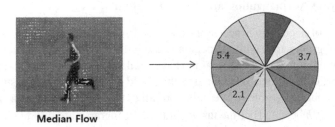

Median Flow

Fig. 2. Weight computation for each bin from median flow

W_i is the weight of each bin for $1 \le i \le N_{bin}$ and $\theta(\bar{v})$ is the direction angle of vector $\bar{v}$ from $-\pi$ to π. Using these weights, initial weighted histograms for blocks $H^I = \left\{ h^I_1, h^I_2, \cdots, h^I_{N_B} \right\}$ are constructed.

2.1.3 Aligning Direction

N_B number of initial histograms were generated in the previous steps. However these histograms have different distributions for the same actions with different direction. For example, stretching left arm to the left side and right arm to the right side have different (but symmetric) histograms. To avoid this issue, direction should be aligned so that we can get the same histograms for the same actions with different directions. Predicting direction of actions could be resolved by comparing total magnitude of each half side in the histogram. Right half is from $-\frac{\pi}{2}$ to $\frac{\pi}{2}$ and left half side is the rest. If sum of magnitude in the right half side is bigger than the left, weight of each bin is switched with the symmetric bin. Then we get the aligned histograms for blocks $H^d = \left\{ h_1^d, h_2^d, \cdots, h_{N_B}^d \right\}$. Figure 3 shows overview of direction aligning.

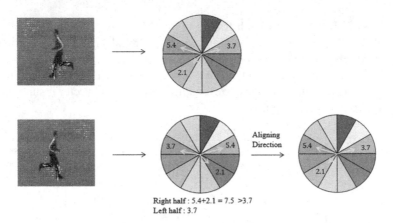

Right half : 5.4+2.1 = 7.5 >3.7
Left half : 3.7

Fig. 3. After aligning, same action with different direction has same distribution

2.2   Histogram Normalization and Concatenation

Normalization of histogram is necessary for scale invariant feature generation since the weight of histogram varies with the person's size in the scene even if the action is same. However, normalizing every histogram to have total weight of 1 could enlarge the noise effect in the actionless block. Introducing a threshold τ in total weight of histogram could help us to sort out this problem. We only normalize the high energy histogram, namely $\|h_i^d\| \geq \tau$, to be $\|h_i^d\| = 1$. That means weights of histograms are divided by their total weight of histogram. The rest low energy histograms ($\|h_i^d\| < \tau$) are not divided by their total weight but by τ to weaker the effect while maintaining the information. Then we get the final histogram $H = \left\{ h_1, h_2, \cdots, h_{N_B} \right\}$. Concatenating these histograms as columns, histogram pattern matrix $\mathbf{H} = \left[h_1 h_2 \cdots h_{N_B} \right]$ for each video which have $N_B \times N_{bin}$ dimension is constructed. Figure 4 shows whole processes in this step. Although $\mathbf{H}$ is a representation of an action, it is hard to work as a feature directly since each actor starts action at different time. Also it has a large amount of noise even if somewhat reduced by median. Section 2.3 deals with this problem.

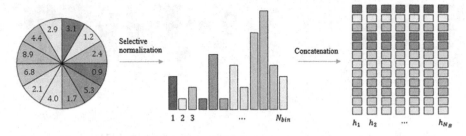

Fig. 4. Histogram normalization and concatenation

2.3   Low-Pass Filtering in Frequency Domain

Handling in frequency domain has two main advantages. First, we can get the shift-invariant property by using only magnitude after Fourier transform. Second, since noise commonly has high frequency, it could be removed by low-pass filtering in the frequency domain. Fourier transform converts the domain from time to frequency.

$$\hat{H}_w(k,l) = \sum_{n=0}^{N_B-1} \sum_{m=0}^{N_{bin}-1} e^{-i(\omega_k n + \omega_l m)} \cdot \mathbf{H}(n,m), \tag{4}$$

$$H_w = \begin{cases} \left|\hat{H}_w(k,l)\right| & if \; k \in \left[-\omega_B, \omega_B\right], \; l \in \left[-\omega_{bin}, \omega_{bin}\right], \\ 0 & otherwise \end{cases} \tag{5}$$

where $\omega_k = \frac{2\pi k}{N_B}$, $\omega_l = \frac{2\pi l}{N_{bin}}$ and ω_B, ω_b are fixed constants determining bandwidth of low-pass filter. H_w is the final feature of action representation.

3   Experimental Results

We evaluated the performance on two datasets, benchmark dataset (KTH, [20]) and our own dataset (Smart class). We used the same experimental setting in each dataset as $N_f = 5, N_{bin} = 32, \tau = 0.1, w_{bin} = 2\pi, w_B = 0.5\pi$. Classification is performed by multi-class SVM [21]. For the optical flow extraction we used the algorithm proposed by Liu. [22]. This could be replaced by any other methods.

3.1   KTH Dataset

Videos in KTH dataset have 6 actions (walking, jogging, running, boxing, hand waving, hand clapping) for 4 scenarios (indoor, outdoor, scale variation, different clothes) with 25 subjects, totally 600 scenes (see Fig. 5).

Fig. 5. KTH dataset

Table 1 shows recognition accuracy and confusion matrix in KTH dataset. We used original split, 16 subjects for training and 9 subjects for testing, suggested in [20]. Our method shows comparable performance with state-of-the-art algorithms in spite of simple process and relatively low computational load.

Table 1. Recognition accuracy and confusion matrix in KTH dataset

Method	Accuracy
Fathi et al. [16]	90.5
Wu et al. [5]	94.5
Laptev et al. [6]	91.8
Kovashka et al. [17]	94.5
Sadanand et al. [8]	98.2
Cai et al. [18]	94.2
Our method	96.3

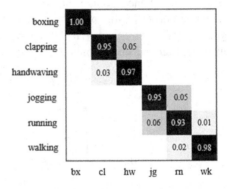

3.2   Smart Class Dataset

Smart class dataset has 6 actions (standing, sitting, lying face down, stretching waist, hand raising, hand waving) with 30 people and 12 sets each, totally 2160 activity videos and additional multi-camera group scenes with 4 cameras.

3.2.1   Single-View

We applied leave-one-and-out method for measuring the accuracy. Smart class dataset and confusion matrix in single-view is shown in Fig. 6. Average precision is 93.17 % that has main confusion on stretching waist and hand raising. It seems that the feature has similar motion pattern in stretching waist or raising hand due to normalization and discarding geometric information.

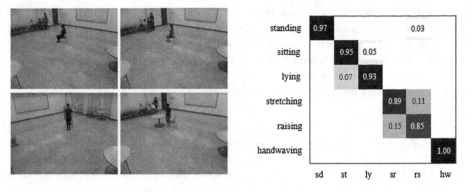

Fig. 6. Smart class dataset in single-view and confusion matrix

3.2.2 Multi-View

Multi-view video has 10 people in the same place and tracking box of each person is given. Figure 7 shows the multi-view group scenes in different views. Each group scene has 4 views and we decide the action by majority vote using SVM score. Specifically, training was performed in single-view videos and we used multi-view scene as testing. First we judged the action of each people in each view, then we had 4 decisions with SVM score for one action. After ignoring the decision which had lower SVM score than threshold, we conducted majority vote and got the final decision of action. Average precision of multi-view scene is 87.04 % which shows discriminative performance despite severe occlusions with objects or other person.

Fig. 7. Multi-view group scenes in different views

4 Conclusion

In this paper, we proposed a novel human action recognition algorithm by extracting a new feature based on optical flow. The proposed feature has property of direction invariance and shift invariance, i.e. it has the same output for the same action with different direction or time duration. It is also resilient to scale variation. In the process, several strategies for generating action specified representation by reducing background noise of optical flow were also presented. Median flow, normalization of histogram and low-pass filtering in frequency domain are comprised in the proposed method. Experimental results show the efficiency of the proposed algorithm and applicability to real-life environment.

Acknowledgement. This work was supported by the Brain Korea 21 Plus Project in 2015 and also by the IT R&D program of MOTIE/KEIT. [10041610, The development of automatic user information(identification, behavior, location) extraction and recognition technology based on perception sensor network(PSN) under real environment for intelligent robot].

References

1. Rahman, M.M., Ishikawa, S.: Robust appearance-base human action recognition. In: ICPR (2004)
2. Wu, X., Xu, D., Duan, L., Luo, J.: Action recognition using context and appearance distribution features. In: CVPR (2011)
3. Laptev, I., Lindeberg, T.: Space-time interest points. In: ICCV (2003)
4. Dollar, P., Rabaud, V., Cottrell, G., Belongie, S.: Behavior recognition via sparse spatio-temporal features. In: Workshop VS-PETS (2005)
5. Wu, X., Xu, D., Duan, L., Luo, J.: Action recognition using context and appearance distribution features. In: CVPR (2011)
6. Yamato, J., Ohya, J., Ishii, K.: Recognizing humanaction in time-sequential images using hidden markov model. In: CVPR (1992)
7. Luo, G., Hu, W.: Learning silhouette dynamics for human action recognition. In: ICIP (2013)
8. Sadanand, S., Corso, J.J.: Action bank: A high-level representation of activity in video. In: CVPR (2012)
9. Wang, H., Klaser, A., Schmid, C., Liu, C.: Action recognition by dense trajectories. In: CVPR (2011)
10. Niebles, J.C., Wang, H., Fei-fei, L.: Unsupervised learning of human action categories using spatialtemporal words. In: BMVC (2006)
11. Klser, A., Marszaek, M., Schmid, C.: A spatiotemporal descriptor based on 3d-gradients. In: BMVC (2008)
12. Bobick, A.F., Davis, J.W.: The recognition of human movement using temporal templates. In: PAMI (2001)
13. Ahad, M., Tan, J., Kim, H., Ishikawa, S.: Motion history image: its variants and applications. In: MVA (2012)
14. Sandhan, T., Choi, J.Y.: Frequencygrams and multi-feature joint sparse representation for action and gesture recognition. In: ICIP (2014)
15. Sandhan, T., Yoo, Y., Yoo, H., Yun, S., Byeon, M.: Multi-task learning with over-sampled time-series representation of a trajectory for traffic motion pattern recognition. In: AVSS (2014)
16. Fathi, A., Mori, G.: Action recognition by learning mid-level motion features. In: CVPR (2008)

17. Kovashka, A., Grauman, K.: Learning a hierarchy of discriminative space-time neighborhood features for human action recognition. In: CVPR (2010)
18. Cai, Q., Yin, Y., Man, H.: Learning spatio-temporal dependencies for action recognition. In: ICIP (2013)
19. Dalal, N., Triggs, B.: Histograms of oriented gradients for human detection. In: CVPR (2005)
20. Schuldt, C., Laptev, I., Caputo, B.: Recognizing human actions: a local svm approach. In: ICPR (2004)
21. Chang, C.C., Lin, C.J.: Libsvm: a library for support vector machine. In: ACM TIST (2011)
22. Liu, C.: Beyound pixels: exploring new representations and applications for motion analysis. In: Doctoral Thesis. Massachusetts Institute of Technology (2009)

Real-Time Human Action Recognition Using CNN Over Temporal Images for Static Video Surveillance Cameras

Cheng-Bin Jin, Shengzhe Li, Trung Dung Do, and Hakil Kim$^{(\boxtimes)}$

Information and Communication Engineering, Inha University, Incheon, Korea
{sbkim, szli, dotrungdung}@vision.inha.ac.kr,
hikim@inha.ac.kr

Abstract. This paper proposes a real-time human action recognition approach to static video surveillance systems. This approach predicts human actions using temporal images and convolutional neural networks (CNN). CNN is a type of deep learning model that can automatically learn features from training videos. Although the state-of-the-art methods have shown high accuracy, they consume a lot of computational resources. Another problem is that many methods assume that exact knowledge of human positions. Moreover, most of the current methods build complex handcrafted features for specific classifiers. Therefore, these kinds of methods are difficult to apply in real-world applications. In this paper, a novel CNN model based on temporal images and a hierarchical action structure is developed for real-time human action recognition. The hierarchical action structure includes three levels: action layer, motion layer, and posture layer. The top layer represents subtle actions; the bottom layer represents posture. Each layer contains one CNN, which means that this model has three CNNs working together; layers are combined to represent many different kinds of action with a large degree of freedom. The developed approach was implemented and achieved superior performance for the ICVL action dataset; the algorithm can run at around 20 frames per second.

Keywords: Video surveillance · Action recognition · Temporal images · Convolutional neural network · Hierarchical action structure

1 Introduction

The ability of a computer to recognize human actions can be important in many real-word applications including intelligent video surveillance, kinematic analysis, video retrieval, and criminal investigation. Based on the types of input video, action recognition can be divided into four classes: surveillance videos, sport videos, movies and user videos, and first-person videos. Different type of videos have different characteristics: Surveillance video [1, 2] usually uses a static camera that records from a side or top view. Therefore, the background of surveillance video is relatively simple, and research objects of surveillance are people or cars. Currently, millions of surveillance cameras are in place throughout the world. This means that more than 800 K video hours are generate per day. The objective of action recognition in the surveillance

© Springer International Publishing Switzerland 2015
Y.-S. Ho et al. (Eds.): PCM 2015, Part II, LNCS 9315, pp. 330–339, 2015.
DOI: 10.1007/978-3-319-24078-7_33

field is understanding of the video. It is necessary to have a program that can automatically label human events. The viewpoint of sport video [3] is the same as that of surveillance video. However, the objects in sport videos are usually fast-moving people. Sport videos need to be segmented manually before performing post-processing. Movies and user videos [4] are recorded with moving cameras; the view is almost always from the front or the side. The problems of jittery video and a dynamic and complicated background make this kind of video more difficult to process than the previous one. First-person videos [5, 6] are becoming popular after Google launched Google Glass. Because this technology employs a moving camera, videos obtained using Google Glass are very dynamic.

However, accurate action recognition is a very challenging task due to large variations in appearance. For example, occlusions, non-rigid motion, scale variation, view-point changes, subtle action, and clothing colors similar to the background color are all important problems. Manual collection of training samples is another difficult task. It requires much human effort and is time consuming. The other challenging task is to create an approach that can process video in real-time and that can be applied in real-world environment applications. The number of approaches to recognizing human action in video has grown at a tremendous rate. Prior approaches can be divided into appearance-based methods, motion-based methods, space-time based methods, and deep learning-based methods this last of which has become a hot topic recently.

Motion history images (MHI) or temporal images [7–9] make up the most popular appearance-based method. The advantages of the MHI method are that it is simple, fast, and it works very well in controlled environments. However, MHI is sensitive to errors of background subtraction. The fatal flaw of MHI is that it cannot capture interior motions and shapes. MHI can only capture silhouettes, but silhouettes tell little about actions. Other appearance-based methods are the active shape model, the learned dynamic prior model, and the motion prior model.

Motion-based methods [10] (generic and parametric optical flow, and temporal activity models) enable an analysis of the temporal information of sub-actions. These methods can also be used to model long-term activities with variable structures of action. However, important questions remain, such as how many sub-action units are meaningful and how is it possible to find these sub-action units for a target activity? Both of these are open problems that need to be solved.

Space-time is a feature-handcraft based method. This method was proposed to handle complex dynamic senses. There are many different descriptors (e.g. HOG, HOF Cuboids, HOG3D, and extended SURF) and detectors (Harris3D, Cuboids, Hessian, and regular dense sampling) [11, 12]. In order to improve performance, spatio-temporal grids [13] and analysis of co-occurrence between action and scene [14] are considered. The spatio-temporal grid method is one that divides interesting regions into many areas; then, it linearly combines extracted descriptors from different areas. Co-occurrence is a method that considers the relationship between action and scene; it gives certain weight to classified results to update the result.

The convolution neural networks (CNN) model [15] is one of a deep learning models. It is a class of supervised machine learning algorithm that can learn a hierarchy of features by building high-level features from low-level ones [16]. Some researchers have started to use CNN to recognize human actions [17, 18]. However, it will be necessary to determine what a good CNN architecture is. This is a question that is still difficult to answer, and a problem that will require further research.

The key contributions of this paper can be summarized as follows:

- This paper proposes a novel model for human detection, human tracking, and recognition of actions in real-time. The model does not make any assumptions (e.g., ground truth of human region, small scale, or viewpoint changes) about the circumstances.
- Hierarchical action structure is described for real-time human action recognition. In this structure, three CNNs work together; layers are combined to represent many different kinds of action with a large degree of freedom.
- Different temporal images are used in the 3 layers of the hierarchical action structure. Experimental results show that these kinds of temporal images are very suitable for use in video surveillance systems, with no need to be concerned about the accuracy or processing time.

The rest of this paper is organized as follows: a definition of the hierarchical action structure is provided in Sect. 2. Different temporal images in 3 layers, and the CNN architecture, are described in Sect. 3. The experimental results for the ICVL (Inha Computer Vision Lab) action dataset are reported in Sect. 4. Section 5 provides the conclusions of the paper.

2   Hierarchical Action Structure

The hierarchical action structure developed for real-time human action recognition is shown in Fig. 1. The structure includes three layers: the action layer, motion layer, and posture layer. The action layer has four classes. These are *nothing*, *texting*, *smoking*, and *others*. The motion layer has classes of *stationary*, *walking*, and *running*. The posture layer has classes of *sitting* and *standing*. There are certain types of common information between the posture layer and the motion layer. For the action layer, the posture layer provides supplementary information. The 3 layers together deliver a complete set of information for human actions.

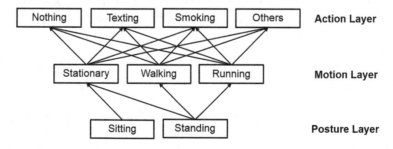

Fig. 1. Hierarchical action structure

The advantages of this structure are that it uses 9 action categories to represent various action combinations. To add a new action, the revise of overall structure is not necessary, rather, it is possible to revise only the corresponding layer and re-train the layer.

3   Human Action Recognition Using CNN

The objective of the paper is to propose a real-time algorithm for recognizing human action in surveillance video; at the same time, the method should not employ any assumptions about the video. In order to process video in real-time, human detection is a precondition of action recognition: this is also a big challenge that is still the subject of much research. First, the approach delineated in this paper performs motion detection using the Gaussian Mixture Model (GMM); after this, the system detects humans in the motion region using a Histogram of Gradient (HOG) [19]. To increase system speed, tracking by detection technique is employed in developed algorithm. Occlusion and difficult-to-detect humans those detected in previous frames but lost in the current frame are detected using the Kalman filter. The algorithm flow chart is shown in Fig. 2.

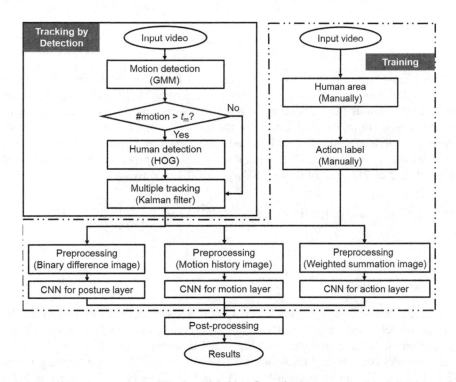

Fig. 2. Algorithm flow chart

3.1   Temporal Images

In the training stage, manually cropped human images and action labels are used as training data. Every layer of the structure has one independent CNN that requires different temporal images. Binary Difference Images (BDI), Motion History Images (MHI), and Weighted Summation Images (WSI) are used in the 3 layers. BDI is the specific form of the temporal images. It is given by Eq. (1)

$$b(x, y, t) = \begin{cases} 1, & if f(x, y, t) - f(x, y, t_0) > threshold \\ 0, & otherwise \end{cases} \tag{1}$$

As can be seen in its name, BDI, is binary image. Pixels in the image are set at 1 if the difference from another image is bigger than the *threshold*. x and y are indexes in the image; $f(x,y,t)$ is the current frame; $f(x,y,t_0)$ is the first frame of the input video. MHI is defined in Eqs. (2)–(4) [8]

$$d(x, y, t) = \begin{cases} 1, & if f(x, y, t) - f(x, y, t - 1) > threshold \\ 0, & otherwise \end{cases} \tag{2}$$

$$h_\tau(x, y, t) = \begin{cases} \tau_{max}, & if\ d(x, y, t) = 1 \\ max(0, h_\tau(x, t, t - 1) - \Delta\tau), & otherwise \end{cases} \tag{3}$$

$$\Delta\tau = \frac{\tau_{max} - \tau_{min}}{n} \tag{4}$$

From Eq. (3) the $h_\tau(x,y,t)$ (MHI) is generated from the difference between the current and previous frame $f(x, y, t-1)$. For every frame, MHI is calculated from the result of the previous MHI. Therefore, this value does not need to be calculated again for the whole set of frames. In Eq. (4), n is the number of the frames to be considered. WSI is the weighted summation of BDI and MHI, given by Eq. (5). There is also one constraint: $w_{BDI} + w_{MHI} = 1$. The temporal images in this paper are constructed using values of τ_{max} of 255, τ_{min} of 0, n of 10, w_{BDI} of 0.4, and w_{MHI} of 0.6.

$$s(x, y, t) = \mathbf{w}^T \cdot \begin{bmatrix} \tau_{max} \cdot b(x, y, t) \\ h_\tau(x, y, t) \end{bmatrix} \cdot \mathbf{w} = \begin{bmatrix} w_{BDI} \\ w_{MHI} \end{bmatrix} \tag{5}$$

3.2   CNN Architecture

According to the desired objectives, a variety of CNN architectures can be devised. To keep the model simple, a light CNN architecture is developed for human action recognition on the ICVL action dataset. This model is shown in Fig. 3.

This model consists of 2 convolutional layers (C1 and C3), 2 subsampling layers (S2 and S4), and 2 full connection layers (F5 and F6). The last full connection layer (F6) is fully connected to the action categories via softmax. The number of kernels in

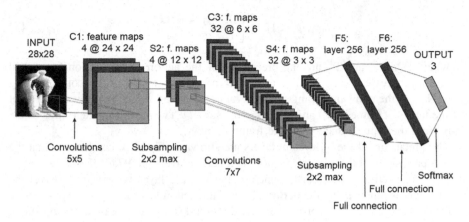

Fig. 3. Architecture of CNN

the two convolutional layers are 4 and 32; the stride of the convolution is 1; the kernel sizes are 5 × 5 and 7 × 7. The subsampling layer uses 2 × 2 max pooling; the stride is 2. Two full connection layers together use 512 neurons.

4 Experimental Results

For this section, experiments were performed to evaluate the proposed method on the ICVL action dataset. The dataset consists of surveillance video data recorded at Inha University. It consists of 158 videos using 11 different indoor and outdoor cameras with a resolution of 1280 × 640 at 20 fps. The durations of the videos are from 1 min to 6 min; each frame has 3 labels for the action, motion, and posture layers. Different training and test data are used in the proposed method; statistics for the data used in the experiments are provided in Table 1.

Table 1. Number of videos in the training and test from the ICVL action dataset

Camera/Data	C01	C02	C03	C04	C05	C06	C07	C08	C09	C10	C11	Tot.
Tra.	13	15	21	12	23	8	17	14	23	0	0	146
Tes.	1	1	2	1	1	1	1	1	1	1	1	12
Tot.	14	16	23	13	24	9	18	15	24	1	1	158

The performances of the 3 different layers are evaluated using frame-by-frame metrics. The metric is calculated according to:

$$P^d = 1 - \sum_{j=1}^{N_c} \sum_{i=1}^{N_{total}} \frac{N_j}{N_{total}} I\big(y_i, \varphi^d(x_i)\big), \ d \in \{A, M, P\} \tag{6}$$

$$I\left(y_i, \varphi^d(x_i)\right) = \begin{cases} 0, & y_i = \varphi^d(x_i) \\ 1, & otherwise \end{cases} \qquad (7)$$

where P^A, $P^{M,}$ and P^P are the precisions of the action layer, motion layer, and posture layer, respectively. N_c is the number of action classes in the corresponding layer; N_{total} is the number of frames in the evaluated videos. $I(y_i, \varphi^d(x_i))$ gives a value of 1 if the labels of frames i (y_i) and $\varphi^d(x_i)$ are the same. $\varphi^d(x_i)$ represents the results of the CNN from the input frame x_i; i is the frame number.

The precisions of the 3 different layers are shown in Fig. 4. It can be seen that the median precisions of the posture, motion, and action layers are 97.77 %, 85.99 %, and 71.29 %, respectively. The performance of the posture layer is very impressive; the motion layer is quite stable; the performance of the action layer is slightly weak. These results demonstrate that the appearance-based method is not good at representing subtle actions and that the movements of *texting* and *smoking* are small.

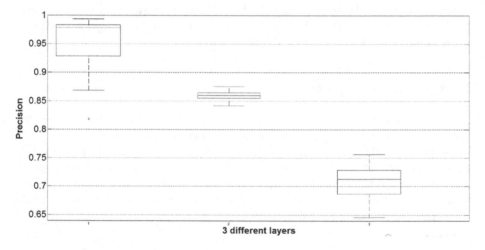

Fig. 4. Precisions of 3 different layers

The confusion matrix for the ICVL action dataset is shown in Fig. 5. The figure shows certain levels of confusion between *running* and *walking*, *texting* and *nothing*, and *smoking* and *nothing*. Some possible explanations for these results are that they are caused by the light architecture of the CNN and that there exists an imbalance in the number of training samples. In the ICVL action dataset, there are many sets of sample data for *standing*, *walking*, *nothing*, and *texting*; however, there are only few samples for other types of action. For example, the dataset has 18,504 training samples for *nothing*, but just 1,106 for *smoking* a sixteen-fold difference. The proportion even bigger than 16 times. However, the multiple layers of the hierarchical action structure can eliminate many misclassifications.

Figure 6 shows certain actions that were correctly recognized and certain actions that were misclassified. Detected bounding boxes, trajectories of objects, object IDs,

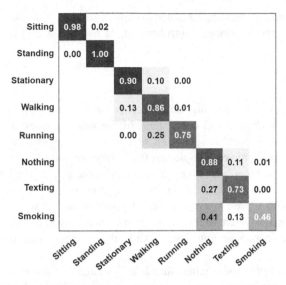

Fig. 5. Confusion matrix for classification results on ICVL action dataset

and 3-layer action results are shown in Fig. 6. The object ID and the trajectory were updated every time after tracking when the detection method failed to detect the object. The top row shows actions that were correctly recognized by the proposed model; the bottom row shows those that were misclassified by the model.

In order to provide an evaluation of the processing time, the experimental environment was established on a computer with an Intel(R) Core™ i7-3770 CPU @ 3.40 GHz and two 4 GB RAMs. The input video was resized to 640 × 360 from the original 1280 × 640. The processing time was tested on 12 videos and calculated as the average processing time. The average processing time for one frame is 46.9319 ms. GMM takes 11.3046 ms, HOG takes 12.6045 ms, the 3 temporal images take 1.2569 ms, the 3 CNNs take 3.6577 ms per human, and the other processes (e.g. initialization,

Fig. 6. Correctly recognized and misclassified results

Kalman filtering, post-processing, and displaying of results, etc.) take 18.2083 ms. As can be seen above, the developed algorithm runs at more than 20 frames per second.

5 Conclusions

This paper has proposed a real-time human action recognition approach that does not use any assumptions about the circumstances of the video in question. The developed approach constructs temporal images from several static images using BDI, MHI, and WSI. Temporal images are very simple and fast. They are quite suitable for use in fixed surveillance video cameras, with no need to consider the processing time or precision. Using the proposed hierarchical action structure, the model can employ as limited number of actions to represent many different kinds of action with a large degree of freedom. Further, employment of this structure makes it easy to add new actions simply by re-training the corresponding layer. And, this method can effectively reduce misclassifications.

In this paper, a light CNN architecture is considered for action recognition. There are other deep architectures, such as Recurrent Neural Networks and Deep Belief Networks, which have achieved promising performance for speech recognition and image recognition. It would be interesting to employ such models for action recognition or to make the current CNN deeper and more complicated.

Acknowledgements. This research was funded by the MSIP (Ministry of Science, ICT & Future Planning), Korea in the ICT R&D Program 2015 (Project ID: 1391203002-130010200).

References

1. Oh, S., Hoogs, A., Perera, A., et al.: A large-scale benchmark dataset for event recognition in surveillance video. In: 2011 IEEE Conference on Computer Vision and Pattern Recognition (CVPR), Providence, Rhode Island ,pp. 3153–3160 (2011)
2. Vahdat, A., Gao, B., Ranjbar, M., et al.: A discriminative key pose sequence model for recognizing human interactions. In: 2011 IEEE International Conference on Computer Vision Workshops (ICCV Workshops), pp. 1729–1736, Barcelona (2011)
3. Lan, T., Wang, Y., Yang, W., et al.: Discriminative latent models for recognizing contextual group activities. IEEE Trans. Pattern Anal. Mach. Intell. **34**(8), 1549–1562 (2011)
4. Kim, I., Oh, S., Vahdat, A., et al.: Segmental multi-way local polling for video recognition. In: Proceedings of the 21st ACM International Conference on Multimedia, MM 2013, pp. 637–640, New York (2013)
5. Pirsiavash, H., Ramanan, D.: Detecting activities of daily living in first-person camera views. In: 2012 IEEE Conference on Computer Vision and Pattern Recognition (CVPR), Providence, Rhode Island, pp. 2847–2854 (2011)
6. Ryoo, M.S., Matthies, L.: First-person activity recognition: what are they doing to me? In: 2013 IEEE Conference on Computer Vision and Pattern Recognition (CVPR), Portland, Oregon, pp. 2730–2737 (2013)

7. Davis, J.W., Bobick, A.F.: The representation and recognition of human movement using temporal templates. In: 1997 IEEE Computer Society Conference on Computer Vision and Pattern Recognition, San Juan, pp. 928–934 (1997)
8. Bobick, A.F., Davis, J.W.: The recognition of human movement using temporal templates. IEEE Trans. Pattern Anal. Mach. Intell. 3(3), 257–267 (2001)
9. Blank, M., Gorelick, L., Schechtman, E., et al.: Actions as space-time shapes. In: 2005 Tenth IEEE International Conference on Computer Vision (ICCV), Beijing, pp. 1395–1402 (2005)
10. Tang, K., Fei-Fei, L., Koller,.D.: Learning latent temporal structure for complex event detection. In: 2012 IEEE Conference on Computer Vision and Pattern Recognition (CVPR), Providence, Rhode Island, pp. 1250–1257 (2012)
11. Wang, H., Schmid, C.: Action recognition with improved trajectories. In: 2013 IEEE International Conference on Computer Vision (ICCV), Sydney, New South Wales, pp. 3551–3558 (2013)
12. Jiang, Z., Lin, Z., Davis, L.S.: A unified tree-based framework for joint action localization, recognition and segmentation. Comput. Vis. Image Underst. 117(10), 1345–1355 (2013)
13. Felzenszwalb, P., McAllester, D., Ramanan, D.: A discriminatively trained, multiscale, deformable part model. In: 2008 IEEE Conference on Computer Vision and Pattern Recognition (CVPR), Anchorage, Alaska, pp. 1–8 (2008)
14. Marszalek, M., Laptev, I., Schmid, C.: Actions in context. In: 2009 IEEE Conference on Computer vision and Pattern Recognition (CVPR), Miami, Florida, pp. 2929–2936 (2009)
15. Krizhevsky, A., Sutskever, I., Hinton, G.E.: ImageNet classification with deep convolutional neural networks. In: Advances in Neural Information Processing Systems, vol. 25 (2012)
16. Ji, S., Xu, W., Yang, M., et al.: 3D convolutional neural networks for human action recognition. IEEE Trans. Pattern Anal. Mach. Intell. 35(1), 221–231 (2013)
17. Toshev, A., Szegedy, C.: DeepPose: human pose estimation via deep neural networks. In: 2014 IEEE Conference on Computer Vision and Pattern Recognition (CVPR), Columbus, Ohio, pp. 1653–1660 (2014)
18. Sun, L., Jia, K., Chan, T., et al.: DL-SFA: deeply-learned slow feature analysis for action recognition. In: 2014 IEEE Conference on Computer Vision and Pattern Recognition (CVPR), Columbus, Ohio, pp. 2625–2632 (2014)
19. Dalal, N., Triggs, B.: Histograms of oriented gradients for human detection. In: 2005 IEEE Conference on Computer Vision and Pattern Recognition (CVPR), San Diego, California, pp. 886–893 (2005)

Scalable Tamper Detection and Localization Scheme for JPEG2000 Codestreams

Takeshi Ogasawara[✉], Shoko Imaizumi, and Naokazu Aoki

Graduate School of Advanced Integration Science,
Chiba University, 1-33 Yayoicho, Inage-ku, Chiba-shi, Chiba 263-8522, Japan
{afka3712,imaizumi}@chiba-u.jp

Abstract. We propose an efficient tamper detection scheme for JPEG2000 codestreams. The proposed scheme embeds information while maintaining the scalability function of JPEG2000 and can detect tampered layers or tampered resolution levels for each decoded image quality level. Marker codes that delimit a JPEG2000 codestream must not be generated in the body data that possesses the image information. We can prevent new marker codes from being generated when embedding information into the JPEG2000 codestream. The experimental results show that our scheme can preserve the high quality of the embedded images.

Keywords: Tamper detection · Reversible data hiding · Marker code · Hash function

1 Introduction

With the rapid growth of communication technology, we can exchange information, including digital images/videos, audio data, and documents, through a network channel. On the other hand, security is becoming a more and more serious issue these days. Many researches have focused on multimedia security, such as copyright and privacy protection [1,2]. Reversible information embedding, which secretly embeds some information into a medium, is frequently exploited for tamper-detection methods to confirm the authenticity of the information [3]. If there has been no tampering with the medium, the embedded information can be correctly extracted, and the original medium can be restored without any distortion. There are different kinds of reversible data embedding techniques, for instance, embedding into the spatial domain [4], embedding using histogram shifting [5], and embedding into binary data [6].

JPEG2000 [7,8], which is an international standard for image/video coding, achieves high coding performance compared to that of JPEG and has efficient functions, e.g., hierarchical decoding and region of interest encoding (ROI). There are many kinds of data embedding schemes for JPEG2000 coded images, such as a scheme exploiting layer structure [9] and another scheme embedding into discrete wavelet transform (DWT) coefficients [10,11]. Meanwhile, a tamper detection scheme for JPEG2000 coded images has been proposed [12]. On tamper

© Springer International Publishing Switzerland 2015
Y.-S. Ho et al. (Eds.): PCM 2015, Part II, LNCS 9315, pp. 340–349, 2015.
DOI: 10.1007/978-3-319-24078-7_34

detection, the embedded information is first extracted, and the image without the embedded information is restored. The extracted information is compared to the restored image information. If they are the same, it is determined that there has been no tampering. Otherwise, it is determined that the image has somehow been tampered. This scheme, however, does not preserve the scalability function of JPEG2000. When tampering can be detected, the previous scheme can identify the tampered position in the spatial domain but cannot localize which portion of the codestream has been tampered.

In this paper, we propose an efficient tamper detection scheme for JPEG2000 codestreams. The proposed scheme preserves the scalability function of JPEG2000 and can localize the tampered portion in a codestream. Furthermore, our scheme can prevent new marker codes from being generated in a JPEG2000 codestream by processing an image as well as some security techniques for JPEG2000 [13–15]. Marker codes that delimit a codestream should not exist within the body data of JPEG2000 packets, which contains image information. If new marker codes are generated in a codestream or existing marker codes are eliminated from a codestream, the JPEG2000 coded image cannot be decoded normally. Hence, we have to consider not generating new marker codes when embedding information into the JPEG2000 codestream. By embedding information reversibly while maintaining the scalability function of JPEG2000 without generating new marker codes, the proposed scheme can detect tampered layers or tampered resolution levels for each decoded image quality level. Moreover, we discuss reducing image degradation by controlling the amount of embedded information.

2 Related Work

2.1 JPEG2000 Codestreams [7,8]

JPEG2000 has a lot of efficient functions and higher coding performance compared to JPEG. JPEG2000 can decode in a hierarchical way and can suppress image distortion of coded images at low bit rate, which is one of the serious issues of JPEG. A simple example of a JPEG2000 codestream structure, where the progression order is the layer-resolution-component-position order (LRCP), is shown in Fig. 1. The structure of the JPEG2000 codestream will be discussed in more detail below.

(1) Quality Layer

Quality layers (hereinafter referred to as "layer") have been hierarchized in accordance with their contribution to the coded-image quality. The left layers in Fig. 1 affect the image quality more than the right layers. Therefore, it is possible to progressively improve the image quality by decoding the codestream from those important layers.

(2) Resolution Level

JPEG2000 performs two-dimensional DWT on the input image and configures resolution levels that achieve spatial scalability. Resolution level 0 contains the LL information, and other resolution levels contain HL, LH, and HH information.

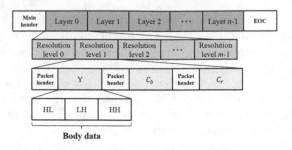

Fig. 1. Example of JPEG2000 codestream structure (LRCP)

(3) Color Components

Color components have color information, such as RGB or YC_bC_r color space. In JPEG2000, RGB signals are generally transformed into YC_bC_r signals prior to the encoding procedure to improve the coding performance.

(4) JP2 Packet

A JPEG2000 packet (hereinafter referred to as "JP2 packet") is the smallest unit in a JPEG2000 codestream. A JP2 packet is composed of the body data, which has the coded image information, and a packet header. In the case of a color image, JP2 packets are generated with respect to each color component. Each JP2 packet contains the information of its layer, resolution level, color component, and precincts, which is needed to decode the codestream. A codestream is decoded in order from the first JP2 packet to the last one; thus, it is possible to arbitrarily define the priorities of the SNR, spatial resolution, color components, and spatial domain by designating the order of the JP2 packets on coding the image. The order that the JP2 packets are sorted into in a codestream is referred to as the progression order.

2.2   JPEG2000 Marker Code

The marker codes of JPEG2000 are represented with two bytes and partition a JPEG2000 codestream. In particular, those in the range $FF90_h$ to $FFFF_h$ have a special meaning to delimit JP2 packets, tiles, and so forth. For example, $FF90_h$ means SOT (start of tile) and $FF91_h$ means SOP (start of packet header). They should never exist within the body data of JP2 packets. When a false marker code is generated in a codestream by embedding information or performing encryption, the codestream cannot be decoded properly. We thus have to consider not generating those marker codes in the range $FF90_h$ to $FFFF_h$ when embedding information.

3   Proposed Scheme

We propose an efficient tamper detection scheme that maintains the scalability function of JPEG2000, and we describe it in the order of the conditions for avoiding marker codes, embedding process, and tamper detection. Note that although we only refer to layers and resolution levels in this paper, the proposed scheme can easily be extended to deal with precincts and color components.

3.1 Conditions for Avoiding Marker Codes

As described in Sect. 2.2, marker codes should not exist in the body data of each JP2 packet. We ensure that the embedding procedure does not generate marker codes in the range $FF90_h$ to $FFFF_h$ in the body data when embedding information into a JPEG2000 codestream. Therefore, the proposed scheme should embed the information in accordance with the following conditions.

Case 1 The byte followed by the target byte is FF_h.

In this case, the target byte should be below 90_h and must not be transformed to 90_h or above. Hence, the embedding process can be performed on the lower half byte of the target byte.

Case 2 The byte followed by the target byte is below FF_h, and another byte following the target byte is also below 90_h.

In this case, the target byte can be transformed to any value. Therefore, the embedding process can be performed on the target byte without any restriction.

Case 3 The byte followed by the target byte is below FF_h, and another byte following the target byte is 90_h or above.

In this case, the target byte should not be FF_h and must not be transformed to FF_h. We thus cannot perform the embedding process on the target byte.

3.2 Information Embedding Procedure

We explain the embedding procedure in the proposed scheme. In this section, we assume the target is the JPEG2000 codestream as shown in Fig. 1. We define

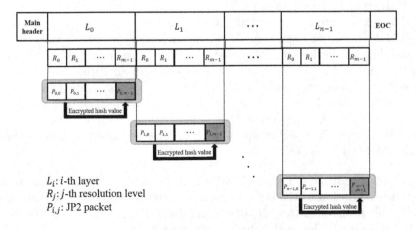

L_i: i-th layer
R_j: j-th resolution level
$P_{i,j}$: JP2 packet

Fig. 2. Layer-based embedding

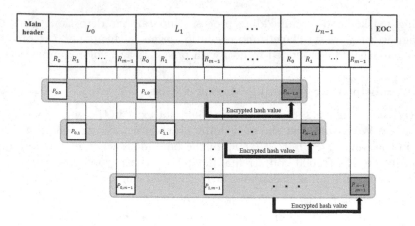

Fig. 3. Resolution level-based embedding

the number of layers and the number of resolution levels of the codestream as n and m, respectively. Each JP2 packet with the i-th $\{i = 0, 1, ..., n-1\}$ layer and the j-th $\{j = 0, 1, ..., m-1\}$ resolution level is represented as $P_{i,j}$.

Step 1. Set $i = 0$.

Step 2. Extract and combine the JP2 packets of the i-th layer, and calculate the hash value using a one-way hash function.

Step 3. Encrypt the hash value with an encryption key.

Step 4. Increment i by 1 ($i = i + 1$).

Step 5. Repeat Steps 2 to 4 until $i = n - 1$.

Step 6. Set $j = 0$.

Step 7. Extract and combine the JP2 packets of the j-th resolution level from each layer, and calculate the hash value using a one-way hash function.

Step 8. Encrypt the hash value with an encryption key.

Step 9. Increment j by 1 ($j = j + 1$).

Step 10. Repeat Steps 7 to 9 until $j = m - 1$.

Step 11. As shown in Figs. 2 and 3, embed each encrypted hash value into the body data of the lowest JP2 packet $P_{i,m-1}$ in layer i or that of the lowest JP2 packet $P_{n-1,j}$ in resolution level j using the algorithm based on pairwise logical computation (PWLC) data hiding [6] that embeds information into binary data. Here, the target body data is transformed into binary data before the embedding process. If it is impossible to embed the hash value into the lowest JP2 packet due to lack of length, embed it into the lowest embeddable JP2 packet.

Moreover, the proposed scheme can embed the shortened hash values to suppress the degradation of the embedded-image quality. For example, we assume that an a-bit hash value ($h = \{h_x | x = 1, 2, ..., a\}$) is shortened to 4 bits. In this case, the shortened 4-bit hash value is generated by

$$h_1' = h_1 \oplus h_5 \oplus \cdots \oplus h_{a-3} \tag{1}$$

$$h_2' = h_2 \oplus h_6 \oplus \cdots \oplus h_{a-2} \tag{2}$$

$$h_3' = h_3 \oplus h_7 \oplus \cdots \oplus h_{a-1} \tag{3}$$

$$h_4' = h_4 \oplus h_8 \oplus \cdots \oplus h_a. \tag{4}$$

3.3   Tamper Detection Procedure

We explain the tamper detection procedure in accordance with the three kinds of decoding cases.

Decoding Selected Layers. We first explain the tamper detection procedure in the case of decoding the selected layer(s). We define the number of the layers to be verified as t. $P_{l,r}$ indicates a JP2 packet, where the layer is $l\{l = 0, 1, ..., t - 1\}(t < n)$ and the resolution level is $r\{r = 0, 1, ..., m - 1\}$. Figure 4 shows an example of tamper detection for the codestream where layer 0 and 1 are decoded $(t = 2)$.

Step 1. Set $l = 0$

Step 2. Extract the encrypted hash value from the embedded JP2 packet of layer l using PWLC data hiding, and restore the JP2 packet without the encrypted hash value.

Step 3. Decrypt the encrypted hash value using the corresponding encryption key (hereinafter referred to as "extracted hash value").

Step 4. Extract and combine the JP2 packets of layer l, and calculate its hash value using a one-way hash function (hereinafter referred to as "restored hash value").

Step 5. Compare the restored hash value in Step 4 to the extracted hash value in Step 3. If they are identical, the layer is identified as authentic without any tampering. Otherwise, the layer is identified as unauthentic and has been tampered.

Step 6. Increment l by 1 $(l = l + 1)$.

Step 7. Repeat Steps 2 to 6 until $l = t - 1$.

In the case where we embed the shortened hash values into the JP2 packets, the restored hash values are first shortened to the same length as the extracted ones and are compared to them.

Decoding Selected Resolution Levels. Next, we show the tamper detection procedure in the case of decoding the selected resolution level(s). We define the numbers of the resolution levels to be verified as s. $P_{l,r}$ indicates a JP2 packet, where the layer is $l\{l = 0, 1, ..., n - 1\}$ and the resolution level is $r\{r = 0, 1, ..., s - 1\}(s < m)$. Figure 5 shows an example of tamper detection for the codestream where resolution level 0 and 1 are decoded $(s = 2)$.

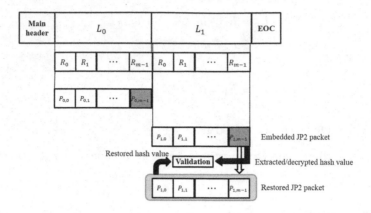

Fig. 4. Tamper detection process (decoded layers are 0 and 1)

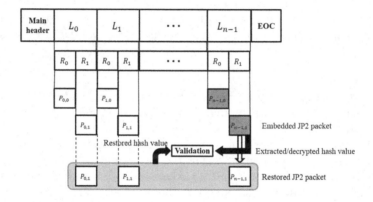

Fig. 5. Tamper detection process (decoded resolution levels are 0 and 1)

Step 1. Set $r = 0$

Step 2. Extract the encrypted hash value from the embedded JP2 packet of resolution level r using PWLC data hiding, and restore the JP2 packet without the encrypted hash value.

Step 3. Decrypt the encrypted hash value using the corresponding encryption key.

Step 4. Extract and combine the JP2 packets of resolution level r, and calculate its hash value using a one-way hash function.

Step 5. Compare the restored hash value in Step 4 to the extracted hash value in Step 3. If they are identical, the resolution level is identified as authentic without any tampering. Otherwise, the resolution level is identified as unauthentic and has been tampered.

Step 6. Increment r by 1 ($r = r + 1$).

Step 7. Repeat Steps 2 to 6 until $r = s - 1$.

In the case where we embed the shortened hash values into the JP2 packets, the restored hash values are first shortened to the same length as the extracted ones and are compared to them.

Decoding Entire Codestream. Finally, we describe the tamper detection procedure in the case of decoding the entire codestream.

Step 1. Extract all of the encrypted hash values from the embedded JP2 packets of each layer and each resolution level using PWLC data hiding, and restore the JP2 packet without the encrypted hash value.

Step 2. Decrypt each encrypted hash value using the corresponding encryption key.

Step 3. Extract and combine the JP2 packets with respect to each layer and each resolution level, and calculate their hash values using a one-way hash function.

Step 4. Compare the restored hash value in Step 3 to the extracted hash value in Step 2 with respect to each layer and each resolution level. If they are identical, the layer or the resolution level is identified as authentic without any tampering. Otherwise, the layer or the resolution level is identified as unauthentic and has been tampered.

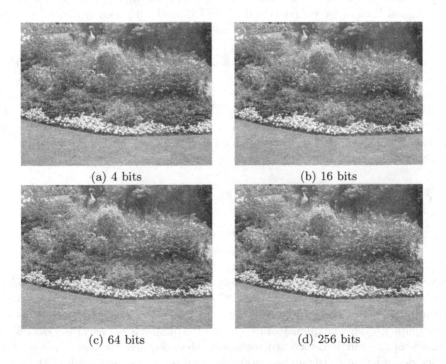

(a) 4 bits

(b) 16 bits

(c) 64 bits

(d) 256 bits

Fig. 6. Embedded images (Flower garden)

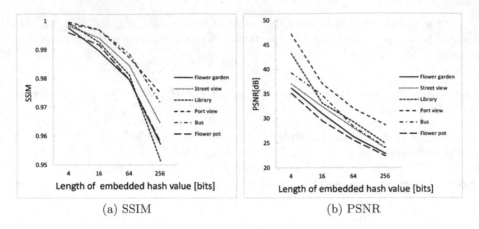

(a) SSIM   (b) PSNR

Fig. 7. Results of image quality assessment

4 Experimental Results

We performed some experiments on the six IHC standard images [16], which
have been transformed into grayscale and resized to 864 × 1152, to evaluate the
image quality of the embedded images generated by using the proposed scheme.
The original images were first coded by JPEG2000 with four layers and three
resolution revels. We used SHA-256, which generates 256-bit hash values in our
experiments. Note that this scheme can also be easily applied to color images.

Figure 6 depicts the embedded images of the flower garden, where we embedded 4, 16, 64, or 256 bits into each target JP2 packet. We evaluated the quality
of those embedded images using SSIM [17] and PSNR as shown in Fig. 7. Here,
the SSIM values and the PSNR values in Fig. 7 are the average values after performing the experiments ten times. It has been confirmed that our scheme can
suppress the degradation of the embedded image quality and control the image
quality by shortening the hash values.

5 Conclusion

We proposed an efficient tamper detection scheme for JPEG2000 codestreams.
Although the conventional scheme does not preserve the scalability function of
JPEG2000 and cannot localize the tampered portion(s) in the codestream, the
proposed scheme embeds information into JPEG2000 codestreams using PWLC
data hiding while preserving the JPEG2000 scalability function and can detect
the tampered areas, such as the layers or/and resolution levels, for each decoded
image quality level. The quality of the embedded images is not seriously degraded
in this scheme. Furthermore, we confirmed that our scheme can suppress the
quality degradation of the embedded images by reducing the length of the hash
values embedded into the JP2 packets.

Acknowledgments. This work was supported by JSPS KAKENHI Grant Number 26820138.

References

1. Wu, Y., Ma, D., Deng, R.H.: Progressive protection of JPEG2000 codestreams. In: Proceedings of IEEE ICIP, pp. V-3447–V-3450 (2004)
2. Schneider, M., Chang, S.-F.: Robust content based digital signature for image authentication. In: Proceedings of IEEE ICIP, pp. III-227–III-230 (1996)
3. Han, S., Fujiyoshi, M., Kiya, H.: A Reversible image authentication method without memorization of hiding parameters. IEICE Trans. Fundam. **E92–A**(10), 2572–2579 (2009)
4. Jin, H.L., Fujiyoshi, M., Kiya, H.: Lossless data hiding in the spatial domain for high quality images. IEICE Trans. Fundam. **E90–A**(4), 771–777 (2007)
5. Ni, Z., Shi, Y.-Q., Ansari, N., Su, W.: Reversible data hiding. IEEE Trans. Circuits Syst. Video Technol. **16**(3), 354–362 (2006)
6. Tsai, C.-L., Chiang, H.-F., Fan, K.-C., Chung, C.-D.: Reversible data hiding and lossless reconstruction of binary images using pair-wise logical computation mechanism. Patten Recognit. **38**(11), 1993–2006 (2005)
7. Information Technology – JPEG 2000 Image Coding System - Part1: Code Coding System ISO/IEC IS-15444-1 (2004)
8. Taubman, D.S., Marcellin, M.W.: JPEG2000: Image Compression Fundamentals. Standards and Practice. Kluwer Academic Publishers, Norwell (2002)
9. Ando, K., Kobayashi, H., Kiya, H.: A method for embedding binary data into JPEG2000 bit streams based on the layer structure. In: Proceedings of EURASIP EUSIPCO, pp. III-89–III-92 (2002)
10. Suhail, M.A., Obaidat, M.S.: On the digital watermarking in JPEG2000. In: Proceedings of IEEE ICECS, pp. II-871–II-874 (2001)
11. Chen, T.-S., Chen, J., Chen, J.-G.: A simple and efficient watermark technique based on JPEG 2000 Codec. Multimedi. Syst. **10**(1), 16–26 (2004)
12. Sun, Q., Chang, S.F.: A secure and robust digital signature scheme for JPEG2000 image authentication. IEEE Trans. **7**(3), 480–494 (2005)
13. Ikeda, H., Iwamura, K.: Selective encryption scheme and mode to avoid generating marker codes in JPEG2000 code streams with block cipher. In: Proceedings of IEEE WAINA, pp. 593–600 (2011)
14. Kiya, H., Imaizumi, S.: Partial-scrambling of image encoded using JPEG2000 without generating marker codes. In: Proceedings of IEEE ICIP, pp. III-205–III-208 (2003)
15. Wu, H., Ma, D.: Efficient and sucure encryption schemes for JPEG2000. In: Proceedings of IEEE ICASSP, pp. V-869–V-872 (2004)
16. IHC standard images. http://www.ieice.org/iss/emm/ihc/en/image/image.php
17. Wang, Z., Bovik, A.C., Sheikh, H.R., Simoncelli, E.P.: Image quality assessment: from error visibility to structural similarity. IEEE Trans. Image Process. **13**(4), 600–612 (2004)

Developing a Visual Stopping Criterion for Image Mosaicing Using Invariant Color Histograms

Armagan Elibol and Hyunjung Shim$^{(\boxtimes)}$

School of Integrated Technology, Yonsei University, Incheon, Republic of Korea
{aelibol,kateshim}@yonsei.ac.kr

Abstract. For over a decade, image mosaicing techniques have been widely used in various applications *e.g.*, generating a wide field-of-view image, 2D optical maps in remote sensing or medical imaging. In general, image mosaicing combines a sequence of images into a single image referred to as a mosaic image. Its process is roughly divided into the iterative image registration and blending. Unfortunately, the computational cost of iterative image registration increases exponentially given a large number of images. As a result, mosaicing for a large scale scene is often prohibitive for real-time applications. In this paper, we introduce an effective visual criterion to reduce the number of image mosaicing iterations while retaining the visual quality of the mosaic. We analyze the change in invariant color histograms of the mosaic image over iterations and use it to determine a termination condition. Based on various experimental evaluations using four different datasets, we significantly improve the computational efficiency of mosaicing algorithm.

Keywords: Image mosaicing · Visual quality · Optical mapping · Invariant color histogram

1 Introduction

Image mosaicing is a class of techniques that register overlapping images and combine them into a larger image [12]. Since mosaicing is effective to create a wide-field-of-view image (i.e., a mosaic image) from a set of images and/or video, the resultant mosaics have been very useful for different scientific studies such as geology [5,9], biology [10] or archaeology [1,11]. Especially with the rapid development of mobile platforms, it becomes possible to obtain optical data of areas beyond the human reach. Mosaics of these areas can help revealling locations of areas of interest or visualize temporal changes in the morphology of bio-diversity of the terrain. For that, mosaics are analyzed by a human expert and provide the global perspective on the area of interest.

In general, image mosaicing is composed of two main phases: iterative image registration for aligning image pairs and image blending for obtaining the final

© Springer International Publishing Switzerland 2015
Y.-S. Ho et al. (Eds.): PCM 2015, Part II, LNCS 9315, pp. 350–359, 2015.
DOI: 10.1007/978-3-319-24078-7_35

mosaic. An image registration process is composed of a pairwise and global registration. While pairwise registration is to identify the transformation between two overlapping images in the sequence, global registration extracts the best possible transformation parameters of each image with respect to a common mosaic coordinate frame. Image blending imposes the smooth transition along the seam in a final mosaic image after global registration and this improves the final quality of the mosaic. The blending is necessary because photometric differences are the main source of seams and they can occur even under the perfect geometric alignment.

Image mosaicing is accomplished via iterating pairwise image registration and global registration (updating the estimate of camera trajectory) using possible overlapping image pairs. Considering time-consecutive images, they generally present significant overlaps. While registering them, their registration parameters can serve as an initial estimate of camera trajectory. However, this initial estimate suffers from error accumulation. This is because the absolute homography, a planar transformation between an input frame and global frame, is derived from multiple relative homographies, a planar transformation between two input frames. When computing each relative homography, we purely rely on correspondences, which vary upon the performance of feature descriptors and matching algorithm. Consequently, each relative homography potentially hides the error caused by incorrect correspondences. Since the absolute homography aggregates multiple relative homographies, the errors from each homography are accumulated in the absolute homography.

Non-consecutive overlapping image pairs can be predicted by this coarse estimate. Registering non-consecutive overlapping image pairs helps improve the trajectory and mosaic. Once overlapping image pairs are identified, global registration methods can be employed in order to find the best transformation parameters between image coordinate frame and a global frame. Note that we can choose an arbitrary image frame to fix the global coordinate system. In our implementation, we choose the first frame as the global frame. Global registration is done by minimizing an error defined by the distance of correspondences between image pairs. This step requires the non-linear optimization, which comes with high computational cost. This cost increases drastically if we are given a large number of input images to create a huge mosaic.

In this paper, we aim to obtain a mosaic image using a reduced number of overlapping image pairs with retaining the visual quality as well as possible to the one using all image pairs. In this way, we can reduce the computational cost introduced by global registration as well as the cost of identifying and registering overlapping image pairs. In [4], the importance of overlapping image pairs have been evaluated by using a weighted shortest path algorithm. Although the importance of the overlapping image pairs were evaluated through their shortest alternative paths and final mosaics were nearly identical to their counterpart ones, the visual quality of image registration and intermediate mosaics were not analyzed. In this paper, we propose to use a deformation and viewpoint invariant color histogram [2] (referred to as an invariant histogram for the rest of paper.)

to measure the changes in visual quality of mosaic after each iteration of the image mosaicing process. The important property of the invariant histogram is that it is invariant under any mapping of the surface that is locally affine. This property is particularly beneficial to measure the image similarity under a wide class of viewpoint changes or deformations. Since images are warped with different transformation parameters to compose the mosaic, the change in the invariant histogram is caused by the misregistration between images in our application. Therefore, we find that the change in invariant histogram is an adequate measure to evaluate our mosaicing process. The proposed method can be integrated into various existing frameworks in image mosaicing to improve their computational efficiency.

2   Invariant Histogram Based Mosaic Image Quality Monitoring

Standard color histograms are sensitive to changes in the viewpoint. Domke and Aloimonos [2] proposed a new color histogram that is invariant to an arbitrary transformation of locally affine surface. They weight pixels using gradients of different color channel. In our context, individual image is warped in global frame to form a mosaic assuming the target surface being locally affine. If the alignment between images remains same, applying an arbitrary transformation does not change the invariant histogram [2]. Our proposal is to generate the intermediate mosaics and compare its invariant histogram with that of previous iteration. If the ratio of change is lower than a threshold, we terminate mosaicing iterations. Our method can be interpreted as adding constraint to image mosaicing framework by monitoring invariant histograms of the mosaics produced at each iteration. A standard image mosaicing pipeline combined with our method is illustrated in Fig. 1. To compare histograms of two images a and b, we employ the same metric in [2]. For an image a and b, computation of differences between their histograms is given in Eq. 1.

$$d(\mathbf{h}^a, \mathbf{h}^b) = \frac{\sum_c (\mathbf{h}_c^a - \mathbf{h}_c^b)^2}{\sum_c (\mathbf{h}_c^b)^2} \tag{1}$$

where $\mathbf{h}_c$ denotes the histogram value for color channel c and computed as follows:

$$\mathbf{h}_c = \sum_{s, s_c = c} |f_x(s) g_y(s) - f_y(s) g_x(s)| \tag{2}$$

where f and g denote derivatives in two color channels [2].

3   Experimental Results

We have conducted various experiments on four different datasets. The first experiment is to measure how invariant histogram varies upon misregistration

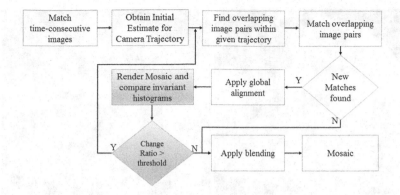

Fig. 1. A mosaicing pipeline of proposed method.

in the mosaic and to monitor the value of the metric given in Eq. 1. For that, we use 33 images of 384×288 pixels cropped from high resolution mosaic. We register images to the mosaic directly in order to obtain their image-to-mosaic planar transformations. Given these transformation parameters, the mosaic is generated by a bottom-up strategy. This mosaic serves a ground-truth as illustrated in Fig. 2. For image registration, we extract the Scale Invariant Feature Transform (SIFT) [8] features and apply Random Sample Consensus (RANSAC) eliminate outliers and estimate the planar transformation. To analyze the robustness of proposed method, we generate the misalignment in image pairs and report the effects of misalignment in the quality of mosaic. To simulate misalignments, we add a Gaussian random noise with zero mean and several levels of standard deviation to the translation parameters both x and y direction. Then, we obtain misaligned mosaics due to the erroneous parameters. The invariant histograms of misaligned mosaics were compared with the one of ground truth mosaic by using Eq. 1. For each variance level of noise, we randomly draw 1000 samples of noise. From this experiment, we observed how the value has changed and how the registration errors have evolved over the significance of noise. Furthermore, to quantify the errors in camera trajectory, we register images pairwise. A totally 528 image pairs were registered and the total number of correspondences over these pairs becomes $142,317$. For each noisy transformation set, a symmetric transfer error [7] is computed.

We summarize our results in Table 1. Numbers given in the table are statistically computed over 1000 trials for each noise level. For higher level of noise, mosaics that have the maximum symmetric transfer error within trials are illustrated in Fig. 2. We find that starting from the noise level of 10 pixels, a visual disturbance on mosaic can be easily recognizable. This provides some insights for choosing a threshold. For the experiments with real image sequences, we terminate the iteration if the change between histograms is smaller than or equal 10^{-4} in two consecutive iterations. Taking into account the mosaics in Fig. 2 and symmetric transfer errors in Table 1, it can be concluded that symmetric transfer error may not provide fully accurate information about the visual quality of the mosaics. However, the noise level of parameters is strongly correlated with

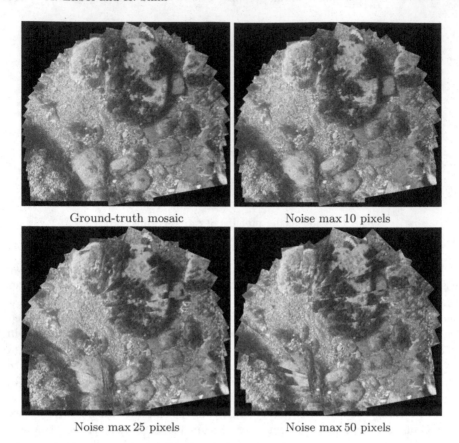

Ground-truth mosaic   Noise max 10 pixels

Noise max 25 pixels   Noise max 50 pixels

Fig. 2. Mosaics obtained with additive noise on the translation parameters of their planar image-to-mosaic transformations

Table 1. Change on invariant histograms and computed symmetric transfer errors with different levels of noise. Change on histograms is computed by using Eq. 1

Noise Level (in pixels)	Change on histogram			Symmetric Transfer Error (in pixels)		
	Mean	Standard Deviation	Maximum	Mean	Standard Deviation	Maximum
2.50	1.18E-05	7.28E-06	5.66E-05	2.74	0.67	8.26
5.00	1.45E-04	9.46E-05	6.18E-04	5.29	1.25	11.18
10.00	1.02E-03	4.37E-04	2.89E-03	10.48	2.47	17.28
25.00	4.84E-03	2.12E-03	1.44E-02	26.09	6.16	36.63
50.00	1.62E-02	7.08E-03	4.79E-02	52.24	12.30	70.34

to the visual errors in mosaics. On the other hand, it should be noted that the noise in our experiments was only added to the translation parameters. Having small noise on the rotation and scale parameters can provoke more noticeable errors on the final mosaic.

Finally, we have evaluated the computational performance of our method on three datasets (referred as Underwater Dataset I (UWDI), Underwater Dataset

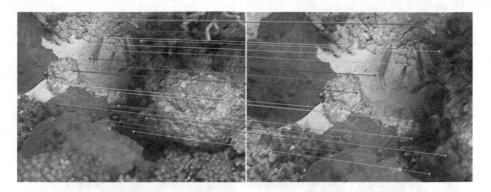

Fig. 3. Image pairs and established correspondences between them. This is the image pair where the symmetric transfer error had a maximum value in the UWDI.

II (UWDII), and aerial). They are extracted from a high-resolution image using real trajectory parameters of different Unmanned Vehicle (UVs). The UWDI is composed of 555 images of 512×384 pixels. Total number of successfully registered (An image pair is considered successfully matched if it has a minimum of 20 inliers.) overlapping image pairs is $18,392$ and total number of correspondences is $7,992,010$. The UWDII consists of 460 images of 572×380 pixels. This dataset is relatively sparse, having only $1,897$ overlapping image pairs, and presents two non-overlapping time-consecutive image pairs. Such properties of dataset falls apart traditional methods, which requires overlap between time-consecutive images. The total number of correspondences is $828,947$. The aerial

Table 2. Summary of results obtained using proposed method during the image mosaicing process. Strategy 'Without' represents the framework in the Fig. 1 without proposed steps.

Dataset	Strategy	Successful Pairs	Unsuccessful Pairs	Avg. Error in pixels	Std. Deviation Pairs in pixels	Maximum in pixels
UWDI	with proposed method	3,094	0	10.11	6.59	55.22
	without	18,320	2,902	7.34	3.06	31.14
	The approach in [6]	14,646	424	7.66	3.18	31.31
	AGA	18,392	135,343	7.33	3.06	31.14
UWDII	with proposed method	1,091	13	1.27	2.38	55.01
	without	1,877	976	0.83	0.33	4.70
	AGA	1,897	103,673	0.83	0.33	4.89
Aerial Dataset	with proposed method	980	0	1.12	0.43	4.80
	without	3,260	264	1.07	0.42	5.02
	The approach in [6]	3,792	202	1.06	0.42	4.94
	AGA	4,299	30,417	1.06	0.42	4.94

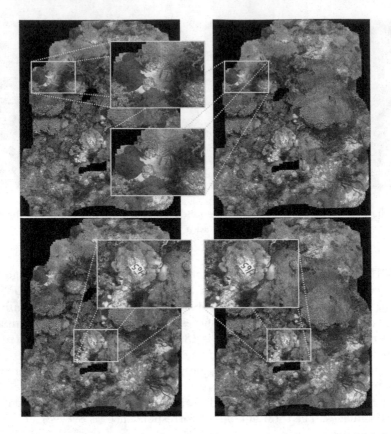

Fig. 4. Mosaics obtained with (left) and without using the proposal (right). (Top) Zoomed region where the symmetric transfer error is maximum. (Bottom) Small misalignment on final mosaics.

dataset comprises 264 images of 387×288 pixels having $4,299$ matched image pair and the total number of correspondences is $432,086$. Our termination criterion is integrated into the image mosaicing method in [3] because this mosaicing algorithm allows to handle randomly ordered image sequence. In this way, we can manage the case when there are non-overlapping time-consecutive images like in the UWDII. Table 2 presents the summary of the results. The second column corresponds to the tested method. The third column shows the total number of successfully matched image pairs. The fourth column contains the total number of image pairs that were not successfully matched and we denote them as *unsuccessful pairs*. The last three columns correspond to the average symmetric transfer error, the standard deviation, and maximum error calculated using all the correspondences identified by All-against-all (AGA) matching strategy. For the UWDI, global registration is carried out using five points (four corners and the center of the image). Since the UWDI and aerial dataset provide an overlap between time-consecutive images, we make a comparison with the method in [6].

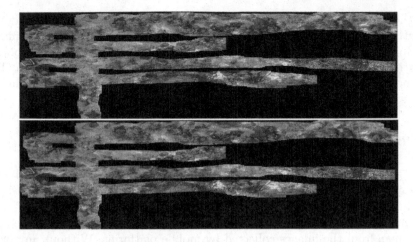

Fig. 5. Obtained mosaics of the UWDII with (top) and without using the proposal (bottom).

Fig. 6. Obtained mosaics of the aerial dataset with and without using the proposal.

Based on our experiments, we find that the maximum symmetric transfer error usually appears on overlapping image pairs with a big change on scale. Since their scale varies, one of them may not be visible in a final mosaic. Therefore, the visual quality of the final mosaic does not reflect the maximum symmetric transfer errors entirely as seen in Fig. 3. From the results presented in Table 2, mosaics can be obtained with a small number of image matching attempts without disturbing the final visual quality. Figs. 4, 5, and 6 show the obtained mosaics with and without using our proposal.

Although the computational times are not reported here, our method significantly reduces the total number of image mosaicing iterations and image matching attempts. The bottleneck of proposed method is the rendering phase, generating mosaic at each iteration and computing the invariant histogram. The time spent for rendering step can be reduced by applying the multiscale image analysis.

4   Conclusion and Future Work

Lately, great advancements in the mobile robotic platforms make it possible to obtain optical data from areas unreachable by humans. In most of the cases, a single image is not sufficient to provide an overview of the area of interest. To this end, Image mosaicing has been an indispensable tool for creating a large-area optical map from the images collected by mobile platforms. Without any prior on camera trajectory, a common mosaicing strategy is to apply the AGA image matching and then to perform global registration. This approach is exhaustive as it also attempts to register images that do not overlap. Therefore, its algorithmic complexity grows quadratically with the total number of images, which limits its usage in a small scale dataset.

Our experiments showed that invariant color histograms can be used as a visual stopping criterion during image mosaicing process. Also, we find that symmetric transfer error may not be an accurate indicator of visual quality of final mosaic, especially when camera trajectory provides scale changes and high overlapping area between both consecutive and non-consecutive images. Another important point can be stressed that identifying all overlapping image pairs may not be necessarily improving the visual quality of mosaic although it improves the camera trajectory estimate. In the future, we plan to extend invariant histograms based stopping criterion for mosaicing with low-overlapping image pairs.

Acknowledgments.. Authors would like to thank Underwater Vision Laboratory of Computer Vision and Robotics Institute of University of Girona for providing high-resolution test images and real trajectory parameters. This research was supported by the MSIP (Ministry of Science, ICT and Future Planning), Republic of Korea, under the IT Consilience Creative Program (NIPA-2014-H0201-14-1002) supervised by the NIPA (National IT Industry Promotion Agency). Aerial High-resolution image was retrieved from https://unsplash.com/stevenlewis on the 27th of April, 2015.

References

1. Bingham, B., Foley, B., Singh, H., Camilli, R., Delaporta, K., Eustice, R., Mallios, A., Mindell, D., Roman, C., Sakellariou, D.: Robotic tools for deep water archaeology: Surveying an ancient shipwreck with an autonomous underwater vehicle. J. Field Rob. **27**(6), 702–717 (2010)
2. Domke, J., Aloimonos, Y.: Deformation and viewpoint invariant color histograms. In: BMVC, pp. 509–518 (2006)

3. Elibol, A., Gracias, N., Garcia, R.: Fast topology estimation for image mosaicing using adaptive information thresholding. Rob. Auton. Syst. **61**(2), 125–136 (2013)
4. Elibol, A., Gracias, N., Garcia, R., Kim, J.: Graph theory approach for match reduction in image mosaicing. J. Opt. Soc. Am. A. **31**(4), 773–782 (2014). http://josaa.osa.org/abstract.cfm?URI=josaa-31-4-773
5. Escartin, J., Garcia, R., Delaunoy, O., Ferrer, J., Gracias, N., Elibol, A., Cufi, X., Neumann, L., Fornari, D.J., Humpris, S.E., Renard, J.: Globally aligned photomosaic of the lucky strike hydrothermal vent field (Mid-Atlantic Ridge, 3718.5'N): Release of georeferenced data, mosaic construction, and viewing software. Geochem. Geophys. Geosyst. **9**(12), Q12009 (2008)
6. Gracias, N., Zwaan, S., Bernardino, A., Santos-Victor, J.: Mosaic based navigation for autonomous underwater vehicles. IEEE J. Oceanic Eng. **28**(4), 609–624 (2003)
7. Hartley, R., Zisserman, A.: Multiple View Geometry in Computer Vision, 2nd edn. Cambridge University Press, Harlow (2004)
8. Lowe, D.: Distinctive image features from scale-invariant keypoints. Int. J. Comput. Vision **60**(2), 91–110 (2004)
9. Park, J.Y., Choi, J.Y., Jeong, E.Y.: Applying an underwater photography technique to nearshore benthic mapping: A case study in a rocky shore environment. J. Coastal Res. SI **64**, 1764–1768 (2011)
10. Pizarro, O., Williams, S.B., Jakuba, M.V., Johnson-Roberson, M., Mahon, I., Bryson, M., Steinberg, D., Friedman, A., Dansereau, D., Nourani-Vatani, N., Bongiorno, D., Bewley, M., Bender, A., Ashan, N., Douillard, B.: Benthic monitoring with robotic platforms - the experience of Australia. In: IEEE International Underwater Technology Symposium (UT), pp. 1–10 (2013)
11. Scaradozzi, D., Sorbi, L., Zoppini, F., Gambogi, P.: Tools and techniques for underwater archaeological sites documentation. In: Oceans - San Diego 2013, pp. 1–6 (2013)
12. Szeliski, R.: Image alignment and stitching: A tutorial. Found. Trends® Comput. Graph. Vis. **2**(1), 1–104 (2006)

Intelligent Reconstruction and Assembling of Pipeline from Point Cloud Data in Smart Plant 3D

Pavitra Holi, Seong Sill Park, Ashok Kumar Patil,
G. Ajay Kumar, and Young Ho Chai[✉]

Graduate School of Advanced Imaging Science, Multimedia and Film, Chung-Ang University,
#221 Heuksuk-Dong, Dongjak-Gu, Seoul, Republic of Korea
{pavitraholi,ashokpatil03,ajaygkumar}@hotmail.com
pssil1321@gmail.com, yhchai@cau.ac.kr

Abstract. The laser-scanned data of subsisting industrial pipeline plants are not only astronomically immense, but are withal intricately entwined like a net. The users must identify 3D points corresponding to each pipeline to be modelled in immensely colossal laser-scanned data sets. To accurately identify the 3D points corresponding to each pipeline, the users need to have some cognizance of direction and design of the pipelines. In addition, manually identifying each pipeline from gigantic and intricate scanned data is proximately infeasible, time-consuming and cumbersome process. In order to simplify and make the process more facile for reconstruction process an intelligent way of reconstruction and assembling of pipeline from point cloud data in Smart Plant 3D (SP3D) is proposed. The presented results shows that the proposed method indeed contribute automation of 3D pipeline model.

Keywords: Point cloud · Segmentation · Cylinder detection · SP3D automation

1 Introduction

In recent years the advancement in laser technology, incremented the demand for reconstruction of pipeline of an industrial plant from the laser-scanned point cloud data [1]. Conventional method of engendering 3D models of pipeline is being done utilizing commercial software's such as Leica Cyclone [2],Smart Plant 3D [3] and AutoCAD Plant 3D. The data obtained from laser scan is sizably voluminous, manually detecting explicit geometric information available from point cloud data to reconstruct pipe-line is time consuming and intricate [4].

So far, several research has been done on reconstruction of point cloud data automatically like Bosche [5] and Rabbani et al. [6]. To make the reconstruction process simple and more facile an automation of intelligently reconstruction and assembling of pipeline from point cloud data in SP3D is proposed.

The rest of the paper is organized as follows: The pre-processing of point cloud data is explained in Sect. 2. Segmentation process with classification and recognition of point cloud is described in Sect. 3. In Sect. 4 detection of cylinder parameters using Hough

© Springer International Publishing Switzerland 2015
Y.-S. Ho et al. (Eds.): PCM 2015, Part II, LNCS 9315, pp. 360–370, 2015.
DOI: 10.1007/978-3-319-24078-7_36

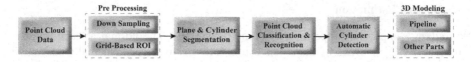

Fig. 1. Workflow of automatic reconstruction and assembling of pipeline process.

transform is explained. We show reconstruction and assembling of pipeline automatically using SP3D in Sect. 5. Section 6 concludes our paper.

2 Pre-processing of Point Cloud

The laser-scanned data captured from an industrial plant contains sizably voluminous accumulation of 3D point cloud data. The data is uneven in density with a lot of noise, which will greatly affect the computational time for reconstruction process. Thus a pre-processing of point cloud data is required to efficiently apply subsequent algorithms to reconstruct the pipelines.

Down sampling and grid-based region of interest has been applied in pre-processing step to accelerate the reconstruction process. Firstly, a 3D voxel grid is generated for the immensely colossal point cloud data. Then all points present inside each voxel are approximated with their centroid. Figure 2, shows a part of a sizably voluminous pipeline plant data filtered with voxel grid algorithm based on Point Cloud Library (PCL) [7].

(a)   (b)

Fig. 2. Down sampling of point cloud data. (a) Original (330227 points). (b) Down sample (116880 points).

3 Segmentation and Classification of Point Cloud

Segmentation commences with pre-processed laser scanned data. It is an essential part in the process of automation of point cloud data. This process mainly consists of three steps: normal estimation, region growing using smoothness constraint and random sample consensus (RANSAC) method with appropriate threshold values for objects classification like pipeline, plane and other parts.

3.1   Normal Estimation

Normals are important properties of geometric surface. The normal at each point is estimated by [7] analysis of the eigenvectors and eigenvalues (Principal Component Analysis) of a covariance matrix created from the nearest neighbors of the query point. More specifically, for each point P_i, assembled the covariance matrix C using Eqs. (1) and (2).

$$C = \frac{1}{k} \sum_{i=1}^{k} \cdot \left(P_i - p\right) \cdot \left(P_i - p\right)^T \tag{1}$$

$$C \cdot \vec{v_j} = \lambda_j \cdot \vec{v_j}, \quad j \in \{0, 1, 2\} \tag{2}$$

Where k is the number of point neighbors considered in the neighbor of P_i, 3D centroid of the nearest neighbors is presented by p, λ_j is the j-th Eigen value of the covariance matrix, and $\vec{v_j}$ the j-th Eigen vector. The normal estimated by PCA are not consistently oriented, by knowing viewpoint V_p, all normal $\vec{n_i}$ are oriented consistently towards the view point, by satisfying Eq. (3).

$$\vec{n_i} \cdot \left(V_p - P_i\right) > 0 \tag{3}$$

3.2   Region Growing

The Fig. 3, shows flow of region growing process [6], the purpose of this process is to merge the points close enough in terms of smoothness constraint. The output is a set of clusters of point cloud that are considered to be a part of same surface.

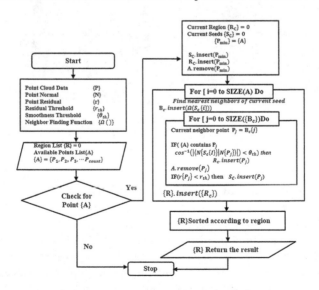

Fig. 3.   Flow chart of region growing process.

This algorithm sorts the points by their residual value and it begins region growing from the point which has minimum residual value. Then the picked point added to the set and named as seed. Once seeds set become empty the process has grown the region and process starts from beginning. Then by assigning proper parameter values, the cluster has been owed, which has its own color as shown in Fig. 4.

Fig. 4. Result of segmentation. (a) Input data. (b) Segmented data.

3.3 Pipe and Plane Separation

After completing segmentation using smoothness constraint, clusters of point cloud data is obtained, which include many different objects like plane, cylinder and some unwanted parts. In order to extract required data RANSAC algorithm [7] is applied. By applying RANSAC robust estimator iteratively and setting proper threshold values, objects like plane and pipeline are differentiated as in Fig. 5.

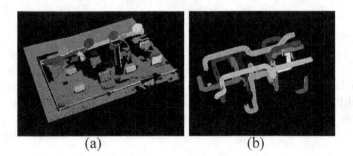

Fig. 5. Objects separation using RANSAC. (a) Plane. (b) Cylinders.

3.4 Point Cloud Classification and Recognition

In order to reconstruct 3D point cloud data, distinguishing of the segmented data should be done effectively. The ability to classify and recognize the objects in segmented data make the reconstruction process easier. Figure 6, shows the overview of classification and recognition step.

This process begins with considering the segmented cloud as input; each cloud will be processed in order to extract the features which uniquely describe the shape of the object. Once the feature extracted, it has to be manually labelled to train support vector machine (SVM) a machine learning algorithm. During classification, given a set of training examples, the SVM classifier endeavors to find optimally disuniting hyper planes which group sets of trained examples belonging to the same class and separates those from different classes. Depending on the acceptable threshold, cloud will be presaged as one of the trained labels.

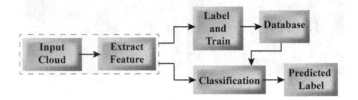

Fig. 6. Overview of classification and recognition process.

Different segmentation provides different information about the geometry of the object [8, 9]. In this process required cloud features are extracted based on normal with specific search radius parameters using multiple region growing segmentation. Figure 7, shows extracted elbow cloud part features from input cloud data.

(a)   (b)   (c)

Fig. 7. Feature extraction. (a) Input cloud. (b) Segmented cloud. (c) Extracted cloud.

Extracted cloud features are trained using SVM. The LIBSVM package [10] with RBF kernel by providing gamma = 0.5, and probability parameter value = 1, has been used to train cloud features as shown in Fig. 8. In classification stage each segmented point cloud data is compared with already trained data, the similar point

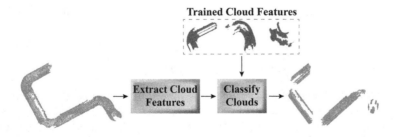

Fig. 8. Classified objects from input cloud data.

cloud features data which matches with the trained data will be removed from the inputted cloud data.

4   Detection of Cylinder Parameter Using Hough Transform

After segmentation and classification process, the point cloud data mainly consists of cylinders. These cylinder parameters in point cloud data are extracted using efficient Hough transform method [11]. The five parameters of cylinder are, (θ, φ) gives the axis direction in spherical coordinate, r is the radius, $P(u, v)$ gives the position in terms of u and v along with axial direction $n = (cos\theta * sin\varphi, sin\theta * sin\varphi, cos\varphi)$ from the cylinder's local coordinate system as shown in Fig. 9. The Hough transform algorithm consists of two sequential steps: first step is orientation estimation and second is position and radius estimation.

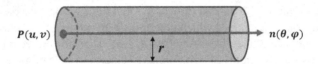

Fig. 9. Parameters of cylinder.

4.1   Orientation Estimation

In orientation estimation the Gaussian sphere of the input point cloud data is used and a 2D Hough transform is performed for finding vigorous hypothesis for the direction of cylinder axis. This is mainly predicated on the observation that, for cylinders the normal form a great circle on the Gaussian sphere [12], this great circle results from intersection of the unit sphere with plane passing through origin. Thus normal vector of this plane is same as cylinder axis.

In this process initially normal [7] are calculated for all point clouds Fig. 10(a), then make a sampled Hough space to represent Hough Gaussian sphere Fig. 10(b), by using approximate uniform sampling method. In this method each points are placed with equal distance, so that they represent equal area on sphere surface, after that each points in point cloud data use the spherical coordinates of its normal obtained, to derive matrix R using Eq. (4).

$$R = \begin{pmatrix} sin^2\varphi - cos\varphi cos^2\theta & -(1 + cos\varphi)\, cos\theta sin\theta & sin\varphi cos\theta \\ -(1 + cos\varphi)\, cos\theta sin\theta & cos\varphi\, cos^2\theta - cos\varphi + cos^2\theta & sin\varphi sin\theta \\ sin\varphi cos\theta & sin\varphi sin\theta & cos\varphi \end{pmatrix} \quad (4)$$

Finally found the cell in Hough Gaussian sphere whose accumulator values, greater than threshold are obtained, which results required cylinder axis. Figure 10, shows the result of orientation estimation.

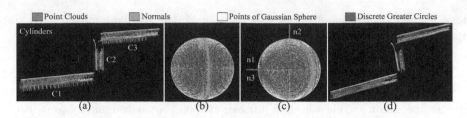

Fig. 10. Orientation estimation. (a) Point cloud with normal. (b) Normals on Gaussian sphere. (c) Estimated orientation on Gaussian sphere. (d) Computed cylinder axis.

4.2 Position and Radius Estimation

The remaining three unknown parameters, position $P(u, v)$ and radius r are estimated using 3D Hough transform. This process begins with orientation obtained from previous step, then take some of its neighbor points on Gaussian sphere and begin projecting all the points to the plane with a normal direction equal to the cylinder axis, by applying singular value decomposition [13] which determines, an orthonormal coordinate frame for projection, consisting three basis vectors (u, v, n). Then for projected point in 3D dataset, generate sampled circle [14] with radius r centered at each point. For each value of radius r in a user-specified radius range, increment cells in the Hough space (voting grid) using Eq. (5) for each projected point given by (u_p, v_p).

$$\left(r \cos \omega + u_p, r \sin \omega + v_p \right) \quad 0 \leq \omega < 2\pi \tag{5}$$

By finding peak in Hough space gives us cylinder orientation and radius. Then transform the position to world coordinate system by using Eq. (6).

$$T = \begin{pmatrix} u_x & u_y & u_z \\ v_x & v_y & v_z \\ n_x & n_y & n_z \end{pmatrix} \tag{6}$$

Figure 11, shows the position and radius estimation. By using Hough transform algorithm found all parameters of cylinders needed to generate a 3D pipe model. The extracted cylinder parameters with segmentation information are saved as specification data file.

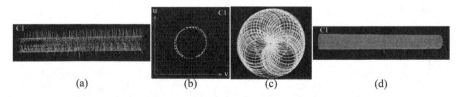

Fig. 11. Position and radius estimation. (a) Orientation data. (b) Voting grid. (c) Generated circles and peak inside Hough sphere. (d) Radius of cylinder.

5   Reconstruction and Assembling of Pipeline in SP3D

The process of reconstruction and assembling of pipeline is shown in Fig. 12. The saved specification data files of cylinders are read and stored in SQL [15] database to generate pipeline automatically in SP3D.

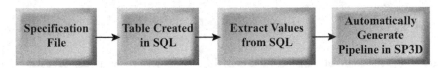

Fig. 12.   Generating and assembling process of pipeline in SP3D.

5.1   SP3D Interfacing

One of the important process in assembling of pipeline is the proper specification list Fig. 13(a) (Excel file) used to create plant objects. This list is provided by the industrial plants to interlink SP3D and .NET automation program. The given specification list creates catalog in SP3D using Bulk-Load utility as shown in Fig. 13.

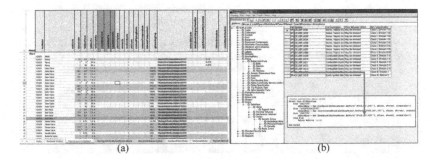

Fig. 13.   Specification file and Catalog file. (a) Specification list (Excel file). (b) Accessing of catalog objects in automation program.

After creation of catalog list, a SQL database table has to be created as shown in Fig. 14(a). The saved specification data files from the cylinder parameter detection process are uploaded into the table created in SQL. These specification database tables are used to extract pipeline information using .NET automation program as shown in Fig. 14(b).

Column Name	Data Type	Allow Nulls
EID	int	☐
Pipe_name	varchar(50)	☐
Diameter	float	☐
posX1	float	☐
posY1	float	☐
posZ1	float	☐
posx2	float	☐
posy2	float	☐
posz2	float	☐

(a)   (b)

Fig. 14. Specification database. (a) Table in SQL. (b) Extracted values from SQL database.

5.2 Experimental Result

After interfacing .NET automation program with SP3D, specification values from SQL database are extracted to generate all pipes automatically as shown in Fig. 15(b), and by using segmentation information in .NET automation program, position and direction of elbows are calculated which has to be inserted between generated pipes from automation algorithm, which intelligently assembles required pipes to generate pipeline as shown in Fig. 15(c).

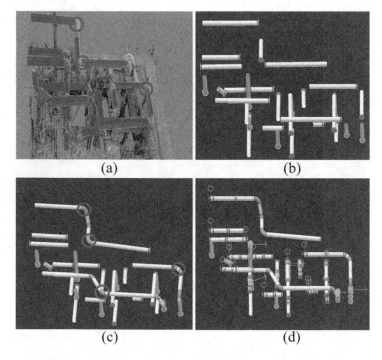

(a)   (b)

(c)   (d)

Fig. 15. Intelligently assembling of pipeline. (a) Input point cloud. (b) Automatically generated pipes in SP3D. (c) Intelligently assembled pipeline. (d) Inserted objects on generated pipelines.

From the results it can optically discern that generated pipelines are reconstructed and assembled from the point cloud data. This process of reconstruction of pipeline plant is simple and efficient. From the developed automation program we can withal insert other tools and objects as required for the reconstruction of pipeline plant like valves, flanges, supports and other equipment's as shown in Fig. 15(d).

6 Conclusion

In this paper an intelligent process of reconstruction and assembling pipelines in SP3D is presented. The proposed method consists of pre-processing of point cloud data, segmentation, classification, recognition and automatic cylinder detection along with its parameters. The extracted parameters are preserved to form specification data file. By utilizing SQL database and catalog list from SP3D in automation program, we successfully generated and intelligently assembled pipeline in SP3D. The proposed method is advantageous, which detects pipes and assembles them automatically. This process enabled us to engender wide variety of objects. The results shows that the automatically assembled pipelines are accurate and homogeneous to the inputted point cloud data acquired from the laser scanning techniques.

Acknowledgements. This work was supported by Commercializations Promotion Agency for R & D Outcomes – Grant funded by the Ministry of Science, ICT and future Planning - 2013A000019.

References

1. Wehr, A., Lohr, U.: Airborne laser scanning—an introduction and overview. ISPRS J. Photogramm. Remote Sens. **54**, 68–82 (1999)
2. Leica Cyclone., Heerberg, Switzerland, Leica Geosystems AG
3. Smart Plant 3D., Huntsville, AL, Intergraph Corporation
4. Son, H., Kim, C., Kim, C.: Fully automated as-built 3D pipeline segmentation based on curvature computation from laser-scanned data. In: Proceedings of Computing in Civil Engineering–Proceedings of the 2013 ASCE International Workshop on Computing in Civil Engineering, pp. 765–772 (2013)
5. Bosché, F.: "Model Spell Checker" for Primitive-based As-built Modeling in Construction. University of Texas at Austin, Austin (2003)
6. Rabbani, T., van den Heuvel, F., Vosselmann, G.: Segmentation of point clouds using smoothness constraint. Inter. Arch. Photogramm. Remote Sens. Spat. Inf. Sci. **36**, 248–253 (2006)
7. Rusu, R.B., Cousins, S.: 3D is here: point cloud library (pcl). In: 2011 IEEE International Conference on Robotics and Automation (ICRA), pp. 1–4. IEEE (2011)
8. Golovinskiy, A., Kim, V.G., Funkhouser, T.: Shape-based recognition of 3D point clouds in urban environments. In: 2009 IEEE 12th International Conference on Computer Vision, pp. 2154–2161. IEEE (2009)
9. Huang, J., You, S.: Detecting objects in scene point cloud: a combinational approach. In: 2013 International Conference on 3D Vision-3DV 2013, pp. 175–182. IEEE (2013)

10. Chang, C.-C., Lin, C.-J.: LIBSVM: a library for support vector machines. ACM Trans. Intell. Syst. Technol. (TIST) 2, 27 (2011)
11. Rabbani, T., Van Den Heuvel, F.: Efficient hough transform for automatic detection of cylinders in point clouds. In: ISPRS WG III/3, III/4 3, pp. 60–65 (2005)
12. Do Carmo, M.P.: Differential Geometry of Curves and Surfaces. Prentice-Hall, Upper Saddle River (1976)
13. Golub, G.H., Van Loan, C.F.: Matrix computations. JHU Press, Baltimore (2012)
14. Kimme, C., Ballard, D., Sklansky, J.: Finding circles by an array of accumulators. Commun. ACM 18, 120–122 (1975)
15. Microsoft SQL Server: Microsoft Corporation, Redmond (2008)

A Rotational Invariant Non-local Mean

Rassulzhan Poltayev and Byung-Woo Hong[✉]

Computer Science Department, Chung-Ang University, Seoul, Korea
hong@cau.ac.kr

Abstract. The image restoration and noise reduction are used to improve image quality and to develop more robust and high performance algorithms to solve denoising problems in the image processing. Every approach has their own limitations and more practicable properties with the specific conditions. This paper considers rotational transformation with non-local means denoising algorithm. The non-local means denoising algorithm use repetitive patterns in the image for noise reduction. Therefore, the affine transformations will extend search space of the problem and will cause of more qualitative results.

Keywords: Non-local means · Image denoising · Rotation-invariant · Image neighborhood · Euclidean distance

1 Introduction

Image noise is a random usually unwanted variation in brightness or color information in an image. Image noise arising from noisy sensor or channel transmission errors usually appears as discrete isolated pixel variation that are not spatially correlated (Pratt, 1978). Images always is subject to noise and unwanted variation of illumination and color intensity values, which are needed for their representation. There are several applications as medical image analysis, segmentation, object detection where the noise reduction is playing one of the important step which has a high impact to the performance of any algorithm. In that manner, the noise reduction is widely used concept.

The main goal of noise reduction is to restore the original image as much as possible by using denoising methods. Digital images are very sensitive to different kind of noises, which are affected to the quality of images.

The image denoising has been the most important and widely studied problems in the image processing and computer vision. The researchers have proposed different kind of algorithms to achieve the aim of denoising. So far image denoising methods can basically divided into two categories: spatial methods and transform domain filtering methods. Essentially, all the existing methods rely on some explicit or implicit assumption about the noise free signal for separating them from a noise.

© Springer International Publishing Switzerland 2015
Y.-S. Ho et al. (Eds.): PCM 2015, Part II, LNCS 9315, pp. 371–380, 2015.
DOI: 10.1007/978-3-319-24078-7_37

If the recovering of original image from an observed noisy image is the main problem, then the general solution is to find original image and to estimate the noise. The noisy image is combination of an image with the noise (1):

$$v(i) = u(i) + n(i) \tag{1}$$

where $v(i)$ is a noisy image (observed image); $u(i)$ is an original image; $n(i)$ is an image noise. Therefore, the denoising process is to estimate original clean image $u(i)$ from the noisy image $v(i)$. One of the popular method for the noise reduction is the non-local means (NLMs) denoising algorithm, which is developed by Buades et al. [1,3].

The non-local means denoising filter is related to universal denoising algorithm, what has been proposed to process image sequences and relies on the principle that the image sequence contains repeated patterns. The detection of the repeated patterns can be used to reduce a noise in images. Such an approach is already popular in texture synthesis, image in-painting, video completion, and has also been explored for image restoration. The underlying idea of this method is to estimate, $u(i)$ using a weighted average of all pixels neighborhoods. The paper is considered non-local means and propose a methods by using synthetic images show the differences between standard non-local means and new proposing method to improve estimation of an noise and to improve quality of an image.

2   Related Works

There are numerous denoising algorithms to estimate $u(i)$ from $v(i)$, and the most of those methods can be categorized by different parameters. However, the main issue of the noise reduction is modeling of that noise. The weak point of any algorithm is the sufficiency of the image model. By this way, all denoising algorithms are based on:

- a noise model;
- a image smoothness model;

The another attribute which has impact on the result of denoising is filter parameter h. The parameter measures a degree of filtering applied to the image. In such way, most results of denoising algorithms can be represented as decomposition of an observed image with noise.

$$v = D_h v + n(D_h, v) \tag{2}$$

where D_h is denoising method with h parameter;

- $D_h v$ is smoother than v
- $n(D_h, v)$ is the noise estimated by the method

The represented approaches for solving the problem of noise reduction are include the local methods, as linear filters and anisotropic diffusion; and global methods like total variation or wavelet shrinkage; and discrete methods, for example, Markov Random fields image denoising and discrete universal denoising algorithm.

As in a general problem with reverse tasks, there are many denoising algorithms based on prior solutions. In general, only prior knowledge consider the original natural images. In this case, small noises after prior skills should represent natural image statistics.

One of the important point in the natural image statistics is considering repetitive local patterns. This observation become base of two different approaches: non-local means and block matching and 3D filtering denoising algorithms. In the NLMs the kernel of the noise reduction created by patches of noise corrupted images, so the kernel operator of the image denoising affected from those patches.

The NLMs filter is an evolution of the Yaroslavsky filter (Yaroslavsky, 1985) which averages similar image pixels according to their intensity distance [4]. Some filters, like the SUSAN or the bilateral filters are based in the same principle [5]. The main differences of the NLMs with these methods is that the similarity between pixels is more robust in front of the noise level by using region comparison rather than pixel comparison and that pattern redundancy is not restricted to be local. Pixels far from the pixel being filtered are not penalized due to its distance to the current pixel, as happens with the bilateral filter. The conventional algorithms consider translation transformation of neighborhood vectors. However, they do not consider other kinds of symmetry, especially images with explicit symmetrical structures, such as texture images. If transformed neighborhood patterns were also included in similarity comparison, and better use of the repetitive structural characteristics will improve performance of the denoising algorithms.

3   Rotational Invariant Non-local Mean

The non local means assumes that an image contains some similar repetitive structures, where the self-similarity property can be used to suppress noise.

The basic principle of the non-local means filter is that for a certain pixel i and any other pixel j in the image, if i and j have similar neighboring pixels then pixel j has a larger weight in determining of the value i. The pixel i and neighborhood of pixel i are called the comparison window. The procedure is very time consuming in order to compute all the pixels in the image. Therefore a certain area around pixel i is considered which is called the search window. In a noisy image I, $(v(i), i \in I)$ denotes all the pixel values; after denoising the corresponding pixel value will be equal to $u(i)$:

$$u(i) = \sum_{j \in I} w(i,j)v(j) \tag{3}$$

where $w(i,j)$ is the weight of all other pixels in the image I, denoting the similarity of pixels i and j, w ranges $0 \leq w \leq 1$ and $\sum_j w(i,j) = 1$. The weight is based on the similarity of $v(N_i)$ and $v(N_j)$, which denote the two neighborhoods of pixel i and j:

$$w(i,j) = \frac{1}{Z(i)} exp(-\frac{||v(N_i) - v(N_j)||^2_{2,a}}{h^2}), \tag{4}$$

$$E(i,j) = ||v(N_i) - v(N_j)||^2_{2,\sigma} \tag{5}$$

$$Z(i) = \sum_j exp(-\frac{||v(N_i) - v(N_j)||^2_{2,\sigma}}{h^2}) \tag{6}$$

where $E(i,j)$ is the Gaussian weighted Euclidean distance and $Z(i)$ is the normalization factor; therefore, $\sum w(i,j) = 1$. The parameter h controls the decay of the exponential function, and thus adjust the filtering effect, which depends on the image to be filtered.

The NLMs is very effective and simple approach, what make it one of the popular. However, it is computational expensive algorithm. There are several methods have been proposed to improve the performance and quality of that algorithm. By considering different objects, we will see that they have rotational symmetry, where the original non-local means algorithm does not work well in the filtering of such images. Therefore, the rotational symmetry will extend search area for the image noise reduction. The rotations help us to consider different transformation of patches in order to find the most similar neighborhood pixels. Hence, we are going to consider two different ways of transformations with following rotation angles: $45°$ and $90°$; to find affine transformation of sub images, 4 and 8 different kind of transformations can be considered in order to find the most similar one. Therefore, the following rotation angles has been considered: $[0; 90; 180; 270]$ and $[45; 135; 225; 315]$. The problem is to find most similar patch by using Euclidean distance.

Firstly, the algorithm calculate possible rotations of selected patch. Next step is looking for neighborhood patch, which has most closest distance for one of their transformations. After finding the most similar rotation element, a rotational invariant patch is going to be calculated for the following neighborhood vector. Then the weighted average of all similar patches inside of the search window of image should be calculated as in conventional non-local means, and because of this reason the method will reduce noise ratio by using weighted average.

According to this modification the pixel with the similar gray level to $v(N_i)$ will have larger weights on average, even if it is represented by rotational transformation. The following modification is described in the Listing 1.

Algorithm 1. Calculate Rotation_L2_Filter(*p*)

$r \Leftarrow rotations(p)$
$result \Leftarrow 0$
while $q \neq LAST$ **do**
  $dist = MAX\_INT$
  $angle = 0$
  while $p \neq LAST\_ROTATION$ **do**
    $d = l2norm(p, q)$
    if $|d| < dist$ **then**
      $dist \Leftarrow d$
      $angle \Leftarrow current$
    end if
  end while
  $weight = e^{\frac{-dist^2}{h^2}}$
  $result = result + weight * rotate(q, -angle)$
  $sum = sum + weight$
end while
$result = result/sum$
return result;

The basic idea of a rotational invariant NLMs can be represented in a simple generic algorithm:

1. Estimate the angle of rotation between the patches.
2. To each pixel in the first patch, find the position of the corresponding pixel in the second patch by rotating vector on estimated angle.

Hence, the different kind transformation of patches give the most of appropriate patterns and extend search space of the problem to improve quality of the image.

4   Experimental Results

Before comparing the experimental results, the objective image quality evaluation metrics [6] should be considered. The performance of each filter is evaluated quantitatively for images with impulse noise using the following objective quality metrics. Definition: $x_{j,k}$ denotes samples of original image, $y_{j,k}$ denotes the sample of filtered image. M and N are number of pixels in row and column directions, respectively.

The mean-square-error (MSE) is the simples, and the most widely used, image quality measurement. This metrics is frequently used in the signal processing and it is defined as follows (7):

$$MSE = \frac{1}{MN} \sum_{j=1}^{M} \sum_{k=1}^{N} (x_{j,k} - y_{j,k})^2 \tag{7}$$

where $x_{j,k}$ represents original image and $y_{j,k}$ represents the restored image; and j and k are the pixel position of the $M \times N$ image. MSE is zero when $x_{j,k} = y_{j,k}$.

The peak signal noise ratio (PSNR) is most commonly used as a measure of the quality of the restored image. The small value of PSNR means that image is poor quality. The PSNR is defined as (8):

$$PSNR = 10 \log \frac{(2^n - 1)^2}{MSE} = 10 \log \frac{255^2}{MSE} \qquad (8)$$

In this way, the effectiveness of the proposed algorithm can be obtained from calculating the difference between PSNR and MSE of the original and denoised images. The non-local means algorithm has three parameters and the filter results depend highly on their settings. The first parameter, $R_{search\ area}$, is the radius of a search window. Despite, that the original method use all the pixels in the image to find the weighted average of every pixel, the algorithm use specific size of window in order to reduce the number of computations, as shown in following Eq. (9).

$$u(i) = \sum_{j \in R_{search\ area}} w(i,j)v(j) \qquad (9)$$

The second parameter, $R_{neighborhood}$ is the radius of the neighborhood window used to find the similarity between two pixels. If the value of $R_{neighborhood}$ is increased the similarity measure will be more robust, but fewer similar neighborhoods will be found, Table 1.

Table 1. PSNR comparison table for different radius of the neighborhood window while R of search window is equal to 5

$R_{neighborhood}$	NLM	NL+Rotation
1	24.05	24.05
2	**23.48**	**23.66**
3	22.79	22.85
4	22.93	22.82

The third parameter, h, is related to the smooth parameter of the exponential curve to control the degree of smoothing. If h is too small, little noise will be removed while if h is set too high, the image will become more blurry.

In the experiments, $R_{search\ area}$ set to 5, which seems a reasonable value for most of the images. The best settings for $R_{neighborhood}$ and h under different noise levels had different combinations, Table 2, Fig. 2.

Table 2. PSNR comparison table for different smoothing parameter

h(smooth param)	NLMs	NLMs+Rotation
2	22.14	22.14
3	22.15	22.16
5	22.34	22.39
10	22.78	22.75
15	22.93	22.94
20	23.16	23.24
22	**23.28**	**23.40**

22.14dB	$PSNR_{orig} = 23.48\ dB$	Noise
22.14dB	$PSNR_{modified} = 23.66\ dB$	Noise

Fig. 1. The denoising results of two different algorithms, where the standard deviation of noisy image is $\sigma = 20$. The first line of images represent the noisy image, denoised image by using conventional non-local means filter and the noise of image which has been estimated. The second row shows denoising by using the rotational-invariant non-local means.

The proposing of using rotational invariant patterns show the effectiveness of such method according to comparison results of the PSNR values in the previous tables. The one of denoising examples is showing in following figure, Fig. 1:

Original images

Noisy images $PSNR = 22.14\ dB$

Denoised[NLMs]:

$PSNR_{h=2} = 22.14\ dB$ $PSNR_{h=10} = 22.78\ dB$ $PSNR_{h=20} = 23.16\ dB$

Denoised[rotational NLMs]:

$PSNR_{h=2} = 22.14\ dB$ $PSNR_{h=10} = 22.75\ dB$ $PSNR_{h=20} = 23.24\ dB$

Fig. 2. The denoising results comparison between conventional and proposed algorithm according to different smoothing parameter values.

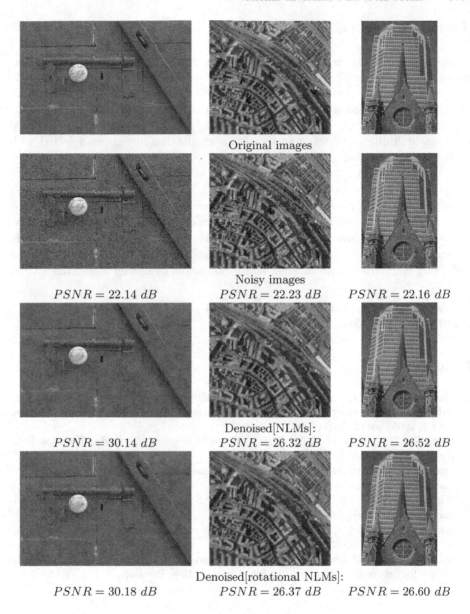

Original images

| $PSNR = 22.14\ dB$ | $PSNR = 22.23\ dB$ | $PSNR = 22.16\ dB$ |

Noisy images

Denoised[NLMs]:

| $PSNR = 30.14\ dB$ | $PSNR = 26.32\ dB$ | $PSNR = 26.52\ dB$ |

Denoised[rotational NLMs]:

| $PSNR = 30.18\ dB$ | $PSNR = 26.37\ dB$ | $PSNR = 26.60\ dB$ |

Fig. 3. The denoising results for the "Door", "Satellite", "Church" test images. Comparison between conventional and rotation-invariant non-local means denoising algorithm.

5   Discussion and Conclusion

In this paper, we have investigated the idea of improving a non-local means denoising algorithm by considering rotational transformations of patches. The method operates by extending a search space of the problem, and considering rotational transformations of the patches in order to calculate the weights for non-local averaging by applying the rotation transformations to the observed patches. In addition, by using such estimation, patches have been selected with higher level of similarity.

The experimental results are compared several denoising parameters to find one of their optimal combination, Fig. 3. Firstly, the smooth parameter is depending on the variation of the noise. The effectiveness of the algorithm is changing by manipulation the radius of the neighborhood and search window. Hence, a rotational invariant non-local means is computationally time expensive, because of extending the search area of the problem, but more robust for noise reduction compare to conventional non-local means denoising algorithm.

Acknowledgments. This work was supported by NRF-2011-0007898 and NRF-2014R1A2A1A11051941.

References

1. Buades, A., Coll, B., Morel, J-M.: A non-local algorithm for image denoising. In: IEEE Computer Society Conference on Computer Vision and Pattern Recognition, CVPR 2005, vol. 2. IEE (2005)
2. Gonzalez, R.C., Woods, R.E.: Digital Image Processing, 2nd edn. Prentice Hall, Upper Saddle River (2002)
3. Buades, A., Coll, B., Morel, J.M.: A review of image denoising algorithms, with new one. Multiscale Model Sim **4**, 490–530 (2006)
4. Yaroslavsky, L.P.: Digital Picture Processing. Springer Series in Information Sciences, 9th edn. Springer-Verlag, Berlin (1985)
5. Smith, S.M., Brady, J.M.: SUSAN - a new approach to low level image processing. Int. J. Comput. Vis. **23**, 45–78 (1997)
6. Sheikh, M.A.A., Mukhopadhyay, S.: Comparative analysis of noise reduction techniques for image enhancement. In: Int. J. Innov. Res. Develop., vol. 2.12 (2013)

Adaptive Layered Video Transmission with Channel Characteristics

Fan Zhang[1], Anhong Wang[1(✉)], Xiaoli Ma[2], and Bing Zeng[3]

[1] Institute of Digital Media & Communication,
Taiyuan University of Science and Technology, Taiyuan 030024, China
837299874@qq.com, wah_ty@163.com
[2] Georgia Institute of Technology, Atlanta 30314, USA
xiaoli.ieee@gmail.com
[3] University of Electronic Science and Technology of China,
Chengdu 611731, China
eezeng@uestc.edu.cn

Abstract. In wireless video transmission, the layered video transmission combining with a layered video coding can gracefully accommodate the receivers' heterogeneity. In this paper, we propose a new layered video transmission scheme based on the wireless channel characteristics of orthogonal frequency division multiplexing (OFDM) physical (PHY) layer, leading to an adaptive layered video transmission (ALAVIT). In our scheme, scalable video coding (SVC) is exploited to generate the layered video bit-streams; and the resulted base layer (BL) and enhancement layer (EL) bits are modulated differently to obtain their individual symbols. According to the estimated channel characteristics described by the H parameters of a connected wireless receiver, subcarriers with good channel quality and more power are allocated to BL symbols for the protection of these important bits. As compared to the state-of-the-art PHY layer techniques such as s-mod and MixCast, our ALAVIT scheme is able to provide a better performance.

Keywords: Wireless video transmission · OFDM PHY layer · SVC · Subcarrier allocation · Power allocation · Modulation adjustment

1 Introduction

Video service has been a business hotspot in the current wireless communication systems. However, how to transmit video data effectively over wireless networks still remains as a challenging problem.

In traditional wireless transmission schemes, PHY layer equally treats bits from the upper layer and tries its best to transmit these bits efficiently, just aiming to maximize the throughput. However, different bits from the same bit-stream may have different importance. For example, in a layered video coding system such as SVC, video sequence is encoded into a BL and several ELs, where the loss of BL would cause a serious degradation in video quality. Hence, it is necessary to give the BL bits a stronger protection.

© Springer International Publishing Switzerland 2015
Y.-S. Ho et al. (Eds.): PCM 2015, Part II, LNCS 9315, pp. 381–389, 2015.
DOI: 10.1007/978-3-319-24078-7_38

Recently, some PHY-layer techniques have been developed with the consideration of protecting the important bits. For instance, Digital Video Broadcast-Terrestrial (DVB-T) standard [1] combines a layered video coding scheme with a layered transmission scheme. Typically, the layered transmission method involved in DVB-T is hierarchical modulation (h-mod) [2]. Through this layered transmission technique, the strong receivers (such as the receivers near to the base station) can recover full-quality video using the received BL and EL bits, while the weak receivers (those far away from the base station) only obtain a basic quality with the received BL bits. Consequently, this accommodates gracefully the receivers' heterogeneity in wireless networks. However, h-mod requires specialized hardware and has performance limits.

For the same purpose, Cai et al. proposed the scalable modulation (s-mod) to provide differentiated services in PHY layer [3]. In s-mod, bits of different layers are mapped to constellations owning different minimum Euclidean distances, realizing different levels of protection for different layers. However, when the constellation points are dense, s-mod suffers from severe degradation of bit error rate (BER). Lately, Cui et al. proposed MixCast modulation [4]. MixCast mixes BL and EL bits into arbitrary number of wireless symbols by using the weighted sum operation. The weights for BL bits have larger absolute values than the weights for EL bits, and thus important BL bits are protected. However, the error propagation of demodulation process may lead to severe degradation in video quality.

More recently, some layered video transmission schemes that combine SVC with OFDM PHY layer are proposed. For instance, Yang et al. proposed a scalable video transmission scheme based on multi-antenna OFDM and SVC [5]. Kim et al. proposed a resource allocation algorithm for the OFDM-based cellular system which uses SVC as the encoder [6]. However, all these schemes do not optimize the resource allocation from the aspect of the BL's BER.

Although the utilization of frequency diversity in OFDM is well discussed in recent years, our contributions focus on two aspects: (1) we propose a cross-layer layered video transmission scheme over OFDM PHY layer which allocates subcarriers and power to video layers with different priority according to the channel characteristics, (2) we minimize the BER of BL so that the important BL bits can be protected strongly, making sure that each receiver can obtain basic video quality. Hence, a novel OFDM PHY layer video transmission scheme based on the channel characteristics is proposed, leading to an adaptive layered video transmission, abbreviated as ALAVIT. At the sending side, BL and EL bits are first modulated using different modulation schemes to obtain individual wireless symbols. Then, according to the estimated channel characteristics described by the H parameters, subcarriers with good channel quality are allocated to BL symbols, while the rest subcarriers to EL symbols for the sake of protecting the important BL bits. In addition, more power is allocated to BL using the water-filling algorithm. Experimental results show the advantages of our scheme as compared with s-mod and MixCast schemes.

2 Review of Scalable Video Coding

As an extension of H.264/AVC standard, SVC has attracted great attention due to its superiority for layered video transmission over heterogeneous networks. SVC can recover video quality proportional to the supported bit rates, and it provides three kinds of scalability [7]. For each type of scalability, the video is encoded into one BL and several ELs. The BL provides the basic video quality, while better video quality can be achieved with the ELs. However, when the BL is damaged or lost, the ELs themselves cannot improve the video quality or even cannot decode the video bit-stream. This means that more protection should be provided to the BL bits. Considering this, we try an alternative PHY layer technique for the transmission of SVC bit-streams, which exploits the adaptive allocations of subcarriers and power for the video transmission over OFDM PHY layer.

3 Adaptive Layered Video Transmission with Channel Characteristics

3.1 The Framework of Our Transmission System

The purpose of our work is to provide different levels of protection to bits with different priorities on the base of the channel characteristics of OFDM PHY layer. The framework of our proposed scheme is shown in Fig. 1. The sender transmits N complex symbols S_n ($n \in [1, N]$) in parallel, where the transmission process is composed of three parts, i.e. subcarrier allocation, power allocation, and OFDM modulation [8]. First, we use different modulation schemes to modulate BL and EL bit-streams by considering their different priorities. Next, subcarriers and power are allocated according to the CSI from the robust feedback channel. Finally, an inverse fast Fourier transform (IFFT) of order N is used to obtain subchannel division, and after IFFT, a cyclic prefix (CP) is inserted to remove the intersymbol interference (ISI).

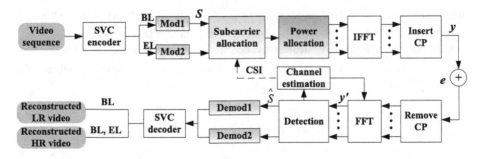

Fig. 1. The framework of our scheme.

At the receiver side, the received complex signal y'_n, i.e. the output of the fast Fourier transform (FFT) at the n-th subcarrier, can be described as:

$$y'_n = P_n \cdot H_n \cdot S_n + e_n \tag{1}$$

where P_n is the power allocated to the n-th subcarrier, H_n is corresponding channel function gain—the so-called H parameters, and e_n is the additive noise. The recovering procedure is merely reversed, where the detection process is to find the optimal solution to the following formula:

$$\hat{S} = \arg\min_{S \in C} \left\| y' - P \cdot H \cdot S \right\|_2 \tag{2}$$

where C is the set of constellation points.

Note that, in our scheme, the estimated CSI is returned from a feedback channel, which is usually available in current communications.

3.2   Subcarrier Allocation

Due to the fact that different subcarriers may face different channel conditions, different subcarriers should be allocated to different data with the purpose of protecting important data. Our goal is to minimize the BER of BL so that the important BL bits can be protected strongly. Our scheme is based on the fact that BER related to the n-th subcarrier can be approximated as follows [9]:

$$BER_n \cong \frac{2(\sqrt{M_n} - 1)}{\sqrt{M_n} \cdot \log_2 M_n} \cdot erfc \sqrt{\frac{P_n \cdot 3|H_n|^2}{2(M_n - 1)E_n}} \tag{3}$$

where M_n is the constellation size and E_n is the noise variance of the n-th subcarrier. From Eq. (3), we find that the performance of each subcarrier depends on both $|H_n|^2$ and P_n. The larger the values of $|H_n|^2$ and P_n, the smaller the BER. Therefore, in our scheme, subcarriers that have larger value of $|H_n|^2$ are deliberately arranged to transmit BL symbols, while EL symbols use subcarriers that have relatively small $|H_n|^2$.

Assuming that there are N_s strongest subcarriers (corresponding to N_s largest values of $|H_n|^2$) allocated to transmit BL bits, and EL bits use the remaining N_w subcarriers, then the BER of BL can be calculated as:

$$BER_{BL} = \frac{1}{N_s} \sum_{n=1}^{N_s} BER_n \tag{4}$$

In order to obtain the values of N_s and N_w, the allocation of subcarriers is deliberately set to a restraint that the transmission rates of our scheme are equal to the ones used in the reference schemes. Consequently, the values of N_s and N_w are approximated by Eqs. (5) and (6) respectively, with $N_s + N_w \leq N$:

$$N_s \approx \frac{N \log_2(M_{BL\_ref})}{\log_2(M_{BL})} \tag{5}$$

$$N_w \approx \frac{N \log_2(M_{EL\_ref})}{\log_2(M_{EL})} \tag{6}$$

where M_{BL_ref} and M_{EL_ref} respectively represent the constellation sizes of BL and EL in the referenced scheme, while M_{BL} and M_{EL} are the corresponding constellation sizes of our scheme. Here, the referenced scheme is the s-mod in [3] or the MixCast in [4].

Specially, subcarriers are first ordered so that $k_1 \geq k_2 \geq \ldots\ldots \geq k_N$, with $k_m = |H_{o(m)}|^2$, where the index ordering is taken into account by the ranking function $n = o(m)$. According to the index, BL bits are assigned to N_s subcarriers which have N_s largest values of $|H_n|^2$, while EL bits to N_w subcarriers with relatively small $|H_n|^2$. The principle of subcarrier allocation is illustrated in Fig. 2 where the green represents the subcarriers allocated to BL and the yellow shows the ones selected to transmit EL data (only eight subcarriers are included here for simplicity, and the second row is the values of $|H_n|^2$). As shown in Fig. 2, four subcarriers with largest $|H_n|^2$ are used to transmit the BL data, and the rest four subcarriers are allocated to transmit the EL data.

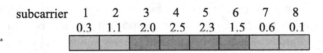

subcarrier	1	2	3	4	5	6	7	8
	0.3	1.1	2.0	2.5	2.3	1.5	0.6	0.1

Fig. 2. Illustration of the subcarrier allocation.

3.3 Power Allocation

The power allocation follows the idea of the water-filling algorithm [10], which results in an allocation of more power to the good channels with large $|H_n|^2$ and less power to the small $|H_n|^2$ ones. By this method, BER of the important BL bits will decrease because more power is allocated to them, which can also be verified by Eq. (3). The power allocated to n-th subcarrier P_n is:

$$P_n = \begin{cases} a - \dfrac{E_n}{|H_n|^2} & if \quad \dfrac{E_n}{|H_n|^2} < a \\ 0 & if \quad \dfrac{E_n}{|H_n|^2} \geq a \end{cases} \tag{7}$$

where a is a threshold guaranteeing $\sum_{n=1}^{N} P_n = P$, P is the power budget, so we set $a = \frac{1}{N}(P + \sum_{n=1}^{N} \frac{E_n}{|H_n|^2})$, where N denotes the total number of subcarriers.

3.4   Modulation Adjustment

In our scheme, we also adjust modulation scheme in order to provide different protection to the different-importance data. Namely, error resilient modulation schemes are applied to the BL bits. On the other hand, by adjusting modulation schemes and the number of subcarriers, we can flexibly adjust the transmission rates. For example, at a given video bit rate, we may choose binary phase shift keying (BPSK) to modulate BL and quadrature phase shift keying (QPSK) to modulate EL. If the total number of subcarriers is 64 and the numbers of subcarriers allocated to BL and EL are 16 and 48, respectively, then, the transmission rate ratio of BL and EL is 1:6.

4   Experimental Results

We first compare the PHY layer performance between s-mod and ALAVIT. These two schemes are under the same transmission rate ratio. For both BL and EL, s-mod uses QPSK, while our ALAVIT uses 16 quadrature amplitude modulation (16QAM), and the numbers of subcarriers allocated to BL and EL are equal. We measure BERs of different layers after demodulation without using any error correcting coding and power allocation. The experimental results shown in Fig. 3 demonstrate that the BER performances of both BL and EL of our scheme are better than the corresponding ones of s-mod. This benefit comes from the exploration of subcarrier allocation adopted in our scheme. The reason why EL's BER of our proposed scheme is better is that the constellation of s-mod is obtained by superimposing two individual constellations. The superimposition gives very weak protection for its EL.

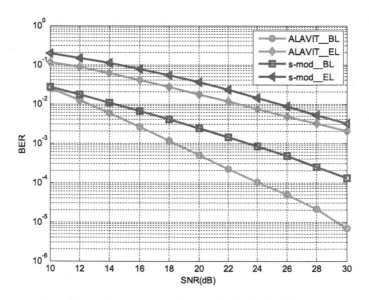

Fig. 3. BER comparisons between s-mod and ALAVIT.

Peak signal to noise ratio (PSNR) reflects the distortion of the video. Thus, we use it as a video quality metric. CIF@30 Hz video *"Foreman"* is used to perform the experiments.

The Joint Scalable Video Model (JSVM) is used as codec, where the encoder is configured with two-layer coarse granular scalability (CGS) [7]: one BL and one EL. The bit rates of the BL and the EL for "Foreman" are 157.82 Kbps and 573.16 Kbps, respectively, which are similar to that of s-mod in [3]. At the receiver side, if there are any bit errors, the decoder discards the frame. Figure 4 shows the PSNR comparisons of s-mod and ALAVIT at SNR = 14 dB without power allocation. From Fig. 4, it is obvious that PSNRs of our ALAVIT scheme are better than that of s-mod, with gain of around 4 dB.

Figure 5 compares the PSNR performances of MixCast and ALAVIT approach at SNR = 14 dB without using power allocation. Video *"Football"* with CIF@30 Hz is used to perform the experiment. We set the same experiment conditions as MixCast in [4]. From Fig. 5, due to the error propagation in the MixCast demodulation process, MixCast suffers from severe degradation in video quality. Nevertheless, PSNRs of our ALAVIT are quite stable, meaning that our scheme will not cause annoying experiences for video users.

Figure 6 shows the BER performances with or without power allocation, which indicates that BL's BER decreases with the help of power allocation. However, the BERs of EL remain nearly the same. This is because less power is allocated to EL. On the base of these results, we can find that subcarrier allocation plays a more dominant role than power allocation in protecting the important data.

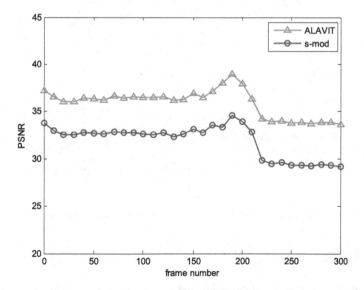

Fig. 4. PSNR comparisons between s-mod and ALAVIT.

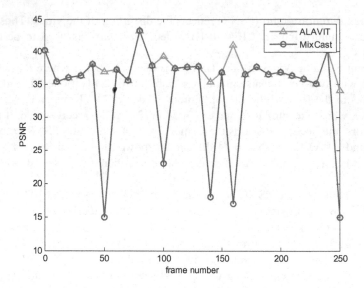

Fig. 5. PSNR comparison between MixCast modulation and ALAVIT.

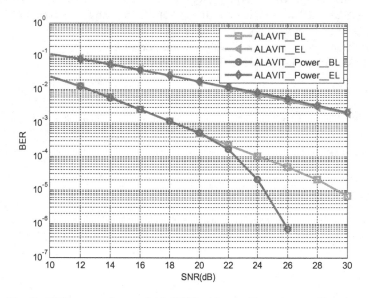

Fig. 6. BER comparisons of ALAVIT with or without power allocation.

5 Conclusion

In this paper, we propose an adaptive layered video transmission scheme using a novel OFDM PHY layer technique. Through the special allocations of subcarriers and power, our scheme provides a general framework for adaptive layered video transmission. The proposed scheme can be adjusted in order to protect important data by selecting different modulation, subcarriers, or power. The simulation results show that our scheme has achieved superior performance when compared with the state-of-the-art s-mod and MixCast which also use PHY layer technique.

Acknowledgements. This work has been supported in part by National Natural Science Foundation of China (No. 61272262 and No. 61210006), International Cooperative Program of Shanxi Province (No. 2015031003-2), The Program of "One hundred Talented People" of Shanxi Province, Research Project Supported by Shanxi Scholarship Council of China (2014-056) and Program for New Century Excellent Talent in Universities (NCET-12-1037).

References

1. ETSI, "Digital Video Broadcasting (DVB)," 2006
2. Wang, S., Yi, B.K.: Optimizing enhanced hierarchical modulations. In: Proceedings of IEEE Global Telecommunications Conference, pp. 1–5, New Orleans, November 2008
3. Cai, L., Xiang, S., Luo, Y.: Scalable modulation for video transmission in wireless networks. IEEE Trans. Vehicular Technol. **60**(9), 4314–4323 (2011)
4. Cui, H., Luo, C., Chen, C.W.: MixCast modulation for layered video multicast over WLANs. In: Proceedings of VCIP, pp. 1–4, Tainan (2011)
5. Yang, Z., Zhao, Y.: Scalable video multicast over multi-antenna OFDM systems. Wirel. Pers. Commun. **70**(4), 1487–1504 (2013)
6. Kim, D., Fujii, T., Lee, K.: A resource allocation algorithm for OFDM-based cellular system serving unicast and multicast services. EURASIP J. Wirel. Commun. Netw. **26**(1), 1–16 (2013)
7. Schwarz, H., Marpe, D., Wiegand, T.: Overview of the scalable video coding extension of the H. 264/AVC standard. IEEE Trans. Circuits Syst. Video Technol. **17**(9), 1103–1120 (2007)
8. Wang, Z.D., Giannakis, G.B.: Wireless multicarrier communications. IEEE Signal Process. Mag. **17**(3), 29–48 (2000)
9. Sklar, B.: Digital Communications. Prentice Hall PTR, Upper Saddle River (2001)
10. Yu, W., Rhee, W., Boyd, S.: Iterative water-filling for Gaussian vector multiple-access channels. IEEE Trans. Inf. Theory **50**(1), 145–152 (2004)

An Accurate and Efficient Nonlinear Depth Quantization Scheme

Jian Jin[1,2], Yao Zhao[1,2(✉)], Chunyu Lin[1,2], and Anhong Wang[3]

[1] Institute of Information Science, Beijing Jiaotong University, Beijing, China
yzhao@bjtu.edu.cn
[2] Beijing Key Laboratory of Advanced Information Science
and Network Technology, Beijing, China
[3] School of Electronic Information Engineering,
Taiyuan University of Science and Technology, Taiyuan, China

Abstract. As known, depth information exists as floating distance data, when firstly captured by depth sensor. In view of storage and transmission, it is necessary to be quantized into several depth layers. Generally, it is mutually contradictory between the efficiency of depth quantization and the accuracy of view synthesis. Actually, since 3D-warping rounding calculation exists during view synthesis, depth changes within a certain range will not cause different warped position. This phenomenon provides a good way to quantize depth data more efficiently. However, 3D-warping rounding calculation can also bring additional view synthesis distortion, if the warped-interval and image-resolution-interval are misaligned. Hence, to achieve efficient depth quantization without introducing additional view synthesis distortion, an *accurate and efficient nonlinear-depth quantization* scheme (AE-NDQ) is presented in which the alignment between warped-interval and image-resolution-interval is taken into consideration during the depth quantization. Experimental results show, compared with the *efficient nonlinear-depth-quantization (E-NDQ)*, AE-NDQ needs almost the same bits to represent the depth layers but maintains more accurate on view synthesis. For the traditional *8-bits nonlinear-depth-quantization (NDQ)*, AE-NDQ needs less bits to represent the depth layers, while has the same accuracy of the synthesized view.

Keywords: 3D-TV · Depth-image-based rendering (DIBR) · Depth quantization

1 Introduction

In three-dimensional television (3D-TV), multiple view plus depth (MVD) is the main data format, which is being investigated passionately. Arbitrary synthesized virtual views can be generated using depth-image-based rendering (DIBR) technique. Actually, depth information is used as the geometrical information to calculate the disparity between reference view and virtual view aiming at achieving pixels shifting. Normally, the original depth distance data exists as floating data. In view of storage and transmission, it is imperative to quantize the floating depth data into discrete values. There exist several methods for depth quantization.

© Springer International Publishing Switzerland 2015
Y.-S. Ho et al. (Eds.): PCM 2015, Part II, LNCS 9315, pp. 390–399, 2015.
DOI: 10.1007/978-3-319-24078-7_39

The straight forward method is to divide the scene's depth range into equidistant layers as shown in Fig. 1(a), which is known as *linear-depth-quantization (LDQ)* and this method is also very common in computer graphics. However, it has a drawback that the widths of warped-intervals in the virtual view are not the same, e.g., the nearer depth layer corresponds to a wider warped-interval, while the farther one represents to a narrower warped-interval. This results in a perceived depth resolution that is much lower for close objects than for scene parts that are farther away. Therefore, for the physical scene, *NDQ* is more suitable, since *NDQ* can make sure the quantized layers corresponding to the same width warped-interval and the rendering of whole scene in the virtual view has the same resolution as shown in Fig. 1(b). *NDQ* can be derived from the theory of "plenoptic sampling" by Jin-Xing Chai et al. in 2000 [1], which is expressed much simpler by Feladmann et al. [2] in 1-D parallel model [3]. It is also explicitly derived the dependencies between the depth range of the scene, the image resolution and the efficient depth scaling. What's more, the minimum required number of depth layers are determined which is treated as the rudiment of efficient nonlinear-depth-quantization (*E-NDQ*) in this paper. And this work was further discussed by Christoph Fehn et al. in 2004 [4]. *E-NDQ* has taken the view synthesis into consideration which can roughly preserve the warped-interval equaling to the image-resolution-interval in virtual view, while cannot ensure their alignment. Once misalignment happens, that means this kind of quantization will introduce view synthesis distortion and it is not the optimal in terms of view synthesis. Now, a variant version named *8-bits NDQ* is adopted in 3-D video [5] as the current standard which is widely used in depth representation. Compared with *E-NDQ*, *8-bits NDQ* uses 256 layers to quantize the physical scene, where 256 layers are many enough to represent the depth information relatively accurate.

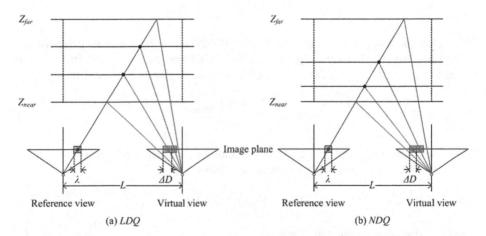

(a) *LDQ*   (b) *NDQ*

Fig. 1. Illustration of *LDQ* and *NDQ*, the red block and blue block are the warped-interval and image-resolution-interval, and their lengths are represented as λ and ΔD respectively (Color figure online).

In this paper, an accurate-efficient nonlinear-depth-quantization (*AE-NDQ*) is explored, which inherits the efficiency of *E-NDQ*. Besides, since 3D-warping rounding calculation is considered so that the warped-interval and the image-resolution-interval are aligned very well and the view synthesis distortion can also be eliminated fundamentally. The rest of this paper is organized as follows. *E-NDQ* is briefly reviewed in Sect. 2. Based on this, *AE-NDQ* is proposed in Sect. 3. Experimental results are shown in Sect. 3. Section 4 concludes this paper.

2 Efficient Nonlinear-Depth-Quantization: A Brief Review

In this section, the theory of *E-NDQ* is given out firstly. Then, a corresponding analysis is following.

2.1 Theory of *E-NDQ*

Note that 1-D parallel model is chosen as target, which means the physical scene is captured by two cameras with the co-planar and parallel camera vectors and share the same focal length f_x, only horizontal disparity is considered. L is the baseline, which represents the distance between reference view and virtual view. The physical scene is bounded by a depth range Z_{far} and Z_{near}. Correspondingly, the disparity range is represented with $D(Z_{far})$ and $D(Z_{near})$. So, we can obtain such formulas by [6]:

$$D(Z_{far}) = \frac{f_x \cdot L}{Z_{far}} \tag{1}$$

$$D(Z_{near}) = \frac{f_x \cdot L}{Z_{near}} \tag{2}$$

Firstly, assuming that N is the total numbers of depth layers, the screen disparity values D are provided by 3D-warping process, which is desired as equidistant interval, the warped-interval $\Delta D = |D_{v+1}-D_v|$ (with $v \in [0,..,N-1]$) can simply be calculated as:

$$\Delta D = \frac{D(Z_{far}) - D(Z_{near})}{N - 1} \tag{3}$$

Then, each screen disparity value D_v corresponds to a certain depth layer Z_v (with $v \in [0,..,N-1]$). So, the D_v can be represented as follows:

$$D_v = D(Z_{far}) - v \cdot \Delta D \tag{4}$$

Finally, a conditional equation is given as:

$$\frac{1}{Z_v} = \frac{1}{Z_{near}} \cdot \left(\frac{v}{N - 1}\right) + \frac{1}{Z_{far}} \cdot \left(1 - \frac{v}{N - 1}\right) \tag{5}$$

Since N is the depth levels, it can be set according to practical needs. So, Z_v could be determined by Eq. (5). Therefore, any points in the scene with the physical depth value Z could be quantized to a level, if it meets the relationship as follow:

$$\frac{1}{Z_v} - \frac{0.5}{N-1} \cdot \left(\frac{1}{Z_{near}} - \frac{1}{Z_{far}}\right) \leq \frac{1}{Z} \leq \frac{1}{Z_v} + \frac{0.5}{N-1} \cdot \left(\frac{1}{Z_{near}} - \frac{1}{Z_{far}}\right) \tag{6}$$

2.2   Analysis of E-NDQ

In the derivation of E-NDQ, the warped-interval ΔD can maintain a constant. If ΔD equals to the image-resolution-interval λ, it would be ideally considered that the depth quantization achieves the highest efficiency and the numbers of depth layers N may achieve the minimum N_{min}. In this subsection, we will explore the expression of N_{min}.

As shown in Fig. 2, if the physical depth value Z_v and Z_{v+1} are given to point a (with the coordinate x_{ref-a}) in the reference view, they will be warped to point b (with x_{vir-b}) and point c (with x_{vir-c}) in the virtual view respectively and we can obtain:

$$x_{ref-a} - x_{vir-b} = \frac{f_x \cdot L}{Z_v} \tag{7}$$

$$x_{ref-a} - x_{vir-c} = \frac{f_x \cdot L}{Z_{v+1}} \tag{8}$$

Since Z_v and Z_{v+1} are the neighboring depth levels, point b and c are the neighboring pixels. So,

$$x_{vir-c} - x_{vir-b} = \Delta D \tag{9}$$

If ΔD equals to λ, we can express N_{min} as:

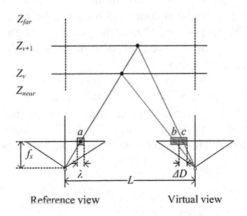

Fig. 2. Illustration of the E-NDQ's derivation.

$$N_{min} = \frac{f_x \cdot L}{\lambda} \cdot \left(\frac{1}{Z_{near}} - \frac{1}{Z_{far}} \right) + 1 \qquad (10)$$

where the f_x, L, Z_{far} and Z_{near} could exist as arbitrary positive real number, λ could be full-pixel, half-pixel and quarter-pixel. That means N_{min} could be a float data. If that happens, the float N_{min} will be rounded into integer, which leads to the warped-interval cannot equal to image-resolution-interval strictly thus causing their misalignment as shown in Fig. 3(a). Even though N_{min} is accidentally an integer, it still cannot make sure their alignment e.g., the malposition between them can also cause the misalignment, as shown in Fig. 3(b).

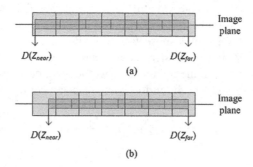

Fig. 3. Illustration of two kinds of misalignment between warped-interval (red block) and image-resolution-interval (blue block), (a) The misalignment caused by different lengths. (b) The misalignment caused by malposition (Color figure online).

Once misalignment exists, it will cause that some points which should be exhibited in the right position are directly pulled to the wrong one by 3D-warping rounding calculation as shown in Fig. 4. For instance, all the points between these two dotted red lines should be warped to the red block and be represented as point a. After 3D-warping rounding calculation, it will be rounded to the blue block, which means that all the points between these two dotted red lines will be warped to the blue block and will be exhibited in this position. In fact, the points in the green region should be exhibited in the dotted blue block instead of the right neighboring blue block and this will cause view synthesis distortion.

Based on the obviation above, E-NDQ has more contributions on the efficiency of depth representation compared with its low performance on the accuracy of view synthesis. However, if the warped-interval and the image-resolution-interval are aligned quite well, in other words, if we quantize the scene using the blue lines, the view synthesis distortion will be eliminated and the accuracy of view synthesis can also be improved.

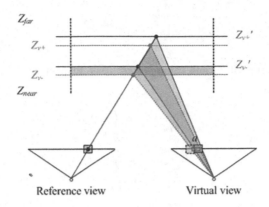

Fig. 4. Quantizing the scene using *E-NDQ*, some dotted red lines are generated. After 3D-warping rounding calculation, it will be represented by its center point *a*. Quantizing the scene based on the blue block's left and right bounds, the blue lines are generated (Color figure online).

2.3   Accurate-Efficient Nonlinear-Depth-Quantization

In this section, we focus on the alignment between warped-interval and image-resolution-interval. Figure 5(a) shows their original relationship, which is directly generated using *E-NDQ*. The ideal case is shown in Fig. 5(c). To achieve that we first redefine the depth range Z_{near}' and Z_{far}' so that $D(Z_{near}')$ and $D(Z_{far}')$ can be the center of image-resolution-interval. For conciseness, integer screen resolution is considered (half or quarter screen resolution is the same). $D(Z_{near}')$ and $D(Z_{far}')$ are redefined as:

$$D\left(Z_{far}'\right) = \lambda \cdot \left\lfloor \frac{D\left(Z_{far}\right)}{\lambda} \right\rfloor \qquad (11)$$

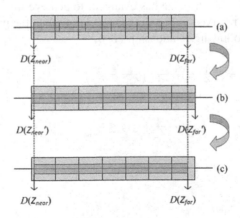

Fig. 5. Illustration of three different relationships between warped-interval and image-resolution-interval.

$$D(Z'_{near}) = \lambda \cdot \left\lceil \frac{D(Z_{near})}{\lambda} \right\rceil \tag{12}$$

where $\lfloor \bullet \rfloor$ and $\lceil \bullet \rceil$ denote floor and ceiling operations, respectively. Z_{near}' and Z_{far}' are written as following:

$$Z'_{far} = \frac{f_x \cdot L}{D(Z'_{far})} \tag{13}$$

$$Z'_{far} = \frac{f_x \cdot L}{D(Z'_{far})} \tag{14}$$

Then, E-NDQ is used. Note that formula (5) can be re-expressed as:

$$\frac{1}{Z'_v} = \frac{1}{Z'_{near}} \cdot \left(\frac{v}{N-1} \right) + \frac{1}{Z'_{far}} \cdot \left(1 - \frac{v}{N-1} \right) \tag{15}$$

And N_{min}' is calculated by

$$N'_{min} = \frac{f_x \cdot L}{\lambda} \cdot \left(\frac{1}{Z'_{near}} - \frac{1}{Z'_{far}} \right) + 1 = \frac{D(Z'_{near}) - D(Z'_{far})}{\lambda} + 1 \tag{16}$$

Since $D(Z_{near}')$ and $D(Z_{far}')$ are the integer screen resolution, undoubtedly, N_{min}' will be an integer. So, the scene will be quantized by (17):

$$\frac{1}{Z'_v} - \frac{0.5}{N'_{min}-1} \cdot \left(\frac{1}{Z'_{near}} - \frac{1}{Z'_{far}} \right) \le \frac{1}{Z} \le \frac{1}{Z'_v} + \frac{0.5}{N'_{min}-1} \cdot \left(\frac{1}{Z'_{near}} - \frac{1}{Z'_{far}} \right) \tag{17}$$

After this, the warped-interval and image-resolution-interval are aligned as shown in Fig. 5(b). However, the depth range has changed, to achieve the optimal, the original depth range shown be reserved as Fig. 5(c). Considering that, the physical scene should be quantized by the following relationship (18):

$$\begin{cases} \frac{1}{Z'_{far}} \le \frac{1}{Z} \le \frac{1}{Z'_v} - \frac{0.5}{N'_{min}-1} \cdot \left(\frac{1}{Z'_{near}} - \frac{1}{Z'_{far}} \right), & v = 0 \\ \frac{1}{Z'_v} - \frac{0.5}{N'_{min}-1} \cdot \left(\frac{1}{Z'_{near}} - \frac{1}{Z'_{far}} \right) \le \frac{1}{Z} \le \frac{1}{Z'_v} + \frac{0.5}{N'_{min}-1} \cdot \left(\frac{1}{Z'_{near}} - \frac{1}{Z'_{far}} \right), & 1 \le v \le N'_{min} - 2 \\ \frac{1}{Z'_v} + \frac{0.5}{N'_{min}-1} \cdot \left(\frac{1}{Z'_{near}} - \frac{1}{Z'_{far}} \right) \le \frac{1}{Z} \le \frac{1}{Z'_{near}}, & v = N'_{min} - 1 \end{cases} \tag{18}$$

3   Experimental Results

In this section, we present detailed simulation results for the proposed algorithm. Before that, the design of experiments is firstly introduced, which contains three main steps:

(1) The *8-bits* gray-scale depth image Q_0 is dequantized by [7],

$$Z = \frac{1}{\frac{v}{255} \cdot \left(\frac{1}{Z_{near}} - \frac{1}{Z_{far}}\right) + \frac{1}{Z_{far}}} \tag{19}$$

where v is the depth value and then the physical depth distance value Z of the scene is achieved and regarded as the ground truth, since *8-bits NDQ* uses many enough depth levels to represent the depth information accurately.

(2) Use *E-NDQ* and the proposed method *AE-NDQ* to quantize the scene. And the quantized depth images are Q_1 and Q_2 respectively.

(3) Q_0, Q_1 and Q_2 are used to direct the view synthesis. And V_0, V_1 and V_2 are the warped virtual views.

Then, two main performance metrics are given: (1) the efficiency of depth quantization, (2) the accuracy of view synthesis. The former metric is described by the total number of depth layers and the latter one is evaluated by the virtual view's Luma Peak Signal to Noise Ratio (PSNR) and the Mean Structural Similarity Index Metric (MSSIM) [8] objectively and the warped virtual views are also exhibited in Fig. 6.

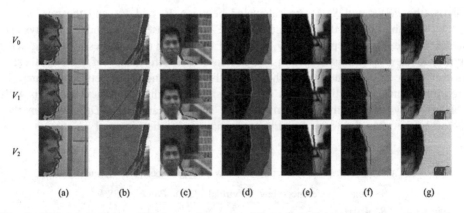

Fig. 6. Illustration of the subjective performance of these virtual views in these methods, (a) to (g) are *Bookarrival*, *Mobile*, *Lovebird1*, *Café*, *Kendo*, *Balloons*, and *Newspaper* respectively. The green parts are the holes.

3.1   Efficiency Evaluation on Depth Quantization

As shown in Table 1, N_{min} is calculated by formula (10) which represents the required minimum number of depth layers in *E-NDQ*. The experimental results have shown that N_{min} can be either a float or an integer, which proved the analysis in Sect. 2. B. To make sure the warped-interval maintaining equidistant, it will be rounded into integer *round* (N_{min}). After this, the *round* (N_{min}) will be used to quantize the physical scene. When the proposed method *AE-NDQ* is applied, the required minimum number of depth layers N_{min}' is calculated by formula (16). The experimental results have also proved N_{min}' is an integer. What's more, the difference between *round* (N_{min}) and N_{min}' is no more than 1, which means, the proposed method has almost the same efficiency as *E-NDQ* in view of depth quantization. That's because the efficiency of *E-NDQ* is main inherited. Besides, only 4 to 6 bits are required to represent all the needed depth layers.

Table 1. The required minimum numbers of depth layers in *E-NDQ* and *AE-NDQ*.

Sequences	N_{min}	Round(N_{min})	N_{min}'
Bookarrival	21.13	21	22
Mobile	13.44	13	14
Lovebird1	35.50	36	36
Café	61	61	62
Kendo	25	25	26
Balloons	25	25	26
Newspaper	36	36	37

3.2   Accuracy Evaluation on View Synthesis

Firstly, the *PSNR* and *SSIM* are calculated between V_0 and V_1, and the results are shown in Table 2. The data suggests that if *E-NDQ* is used to quantize the physical scene, the generated depth map Q_1 is not fit to direct view synthesis, since it will introduce view synthesis distortion and decrease the performance of view synthesis. However, the experimental result shows that the *PSNR* and *SSIM* between V_0 and V_2 are infinite and 1 respectively. That's means using Q_2 to direct the view synthesis will not bring any view synthesis distortion at all. That's because *AE-NDQ* maintains the alignment between the warped-interval and image-resolution-interval.

Table 2. Accuracy evaluation on view synthesis between V_0 and V_1.

Sequences (reference view → virtual view)	PSNR	SSIM
Bookarrival (8 → 9)	23.3883	0.8600
Mobile(3 → 4)	25.8789	0.9455
Lovebird1(4 → 5)	22.6380	0.8219
Café(2 → 3)	28.1909	0.9733
Kendo(1 → 2)	Inf.	1
Balloons(1 → 2)	Inf.	1
Newspaper(2 → 4)	Inf.	1

Besides, we also exhibit the subjective performance of these virtual views in these methods as shown in Fig. 6. It can be seen that the holes' size and positions between V_0 and V_2 are the same, which are different from V_1. This also proves the proposed method has the same performance with *8-bits NDQ* in term of the accurate view synthesis.

4 Conclusion

In this paper, we propose the *AE-NDQ*. In the stage of *AE-NDQ* design, the efficiency of *E-NDQ* is partly inherited. Only 4-6 bits can represent all the depth layers. Besides, we also take view synthesis into consideration, especially the effect of 3D-warping rounding calculation. Inspirited by this, we try to align warped-interval with image-resolution-interval very well. After that, the view synthesis distortions in the virtual view which is caused by the depth quantization are all eliminated. The experimental results also prove that the depth image, which is quantized using *AE-NDQ*, can be used to render the virtual view as accurately *8-bits NDQ* does. Therefore, *AE-NDQ* is more suitable to be used in the physical scene quantization for its using less bits to represent depth layers.

Acknowledgement. This work was supported by National Natural Science Foundation of China (no. 61210006, no. 61402034, no. 61202240, no. 61272051 and no. 61272262), Beijing Natural Science Foundation (4154082) and SRFDP (20130009120038), the Fundamental Research Funds for the Central Universities (2015JBM032), International Cooperative Program of Shanxi Province (No. 2015031003-2), the Program of "One hundred Talented People" of Shanxi Province, and Shanxi Scholarship Council of China (2014-056).

References

1. Chai, J.-X., Tong, X., Chan, S.C., Shum, H.-Y.: Plenoptic Sampling. In: Proceedings of ACM SIG-GRAPH 2000, New Orleans, LA, USA, pp. 307–318, July 2000
2. Feldmann, I., Schreer, O., Kauff, P.: Nonlinear depth scaling for immersive video applications. In: Proceedings of International Workshop on Image Analysis for Multimedia Interactive Services 2003, London, UK, pp. 433–438, April 2003
3. Tian, D., Lai, P.-L., Lopez, P., Gomila, C.: View synthesis techniques for 3D video. Appl. Digital Image Process. XXXII, Proc. SPIE **7443**, 74430T–74430T (2009)
4. Fehn, C., Hopf, K., Quante, B.: Key technologies for an advanced 3D TV system. Int. Soc. Optics Photonics, Optics East (2004)
5. Muller, K., Merkle, P., Wiegand, T.: 3-D video representation using depth maps. Proc. IEEE **99**(4), 643–656 (2011)
6. Zhao, Y., Zhu, C., Chen, Z., Tian, D., Yu, L.: Boundary artifact reduction in view synthesis of 3D video: from perspective of texture-depth alignment. IEEE Trans. Broadcast. **57**(2), 510–522 (2011)
7. Zhao, Y., Zhu, C., Chen, Z., Yu, L.: Depth no-synthesis-error model for view synthesis in 3-D video. IEEE Trans. Image Process. **20**(8), 2221–2228 (2011)
8. Wang, Z., Bovik, A.C., Sheikh, H.R., Simoncelli, E.P.: Image quality assessment: from error visibility to structural similarity. IEEE Trans. Image Process. **13**(4), 600–612 (2004)

Synthesis-Aware Region-Based
3D Video Coding

Zhiwei Xing[1], Anhong Wang[1(✉)], Jian Jin[2], and Yingchun Wu[1]

[1] Institute of Digital Media & Communication,
Taiyuan University of Science and Technology, Taiyuan 030024, China
{zhiweix1990, wah_ty}@163.com, yingchunwu3030@sina.com
[2] Beijing Jiaotong University, Beijing 100044, China
jianjin@bjtu.edu.cn

Abstract. In a depth-image-based rendering (DIBR)-based 3D video system, the original 3D video is commonly compressed from the point of the video itself, paying little attention to its contribution to the synthesized virtual view. This paper first proposes a method to divide the original video into different regions according to its contribution to the synthesized view and then proposes to measure the regions using compressive sensing with different measurement rates, leading to a synthesis-aware region-based 3D video coding approach. Experimental results show that our approach can achieve better synthesized quality under the same equivalent measurement rate. Our approach is suitable for the applications when the virtual view is more important than the original views.

Keywords: 3D video coding · Region-division · Synthesis-aware coding · DIBR · Compressive sensing

1 Introduction

Three-dimension (3D) video technique has generated increasing attraction in that it offers us an immersive experience. Today, 3D video has two formats: two/multi-view stereo and multi-view plus depth. The latter one is the most widely used format because it is capable of decreasing the number of views needed for synthesizing free virtual viewpoints by using depth information via the depth-image-based-rendering (DIBR) [1–3] technique.

In general, DIBR has three steps: (1) warping, being implemented to get the virtual views through geometry transformation by using the original views and their depth maps with corresponding parameters of cameras; (2) merging, which will blend the two warped views into one virtual view; (3) hole-filling, which involves wiping off the cracks and holes appearing in the virtual view.

Recently, the 3D High Efficient Video Coding (3D-HEVC) [4, 5] has emerged as a standard algorithm for 3D video coding. It is capable of efficiently compressing 3D video by making use of the inter-view correlation and intra-view correlation of the original texture videos, and even the correlation between texture video and depth video. However, when 3D-HEVC compresses the original 3D videos, it is only from the

© Springer International Publishing Switzerland 2015
Y.-S. Ho et al. (Eds.): PCM 2015, Part II, LNCS 9315, pp. 400–409, 2015.
DOI: 10.1007/978-3-319-24078-7_40

viewpoint of the video itself, ignoring different regions' contribution to the virtual view synthesized by DIBR algorithm. In some applications, it is just the virtual videos (rather than the original video) that may be the most needed for the viewer. For example, the original views captured by cameras may be unsuitable to watch directly when the baseline length is longer than human's pupil distance, typically 65 mm. This may not cause the viewer to experience an obvious stereo feeling or even make them feel uncomfortable. Hence, some middle virtual views are expected to be synthesized by the original views, even at the cost of sacrificing the quality of the original views under the constraint of limited bandwidth for transmitting the original views. A framework of 3D video coding using view synthesis prediction is proposed in [6], which compresses the color and depth data simultaneously by designing four types of the view synthesis prediction method according to the view position. However, the procedure maybe very complicated, as a new virtual view needs to be synthesized both in the encoder and decoder.

Some regions of the original views are invisible in the virtual views, which may be occluded by the foreground or out of the virtual views' visual range, so these regions are not as significant as the other regions, which can be seen in virtual views. Hence, from the point of compression, these invisible pixels should pay less attention than significant regions that are visible in the virtual views. In the meantime, there are many repeated regions that exist in both original views; therefore, they should be identified in advance and coded only once in order to save the bitrate.

Based on the above analysis, in this paper, we propose an approach of 3D coding which takes into account the view synthesis process and region division. In our approach, we first implement region-division in original views to both the left and right views, according to the estimated significance to the virtual view. Then we deliberately perform an unequal measurement through block-based compressed sensing (BCS) for significant and non-significant regions. Specifically, we arbitrarily chose one view as the dominant view, in which, almost all regions (except the regions of occlusion and out of visual range of virtual view) are regarded as significant and measured using a high measurement rate. For the other original view, we only pick the regions corresponding to the disocclusion of the virtual view (that is invisible through the dominant view but can be seen by virtual view and other views) as significant regions. The non-significant regions are measured by a lower CS-rate. The experimental results show that our approach can obtain a better synthesized view under the same equivalent measurement rate due to the exploration of this unequal measurement to regions.

The rest of this paper is organized as follows. Section 2 provides a brief review of compressive sensing. Section 3 presents the details of our methods. Section 4 reports the simulation results, and Sect. 5 is the conclusion.

2 Review of Block-Based Compressive Sensing

In our work, the CS is used to flexibly measure the different regions of the original videos through adjusting the CS-rate, therefore, we will briefly review the principle of CS and the block-based CS approach in this section.

According to CS [6] theory, for a sparse signal x in the transform domain Ψ, we can obtain a sampled vector y by using measurement matrix Φ, which is usually chosen to be orthonormal, such that $\Phi\Phi^T = I$. Then, the sampling projection y is obtained by a linear transformation:

$$y = \Phi x \tag{1}$$

In the reconstruction of CS, assuming a sparse set of transform parameters, $x' = \Psi x$. The ideal recovery is to find the x' with the smallest l_0 norm consistent with y:

$$x' = \arg\min_{x'} \parallel x' \parallel_{l_o}, \text{s.t.} y = \Phi\Psi^{-1}x' \tag{2}$$

where Ψ^{-1} represents the inverse transform, several procedures have been proposed to optimize l_0 [7]. In practice, such an l_0-constrained optimization is suffering from its computational infeasibility. Thus, CS turns to solve an l_1-based convex optimization [8].

When applied to 2-D images, however, CS faces several challenges; e.g., the huge measurement matrix and the very complex reconstruction process. To address these challenges, block-based compressed sensing (BCS) [9, 10] sampling is proposed. Typically, an image is divided into several non-overlap blocks, and each block is measured by the same measurement matrix Φ_B. This is equal in measurement to the whole image with a diagonal matrix Φ:

$$\Phi = \begin{bmatrix} \Phi_B & & \\ & \ddots & \\ & & \Phi_B \end{bmatrix} \tag{3}$$

Since the size of Φ_B is smaller than Φ, the memory required to restore the measurement matrix is greatly reduced and the reconstruction is sped up. In addition, BCS-SPL combines a Wiener filter to the SPL algorithm, thus offering a much smoothed reconstruction.

3   Synthesis-aware Region-Based 3D Coding with BCS

Based on the observations in the former sections, we want to perform coding of different accuracy to different regions in the original videos according to their attribution to the virtual views.

Figure 1 shows the flow chart of our proposed approach. Firstly, we perform region-division to the original views from the viewpoint of synthesizing the virtual view. Secondly, the regions are measured by BCS-SPL with different measurement rates. Finally, the original views are recovered by BCS-SPL reconstruction and used to synthesize the virtual view. For simplification, we just synthesize one virtual view that is located in the middle of two original views; the extension to the arbitrary view is straight.

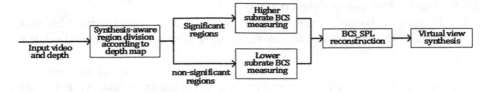

Fig. 1. The flow chart of the proposed approach

3.1 Region-Division of Original Videos

We conduct region-division to the text video according to the boundary of regions in the depth map, considering that the depth map provides geometry information. Specifically, we first detect the boundary of the depth map by using a threshold derived from the synthesis process and then perform region division to the corresponding texture video according to the depth video's division.

3.1.1 Calculate the Threshold to Detect Boundaries

We need to distinguish cracks and holes as cracks are with small size and are usually interpreted by the neighbor pixels, while holes have a large size and are filled by the other original virtual view. Based on the principle of 3D warping, the threshold for distinguishing cracks from holes can be calculated by Eq. (4) when assuming that the size of holes is greater than 2 pixels,

$$\Delta Z = \frac{255}{\frac{L}{2} \cdot f_x \cdot \left(\frac{1}{Z_{near}} - \frac{1}{Z_{far}} \right)} \tag{4}$$

where f_x is the focal length, L is the length of baseline, Z_{near} and Z_{far} is the nearest and farthest depth range of the real scene.

3.1.2 Boundaries Detection Process

After obtaining the threshold, the boundary detection is implemented. Two masks Bl and Br are used to represent the left boundary and right boundary of each original view, respectively. If the depth difference ΔZ_d between two adjacent pixels is larger than the threshold, the corresponding pixel location $Bl(x, y)$ is set to 1, otherwise to 0, as shown in Eq. (5):

$$Bl(x, y) = \begin{cases} 1, \Delta Z_d = d(x, y) - d(x, y + 1) \geq \Delta Z \\ 0, otherwise \end{cases} \tag{5}$$

The right boundary Br is obtained in the same way in Eq. (6):

$$Br(x, y) = \begin{cases} 1, \Delta Z_d = d(x, y - 1) - d(x, y) \leq -\Delta Z \\ 0, otherwise \end{cases} \tag{6}$$

3.1.3 Region Division of Original Views

After obtaining the boundary information, we implement the region division and pick out the regions that are required for the virtual view from the region-divided original views.

Figure 2 illustrates a simple model for the region division, where AL represents the background of all the visual range from V_1 to V_3, MN is the foreground, V_1 and V_3 are original views and V_2 is the virtual view, and L is the length of baseline. From Fig. 2 we can see that, for the virtual view V_2, the regions AB and EF in the left original view V_1 are redundant, as region AB is out of the visual range of virtual view, and region EF is occluded by foreground MN. So, we deem them as a non-significant region. Regions HI and JK in V_3 are significant regions since they cannot be obtained in V_1 but can be seen in V_2.

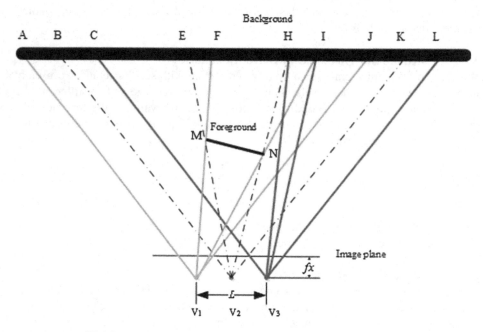

Fig. 2. Region-division approach of original left and right view

Next, we will illustrate how to get each region orderly, from left to right.

Let L_{AB} represent the length of region AB. Since the flat background plane is in parallel to the image plane, we can get,

$$L_{AB} = \frac{L}{2} \cdot f_x \cdot \left(\frac{d_A}{255} \cdot \left(\frac{1}{z_{near}} - \frac{1}{z_{far}} \right) + \frac{1}{z_{far}} \right) \tag{7}$$

where d_A represents the depth value of point A.

Region EF is also insignificant for virtual V_2 as it's occluded by foreground MN. Let L_{EF} represent the length of region EF. According to the similarity between triangle V_1MV_2 and triangle FME, we can get,

$$L_{EF} = \frac{L}{2} \cdot f_x \cdot \frac{d_M - d_F}{255} \cdot \left(\frac{1}{z_{near}} - \frac{1}{z_{far}} \right)$$

(8)

After these two regions AB and EF in V_1 are conformed, we can find out the other significant regions in V_1, meaning the region BE and region MJ.

For region HI, it is also a significant region, as V_1 is unable to see it, but in V_2 and V_3, it is visible. Therefore, this region must be found in V_3. To find this region, we first warp point I in V_1 to V_3, and then the length of HI in V_3 can be obtained,

$$L_{HI} = \frac{L}{2} \cdot f_x \cdot \frac{d_N - d_I}{255} \cdot \left(\frac{1}{z_{near}} - \frac{1}{z_{far}} \right)$$

(9)

And then the region JK is a significant region in V_3, as it's out of the visual range of V_1 but visible in both virtual view V_2 and the original V_3. The length of JK can be calculated,

$$L_{JK} = \frac{L}{2} \cdot f_x \cdot \left(\frac{d_J}{255} \cdot \left(\frac{1}{z_{near}} - \frac{1}{z_{far}} \right) + \frac{1}{z_{far}} \right)$$

(10)

After warping V_1 to V_3, the point J is the right edge of V_3'; therefore it can be located easily.

The other regions of V_3 are insignificant to the virtual view V_2, and they can be easily identified. Until now we have found out all the regions that the virtual view V_2 needs.

3.2   Rate Allocation for Different Regions with BCS_SPL

In this section, we will measure different regions using different CS-rates according to its contribution to the virtual view. To the significant regions, a higher CS-rate is allocated, while to the non-significant ones, a lower CS-rate is used. We assume that most regions of the left view are significant regions, since it is regarded as the dominant view, and only a part of regions in the right view are significant, according to our analysis in Sect. 3.1.

In our experiment, two CS-rates are used to measure the region-divided original views, as shown in Eq. (11). As BCS_SPL is applied on a block basis, we let the CS-rate be $R1$ when more than 50 % of the current block pixels are significant pixels; otherwise the rate is set to $R2$, as below:

$$subrate = \begin{cases} R1, \text{ to significant regions} \\ R2, \text{ to insignificant regions} \end{cases}$$

(11)

After allocating the different CS-rates, we can obtain an equivalent measure rate *subrate_equal* of each original frame, which can be obtained by Eq. (12):

$$subrate\_equal = \frac{1}{col * row} * \sum region\_size * subrate \qquad (12)$$

where *col* and *row* represent the current frame's width and height, respectively, *region_size* represents the size of each regions, and subrate represents the corresponding sub-sampling rate.

After getting all the CS-rates that are needed in CS measuring, we can implement CS measurement and the reconstruction process, and finally the virtual view synthesis is implemented, as shown in Fig. 1.

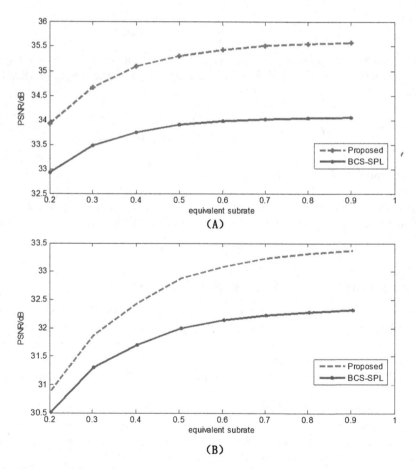

Fig. 3. The comparison PSNR of Y component for each sequence in our approach and BCS_SPL: (A) Kendo, (B) Book arrival

4   Experimental Results

In order to evaluate the performances of our method, we conduct experiments to show the subject and object performance comparisons between our method and the BCS_SPL method, which uses equivalent CS-rate for the two original views. For the two original videos of BCS-SPL methods, we use the equivalent subrate, which is the average of the two *subrate_equals* from the two region-divided original views. The block size in BCS_SPL is set to 16*16 in our experiment.

For each sequence, we test 10 frames and use eight groups of R1 and R2 CS-rates, as shown in Eq. (10). Figure 3 shows the comparison PSNR of Y component for each sequence. We can see that our proposed region-aware approach demonstrates a great

(A)   (B)

Fig. 4. The visual quality comparison of the virtual views in our approach and BCS-SPL: (A) Kendo, (B) Book arrival. The picture on the top is the original texture video, the middle one is the virtual view synthesized using undivided original views, and the picture at the base is our proposed approach.

performance improvement compared to the BCS-SPL method without region division, and the average PSNR increase of the Y component is nearly 1 dB for the test sequences.

The visual quality comparison for the virtual views is shown in Fig. 4. It is obvious that our proposed approach can efficiently improve the texture video with respect for both PSNR and the visual quality of synthesized views.

5 Conclusion

In this paper, we propose a region based coding approach of 3D video coding, which takes virtual view rendering and region division into consideration. Firstly, we implement the region-division into the original views, according to the estimated significance to the virtual view. Then, we perform an unequal measurement through block-based compressed sensing (BCS) for significant and non-significant regions. The significant regions are measured by a higher CS-rate and non-significant regions by a lower CS-rate. Experimental results show that our approach can achieve better synthesized quality under the same equivalent measurement rate. Our approach is appropriate when the virtual view is more crucial than the original views.

Acknowledgements. This work has been supported in part by National Natural Science Foundation of China (No. 61272262 and No. 61210006), International Cooperative Program of Shanxi Province (No. 2015031003-2), The Program of "One hundred Talented People" of Shanxi Province, Research Project Supported by Shanxi Scholarship Council of China (2014-056) and Program for New Century Excellent Talent in Universities (NCET-12-1037).

References

1. Tian, Dong, Lai, Po-Lin, Lopez, Patrick, Gomila, Cristina: View synthesis techniques for 3D video. Proc. SPIE Appl. Digital Image Process. XXXII **7443**, 74430T-1–74430-11 (2009)
2. Ndjiki-Nya, P., Köppel, M., Doshkov, D., Lakshman, H.: Depth image-based rendering with advanced texture synthesis for 3D video. IEEE Trans. Multimedia **13**(3), 453–465 (2011)
3. Nguyen, Q.H., Do, M.N., Patel, S.J.: Depth image-based rendering with low resolution depth. In: 2009 16th IEEE International Conference on Image Processing (ICIP), pp. 553–556, November 2009
4. Müller, K., Schwarz, H., Marpe, D., Bartnik, C.: 3D high-efficiency video coding for multi-view video and depth data. IEEE Trans. Image Process. **22**(9), 3366–3378 (2013)
5. Balota, G., Saldanha, M., Sanchez, G., Zatt, B., Porto, M., Agostini, L.: Overview and quality analysis in 3D-HEVC emergent video coding standard. In: 2014 IEEE 5th Latin American Symposium on Circuits and Systems (LASCAS), pp. 1–4 (2014)
6. Lee, C., Ho, Y-.S.: A framework of 3D video coding using view synthesis prediction. Picture Coding Symposium (PCS), pp. 9–12 (2012)
7. Duarte, M.F., Eldar, Y.C.: Structured compressed sensing: from theory to applications. IEEE Trans. Signal Process. **59**(9), 4053–4085 (2011)

8. Do, T.T., Gan, L., Nguyen, N., Tran, T.D.: Sparsity adaptive matching pursuit algorithm for practical compressed sensing. In: Proceedings of the 42th Asilomar Conference on Signals, Systems, and Computers, pp. 581–587, Pacific Grove, California, October 2008
9. Mun, S., Fowler, J.E.: Block compressed sensing of images using directional transforms. In: Proceedings of the International Conference on Image Processing, pp. 3021–3024 Cairo, Egypt, November 2009
10. Gan, L.: Block compressed sensing of natural images. In: Proceedings of the International Conference on Digital Signal Processing, pp. 403–406, Cardiff, UK, July 2007

A Paradigm for Dynamic Adaptive Streaming over HTTP for Multi-view Video

Jimin Xiao[1]([✉]), Miska M. Hannuksela[2], Tammam Tillo[1],
and Moncef Gabbouj[3]

[1] Department of Electrical and Electronic Engineering,
Xi'an Jiaotong-Liverpool University (XJTLU), 111 Ren Ai Road, SIP,
Suzhou 215123, Jiangsu Province, People's Republic of China
{jimin.xiao,tammam.tillo}@xjtlu.edu.cn
http://www.mmtlab.com
[2] Nokia Researcher Center, Tampere, Finland
miska.hannuksela@nokia.com
[3] Tampere University of Technology, Tampere, Finland
moncef.gabbouj@tut.fi

Abstract. HTTP-based delivery for Video on Demand (VoD) has been
gaining popularity within recent years. With the recently proposed
Dynamic Adaptive Streaming over HTTP (DASH), video clients may
dynamically adapt the requested video quality and bitrate to match their
current download rate. To avoid playback interruption, DASH clients
attempt to keep the buffer occupancy above a certain minimum level.
This mechanism works well for the single view video streaming. For
multi-view video streaming application over DASH, the user originates
view switching and that only one view of multi-view content is played
by a DASH client at a given time. For such applications, it is an open
problem how to exploit the buffered video data during the view switching
process. In this paper, we propose two fast and efficient view switching
approaches in the paradigm of DASH systems, which fully exploit the
already buffered video data. The advantages of the proposed approaches
are twofold. One is that the view switching delay will be short. The second advantage is that the rate-distortion performance during the view
switching period will be high, i.e., using less request data to achieve comparable video playback quality. The experimental results demonstrate the
effectiveness of the proposed method.

Keywords: DASH · Multi-view · HTTP · View switch · Inter-view
prediction

1 Introduction

Video streaming has become a key application for mobile devices, such as mobile
phones, tablets, etc. The video streaming traffic continues to grow rapidly. It

This work was supported by National Natural Science Foundation of China
(No.61210006, No.60972085).

© Springer International Publishing Switzerland 2015
Y.-S. Ho et al. (Eds.): PCM 2015, Part II, LNCS 9315, pp. 410–418, 2015.
DOI: 10.1007/978-3-319-24078-7_41

is expected that the consumer Internet video traffic will account for 69 % of the entire consumer Internet traffic by 2017 [1]. Meanwhile, HTTP-streaming has been gaining popularity in recent years. Instead of relying on RTP/UDP for multimedia communications due to its low end-to-end delay, many content providers have resorted to using HTTP/TCP transport for media delivery when the delay constraints allow it. The success of HTTP-based video streaming is attributed to many reasons. First, HTTP/TCP is not affected by firewall/NAT traversal issues that exist in traditional RTP/UDP-based streaming systems. Second, TCP congestion control mechanisms and reliability requirement do not necessarily hurt the performance of video streaming, especially if the video player is able to adapt to large throughput variations.

Dynamic Adaptive Streaming over HTTP (DASH) [2], refers to the video streaming methodology that the video client can dynamically adjust the quality (rate) of the requested video data based on the observed available download rate throughout the streaming process. At the DASH server, stored video segments are encoded with a various versions of video quality (rate), and different versions of segments are aligned for switching. All DASH clients need to do is choosing a most appropriate segment version whose rate matches the DASH client's available rate on a moment-by-moment basis. The chosen segment version changes with the variation of the network condition. One benefit of such approach is that the client can control its playback buffer size by dynamically adjusting the rate at which new fragments are requested.

How to adjust the request bitrate based on the observed network condition, which is the rate adaptation method, is important for DASH users' Quality of Experience (QoE). On one hand, if the request bitrate is too low, the displayed video quality would be low accordingly. On the other hand, if the request bitrate is too high, buffer starvation is inevitable, which is in fact one of the most annoying degradations of the users QoE [3]. In the past five years, a lot of research works on DASH rate adaptation algorithm have been carried out. Adaptation methods can be roughly divided into throughput-based group and buffer based group [4]. The throughput-based methods decide the rate based on the estimated throughput only, such as [5] and [6]. The buffer-based methods decide the rate based on the buffer characteristics as well as the throughput, examples of such systems include [7] and [8]. In [4], a comprehensive evaluation of various rate adaptation methods is conducted.

In the past decade, multi-view video has attracted plenty of research interests both from industry and academia [9]. However, all the existing DASH researches are tailored for the conventional 2D video. In this paper, we are going to study delivering multi-view video over DASH systems. More specifically, our focus will be on how to enable seamless interactive view switching at the DASH client side. It is known that enabling users to interactively navigate through different viewpoints is an important functionality in multi-view streaming systems. In [10], coarse and fine quality layers of several views are grouped and pre-encoded. During actual streaming, two views of high quality plus a subset of views of low quality, selected based on user behavioral prediction, are sent to the client to enable interactive view switching. [11] and our previous paper [12] present solutions based on redundant coded pictures. The contribution of this paper is manyfold. First, to the best of our knowledge, it is the first work to tackle the problem of user-driven view switching of multi-view content over DASH systems. Second, we propose two different solutions which enable timely and efficient view switching at the DASH client side, among which one has high rate-distortion performance while the storage requirement for the DASH server is tremendous, the other one has an intermediate storage requirement with slightly lower rate-distortion performance. This allows the system designer to choose the suitable solution based on the storage and bandwidth resources.

The rest of the paper is organized as follows. In Sect. 2 the proposed DASH-based multi-view video streaming approach is presented in details. In Sect. 3 some experimental results validating the proposed approach are given. Finally, some conclusions are drawn in Sect. 4.

2 DASH-based Multi-view Video Streaming Approach

2.1 Problems

Before describing the proposed DASH-based multi-view video streaming approach, let us first check the problems existing in conventional DASH-based multi-view video streaming systems. Figure 1 describes an illustrative example of such system, where two different viewpoints of the video are stored in the DASH server. Each viewpoint video is segmented into segments (for example 2 − 10 seconds). Each segment is encoded with different quality (rate). The DASH client sends the video request based on its network conditions, and buffers the received video data, which is the same as conventional 2D DASH systems. Problem arises when the

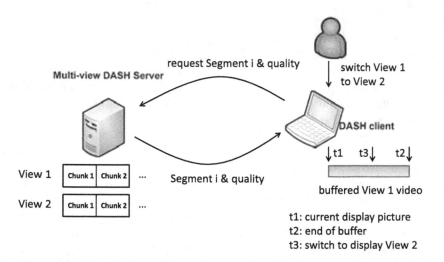

Fig. 1. DASH-based Multi-view Video Streaming; two viewpoints of the same video are stored in the server; the user request to switch from View 1 to View 2.

DASH client user requests to switch from the current viewpoint (View 1) to another viewpoint (View 2), also referred to as the target view. Let us assume the time the user initiates the view switching is t_1, and the end of the buffered video data at time t_1 is t_2. The DASH client estimates that it will be able to receive the data for View 2 starting from media time t_3 and hence start to display View 2 instead of View 1 starting from t_3, $t_1 \leq t_3 \leq t_2$. During t_1 to t_3, the client buffers video data of View 2 which starts from t_3. For the extreme case $t_3 = t_1$, the user wants to switch immediately when the decision is made so as not perceive any delay. The problem of such case is that there is no buffered data for View 2, so there will be a buffer starvation problem. For the case $t_3 = t_2$, the DASH client have $t_2 - t_1$ duration to buffer video data for View 2, but user need to wait $(t_2 - t_1)$ to switch to View 2, which causes user impatience since the length of buffered video data $(t_2 - t_1)$ could be quite long (for example, 60 seconds). A more reasonable choice would be $t_1 < t_3 < t_2$. However, for any chosen value of t_3, the buffer video date from t_3 to t_2 will be wasted.

2.2 Solutions

Facing the problems described in Sect. 2.1, we propose to encode video segments using inter-view prediction as well as using simulcast encoding. Let us take the example described in Fig. 1. Besides the simulcast encoded version of View 1 and View 2, View 1 (View 2) should be encoded based on View 2 (View 1) using inter-view prediction. The encoded streams using inter-view prediction are stored in the DASH server as well as the simulcast ones. Thus, in the example described in Fig. 1, from t_3 to t_2 the dash client can request View 2 stream which is encoded using inter-view prediction from View 1. This is because from t_3 to t_2 the View 1 simulcast encoded data is already buffered, it allows the client to decode the View 2 video by downloading the View 2 stream which is generated using inter-view prediction from View 1. In fact, to achieve the same playback video quality for time period t_3-t_2, the resulted bitstream rate using the proposed inter-view prediction scheme is much lower than that of using simulcast, this is because the rate-distortion performance is improved by using inter-view prediction coding. Finally, from t_2, the DASH client start to request simulcast encoded View 2.

There is one important fact of the proposed approach should be considered. The segments of the buffered View 1 data might be with different bitrate and quality due to the nature of the DASH system. Figure 2 describes a case, where each segment of the buffered video data is encoded with different quantization parameter (QP) values (QP_1, QP_2, QP_3, ...). For example, the QP of Segment 1 in the buffered video of View 1 is QP_1. In this case, the DASH client need a View 2 segment, which is predicted from View 1 with this specific QP (QP_1) to avoid mismatch errors. Bearing in mind that the DASH server does not know what quality version (QP) will be requested by clients, so different versions of inter-view prediction based segments should be stored in the server. For example, if there are N quality versions for the simulcast video, the inter-view predicted segments should be encoded and stored based on all the N versions. Meanwhile, to have different rate for the inter-view predicted segments, they should

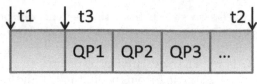

Fig. 2. Different segments of the buffer View 1 data have different encoding quality (QPs).

be encoded with different QPs. Hence, the inter-view predicted segments could be denoted as $C_{i \to j}^{QP_i, QP_j}$, where $i \to j$ denotes that the current encoding View j is predicted from View i, the QP of the view i (j) is QP_i (QP_j). Let us assume N quality versions are available in the DASH server, then N^2 versions are required for segments using inter-view prediction. Since the DASH server cannot predict when viewpoint switching happens, the N^2 versions are needed for the whole video sequence. This requires quite large storage for the DASH server.

We also propose one solution to lower the storage requirement on the DASH server. For the inter-view predicted segments $C_{i \to j}^{QP_i, QP_j}$, instead of encoding view j using all versions of view i with different QPs, view j is encoded using uncompressed view i data. Here we used $C_{i \to j}^{0, QP_j}$ to describe such a segment, where $QP_i = 0$ refers to the uncompressed version of View i. Note that at the DASH client side, the buffered View i data that are used as reference to decode View j is compressed. Therefore, for View j there is mismatch error at the decoder side. Nevertheless, inter-view predicted segments are usually encoded with lower quality than the simulcast ones to avoid buffer starvation problem. Thus, the mismatch error is much smaller compared with the encoding error. Therefore, the rate-distortion performance of $C_{i \to j}^{0, QP_j}$ is much higher than that of the simulcast one. In the following part of the paper, we will call the proposed solution without decoder side mismatch error as MV-DASH, and the proposed solution with decoder side mismatch error as MV-DASH(M). For MV-DASH(M), if N quality versions are required in the DASH server, only N versions are required for segments using inter-view prediction, which is much less storage than the MV-DASH case (N^2).

3   Experimental Results

In the experiment, 4 typical 3D video sequences are used: *Newspaper* [13], *Lovebird1* [13], *Balloons* [14], *Kendo* [14]. We assume for *Newspaper*, the user requests to switch from viewpoint 4 to viewpoint 2; for *Balloons*, the user requests to switch from viewpoint 1 to viewpoint 3; for *Lovebird1*, the user requests to switch from viewpoint 6 to viewpoint 7; for *Kendo*, the user requests to switch from viewpoint 1 to viewpoint 2. Therefore, the test cases cover both narrow

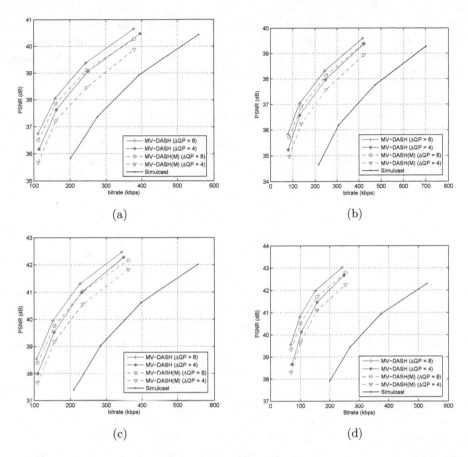

Fig. 3. Rate-distortion performance comparison for Simulcast, MV-DASH, and MV-DASH(M); QP value is set to be 4 and 8 larger than reference views when inter-view prediction is used; (a) *Newspaper* sequence, (b) *Lovebird1* sequence, (c) *Balloons* sequence, (d) *Kendo* sequence.

and wide baselines. For *Newspaper* and *Balloons*, the baseline of view switching is 10 cm, whereas for *Lovebird1* and *Kendo*, it is 5 cm. We compare the rate-distortion performance of the target view using the simulcast and the proposed inter-view prediction methods, including both MV-DASH and MV-DASH(M). The first 100 frames of the video sequences are used for this test. HEVC reference software HM12.1 and HEVC multi-view extension software HTM9.3 are respectively used to simulate the simulcast and inter-view prediction based coding methods. Intra period is set to be 32, the default GOP structure with QP cascading is used [15]. For encoding inter-view predicted views, the QP value is set to be 4 and 8 larger than reference views, the QPs of simulcast coding are $37, 34, 31, 28$. This is because during the period of view switching, there is not much time (which is $t_3 - t_1$ in Sect. 2.2) and not enough network throughput to buffer the target view data so that a sufficient buffer occupancy level is reached. To avoid buffer starvation and re-buffering, a lower quality version are preferred. It should be noted that for the MV-DASH(M) approach, in the encoding stage the reference view is uncompressed, the QP gap is measured at the decoder side.

Table 1. BD-Rate by comparing the different schemes with the simulcast.

Sequence	MV-DASH $(\Delta QP = 8)$	MV-DASH $(\Delta QP = 4)$	MV-DASH(M) $(\Delta QP = 8)$	MV-DASH(M) $(\Delta QP = 4)$
Newspaper	−51.94 %	−45.36 %	−49.37 %	−27.28 %
Lovebird1	−69.72 %	−65.16 %	−68.25 %	−49.74 %
Balloons	−56.93 %	−50.71 %	−55.10 %	−45.34 %
Kendo	−74.03 %	−66.84 %	−72.45 %	−66.69 %
Average	−63.15 %	−57.01 %	−61.29 %	−47.26 %

The rate-distortion performance curves are plotted in Fig. 3 for the target views. Meanwhile, the BD-Rate [16] results by comparing the proposed schemes with the simulcast for the target views are listed in Table 1. In Fig. 3, by looking at the blue curves or the red curves separately, it is concluded that larger QP gap leads to better target view rate-distortion performance for both of the MV-DASH and MV-DASH(M) approaches. For the MV-DASH approach, the average BD-Rate compared with simulcast are −0.63.15 % and −57.01 % for a QP gap of 8 and 4, respectively. Similar results are obtained for the MV-DASH(M) approach. This is reasonable, because large QP gap means a high quality reference view is available for either encoding (MV-DASH) or decoding (MV-DASH(M)) process. On the other hand, by comparing the red curves with the blues ones, it is noted that the rate-distortion performance of MV-DASH(M) is slightly worse than that of MV-DASH. Nevertheless, it is still much better than that of the simulcast. For example, for the case of QP gap 8, the DB-rate of MV-DASH(M) is −61.29 %, which is slightly lower than that of MV-DASH (−63.15 %). Hence, the choice between MV-DASH and MV-DASH(M) is left for the system designer based on the server storage conditions.

The claimed advantage that the view switching delay is short for the proposed approaches is explained as follows. Let us assume the DASH client needs to pre-buffer t_0 length of the target view before starting to display this viewpoint, and we assume this view switching point is t_3. To have the same video quality, the video bitrate using the simulcast is V_0, the video bitrate using one of the proposed approaches is V_1, $V_1 = (1 - \mu)V_0$, here μ is the BD-rate saving. We assume network download bandwidth is B. To pre-buffer t_0 length of video, the simulcast solution takes a time period of t_s, whereas the proposed approach takes a time period of t_p. Hence, we have $t_s B = t_0 V_0$, and $t_p B = t_0 V_1$. Thus, $t_p = (1 - \mu)t_s$, this means that using the proposed approach the view switching delay will be only $1 - \mu$ of the simulcast solution.

4   Conclusions

In this paper, two multi-view video streaming approaches over DASH systems have been proposed. In the proposed solutions, by exploiting the buffer video data, the view switching process is much faster and more efficient than that of conventional methods. Two solutions with different rate-distortion performance and DASH server storage requirement are proposed. Therefore, the system designer can choose appropriate solution by trade off the rate-distortion performance and the DASH server storage requirement. Experimental results demonstrate the effectiveness of the proposed solutions.

References

1. Cisco Systems Inc., Cisco visual networking index: forecast and methodology, 2012–2017. Cisco White Paper, May 2013
2. ISO/IEC JTC 1/SC 29/WG 11 (MPEG), Dynamic adaptive streaming over HTTP, w11578, CD 23001–6, Guangzhou, China, October 2010
3. Dobrian, F., Sekar, V., Awan, A., Stoica, I., Joseph, D.A., Ganjam, A., Zhan, J., Zhang, H.: Understanding the impact of video quality on user engagement. SIGCOMM-Comput. Commun. Rev. **41**(4), 362 (2011)
4. Thang, T.C., Le, H., Pham, A., Ro, Y.M.: An evaluation of bitrate adaptation methods for HTTP live streaming. IEEE J. Select. Areas Commun. **32**(4), 693–705 (2014)
5. Thang, T.C., Ho, Q.-D., Kang, J.-W., Pham, A.T.: Adaptive streaming of audio-visual content using MPEG DASH. IEEE Trans. Consum. Electron. **58**(1), 78–85 (2012)
6. Liu, C., Bouazizi, I., Gabbouj, M.: Rate adaptation for adaptive HTTP streaming. In: Proceedings of the Second Annual ACM Conference on Multimedia Systems (ACM MMSys2011), California, February 2011
7. Muller, C., Lederer, S., Timmerer, C.: An evaluation of dynamic adaptive streaming over http in vehicular environments. In: Proceedings of the 4th Workshop on Mobile Video (MoVid), pp. 37–42, NC, USA, February 2012
8. Miller, K., Quacchio, E., Gennari, G., Wolisz, A.: Adaptation algorithm for adaptive streaming over http. In: Proceedings of the Packet Video Workshop (PV 2012), pp. 173–178, May 2012
9. Chen, Y., Wang, Y.-K., Ugur, K., Hannuksela, M.M., Lainema, J., Gabbouj, M.: The emerging MVC standard for 3D video services. EURASIP J. Adv. Signal Process. **2009**(1), 13 (2009)
10. Kurutepe, E., Civanlar, M.R., Tekalp, A.M.: Client-driven selective streaming of multiview video for interactive 3DTV. IEEE Trans. Circuits Syst. Video Technol. **17**(11), 1558–1565 (2007)
11. Cheung, G., Ortega, A., Cheung, N.: Interactive streaming of stored multiview video using redundant frame structures. IEEE Trans. Image Process. **3**(3), 744–761 (2011)
12. Zhu, L., Hannuksela, M.M., Li, H.: Inter-view-predicted redundant pictures for viewpoint switching in multiview video streaming. In: Proceedings of IEEE International Conference on Acoustics, Speech and Signal Processing (ICASSP), March 2010

13. Electronics and Telecommunications Research Institute and Gwangju Institute of Science and Technology, (2008, April), 3DV Sequences of ETRI and GIST. ftp:// 203.253.128.142/
14. Nagoya University, (2008, Mar.), 3DV Sequences of Nagoya University. http:// www.tanimoto.nuee.nagoya-u.ac.jp/mpeg/mpeg-ftv.html
15. Rusanovskyy, D., Muller, K., Vetro, A.: Common test conditions of 3DV core experiments. ISO/IEC JTC1/SC29 /WG11, M26349, July 2012
16. Bjontegaard, G.: Improvements of the BD-PSNR model. ITU-T SG16 Q.6 Document (2008)

Adaptive Model for Background Extraction Using Depth Map

Boyuan Sun[1,2]($\boxtimes$), Tammam Tillo[1], and Ming Xu[1]

[1] Department of Electrical and Electronic Engineering,
Xian Jiaotong-Liverpool University (XJTLU), 111 Ren Ai Road, SIP,
Suzhou 215123, Jiangsu, People's Republic of China
{boyuan.sun01,tammam.tillo,ming.xu}@xjtlu.edu.cn
http://www.mmtlab.com
[2] Department of Electrical Engineering and Electronics,
University of Liverpool, Liverpool, UK

Abstract. Depth map has attracted great attention for image and video processing in recent years. Depth map gives one more dimensional information about the images besides color (intensity). Depth is independent of color, which is the advantage for extracting the background covered by objects with irregular repetitive motions e.g. rotation. A new algorithm for background extraction using Gaussian Mixture Models (GMM) combined with depth map is presented. The per-pixel mixture model and single Gaussian model are used to model the recent observation in color and depth space respectively. We also incorporate the color-depth consistency check mechanism into the algorithm to improve the accuracy. Our results show much greater robustness than prior state of the art method to handle challenging scenes.

Keywords: Background extraction · Gaussian mixture model · Depth map

1 Introduction

Plenty of computer vision and image and video processing systems like visual surveillance, pedestrian detection and tracking, and 3D video representations, rely on the very fundamental procedure called background subtraction, that classifies the scene into foreground and background regions. These systems' overall performance usually depends on the accuracy and robustness of their background subtraction algorithms. Several algorithms for performing background subtraction have been proposed to effectively generate the background model from temporal sequence of frames.

One simple background model used running Gaussian average approach [1] to model the background independently at each (i, j) pixel location. The model

B. Sun—This work was supported by National Natural Science Foundation of China (No. 61210006, No. 60972085).

Y.-S. Ho et al. (Eds.): PCM 2015, Part II, LNCS 9315, pp. 419–427, 2015.
DOI: 10.1007/978-3-319-24078-7_42

is based on fitting a Gaussian probability density function on the previous n pixel's values. However, this model would fail when the scene contain background motion, such as moving leaf or ripples in the water. Various researchers have proposed other temporal average filters that have a better performance than running Gaussian average. In [3], the median value of recent n frames is regarded as the background model. The main disadvantage of this kind of approach is that a buffer of recent observations is required for the computation. Moreover, the median filter does not have an accurate statistical analysis for the scene. Non stationary backgrounds have been modeled by GMM algorithms [2,4,6]. The Gaussian model's parameters are updated at each new coming frame to identify the changes in the scene. The drawback of this model is the adaptation speed. If the adaptation of the Gaussian model's parameters is fast, the foreground objects with slow movement will be classified as part of the backgrounds. But if the adaptation speed is slow, the model would fail to identify some fast changes of the background such as sudden illumination changes (switch on/off light). Recently, depth information has been widely used in video processing algroithm where the depth map measures the distance between the objects and the camera. Depth data is considered as fourth channel in GMM besides three channels of the color space such as RGB or YUV in [7,8]. This approach gives less strict match condition for the depth data than texture data. But it did not fully exploit the depth information and results can be further improved by processing depth map indenpendent of color data. In [5], the depth information is exploited to identify the background regions which are covered by the object with repetitive motion that the GMM fails to recover. The problem with this approach is that the GMM algorithm and depth map processing algorithm are two independent procedures, which may absorb part of foreground object into background especially at the transition positions between foreground and background regions.

This paper describes an algorithm that fully utilize the depth data combined with GMM algorithm to estimate the backgrounds. It applies the adapting GMM with modified update mechanism and single Gaussian model (SGM) to the expected background appearance and depth values respectively. The result greatly outperforms the original GMM algorithms.

2   Method

2.1   Background Model from GMM

The Gaussian Mixture Model has been widely applied to model the stable background and detect the moving objects. GMM is pixel based algorithm, where each pixel is modeled as a mixture of K Gaussian distributions (K is usually from 3 to 5) [2]. The probability of observing the current pixel value is

$$p(x_t) = \sum_{i=1}^{K} \omega_{i,t} * \eta(x_t, \mu_{i,t}, \sigma_{i,t}^2) \qquad (1)$$

where K is the number of distributions, $\omega_{i,t}$, $\mu_{i,t}$ and $\sigma^2_{i,t}$ are the estimate of the weight, mean value and variance of ith Gaussian distribution respectively at time t. And η is the Gaussian probability density function.

Every new pixel value x_t is checked against the existing K Gaussian distributions, until a match is found. The matched new pixel value is the pixel value that is within 2.5 standard deviation of a distribution [2]. And the parameters will be updated as following [4]

$$\omega_{i,t} = (1 - \alpha)\omega_{i,t-1} + \alpha \tag{2}$$

$$\mu_{i,t} = (1 - \rho)\mu_{i,t-1} + \rho x_t \tag{3}$$

$$\sigma^2_{i,t} = (1 - \rho)\sigma^2_{i,t-1} + \rho(x_t - \mu_{i,t})^2 \tag{4}$$

where α is the learning rate and ρ is

$$\rho = \alpha * \eta(x_t, \mu_{i,t}, \sigma^2_{i,t}) \tag{5}$$

If there is no match to K distributions for current pixel value, the least probable distribution will be replaced with a new distribution that has the current pixel value as its mean, an initially high variance, and low prior weight. The least probable distribution is defined as the distribution with the smallest ratio of ω/σ.

2.2   The Exploit-Ability of Depth Map

Based on the nature of the GMM, the background pixels are the pixels with temporal stable intensity. However, if the foreground object has some certain motions (e.g. rotation) or the moving object that becomes the dominant part of the scene, which means the background is occluded by the foreground object in most frames, the GMM will erroneously classify this foreground object as part of the background. In this scenario, the GMM will be unable to recover the

Fig. 1. The background reference obtained using GMM after 100 frames for Ballet sequence .

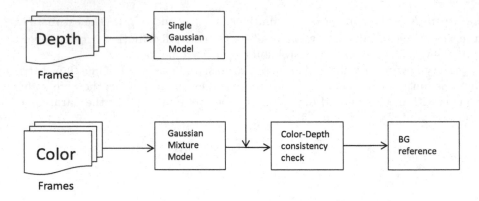

Fig. 2. The framework of proposed algorithm (BG stands for background)

occluded background information. A typical example is shown Fig. 1, where the main body of the dancer is classified as background. The GMM can not model a satisfactory background reference in this case.

In this paper, we propose to solve this kind of problem by exploiting the depth map information as shown in Fig. 2. The depth map measures the distance between the objects and the camera. The regions with larger depth value are far away from the camera and the regions with small depth value are close to the camera. Hence, it is reasonable to assume the regions with large depth value have the high probability to be the background. Base on this assumption, by investigating the depth value of each pixel, the far field regions even if they appear only in a small fraction of the video can be extracted and modeled as background. As illustrated in Fig. 3, it is clear to see that the region (marked in red circle) failed to be recovered by GMM whereas it is distinguishable by the depth map. The associated depth indicates these regions are far away from the camera, which means these regions have high probability to be classified as background.

Since the depth map is less complicated than texture information, we decide to apply single Gaussian model (SGM), which is faster and requires less computations than GMM, to the depth map. The current depth map will be checked against the distribution of the depth model. The matched pixels will be updated using GMM because these pixels have consistent depth. There are two case scenarios for the unmatched pixels as shown in Fig. 4. One is the pixels with smaller depth value, which can be coarsely regarded as foreground and no update for the background model required. Another is the pixels with larger depth value, which has high possibility that the occluded background is revealed.

2.3   Depth-Color Consistency Check

The depth based classification is coarse due to the accuracy limitation of the depth map. In order to refine the classification, the depth-color consistency check

Fig. 3. The color-depth frame comparison: the first row is background reference obtained from GMM, the second and third rows are color and depth image for frame 16, 52 and 71 respectively (color figure online)

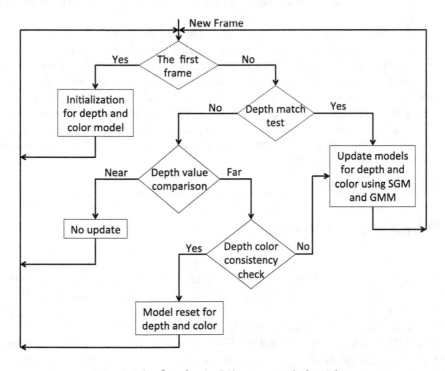

Fig. 4. The flowchart of the proposed algorithm

mechanism is developed, especially for the pixels that could be the revealed as background. These new candidate pixels' value will be checked against the most probable Gaussian distribution of the color. As shown in Fig. 4, if no matched distribution is found, the candidate pixels' values can be regarded as part of the background and the background model needs to reset using the candidates' color and depth information. Otherwise, there are two possibilities. The first is the imperfect depth map that indicate the major changes in depth but actually no such changes and this is related to depth estimation algorithm. The second possibility is the observed background regions share similar color with the foreground objects. Either of these two possibilities will not affect the background model due to the similar color information. These pixels will also be updated using GMM. The detailed description of proposed algorithm is shown in Algorithm 1.

Algorithm 1. Gaussian Mixture Model using Depth Map

1: The empty set of models for color is initialized at the time t_0.
   - Assigning the pixel value of current frame to the mean value μ_{i,t_0} of the first Gaussian model and the rest is set to 0.
   - The variance σ^2_{i,t_0} of all Gaussian model is set to predefined large value, e.g., 900 in this paper.
   - The weight of first Gaussian model is set to 1, and the others are set to 0.
   - The model of depth map is set according to single Gaussian model [1].
2: Compare the current depth observations with the existed model.
3: For the pixels that match to the existing model, the pixels value will be updated by using GMM. The pixels with unmatched small depth value will not be updated and their corresponding color mixture model remains same. The rest pixels will go through the depth color consistency check mechanism.
4: The mixture model of the pixels with consistent depth-color change will be reset using current pixel's depth and color information. The other pixels' model will be updated by GMM.

---

3   Experimental Results

In this section, we tested the proposed depth assisted GMM algorithm on three video sequences. The test sequences include: Microsoft data set *Ballet* (1024 × 768, 100 frames), *Break-dancer* (1024 × 768, 100 frames) and the MPEG-3DTV test sequence *Arrive-book* (1024 × 768 , 50 frames). And the depth maps of the test sequences are obtained from the MPEG depth estimation reference software (DERS) based on graph cuts [9]. The results are compared with the manually marked ground truth which is available in www.mmtlab.com/download.

The results shown in Fig. 5 for proposed method are compared with previous GMM method for whole sequence from test videos. In the *ballet* video, the female

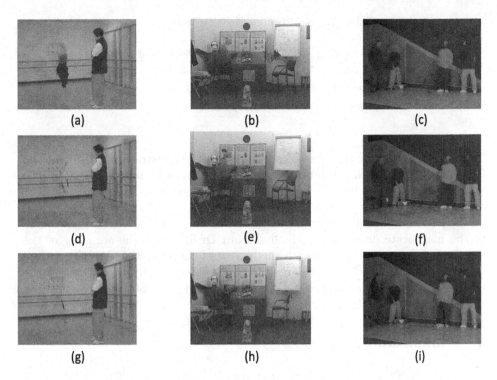

Fig. 5. The experimental results for GMM, Yao's and the proposed method for 3 test sequences

dancer is continuously rotating with small translation on the floor. The male is almost static and he only moved his body very slightly. It is obvious to see the GMM method failed to recover the background that is occluded by the dancing women. For Yao's approach [5], there are some artifacts that can be seen at the edge of the female dancer where the proposed approach is able to recover most occluded background region even for the small part covered by the hand of male dancer shown in Fig. 5(g). For the *book* and the *break-dancer* video, since there is no large background region constantly occluded by the foreground objects, the proposed method still has slightly gain in subjective view. In addition, the moving objects in the scene are quite close to the static background, which makes the modeling process more challenging. As shown in Fig. 5(h) and (i), the proposed method recovers the small part of background occluded by the sitting man and the most part of background covered by dancing man in the front of the scene in *arrive-book* and *break-dancer* video respectively. The noise which appears in Fig. 5(i), e.g., the small black blocks on the ground, is caused by the imperfect depth map information. For the objective assessment of the proposed method, since we are more interested in background similarity between true background and estimated background, PSNR is chosen instead of ROC curve which is more suitable for the measurement of the binary classification system.

Table 1. PSNR results for tested sequences

Method	Ballet	Book	Breakdancer
GMM [4]	19.95 dB	25.18 dB	23.36 dB
Yao's [5]	21.24 dB	26.96 dB	21.66 dB
Proposed	21.53 dB	27.42 dB	23.18 dB

The PSNR results for three tested sequences are demonstrated in Table 1. The PSNR results show that the proposed method is significantly better than the compared alogrithms for ballet and book video sequences. As for break-dancer sequence, the proposed method has gained about 1.5 dB in PSNR over Yao's method althoutgh both of them are less than the GMM method. This is due to the inaccurate depth map estimated from DERS,where the accuracy of the depth map is decreased along with the increased degree of the object's motion. In *break-dancer* sequence, the dancer made huge movement which caused the major changes in depth map especially when dancer's leg moved to the position that is very close to the ground.

4 Conclusion

In this paper, we proposed a depth map assisted Gaussian Mixture Model approach to handle the foreground object with irregular repetitive motions in the video. Through the experimental evaluation, we have shown this approach has a much better performance for objects with irregular repetitive motion in the popular test sequences. For the future works, we will exploit the spatial correlation between the pixels for both depth and color data and then combine all the information to improve the background subtraction algorithm. In addition, the motion information of the texture will be investigated to enhance the performance.

References

1. Wren, C.R., Azarbayejani, A., Darrell, T., Pentland, A.P.: Pfinder: real-time tracking of the human body. IEEE Trans. Pattern Anal. Mach. Intell. **19**(7), 780–785 (1997)
2. Stauffer, C., Grimson, W.E.L.: Adaptive background mixture models for real-timetracking. In: IEEE Computer Society Conference on Computer Vision and Pattern Recognition, vol. 2. IEEE (1999)
3. Lo, B.P.L., Velastin, S.A.: Automatic congestion detection system for underground platforms. In: 2001 Proceedings of 2001 International Symposium on Intelligent Multimedia, Video and Speech Processing, pp. 158–161. IEEE (2001)
4. KaewTraKulPong, P., Bowden, R.: An improved adaptive background mixture model for real-time tracking with shadow detection. In: Remagnino, P., Jones, G.A., Paragios, N., Regazzoni, C.S. (eds.) Video-based surveillance systems, pp. 135–144. Springer, New York (2002)

5. Yao, C., Tillo, T., Zhao, Y., Xiao, J., Bai, H., Lin, C.: Depth map driven hole filling algorithm exploiting temporal correlation information. IEEE Trans. Broadcast. **60**(2), 394–404 (2014)
6. Lee, D.S.: Effective Gaussian mixture learning for video background subtraction. IEEE Trans. Pattern Anal. Mach. Intell. **27**(5), 827–832 (2005)
7. Gordon, G., Darrell, T., Harville, M., Woodfill, J.: Background estimation and removal based on range and color. In: 1999 IEEE Computer Society Conference on Computer Vision and Pattern Recognition, vol. 2. IEEE (1999)
8. Harville, M., Gordon, G., Woodfill, J.: Adaptive video background modeling using color and depth. In: Proceedings of 2001 International Conference on Image Processing, vol. 3, pp. 90–93. IEEE (2001)
9. Tanimoto, M., Fujii, T., Suzuki, K.: View synthesis algorithm in view synthesis reference software 2.0 (vsrs2.0)). ISO/IEC JTCI/SC29/WG11M, 16090 (2009)

An Efficient Partition Scheme for Depth-Based Block Partitioning in 3D-HEVC

Yuhua Zhang[1], Ce Zhu[1(✉)], Yongbing Lin[2], Jianhua Zheng[2], and Yong Wang[1]

[1] University of Electronic Science and Technology of China,
Chengdu, Sichuan, China
yuhua_zhang@126.com, eczhu@uestc.edu.cn,
wangyong_uestc_ee@163.com
[2] HiSilicon Technologies Co. Ltd, Beijing, China
{linyongbing, zhengjianhua}@hisilicon.com

Abstract. In the development of a 3D video extension of High Efficiency Video Coding (HEVC) standard, namely 3D-HEVC, Depth-based Block Partitioning (DBBP) is employed to code texture videos in dependent views by utilizing coded depth information of an independent view. With the DBBP, a proper partition mode is determined with the coded depth information, which divides the current texture block into two regions and thereafter allows for fine-grained motion compensation of foreground and background separately. In the DBBP, the original partition method consists of two steps specifically, i.e. threshold calculation and matched filtering based on down-sampling, which is relatively high complex and redundant with a segment mask generation process. Accordingly, a simple yet more effective partition scheme for the DBBP coding is proposed in this paper, based on the available binary segment mask. While reducing computational complexity significantly, the proposed method also demonstrates bitrate saving for all the dependent texture views and synthesized views under common test conditions (CTC) configuration specified in the 3D-HEVC.

Keywords: 3D-HEVC · Depth-based block partitioning · Partition mode · Binary segment mask · Simple yet more effective

1 Introduction

Recently, 3D video has attracted substantial interests from both industry and academia. Multi-view video plus depth (MVD), as a most promising representation format for 3D video, has been investigated extensively [1, 2]. MVD is generally composed of multiple texture views together with their associated depth maps. With the transmitted texture and its corresponding depth information, virtual views may be generated at the decoder using a well-known Depth Image Based Rendering (DIBR) technique [2, 3], to present audiences a great flexibility in choosing different views with varying depth sensations. To investigate efficient representation and coding methods for MVD, a Joint Collaborative Team on 3D Video Coding Extension Development (JCT-3 V) was

© Springer International Publishing Switzerland 2015
Y.-S. Ho et al. (Eds.): PCM 2015, Part II, LNCS 9315, pp. 428–436, 2015.
DOI: 10.1007/978-3-319-24078-7_43

established by ISO/IEC MPEG and ITU-T VCEG. Based on the newly released 2D video standard - High Efficiency Video Coding (HEVC), JCT-3 V is dedicated to develop a new 3D standard for MVD format, namely 3D-HEVC [4, 5].

Different from the texture video, depth maps usually appear larger areas of nearly constant or only slowly varying values within an object, with sharp edges at the foreground/background boundaries. However, there are always motion and structure correlation between texture and depth videos. Therefore, combined coding of texture and depth videos to exploit inter-component dependencies, can further enhance overall coding performance [6–8].

In general, utilizing depth information for the texture component coding to reduce the texture bitrate appears to be more appropriate. This is based on the fact that the texture bitrate is significantly higher than that of the depth component, as the texture map contains richer and more complex content than the depth map. In the 3D-HEVC, the depth information of independent view (base view) is coded before the texture component of dependent views, which makes it possible to code texture component of dependent views by utilizing the coded depth data. Consequently a novel inter prediction mode for texture video coding was introduced in [9], namely Depth-based Block Partitioning (DBBP), which has been shown to increase coding efficiency for both the texture component and synthesized views.

In the DBBP, the decoder involves five steps: virtual depth identification, segment mask generation, partition determination, bi-segment compensation, and prediction signals merging. First, a so-called virtual depth block of the current coding texture block is addressed from the coded depth picture in the base depth view, utilizing the depth oriented neighboring block based disparity vector (DoNBDV). Second, a binary segmentation mask is obtained from the identified (virtual) depth block in the reconstructed base view, which divides the current texture block into two segments. Third, a block partition mode for the collocated texture block is selected from either $2N \times N$ or $N \times 2N$ in the partition process, also based on the identified depth map. Following that, each of the two partitions is motion compensated and finally recombined based on the segmentation mask. Therein, the original partition determination process consists of two concrete steps, i.e. threshold calculation and matched filtering based on subsamples of the identified depth block in the reconstructed base view. Due to the mean (threshold) calculation and matched filtering which requires certain memory access and computations, the partition method is of relatively high complexity.

To reduce the complexity of the partition determination step, a partition method using three corner subsamples in the identified depth block is presented in [10], at a cost of coding loss. To our observation, both the segment mask generation in the second step and the partition determination in the third step of the DBBP need to perform sampled mean calculation as well as the subsequent binarization for mask generation and partition determination, respectively. In spite of using different sub-sampling rate, there is strong operational redundancy in the two steps. Based on the observation, a simple yet more effective partition scheme is proposed in this paper, just based on the binary segmentation mask obtained in segment mask generation process, by skipping the threshold calculation and matched filtering.

The rest of this paper is organized as follows. In Sect. 2, the detailed DBBP coding in the current 3D-HEVC, particularly the partition determination step, is presented.

Section 3 introduces our proposed efficient partition scheme. Experiment results under the Common Test Conditions (CTC) [11] are shown in Sect. 4. Finally, Sect. 5 concludes the paper.

2 DBBP Coding in 3D-HEVC

In the DBBP coding [12], an appropriate partition mode and a binary segmentation mask are computed from the corresponding depth map. Each of the two partitions (resembling foreground and background) is motion compensated separately and afterwards merged based on the obtained segmentation mask.

2.1 Depth-Based Block Partitioning

DBBP mode is a new inter prediction mode for the dependent texture view, which utilizes the reconstructed depth information of the base view coded before the current view. The DBBP coding generally composes five steps: virtual depth identification, segment mask generation, partition determination, bi-segment compensation, and prediction signals merging. Figure 1 illustrates the five steps.

In the 3D-HEVC, the depth information of the base view is coded before the texture component of the dependent views. Therefore, a virtual depth block can be first identified from the corresponding depth image in the base view using the DoNBDV which is specified in the 3D-HEVC. In the following segment mask generation process, a threshold is calculated first based on the average of all depth samples within the corresponding virtual depth block. Then, a binary segmentation mask $m_D(x,y)$ is generated based on the values of the depth samples and the threshold. When dealing with the depth sample located at coordinator (x,y), its binary mask $m_D(x,y)$ will be set to 1 or 0 comparing its depth value to the threshold.

With the identified virtual depth block, a partition mode also need to be selected from either $2N \times N$ or $N \times 2N$ in the partition determination process, based on threshold calculation and matched filtering. First, a threshold is obtained by calculating the mean value of top-left pixels of all the sub-blocks in the identified virtual depth block, which can be seen as down-sampling with sampling rate 4. An example of used depth samples for 8×8 block in partition process is illustrated in Fig. 2. Second, 4×4 block-wise segmentation map is generated by using the calculated threshold in the first step, and then matched filtering is performed to determine a best matching partition mode. To be specific, a binary matrix is used to represent the matching state, where 0 represents the corresponding pixel in virtual depth block is not larger than the calculated threshold and 1 represents the opposite. Afterwards, two results will be counted for each partition candidate, i.e. $2N \times N$ and $N \times 2N$. Take $N \times 2N$ for example, the two results are: (1) the number of 0-value components in the left plus the number of 1-value components in the right and (2) the number of 1-value components in the left plus the number of 0-value components in the right. By comparing the counted two results for each partition candidate, the one with the largest result value is selected as the optimal partition mode for the current block.

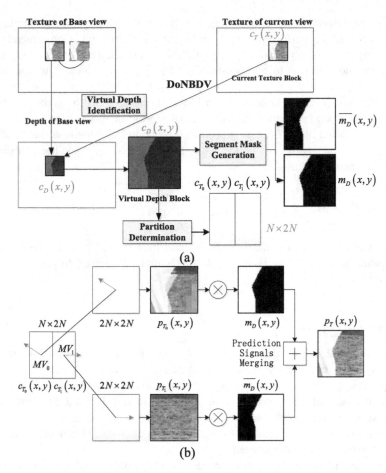

(a)

(b)

Fig. 1. Depth-Based Block Partitioning in 3D-HEVC, (a) steps: virtual depth identification, segment mask generation, partition determination, (b) steps: bi-segment compensation, and prediction signals merging

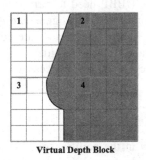

Virtual Depth Block

Fig. 2. Example of deriving the partition mode based on the depth samples.

After the partition mode is determined, full-size $(2N \times 2N)$ motion or disparity compensation will be performed twice, once for each segment. Consequently, two prediction signals $p_{T0}(x, y)$ and $p_{T1}(x, y)$ will be obtained, resulting in two sets of vector information needed to be coded for a DBBP coding block. The approach is based upon this fact that a texture block is typically segmented into foreground and background based on the collocated depth block. These two depth layers can then be compensated independently by their own sets of motion or disparity vectors.

Then, a predicted block $P_T(x, y)$ is produced by combining the resulting prediction signals $p_{T0}(x, y)$ and $p_{T1}(x, y)$ using the segmentation mask $m_D(x, y)$. The combination process is defined as follows:

$$P_T(x, y) = \begin{cases} P_{T0}(x, y), & m_D(x, y) = 0/1 \\ P_{T1}(x, y), & m_D(x, y) = 1/0 \end{cases}, \ x, y \in [0, 2N - 1] \tag{1}$$

DBBP partitions a block into foreground and background regions. To obtain the final predicted block, filtering is applied to the boundary samples to possibly change the intensity values.

2.2 New Observations

As can be seen from above specified description, the partition determination process is of relatively high complexity since it needs to derive a threshold by calculating sub-sample mean value, and then to perform matched filtering by using the derived threshold. In the matched filtering step, a binary matrix is used to represent the matching state and two results need to be counted for each partition candidate. Such design consumes a certain memory access and computations. Therefore, a simplified partition approach is urgently needed.

In the segment mask generation process, a threshold calculation based on the average of all depth samples within the corresponding virtual depth block is performed first. Then, a binary segmentation mask is generated based on sample depth values and the threshold. The design of the segment mask generation process is very similar to the partition determination process. Both the segment mask generation step and the partition determination step of the DBBP need to perform sampled mean calculation as well as the subsequent binarization for mask generation and partition determination, respectively. The objectives of the two processes are consistent, i.e. to find a rough direction of internal boundary according to the identified virtual depth block. So there is strong operational redundancy in the two processes.

3 Proposed Scheme

To reduce the high complexity in the partition process and eliminate its redundancy with the segment mask generation process, an efficient partition scheme based on the binary segmentation mask is proposed.

3.1 Discussions on Similarity in Segment Mask Generation and Partition Determination

As mentioned above, the underlying principle of segment mask generation and partition derivation are the same, i.e. to roughly identify the boundary direction within the current texture block using its derived virtual depth information. An example of the actual vertical object boundary is illustrated in Fig. 3. In order to better segment the foreground and background, clearly, partition candidate $N \times 2N$ should be selected eventually from the two kinds of partition mode $2N \times N$ and $N \times 2N$.

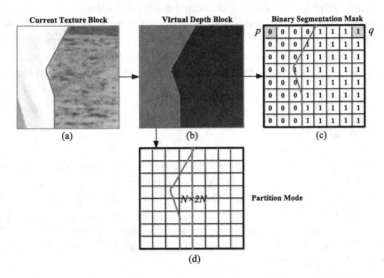

Fig. 3. Example of the vertical directional boundary case in DBBP.

Meanwhile, the threshold in segment mask generation process is calculated based on the average of all depth samples within the corresponding virtual depth block. Then, the binary mask of each sample is generated by comparing its value with the threshold value. In general, the sample values in a same foreground region or a same background region are very similar in the depth map. And the depth value of background region is greater than that of foreground region. So, the segment mask generation method can well separate the foreground and background and identify the boundary direction within the block. For the example in Fig. 3, the final binary segmentation mask will also tend to be of vertical partition.

For a given depth block, the final binary mask and the selected partition mode will present the same boundary direction, referring to Fig. 3(c) and Fig. 3(d). Therefore, utilizing a similar or even the same boundary identification method can obtain the binary mask and partition mode at the same time. To reduce the high complexity in the partition process and eliminate its redundancy with the segment mask generation process at the same time, we propose to derive the partition mode according to the binary mask values obtained in segment mask generation process.

3.2 An Efficient Partition Scheme

Taking the actual vertical object boundary case in Fig. 3(a) for example, the top-left sample p and the top-right sample q will present different binary mask value, as show in Fig. 3(c). Since there are only two candidate partition modes $2N \times N$ and $N \times 2N$ in the DBBP coding, the vertical direction partition candidate $N \times 2N$ is supposed to be selected. Accordingly, for the actual horizontal object border case, the top-left sample p and the top-right sample q will be divided into a same segment and present same binary mask value. And the appropriate partition mode should also be horizontal partition candidate $2N \times N$. Therefore, according to the value of the top-left sample p and the top-right sample q, proper partition mode can be obtained directly.

The proposed partition scheme simply judge two binary mask values ($m_D(0,0)$ for p and $m_D(0, N-1)$ for q), to directly determine the partition mode from either $2N \times N$ or $N \times 2N$ for DBBP coding. If $m_D(0,0)$ is equal to $m_D(0, N-1)$, $2N \times N$ partition mode is selected to be the final partition mode. Otherwise, $N \times 2\,N$ partition mode is selected.

4 Experimental Results

To evaluate the performance of our proposed partition scheme for DBBP coding, experiments are carried out under the CTC configuration required by JCT-3 V. All the experiments were implemented on the 3D-HEVC Test Model (HTM) version 11.0 [13]. Totally eight sequences with a centre-left-right encoding order are tested. Table 1 shows the detailed information of the test sequences and view numbers used for the experiments. To evaluate the coding performance, Bjontegaard metric (BD-rate change in percentage) is employed [14].

Table 1. Test sequences and view numbers used for the experiments

Sequence	Resolution	Frame	Coded view number
Balloons	1024 × 768	300	3-1-5
Kendo	1024 × 768	300	3-1-5
Newspaper_CC	1024 × 768	300	4-2-6
GT_Fly	1920 × 1088	250	5-9-1
Poznan_Hall2	1920 × 1088	250	6-7-5
Poznan_Street	1920 × 1088	250	4-5-3
Undo_Dancer	1920 × 1088	250	5-1-9
Shark	1920 × 1088	300	5-1-9

Table 2 shows the results of the proposed partition scheme under CTC configuration. All the result data were obtained by comparing with the anchor data of HTM11.0.   "$video1$",   "$video2$",   "$video - PSNR/video - bitrate$", "$video - PSNR/total - bitrate$" and "$synth - PSNR/total - bitrate$" indicate the BDBR for the left coded texture view, the right coded texture view, three coded views

Table 2. BD-Rate results for the proposed method under CTC

Sequence	video1	video2	video-PSNR/video-bitrate	video-PSNR/total-bitrate	synth-PSNR/total-bitrate
Balloons	-0.04 %	-0.05 %	-0.01 %	-0.01 %	-0.06 %
Kendo	-0.06 %	0.03 %	-0.01 %	-0.03 %	-0.01 %
Newspaper_CC	-0.06 %	0.05 %	-0.01 %	0.01 %	0.01 %
GT_Fly	-0.06 %	-0.16 %	-0.01 %	0.00 %	-0.03 %
Poznan_Hall2	-0.30 %	-0.05 %	-0.06 %	-0.05 %	-0.03 %
Poznan_Street	0.08 %	0.02 %	0.00 %	0.01 %	0.02 %
Undo_Dancer	-0.01 %	0.06 %	0.01 %	0.00 %	0.01 %
Shark	-0.14 %	-0.03 %	-0.01 %	-0.01 %	0.01 %
1024 × 768	-0.05 %	0.01 %	-0.01 %	-0.01 %	-0.02 %
1920 × 1088	-0.09 %	-0.03 %	-0.02 %	-0.01 %	0.00 %
average	**-0.07 %**	**-0.02 %**	**-0.01 %**	**-0.01 %**	**-0.01 %**

without depth maps, three coded views with depth maps, and six synthesized views, respectively. On average 0.01 % total bit rate saving can be achieved for synthesized views. Moreover, the proposed partition scheme for DBBP can bring 0.07 % and 0.02 % bitrate saving, for view 1 and view 2, respectively.

To explain simplification, computation complexity between the original and the proposed method are compared in Table 3. The number of necessary additions (ADD), binary shifting operations (SHIFT) and binary comparisons (COMP) increases with the block size in the original DBBP scheme. With our proposed partition scheme, the threshold calculation is removed and the number of computations is no longer associated with the block size, which is really less complex compared to the original method. When considering that the DBBP coding is mostly applied on bigger blocks due to its capability of better approximating object boundaries, the effective complexity reduction of the proposed method is significant.

Table 3. Complexity comparison between the original and proposed partition scheme

Block Size	Original Method		Proposed Method	
	Threshold	Matched	Threshold	Matched
8 × 8	3 ADDs	8 + 4 COMPs	NONE	1 COMP
	1 SHIFT	8 ADDs		
16 × 16	15 ADDs	32 + 4 COMPS	NONE	1 COMP
	1 SHIFT	32 ADDS		
32 × 32	63 ADDS	128 + 4 COMPS	NONE	1 COMP
	1 SHIFT	128 ADDS		
64 × 64	255 ADDs	512 + 4 COMPS	NONE	1 COMP
	1 SHIFT	512 ADDS		

5 Conclusion

In the Depth-based Block Partitioning (DBBP) mode, the original partition method consists of two steps, i.e. threshold calculation and matched filtering based on down-sampling, which is relatively high complex and redundant with the segment mask generation process. To deal with this, an efficient partition scheme for DBBP coding is proposed in this paper accordingly. For DBBP coding of any block size, the proposed partition scheme consistently requires only 2 sample values of the segmentation mask and 1 binary comparison operations. It not only reduces the computational complexity, but also achieves slightly better coding performance. Experimental results demonstrate that on average 0.01 %, 0.07 % and 0.02 % BD-rate saving can be obtained for synthesized views, view 1 and view 2, respectively, under CTC configuration.

References

1. Müller, K., Schwarz, H., Marpe, D., Bartnik, C., et al.: 3D high-efficiency video coding for multi-view video and depth data. IEEE Trans. Image Process. **22**(9), 3366–3378 (2013)
2. Zhu, C., Zhao, Y., Yu, L., Tanimoto, M.: 3D-TV system with depth-image-based rendering: architectures, techniques and challenges. Springer, Berlin (2013)
3. Fehn, C.: Depth-image-based rendering (DIBR), compression and transmission for a new approach on 3D-TV. In: Proceedings of SPIE Conference on Stereoscopic Displays and Virtual Reality Systems XI, CA, U.S.A., Vol. 5291, pp. 93–104, January 2004
4. ISO/IEC JTC1/SC29/WG11. Applications and Requirements on 3D Video Coding, Doc. N12035, Geneva, Switzerland, March 2011
5. ISO/IEC JTC1/SC29/WG11. Call for Proposals on 3D Video Coding Technology, Doc. N12036, Geneva, CH, March 2011
6. Lei, J., Li, S., Zhu, C., Sun, M.-T., Hou, C.: Depth coding based on depth-texture motion and structure similarities. IEEE Trans. Circuits Syst. Video Technol. **25**(2), 275–286 (2015)
7. Li, S., Lei, J., Zhu, C., Yu, L., Hou, C.: Pixel-based inter prediction in coded texture assisted depth coding. IEEE Signal Process. Lett. **21**(1), 74–78 (2014)
8. Li, S., Zhu, C., Lei, J.: Depth-texture cooperative clustering and alignment for high efficiency depth intra coding. In: IEEE China Summit & International Conference on Signal and Information Processing (ChinaSIP), Beijing, China, July 2013 (Invited Paper)
9. Jager, F.: Depth-based block partitioning for 3D video coding. In: Picture Coding Symposium (PCS), San Jose, CA (2013)
10. Park, M., Lee, J., Cho, Y., Kim, C.: Partition derivation for DBBP. ITU-T SG 16 WP 3 and ISO/IEC JTC 1/SC 29/WG 11, JCT3 V-I0077, 9th Meeting: Sapporo, JP, 3–9 July 2014
11. Mueller, K., Vetro, A.: Common test conditions of 3DV core experiments. ITU-T SG 16 WP 3 and ISO/IEC JTC 1/SC 29/WG 11, JCT3 V-G1100, 7th Meeting: San José, US, 11–17 January 2014
12. Chen, Y., Tech, G., Wegner, K., et al.: Test model 8 of 3D-HEVC and MV-HEVC. ITU-T SG 16 WP 3 and ISO/IEC JTC 1/SC 29/WG 11, JCT3 V-H1003, 8th Meeting: Valencia, ES (2014)
13. HEVC 3D extension software repository (at HHI). https://hevc.hhi.fraunhofer.de/svn/svn_3DVCSoftware/tags/HTM-11.0
14. Bjontegaard, G.: Calculation of Average PSNR Differences between RD-curves. ITU-T Q.6/SG16 VCEG, VCEG-M33, April 2001

Image Classification with Local Linear Decoding and Global Multi-feature Fusion

Zhang Hong[1,2(✉)] and Wu Ping[1]

[1] College of Computer Science and Technology,
Wuhan University of Science and Technology, Wuhan, China
zhanghong_wust@163.com
[2] Intelligent Information Processing
and Real-Time Industrial Systems Hubei Province Key Laboratory,
Wuhan University of Science and Technology, Wuhan, China

Abstract. Recent years have witnessed a surge of interest in image classification. The combination of deep neural network with feature extraction has improved image classification performance dramatically. In order to improve the performance of image classification, this paper proposes an image classification algorithm based on deep neural network of linear decoder and softmax regression model. First, we learn features of some small image patches with linear decoder; secondly, by convolving and pooling the large images with the learned features, then we obtain the pooled convolved features; thirdly, we use softmax regression model to learn the features for image classification. Experimental results are encouraging and demonstrate the validity and superiority of our method.

Keywords: Linear decoders · Deep neural network · Image classification

1 Introduction

With the vigorous development of Internet and multimedia technology, as well as the rising popularity of social media, the demand of image classification becomes more and more rigorous. Image data gradually become an important medium of information exchange and experience sharing. However, most of the images have no tags. Traditional methods, which mostly rely on manual way to manage and classify images, are time consuming and inefficient. Numerous researches have been devoted to efficient image classification methods with less labeled training data.

The key challenge of image classification is the well-known "semantic gap" issue [1, 2, 11, 12]. It exists between low-level image pixels captured by machines and high-level semantic concepts perceived by human. Recent years have witnessed some important advances of new techniques that attempts to address this grand challenge. One important technique is machine learning. For example, [1] proposed error back propagation algorithm to learn the abstract representation of images. Although this strategy has been proven effective, it still has some shortcomings. Firstly, this method is easy to fall into local minimum and it is difficult to achieve optimal learning outcomes. Secondly, its capacity of expressing complex function is limited. What is more,

© Springer International Publishing Switzerland 2015
Y.-S. Ho et al. (Eds.): PCM 2015, Part II, LNCS 9315, pp. 437–446, 2015.
DOI: 10.1007/978-3-319-24078-7_44

this method cannot be able to help training discriminative classifiers for prediction. To solve these problems, Hinton presented the third generation of neural networks [2] in 2006, namely, the deep learning [13, 19]. It is a branch of machine learning based on a set of algorithms that attempt to model high-level abstractions in data. For example, [3] has proposed convolutional neural networks to extract feature of images. However, this method may require a large amount of labeled data and collecting these are time consuming and labor intensive. Taking into account the problem of limited labeled samples, Poultney, etc. using unlabeled image samples to learn the feature of images [4]. Le proposed a network of sparse autoencoder [5]. It is an unsupervised learning algorithm that applies back-propagation, setting the target values to be equal to the inputs. Nevertheless, in the sparse autoencoder, the sigmoid activation function, whose range is [0, 1], performs poorly on image classification because the input unit is not necessarily within [0, 1]. And the optimizing range of data is still an open question.

Different from above researches, this paper proposes a method of the deep neural network based on local linear decoder and global multi-feature regression. Our framework is shown in Fig. 1. First, we randomly select local areas from original images as the input of the network; then we calculate the parameters learned from the local areas, use these parameters to convolve and pool the original images; furthermore, we propose global multi-feature regression for optimal image classification. It can be seen from Fig. 1 that, in our framework the local network contains input layer, hidden layer and output layer; and the input data needn't be limited or scaled. Our method reduces the complexity of the neural network training and improves the efficiency of data preprocessing.

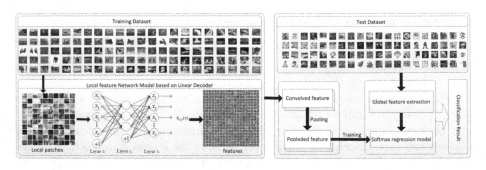

Fig. 1. A framework of deep learning with application to image classification

The rest of this paper is organized as follows. Section 2 introduces our method of convolved feature analysis based on local linear decoder. Section 3 presents the global multi-feature regression for image classification. Section 4 presents the result of experiments. Section 5 concludes this paper and discusses future directions.

2 Convolved Feature Analysis Based on Local Linear Decoder

Motivated by the methods of neural network and sparse autoencoder [6, 12], we present the convolved feature analysis algorithm based on local linear decoder. By analyzing convolved feature, we can reduce the reduancy between feature vectors at neighboring locations and improve the efficiency of the overall representation [7].

2.1 Local Linear Feature Decoder

It can be seen from Fig. 1 that the Linear Feature Decoder (LFD) includes several layers of networks. When convergence threshold or the maximum number of iterations is reached, our neural network will completes intelligent learning of essential feature of image training set. Layer L_1 is the input layer. Layer L_1 processes the input information and transfers it to Layer L_2, i.e., the hidden Layer. After that, the layer handles original image features and passes it to the Layer L_3, i.e., the output layer.

Let $X = \{x_1, x_2, x_3, \ldots, x_i, \ldots, x_m\}$ denote the training data of unlabeled images where x_i represents the i-th input image, m is the number of samples. When the input of Layer L_1 is X, let $Y = \{y_1, y_2, y_3, \ldots, y_i, \ldots, y_m\}$ denote the desired output value of image tags where y_i represents the desired output value of x_i. Let $h_{w,b}(x)$ represents the hypothetical model when X is the input, where w is weight vector and b is bias vector. The process of our local linear feature decoding is to calculate the vectors w and b for each layer. Let $a_i^{(l)}$ denote the activation of unit i in layer l. And $a_j^{(2)}(x_i)$ denote the activation value of the hidden unit when the network is given a specific input of x_i. The average activation value of hidden unit j averaged over the unlabeled training set can be calculated as follows:

$$\hat{\rho}_j = \frac{1}{m}\sum_{i=1}^{m}\left[a_j^{(2)}(x_i)\right] \tag{1}$$

In order to obtain sparse data representation, we add a sparsity parameter ρ with $\hat{\rho}_j = \rho$, and to keep the average activation values of hidden units in a certain range. We add another extra penalty term $\sum_{j=1}^{s_2} KL(\rho \parallel \hat{\rho}_j)$. Then the overall cost function is defined as below:

$$J(w,b) = \left[\frac{1}{m}\sum_{i=1}^{m}\frac{1}{2}\parallel h_{w,b}(x_i) - y_i \parallel^2\right] + \frac{\lambda}{2}\sum_{l=1}^{n_l-1}\sum_{i=1}^{s_l}\sum_{j=1}^{s_{l+1}}\left(w_{ji}^{(l)}\right)^2$$
$$+ \beta\sum_{j=1}^{s_2}\left[\rho\log\frac{\rho}{\hat{\rho}_j} + (1-\rho)\log\frac{1-\rho}{1-\hat{\rho}_j}\right] \tag{2}$$

where the first part is an average sum-of-squares error term, the second part is a weight decay term. It tends to decrease the magnitude of the weights and helps prevent over-fitting. n_l represents number of layers in the network, λ is the weight decay parameter and s_l denotes the number of units in layer l. The third part is a penalty term, and β controls the weight of the sparsity penalty term. We choose the sigmoid function $f(z) = \frac{1}{1+e^{-z}}$ to be the activation function of units. For each unit i in layer l, we compute an "error term" $\sigma_i^{(l)}$ that measures how much that node was "responsible" for any errors in our output. Then we have:

$$
\begin{cases}
\sigma_i^{(2)} = \left(\left(\sum_{j=1}^{s_2} w_{ji}^{(2)} \sigma_j^{(3)} \right) + \beta \left(-\frac{\rho}{\hat{\rho}_i} + \frac{1-\rho}{1-\hat{\rho}_i} \right) \right) f'\left(z_i^{(2)} \right) \\
\sigma_i^{(3)} = -(y_i - \hat{x}_i)
\end{cases}
\tag{3}
$$

where $z_i^{(l+1)}$ denote the total weighted sum of inputs to unit i in layer $l + 1$. Then we can compute the partial derivatives of $J(w, b)$ and use algorithm of gradient descent to calculate the parameter w and b. When the algorithm is converged or the maximum number of iterations is reached, the value of parameter w and b can be determined. And then, the process of feature learning is completed.

2.2   Convolved Feature Analysis Based on Local Network

Each unit in every layer has the weight vector w and bias vector b to be trained in this network, and the number of these vectors depends on the size of the image resolution [14]. The larger the image resolutions, the more the number of vectors of the network need to learn [15, 16]. Visibly, the size of image resolution directly affects the complexity of parameter's learning.

Therefore, this section proposes the convolution algorithm based on locally connected networks. We should cut the unlabeled image training dataset into several local visual areas, and then constitute a regional subset as the input layer of the neural network. With this algorithm we can get the convolved feature matrix of the dataset, and choose this matrix as the input of classifiers.

Let $\Omega = \{(z_1, r_1), (z_2, r_2), \ldots, (z_i, r_i), \ldots, (z_t, r_t)\}$ denote a labeled image training dataset where t is the number of samples, z_i represents the i-th image in the dataset. And the dataset is grouped into k categories, r_i is the category number of z_i. Let $S = \{s_1, s_2, \ldots, s_i, \ldots, s_n\}$ denote an unlabeled images testing dataset where s_i represents the i-th image and n is the number of samples. We assume the image resolution in the dataset Ω and S is $a \times a$, and select local small images from these original images randomly with image resolution of $b \times b$. All of these images that selected from dataset Ω constitute an unlabeled image training dataset $X = \{x_1, x_2, x_3, \ldots, x_i, \ldots, x_m\}$ as shown in Fig. 1, and input it to LFD.

Then, according to Sect. 2.1, we can calculate the weight vector w and bias vector b. Then we set the step length of convolution operation is 1. After completion of the convolution on training and test sets, we can get convolved features matrices $D_{Convoled\_Train}$ and $D_{Convoled\_Test}$, respectively. Its dimension sizes are:

$$k \times (a - b + 1) \times (a - b + 1) \tag{4}$$

where k is the number of units in hidden layer. It is quite clear that the dimension is very high. So, it is not conducive to training classifier. What is more, it is prone to be over-fitting and it will affect the efficiency of image classification. Regarding above these issues, we aggregate statistics of features at various locations [20]. This operation is called pooling [21]. In this paper, we compute the mean value of a particular feature over a region of the image and set the size of the region is $D_{pooling}$ to pool our convolved features over. After that, we can get the pooled feature matrices D_{Pooled_Train} and D_{Pooled_Test}.

3 Global Multi-feature Fusion

According to the pooled feature matrices D_{Pooled_Train} and D_{Pooled_Test} obtained in Sect. 2, we can set the feature matrix D_{Pooled_Train} as training sample in this model. For any input sample (z_i, r_i), the model estimates the probability that $p(r_i = k|z_i)$ for each category. Then we have:

$$h_\theta(z_i) = \begin{bmatrix} p(r_i = 1)|z_i; \theta \\ p(r_i = 2)|z_i; \theta \\ \vdots \\ p(r_i = k)|z_i; \theta \end{bmatrix} = \frac{1}{\sum_{j=1}^{k} e^{\theta_j^T z_i}} \begin{bmatrix} e^{\theta_1^T z_i} \\ e^{\theta_2^T z_i} \\ \vdots \\ e^{\theta_k^T z_i} \end{bmatrix} \tag{5}$$

where $\theta_1, \theta_2, \ldots, \theta_k$ are the model parameters of our model. And we define the cost function of $h_\theta(z_i)$ as follows:

$$J(\theta) = -\frac{1}{t} \left[\sum_{i=1}^{t} \sum_{j=1}^{k} 1\{r_i = j\} \log \frac{e^{\theta_j^T z_i}}{\sum_{l=1}^{k} e^{\theta_l^T z_i}} \right] \tag{6}$$

where $1\{r_i = j\}$ is the indicator function. When $r_i = j$ is a true statement, $1\{r_i = j\} = 1$, otherwise, $1\{r_i = j\} = 0$. In order to avoid the numerical problems associated with softmax regression's over parameterized representation, we add the weight decay term $\frac{\lambda}{2} \sum_{i=1}^{k} \sum_{j=0}^{n} \theta_{ij}^2$ in $J(\theta)$. Further, its partial derivative can be calculated as bellow:

$$\nabla_{\theta_j} J(\theta) = -\frac{1}{t} \sum_{i=1}^{t} [z_i(1\{r_i = j\} - p(r_i = j|z_i; \theta))] + \lambda \theta_j \tag{7}$$

By minimizing $J(\theta)$ with respect to $\theta_1, \theta_2, \ldots, \theta_k$, we will have a working implementation of softmax regression model. Then the model calculates probability value in each category of each sample in (5).

In general, we choose the category with the maximum probability as the prediction result [18]. However, due to unavoidable parameter errors, redundant and over-fitting, the method that only use the features learned in Sect. 2.2 and combine with the softmax regression model [17] will affect the classification results. In order to solve the above problems, this paper adopts the method of multi-step classification. Firstly, for each sample s_i we get its prediction vector c_i as below:

$$c_i = \{c_1, c_2, \ldots, c_k\} \tag{8}$$

where $c_i, i \in [1, k]$ is the prediction value for each class. We put the prediction values in c_i in descending order. For each sample s_i, we select the former k/2 value constitute the candidate set as a preliminary predicted value.

In addition to the local feature matrix learned in LFD, we can also extract global features of images in dataset Ω and S. We set the dimensions of global features extracted from training dataset and testing dataset as p. We can get a global feature matrix $A \in \mathbb{R}^{m \times p}$ from the labeled image training dataset Ω containing m images and a global feature matrix $B \in \mathbb{R}^{n \times p}$ from the unlabeled images testing dataset S. We calculate the Euclidean distance between any two samples z_j and s_i. Then we can obtain the distance matrix D_{ij}, where $z_j \in A$, $s_i \in B$.

The preliminary classification of images was completed. So when calculating the distance between samples, we can only count the distance between test sample and the candidate set C_i. Then, we set the minimum value of D_{ij} corresponding to the sample as the final prediction result of s_j and denoted as L_i. Based on above discussions, the classification process is listed in Fig. 2.

Input: training sets X (unlabeled) and Ω (labeled); testing set S.
Output: classification result L_i for S.

---

1. Select local visual areas from each image in X as the input of the LFD, and calculate weight vector w and bias vector b in (2);

2. Calculate convolved feature matrices $D_{Convoled\_Train}$ and $D_{Convoled\_Test}$ for samples in Ω and S with w and b, and accordingly obtain the pooled feature matrices D_{Pooled_Train} and D_{Pooled_Test};

3. Use matrix D_{Pooled_Train} as the training set to learn classification parameters $\theta_1, \theta_2, \ldots, \theta_k$ in (5) for the softmax regression process; and then obtain prediction vector of c_i in (8) for each testing data s_i in S;

4. Extract global feature matrices from dataset Ω and S, calculate Euclidean distances between any two samples from Ω and S; classify s_i into the class which the nearest neighbor of s_i belongs to.

Fig. 2. Our image classification algorithm based on local linear decoding and global multi-feature fusion

4 Experiments

4.1 Datasets

We test the proposed algorithm of image classification on two datasets: the MSRA-MM [8] and the image dataset collected by ourselves from web pages named as DATASET2[1,2,3]. MSRA-MM is the second version of Microsoft Research Asia Multimedia. It contains about 1 million images. In addition, there are 10 categories including bird, horse, dog, dolphin, elephant, explosion, airplane, autobicycle, car and stream in DATASET2 and each category contains 100 images. Overall, 700 images are selected as the training set, and the rest 300 images are used as an unlabeled test set. Figure 3 shows some image samples where each column represents three image samples randomly selected from a semantic category. As for the MSRA-MM dataset 6000 images are randomly selected as the training set, the rest 3000 images are used as the test set.

Fig. 3. Image samples from DATASET2

4.2 Feature Selection and Analysis Results

Three different types of visual features, i.e., color histogram (in HSV space), color coherence vector (CCV), and Tamura texture are extracted to calculate the global feature matrices A and B.

First, we normalize the image size of MRSA-MM and DATASET2 datasets as 96×96 pixels. And to calculate local linear feature decoder we set the size of local image area as 8×8 pixels. Secondly, we select 100 local image areas from each original image. And then we select 100000 areas from the training set as the input of the LFD. Since there are three channels in RGB images, the number of units in the input layer of L_1 is $8 \times 8 \times 3 = 192$. Similarly, the number of units in the output layer of L_3 is also 192. Besides, we set the number of units in the hidden layer of L_2 as 400 which is an experienced optimal value [10]. We visualize all 400 vectors of w and b in L_2 in Fig. 4.

---

[1] http://image.baidu.com

[2] http://en.wikipedia.org/wiki/Encarta.

[3] http://www.aboutus.org/AnimalBehaviorArchive.org.

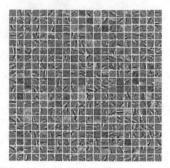

Fig. 4. Visualization of all the values of all parameter w and b

5 Performance Comparison

In order to verify the effectiveness and superiority of the proposed method, we compare softmax regression model with error back propagation [1] and K-Nearest Neighbor algorithms [9]. We use accuracy (ACC) as evaluation metric. Performance comparison results are shown in Fig. 5. From Fig. 5, we have the following observation: the classification of softmax regression have 10.7 percentage points higher than the traditional BP neural network. After determining the classification method, we compare our method with three methods of distance metric: Euclidean Distance, Cosine Distance and Mahalanobis Distance. The Average Precision (AP) is employed as the performance metric for evaluating the classification results.

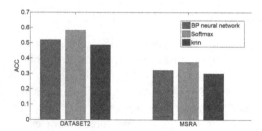

Fig. 5. Performance comparison of different algorithms based on local features

Fig. 6. Precision of results in different algorithms based on multi-feature

It can be seen from Fig. 6 that after combine Euclidean distance metric with softmax regression model, the performance has improved significantly. Our method has 28 percentage points higher than the Softmax + Mahalanobis Distance, 17 percentage points higher than the Softmax + Cosine.

6 Conclusions

In this paper, we propose an image classification algorithm based on local linear decoding and global multi-feature fusion. Our method avoids classification error caused by only use local features learned in this network or the global feature extracted by people. Experimental results show that our method has achieved good classification performance on different images datasets, and outperform the traditional image classification methods obviously. Future study mainly includes further study on optimizing algorithm and how to extend the algorithm to multi-label image classification.

Acknowledgments. This work described in this paper was supported by National Natural Science Foundation of China (No. 61373109, No. 61003127 and No. 61440016), State Key Laboratory of Software Engineering (SKLSE2012-09-31).

References

1. Cigizoglu, H., Kisi, O.: Flow prediction by three back propagation techniques using k-fold partitioning of neural network training data. Nord. Hydrol. **36**(1), 49–64 (2005)
2. Hinton, G.E., Salakhutdinov, R.R.: Reducing the dimensionality of data with neural networks. Science **313**(5786), 504–507 (2006)
3. Krizhevsky, A., Sutskever, I., Hinton, G.E.: Imagenet classification with deep convolutional neural networks. In: Advances in Neural Information Processing Systems, pp. 1097–1105 (2012)
4. Poultney, C., Chopra, S., Cun, Y.L.: Efficient learning of sparse representations with an energy-based model. In: Advances in Neural Information Processing Systems, pp. 1137–1144 (2006)
5. Le, Q.V.: Building high-level features using large scale unsupervised learning. In: IEEE International Conference on Acoustics, Speech and Signal Processing, pp. 8595–8598 (2013)
6. Hinton, G.E.: Learning multiple layers of representation. Trends Cogn. Sci. **11**(10), 428–434 (2007)
7. Kavukcuoglu, K., Sermanet, P., et al.: Learning convolutional feature hierarchies for visual recognition. In: Advances in Neural Information Processing Systems, pp. 1090–1098 (2010)
8. Li, H., Wang, M., Hua, X.S.: MSRA-MM 2.0: a large-scale web multimedia dataset. In: Proceedings of the 2009 IEEE International Conference on Data Mining Workshops, Washington, pp. 164–169 (2009)
9. Keller, J.M., Gray, M.R., Givens, J.A.: A fuzzy k-nearest neighbor algorithm. IEEE Trans. Syst. Man Cybern. **4**, 580–585 (1985)
10. Zhang, R., Zhang, Z.: Effective image retrieval based on hidden concept discovery in image database. IEEE Trans. Image Process. **16**(2), 562–572 (2007)
11. Zhang, Hong, Jun, Yu., Wang, Meng, Liu, Yun: Semi-supervised distance metric learning based on local linear regression for data clustering. Neurocomputing **93**, 100–105 (2012)
12. Karnowski, T.P., Arel, I., Rose, D.: Deep spatiotemporal feature learning with application to image classification. In: Machine Learning and Applications (ICMLA), pp. 883–888 (2010)
13. Wang, Z., Xia, D., Chang, E.Y.: A deep-learning model-based and data-driven hybrid architecture for image annotation. In: Proceedings of ACM, pp. 13–18 (2010)

14. Hörster, E., Lienhart, R.: Deep networks for image retrieval on large-scale databases. In: Proceedings of ACM, pp. 643–646 (2008)
15. Eigen, D., Puhrsch, C., Fergus, R.: Depth map prediction from a single image using a multi-scale deep network. In: Advances in Neural Information Processing Systems, pp. 2366–2374 (2014)
16. Lin, Y., Lv, F., Zhu, S., et al.: Large-scale image classification: fast feature extraction and svm training In: CVPR, pp. 1689–1696 (2011)
17. Zhong, S., Liu, Y., Liu, Y.: Bilinear deep learning for image classification. In: Proceedings of ACM, pp. 343–352 (2011)
18. Wang, J., Yang, J., Yu, K., et al.: Locality-constrained linear coding for image classification In: CVPR, pp. 3360–3367 (2010)
19. Warburton, K.: Deep learning and education for sustainability. Int. J. Sustain. High. Educ. 4 (1), 44–56 (2003)
20. Ciresan, D., Meier, U., Schmidhuber, J.: Multi-column deep neural networks for image classification In: CVPR, pp. 3642–3649 (2012)
21. Giusti, A., Cireşan, D.C., Masci, J., et al.: Fast image scanning with deep max-pooling convolutional neural networks (2013). arXiv:1302.1700. arXiv preprint

Hashing with Inductive Supervised Learning

Mingxing Zhang[1(✉)], Fumin Shen[1], Hanwang Zhang[2], Ning Xie[3],
and Wankou Yang[4]

[1] University of Electronic Science and Technology of China, Chengdu, China
superstar_zhang@hotmail.com, fumin.shen@gmail.com
[2] National University of Singapore, Singapore, Singapore
hanwangzhang@gmail.com
[3] Tongji University, Shanghai, China
ningxie@tongji.edu.cn
[4] Southeast University, Nanjing, China
wankou.yang@yahoo.com

Abstract. Recent years have witnessed the effectiveness and efficiency of learning-based hashing methods which generate short binary codes preserving the Euclidean similarity in the original space of high dimension. However, because of their complexities and out-of-sample problems, most of methods are not appropriate for embedding of large-scale datasets. In this paper, we have proposed a new supervised hashing method to generate class-specific hash codes, which uses an inductive process based on the Inductive Manifold Hashing (IMH) model and leverage supervised information into hash codes generation to address these difficulties and boost the hashing quality. It is experimentally shown that this method gets excellent performance of image classification and retrieval on large-scale multimedia dataset just with very short binary codes.

Keywords: Hashing · Binary code learning · Supervised learning · Image retrieval · Multimedia

1 Introduction

Recently, a great growth of multimedia like images and videos is taking place very day duo to the booming of social network, such as Flickr and YouTube. It is reported in 2010 [1] that Flickr has had over 4 billion images and YouTube received more than 20 h of uploaded videos per minute. How to index and organize these large-scale data effectively but also efficiently is becoming a key challenge for some application like retrieval and classification [2]. Nowadays, various hashing methods have been proposed and seem to offer great promise towards this goal. Hashing methods use a set of hash function that map original features of high dimension into short binary codes, while preserving the Euclidean similarity in the original space. Because of using the short binary code, hashing methods can take fast comparison by Hamming distance and need small storage space for large-scale data.

Generally, various hashing methods can be categorized as unsupervised and (semi-) supervised. Unsupervised hashing methods try to maintain the similarity of original

© Springer International Publishing Switzerland 2015
Y.-S. Ho et al. (Eds.): PCM 2015, Part II, LNCS 9315, pp. 447–455, 2015.
DOI: 10.1007/978-3-319-24078-7_45

features without other information. For example, Locality sensitive hashing (LSH) [3] generates hash codes based on random projection approximating cosine similarity. With the success of LSH, random hash functions have been extended to several similarity measures, such as p-norm distances [3], the Mahalanobis metric [4], and kernel similarity [5, 6]. However, those methods using random projection are data-independent, so several hash tables are required and the generated hash codes relatively have long lengths for high accuracy. In order to learn more compact hash codes, many data-dependent or learning-based hashing methods have been proposed. Those methods aim to learn a set of hash functions rather than randomly selected. For example, PCAH [7] and iterative quantization (ITQ) [8] generate linear hash functions through simple principal component analysis (PCA) projections; Spectral Hashing (SH) [9] learns eigenfunctions that preserve Gaussian affinity; Restricted Boltzmann Machines (RBMs) or Semantic Hashing generates hash codes by using some layers of Restricted Boltzmann Machine; Anchor Graph Hashing (AGH) [10] use Anchor Graphs to capture semantic similarity; and Inductive Hashing on Manifolds [11, 12] takes the intrinsic manifolds structure into consideration.

(Semi-)Supervised hashing methods are designed to preserve some label-based similarity. Specifically, data from same class are semantically similar while data from different classes are different [13]. It is obvious that supervised hashing methods are data dependent. Duo to leveraging label-based information, supervised hashing methods can learn a more class specific coding and be more beneficial to classification or retrieval. LDA hash [14] is a simple but typical supervised method which is based on Linear Discriminant Analysis (LDA). Binary Reconstructive Embeddings (BRE) [15] and Kernel-based Supervised Hashing (KSH) [16] employs a kernel formulation utilizing code inner product for the target hash functions. Two Step Hashing (TSH) [17] uses the spectral method and learns hash function by training classifiers. Semi-Supervised Hashing (SSH) [18] uses a supervised term minimizing the empirical error on the labeled data and an unsupervised term providing effective regularization. Hashing with Decision Trees [19] uses decision trees as hash functions for supervised learning. Minimal Loss Hashing (MLH) [20] learns binary hash functions based on structural SVMs with latent variables. Supervised Discrete Hashing (SDH) [21] introduced a novel discrete optimization algorithm to directly learn the binary codes without continuous relaxation.

Despite the fact that supervised hashing methods can attain higher accuracy than unsupervised hashing methods by leveraging supervised information, supervised hashing methods cause higher computational complexity and slower training mechanisms, especially when it comes to large-scale data. Most supervised hashing approaches are conducted only on datasets of relatively small size. This inefficiency problem makes great disadvantage for the application of existing supervised hashing methods. With the large-scale data like multimedia fast growing on social networks, it is urgent for supervised hashing methods to perform both effectively and efficiently.

With the purpose to address the fore-mentioned problem on large-scale multimedia data, in this paper, we proposed a new supervised hashing method to learn a compact hash code both effectively and efficiently, which is inspired by the idea of Inductive Manifold Hashing (IMH) [11, 12] that allows rapid assignment of new codes to previously unseen data in a manner which preserves the underlying structure of the manifold. We denote our method with Inductive Supervised Hashing (ISH). In our method, we first generate base set of small size and embed the base set to hash codes. Then we train a softmax classifier whose inputs are the distances vector from sample to base set nodes and outputs are the class similarity vector between sample and base set nodes. After that, we can embed a new sample by a weighted combination of hash codes of base set nodes, which weight is the predict similarity vector. The main difference between our method and IMH is that our method conduct supervised learning in the new sample prediction stage while the IMH just do unsupervised prediction which cannot use the large label information sufficiently. The experimental results show that this method gets excellent performance on large-scale multimedia dataset, and is very effective and efficient for image retrieval just with very short codes. The main contributions of this paper are summarized as follows:

1. We propose a new supervised hashing method that can embed large-scale samples into hash codes both effectively and efficiently. First, we use an appropriate hashing method to embed the base set which generated by the cluster algorithm. Then we can embed new samples using hash codes of the base set. Because the size of base set is much smaller than original dataset, we can product hash coding of new samples very efficiently. Second, we train a softmax classifier that produces the weight vector for new samples to combine the hash codes of the base set. As we use the whole labeled dataset to train the classifier, we leverage more class-specific information when embedding new samples into hash codes. So the embedded hash codes can be more effective for some application like classification and retrieval.
2. We use large-scale multimedia datasets on social networks to evaluate our proposed method and other representative methods. Experiments show our method has a better performance than other methods.

2 Related Work

Given a data set $X = \{x_1, x_2, \ldots, x_n\}$ containing n samples, the purpose of hashing is to look for a group of hash function that map the original samples to k-bit binary codes $B = \{b_1, b_2, \ldots, b_n\} \subset \{-1, 1\}^{k \times n}$ in hamming space such that near neighbors in the original space have similar binary codes. In this section, the Inductive Manifolds Hashing method closely related to the proposed method is briefly reviewed.

In order to address the complexities of hashing models and the problems with out-of-sample data, Shen et al. [11, 12] has proposed the Inductive Manifolds Hashing method, which allows rapid assignment of new codes to previously unseen data.

Suppose we have got the manifold-based embedding $Y = \{y_1, y_2, \ldots, y_n\}$ for the all train data $X = \{x_1, x_2, \ldots, x_n\}$. Given a new data point x_q, we generate an embedding y_q which preserves the local neighborhood relationships among its neighbors $N_k(x_q)$ in X by solving the following function:

$$c(y_q) = \sum_{i=1}^{n} w(x_q, x_i) \|y_q - y_i\|^2 \tag{1}$$

where $W_{q,i} = \begin{cases} \exp\left(-\|x_q - x_i\|^2 \big/ \sigma^2\right), & \text{if } x_i \in N_k(x_q) \\ 0 & \text{otherwise} \end{cases}$. Differentiate $c(y_q)$ with

respect to y_q, we obtain

$$\frac{\partial c(y_q)}{y_q}\bigg|_{y_q = y_q*} = 2\sum_{i=1}^{n} w(x_q, x_i)(y_q * -y_i) = 0 \tag{2}$$

which leads to the optimal solution:

$$y_q* = \frac{\sum_{i=1}^{n} w(x_q, x_i) y_i}{\sum_{i=1}^{n} w(x_q, x_i)} \tag{3}$$

Equation (3) provides a simple inductive formulation for the embedding: produce the embedding for a new data point by a (sparse) locally linear combination of the base embeddings. At last, we can use the sign function for Eq. (3) to get the hash codes.

3   The Proposed Method

In this paper, we proposed a new supervised hashing method based on the IMH to learn a binary code more efficiently and effectively. In our method, we first generate base set of small size by clustering on each class. Then we use an ordinary supervised hashing method to embed the base set to hash codes. After that, we use all the labeled samples to train a softmax classifier, whose inputs are the distances vectors from samples to base set nodes and outputs are the similarity vectors that have same dimension as the number of base set nodes. The base set nodes which have some class label with sample have higher similarity scores in output vector. Thus in the test stage, we can embed a new sample by a weighted combination of hash codes of base set nodes, whose weight is the output of softmax, namely the similarity vector.

3.1   Inductive Supervised Learning for Hashing

Suppose we have large-scale dataset $X = \{x_1, x_2, \cdots, x_n\}$, and its class label $L = \{l_1, l_2, \cdots, l_n\}$. As the dataset is very large, it is not suitable to directly use the supervised hashing method on the whole dataset. We can use the idea of Inductive Manifold Hashing that assigns new codes to previously unseen data rapidly using already generated embedding of smaller data. However, in our dataset, we have much label information, so we can product a more effective hashing if we can explore this label information sufficiently.

Given a new data point x_q, instead, we generate an embedding y_q by introducing the label information and the object function can be formulated as follow:

$$c(y_q) = \sum_{i=1}^{n} s(x_q, x_i)\|y_q - y_i\|^2 \tag{4}$$

where $s(x_q, x_i)$ is the similarity between new sample x_q and already known node x_i. If x_q and x_i have same class label, the similarity $s(x_q, x_i)$ is higher, while x_q and x_i come from different classes, the similarity is lower.

Differentiate $c(y_q)$ with respect to y_q, we obtain the optimal solution:

$$y_q* = \frac{\sum_{i=1}^{n} s(x_q, x_i) y_i}{\sum_{i=1}^{n} s(x_q, x_i)} \tag{5}$$

Similarly, Eq. (5) provides an inductive formulation for the embedding. We can further transform Eq. (5) to

$$y_q* = \sum_{i=1}^{n} \frac{s(x_q, x_i)}{\sum_{i=1}^{n} s(x_q, x_i)} y_i = \sum_{i=1}^{n} \alpha_i(x_q) y_i \tag{6}$$

where $\alpha_i(x_q) = \frac{s(x_q, x_i)}{\sum_{i=1}^{n} s(x_q, x_i)}$ is the vector of similarities between x_q and all known node x_i. We can compute $\alpha_i(x_q)$ by using a softmax classifier, which is described in Sect. 3.3 in detail.

3.2   Base Set Embedding

As described above, we produce the embedding for a new sample by a weighted linear combination of the base embeddings of already known nodes. We call these known nodes as base set. In our large-scale dataset $X = \{x_1, x_2, \cdots, x_n\}$, we can generate base set via K-means cluster algorithm. While K-means is unsupervised, in order to preserve the label information, we perform K-means on each sub-data with same label respectively. The label of notes in base set is the same as the label of sub-data samples. If our whole dataset contains t classes, then we can get the base set $B = \{b_{1,1}...b_{m_1 1}, b_{1,2},...b_{m_2,2}, ..., b_{1,t},...b_{m_t,t}\}$ with the label $L_B = \{l_{1,1}...l_{m_1 1}, l_{1,2},...l_{m_2,2}, ..., l_{1,t},... l_{m_t,t}\}$. We need to note that the number of notes in base set from same class is not consistent; it varies according to the number of samples in whole dataset.

After when we get the base set, we can use any hashing method to generate the embeddings $Y = \{y_i\}_{i=1}^{m_1+\cdots+m_t}$ of the base set.

3.3   Hashing with Classifier

When we get the embeddings $Y = \{y_i\}_{i=1}^{m_1+\cdots+m_t}$ of the base set, we can use Eq. (6) to produce the embedding of a new sample. We can compute $\alpha_i(x_q)$ by using a softmax classifier, which is train using all labeled dataset.

For every sample in dataset $X = \{x_1, x_2, \cdots, x_n\}$, we can get a vector of Euclidean distances and a vector of similarities between this sample to all notes in base set respectively. The distances vector is formulated as $d_i = \{\|x_i - b_{1,1}\|_2, ..., \|x_i - b_{m_1}, 1\|_2, ..., \|x_i - b_{1,t}\|_2 ..., \|x_i - b_{m_t,t}\|_2\}$, and the similarities vector is formulated as $l_i = \{1,0\}^{m_1+\cdots+m_t}$ where the similarity is 1 if this sample and the note in base set have same class label and 0 on the contrary.

Then we use the data $D = \{d_i\}_{i=1}^n$ and its label $L_d = \{l_i\}_{i=1}^n$ to train a softmax classifier. For every input, the softmax classifier wants to estimate the probability of each class, which is formulated as follow:

$$h_\theta(d_i) = [p(l_i|d_i;\theta)] = \frac{1}{\sum_{j=1}^{m_1+\cdots+m_t} e^{\theta_j^T d_i}} \begin{bmatrix} e^{\theta_1^T d_i} \\ e^{\theta_2^T d_i} \\ \vdots \\ e^{\theta_{m_1+\cdots+m_t}^T d_i} \end{bmatrix} \tag{7}$$

As we known, original softmax classifier cannot classify the sample with multi-labels, so we modified the loss function to be the sum of negative log likelihood of all probabilities of base set notes having same class label with sample. Our loss function is formulated as follow:

$$c(L_d, D) = -\sum_{i=1}^n l_i^T \log(h_\theta(d_i)) \tag{8}$$

We can get the parameter θ by minimizing Eq. (8) using L-BFGS method and then, for a new sample, we compute $\alpha_i(x_q)$ in Eq. (6). After that, we can get the embedding of this sample by linear combination of the base embeddings according to Eq. (6) and use the sign function to get the final hash codes for this new sample. It is deserved to note that we can just preserve the highest k values in $\alpha_i(x_q)$ to eliminate the influence of notes with less similarity, as well as we can use other appropriate classifier in this stage.

Last, we summarize our proposed Inductive Supervised Hashing method as Algorithm 1.

Algorithm 1 Inductive Supervised Hashing (ISH)

Input: Training data $X=\{x_1,x_2,\cdots,x_n\}$, class label $L=\{l_1,l_2,\cdots,l_n\}$, code length r, base set size m, neighborhood size k

Output: Binary codes $Y=\{y_1,y_2,\cdots,y_n\}$

1) Generate the base set $B = \{b_{1,1},\cdots b_{m_1,1},b_{1,2},\cdots b_{m_2,2},\cdots,b_{1,t},\cdots b_{m_t,t}\}$ by clustering on each class.

2) Compute the similarity matrix S using class label.

3) Embed B into the low-dimensional space using an appropriate hashing method.

4) For every labeled sample in dataset, compute the $D = \{d_i\}_{i=1}^n$ and $L_d = \{l_i\}_{i=1}^n$

5) Use the labeled dataset (D, L_d) to train a classifier like softmax

6) Foe a new sample x_q, use the trained classifier to produce the similarity vector $\alpha_i(x_q)$

7) Obtain the low-dimensional embedding Y for the new sample by Eq (6)

8) Threshold Y at zero.

---

4   Experiment Results

In order to evaluate our proposed ISH method more sufficiently, we use the NUS-WIDE datasets. NUS-WIDE is a large-scale real-world web image dataset comprising over 269,000 images with over 5,000 user-provided tags, and ground-truth of 81 concepts for the entire dataset. The dataset is much larger than the popularly available Corel and Caltech 101 datasets, which contains 161,789 images for training and 107,859 images for testing.

In our experiments, instead of using original data, we use decaf features of images which are extracted from the activation of a deep convolutional network. Five hashing algorithms are compared, including the proposed ISH and three unsupervised methods: PCAH [7], SH [9], AGH [10], as well as the IMH-tSNE [11, 12] method. The provided codes and suggested parameters according to the authors of these methods are used.

For the purpose of demonstrating ISH's capacity of solving the problems with out-of-sample data, we generate the same base set B for IMH-tSNE and ISH on the ILSVRC2012 train data, then compute the embeddings Y for the whole images in the NUS-WIDE dataset. We use the k-means algorithm to obtain the base set B. Because the ILSVRC2012 train dataset contains 1.2 million images, the size of B is set between 1000 and 10000. We use the same embedding method in IMH-tSNE to embed the base set. In our method, we use a similarity matrix S encoding the pairwise semantic affinities among the concepts between two datasets. There are several ways to obtain S, such as mining from the Internet corpus or exploiting large-scale crowd sourcing [22]. Then we use this matrix to produce the label between samples and base set nodes. We use the train set of NUS-WIDE to train our classifier and the test set as query set, which is the same as other methods except no classifier trained. The code length of embedding varies from 16 to 256 bits in all compared method.

Figure 1 shows the performance of our proposed ISH method with respect to the mAP@100 for Hamming ranking with different code length and size of base set. We can see that longer code length and larger size of base set can achieve a better performance in the range of 16 to 256 and 1000 to 10000 respectively.

The comparative results based on mAP@100 and F1-measure with code lengths from 16 to 256 bits are reported in Fig. 2. It can be seen that the proposed ISH using

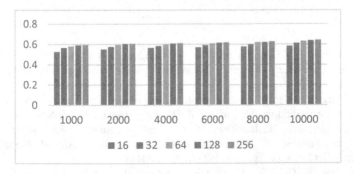

Fig. 1. mAP@100 performance of our proposed ISH with different code length and size of base set

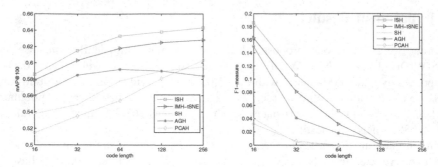

Fig. 2. The comparative results of different methods based on mAP@100 and F1-measure with code lengths

supervised embending for new samples performs best in all cases. IMH-tSNE performs better than unsupervised PCAH, AGH and SH on this dataset, but it is still inferior to ISH. PCAH perform worst in this case.

5 Conclusion

In this paper, we proposed a new supervised compact hashing method based on the IMH to learn a binary code more efficiently and effectively. We first generate base set of small size and embed the base set to hash codes. Then we train a softmax classifier whose inputs are the distances vectors from sample to base set nodes and outputs are the class similarity vector between sample and base set nodes. After that, we can embed a new sample by a weighted combination of hash codes of base set nodes, which weight is the predict similarity vector. Instead of using unsupervised embedding for new samples in IMH, we do a supervised embedding by using the similarity vector for new samples to linearly combine the hash codes of the base set. The experimental results show that this method gets excellent performance on large-scale multimedia dataset, and is very effective and efficient for image retrieval just with very short codes.

References

1. Wang, J., Kumar, S., Chang, S.F.: Semi-supervised hashing for scalable image retrieval. In: 2010 IEEE Conference on Computer Vision and Pattern Recognition (CVPR), pp. 3424–3431. IEEE (2010)
2. Yang, Y., Zha, Z.J., Gao, Y., et al.: Exploiting web images for semantic video indexing via robust sample-specific loss. Multimedia, IEEE Trans. **16**(6), 1677–1689 (2014)
3. Datar, M., Immorlica, N., Indyk, P., et al.: Locality-sensitive hashing scheme based on p-stable distributions. In: Proceedings of the Twentieth Annual Symposium on Computational Geometry, pp. 253–262. ACM (2004)
4. Kulis, B., Jain, P., Grauman, K.: Fast similarity search for learned metrics. Pattern Anal. Mach. Intell. IEEE Trans. **31**(12), 2143–2157 (2009)

5. Kulis, B., Grauman, K.: Kernelized locality-sensitive hashing for scalable image search. In: 2009 IEEE 12th International Conference on Computer Vision, pp. 2130–2137. IEEE (2009)
6. Raginsky, M., Lazebnik, S.: Locality-sensitive binary codes from shift-invariant kernels. In: Advances in Neural Information Processing Systems, pp. 1509–1517 (2009)
7. Wang, J., Kumar, S., Chang, S.F.: Semi-supervised hashing for large-scale search. Pattern Anal. Mach. Intell. IEEE Trans. **34**(12), 2393–2406 (2012)
8. Gong, Y., Lazebnik, S.: Iterative quantization: A procrustean approach to learning binary codes. In: 2011 IEEE Conference on Computer Vision and Pattern Recognition (CVPR), pp. 817–824. IEEE (2011)
9. Weiss, Y., Torralba, A., Fergus, R.: Spectral hashing. In: Advances in Neural Information Processing Systems, pp. 1753–1760 (2009)
10. Liu, W., Wang, J., Kumar, S., et al.: Hashing with graphs. In: Proceedings of the 28th International Conference on Machine Learning (ICML-11), pp. 1–8 (2011)
11. Shen, F., Shen, C., Shi, Q., van den Hengel, A., Tang, Z., Shen, H.T.: Hashing on nonlinear manifolds. Image Process. IEEE Trans. **24**(6), 1839–1851 (2015)
12. Shen, F., Shen, C., Shi, Q., et al.: Inductive hashing on manifolds. In: 2013 IEEE Conference on Computer Vision and Pattern Recognition (CVPR), pp. 1562–1569. IEEE (2013)
13. Yang, Y., Yang, Y., Huang, Z., et al.: Tag localization with spatial correlations and joint group sparsity. In: 2011 IEEE Conference on Computer Vision and Pattern Recognition (CVPR), pp. 881–888. IEEE (2011)
14. Strecha, C., Bronstein, A.M., Bronstein, M.M., et al.: LDAHash: Improved matching with smaller descriptors. Pattern Anal. Mach. Intell. IEEE Trans. **34**(1), 66–78 (2012)
15. Kulis, B., Darrell, T.: Learning to hash with binary reconstructive embeddings. In: Advances in Neural Information Processing Systems, pp. 1042–1050 (2009)
16. Liu, W., Wang, J., Ji, R., et al.: Supervised hashing with kernels. In: 2012 IEEE Conference on Computer Vision and Pattern Recognition (CVPR), pp. 2074–2081. IEEE (2012)
17. Lin, G., Shen, C., Suter, D., et al.: A general two-step approach to learning-based hashing. In: 2013 IEEE International Conference on Computer Vision (ICCV), pp. 2552–2559. IEEE (2013)
18. Wang, J., Kumar, S., Chang, S.F.: Semi-supervised hashing for scalable image retrieval. In: 2010 IEEE Conference on Computer Vision and Pattern Recognition (CVPR), pp. 3424–3431. IEEE (2010)
19. Lin, G., Shen, C., Shi, Q., et al.: Fast supervised hashing with decision trees for high-dimensional data. In: 2014 IEEE Conference on Computer Vision and Pattern Recognition (CVPR), pp. 1971–1978. IEEE (2014)
20. Norouzi, M., Blei, D.M.: Minimal loss hashing for compact binary codes. In: Proceedings of the 28th International Conference on Machine Learning (ICML-11), pp. 353–360 (2011)
21. Shen, F., Shen, C., Liu, W., et al.: Supervised discrete hashing. In: 2015 IEEE Conference on Computer Vision and Pattern Recognition (CVPR), pp. 37–45 (2015)
22. Rohrbach, M., Stark, M., Szarvas, G., et al.: What helps where–and why? semantic relatedness for knowledge transfer. In: 2010 IEEE Conference on Computer Vision and Pattern Recognition (CVPR), pp. 910–917. IEEE (2010)

Graph Based Visualization of Large Scale Microblog Data

Yue Guan, Kaidi Meng, and Haojie Li$^{(\boxtimes)}$

School of Software, Key Laboratory for Ubiquitous Network
and Service Software of Liaoning Province,
Dalian University of Technology, Dalian, China
worm004@hotmail.com, mengkaidi_aileen@sina.com,
hjli@dlut.edu.cn

Abstract. Visualization is an important but tough way to make sense of large scale dataset. In this paper, we propose a graph based method to visualize microblog data. In our scheme, the graph is constructed using the content similarities between data which is more robust than the widely used data relationships. Given a targeted dataset, we first adopt a duplicates removal strategy to reduce the size of the data and a subset is randomly sampled for visualization. Then a multilevel graph layout with a heat map is applied to generate an interactive interface which allows users to move on and scale the layout. In this way, different granularities of summarization information can be immediately presented to users when a certain area is specified in the interface; meanwhile more detailed knowledge on the selected area can be shown in nearly real time by leveraging a hash based microblog retrieval approach. Experiments are conducted on a Brand-Social-Net dataset which contains 3,000,000 microblogs and the experimental results show that, with our visualization method, some meaningful patterns of dataset can be found easily.

1 Introduction

While the time of big data is coming, more and more data is being generated by users in social networks. On one hand, there are enormous values hiding in these user-generated data which leads to the emergence of data mining methods. However, it is recognized that analyzing and understanding a dataset before doing any operation on it is rather helpful. Thus sensemaking is a very important and tough task especially for big data. On the other hand, data can be treated as a kind of resource generated for consuming. Thus users would prefer an innovative interface to consume information more efficiently.

In this paper we propose a graph based method to visualize large scale microblog data. An interface is introduced to make sense of big data. With the interface users can interact with the layout easily to find their interested information. Some visualization methods [1–9] have been proposed these years for big data. Many of them use social relation-ships like retweet or friend/follow to build the relation-ships between data, and they focus on relation-ships among the different facets of data. There are obvious limitations for such kind of methods. Firstly, social relation-ships can be treated as a

© Springer International Publishing Switzerland 2015
Y.-S. Ho et al. (Eds.): PCM 2015, Part II, LNCS 9315, pp. 456–465, 2015.
DOI: 10.1007/978-3-319-24078-7_46

big graph which can be easily destroyed during sampling. While retweet relation-ships are also not reliable because some retweets are just the copy-paste of text content. Secondly, priori knowledge is important to these methods. However, it is usually hard to obtain such priori knowledge from big data in social networks. Some other visualization works [3, 4, 6] use less priori knowledge and they use cluster methods to build a hierarchy structure. The limitation is that the visualization results are highly depended on the cluster methods. It is a great challenge for cluster methods to work with the noisy sparse data in social networks (Fig. 1).

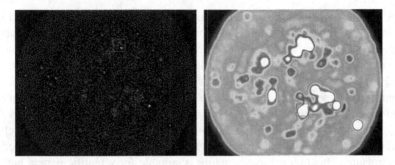

Fig. 1. The visualization result of the proposed method in a large scale microblog dataset. The left one shows nodes in a graph layout, and the right one shows the corresponding 7-color heat map which highlights some clusters in the layout (Color figure online).

In this work we propose a sensemaking method for large scale microblog data. The proposed method uses little priori knowledge, and each microblog is treated as a node of a graph. Similarities between low level textual features are used as constrains for the graph layout. Constrains in our method are more flexible compared to the cluster methods and nodes are not forced being in some cluster or not. Nodes are affected by all other nodes in the graph. The proposed method consists of the following steps. Firstly, we use a preprocess step to filter out a part of noisy and duplicate data. Secondly, a graph is constructed based on the similarities between the low level features of nodes. Thirdly, a graph layout is obtained in a 2d space by using a multilevel layout method. Finally, a hash method is applied to rapidly retrieve similar microblogs from the data set for supporting some interact operations. We also develop an interface to interact with data by moving and scaling. User can also retrieve similar microblogs from a data set by selecting nodes in a certain area. The online interactions are very efficient and can be operated in real-time. The offline part takes less than one hour for pre-processing, feature extraction and graph layout generation in million-level data. The experimental results show some cases of data visualization results which demonstrate the efficiency of our method.

The contribution of this paper is a visualization method for large scale microblog dataset. In this method the relationships among data are constructed using their low-level feature similarities rather than the friend/follow or retweet relationships. A graph layout method which is more flexible than clustering is applied to gather similar data. The rest of the paper is constructed as below. Related work will be surveyed in Sect. 2 in the fields of visualization for big data. Section 3 details the graph construction method which includes preprocess, duplicates removal, feature extraction and edge construction. In Sect. 4 we will introduce our microblog retrieval method which is based on p-stable LSH [11] and Minhash [12]. Section 5 shows the framework of the proposed method and the developed interface. Finally experimental results are shown with two examples in Sect. 6.

2   Related Work

Different visualization methods have been proposed for social network data in recent years. Emden R.Gansner and Yifan Hu [5] proposed a method to visualize a graph as a real map. They constructed a graph based on key words extracted from large data sets, and then mapped the graph into a 2-dimensional space with a Voronoi map. They also proposed a layout generation method [2] for streaming text data, focusing on updating the layout in real-time with the changing text stream. Fernando V. Paulovich et al. [3] proposed a hierarchical node placement strategy for scientific papers. This work firstly uses cluster methods to divide the data into different partitions in different levels, and then builds a cluster-tree with the partitions. Finally the graph layout is calculated from the tree structure. This hierarchical structure is suitable for scientific papers but is too strict for social network data, because there is no 'majors' in microblogs and thus it is impossible to pre-divide them into different clusters. Nan Cao et al. [10] proposed a method called facetAtlas to show different facets in one graph for rich text corpora. Their method use different colors and shapes to act as different relations based on different facets. The limitation is that the visualization result is not intuitive for users.

3   Graph Construction

3.1   Preprocess Method

Social network data is usually of low quality and data preprocess method is important for further steps. In this paper we take Sina Weibo as the evaluation data though it is easy to apply our method to other microblog sources. Microblogs in Sina Weibo usually contain both texts and images data and we preprocess them with text data. "text" and "document" presents this kind of data in the following sections. The ordinary text preprocess method for English social network data contains removing retweet, tag analysis and removing stop words. Other methods are used for processing Chinese short documents in microblogs. Firstly retweets and AT tags like '@somebody' and emoji tags like '[laugh]' are removed. Then all spaces between English words and numbers are deleted to generate some new words. These new words are added into corresponding documents. For example, a text 'WIN 8耗磁盘!' should be converted into 'WIN 8耗磁盘 WIN8'. Then keywords are collected to build up a dictionary for Chinese word segmentation. After a fine-grained segmentation stage all the words in documents are reserved. Finally all the English and Chinese stop words are removed and words which are too long or too short are deleted.

3.2   Feature Extraction

We extract two kinds of features for microblog texts. Firstly TF*IDF feature is extracted for each document. TF*IDF uses Bag of Words model to represent a text and calculate a score for each word in each document. Then a document can be represented as a sparse vector. The score of i-th word in j-th document can be formulated as:

$$X_{ij} = tf_{ij} * idf_i \tag{1}$$

$$Idf_i = \log_2 \left(\frac{N}{df(i)} \right) \tag{2}$$

where tf_{ij} is the frequency of the i-th word in j-th document. N is the total number of microblog documents and df(i) is the number of microblogs where the i-th word appears. TF*IDF is the usually used low level feature in text retrieval task. In graph layout generation for large scale social network, data similarities between TF*IDF features of microblogs are used to determine the distance between them in a 2D space. TF*IDF is a kind of high-dimension feature and people usually reduce its dimension by just keep a small part of words. However, there are at most 140 words in each document in social network thus each single word is very important for a microblog. Here we keep all these words. Since the dimension of TF*IDF is too high for some operations, we suggest another kind of textural feature to be extracted.

We propose to extract Minhash [12] signatures for each document, and use it to remove duplicates and retrieve similar microblogs which will be introduced carefully in the following sections. Minhash is a kind of locality sensitive hashing (LSH) scheme which is a distribution on a set of hash functions F for a set of features. For two features x and y the following equation is satisfied,

$$Pr_{h \in F}[h(x) = h(y)] = sim(x, y) \tag{3}$$

where $sim(x, y) \in [0, 1]$ measures the similarity between x and y. Similarities between Minhash features are used to measure those between TF*IDF features. We firstly generate a set of hash function F, and each of the hash function maps a TF*IDF feature into a 64-bit signature. The distance between two Minhash s_i and s_j is calculated with the hamming distance, and the distance between two TF*IDF x_i, x_j is calculated using the cosine distance. In our experiments the hamming distance is less than 4 means the corresponding microblogs have the same content, and the cosine distance is less than 0.7 means the corresponding Minhash features' hamming distance is less than 21. For short documents in social network, if the TF*IDF distance is less than 0.7 then the corresponding two microblogs can be treated as two similar microblogs which share a same topic in our experiments.

3.3   Duplicates Removal Method

This section shows the duplicates removal method. There are much duplicate data in social network. Some documents or advertisements are very interesting, so people tend to retweet them. However, retweet tags can be removed easily, so the duplicate relationship can only be discovered by comparing document contents. There are many duplicates in microblogs so the size of data can be reduced by removing them. Gurmeet Singh Manku et al. [14] have proposed a good method for detecting near-duplicates for web crawling. The method is applied here to detect near-duplicates in large microblog data.

The method is simple but efficient by using different hash tables to reduce the search space. Minhash features are used here and 4 permuted tables T_0, T_1, T_2, T_3 are constructed for discovering duplicate microblogs whose hamming distance is less than 4 in Minhash features. In each permuted table we search duplicate items for a 64-bit Minhash signature S in the following steps:

(1) Identify all permuted items in T_i whose top 16 bit-positions match the top 16 bit-positions of $\pi_i(s)$. Here π is the select operation.
(2) For each of the permuted items identified in Step 1, check if it differs from $\pi_i(s)$ in at most 3 bit-positions.

This duplicates detection method is proposed for 'long documents' like webpages and there is a little side effect when it is applied to 'short documents' like microblogs. To overcome this, we only applied the method to microblogs longer than 7 words. This is reasonable because the duplicates are usually long documents as they may contain more interesting information than short ones.

3.4 Edge Construction

In this section we build edges for the graph layout. Edges play an important role in force-directed graph layout, which produce attractive forces between nodes. If edges of a graph are linked in good order as a mesh, the layout algorithm will perform well. But the edges of a graph from real-world user-generated data are usually messy. To get a better graph layout result a sparser graph could be constructed. However, the number of edges should be as many as possible to catch enough relationships between microblogs. So it is important to make it balanced between the graph layout's quality and the information quantity that a graph shows. Here a threshold is applied to balance it. It is assumed that two documents in social networks share the same topic if the cosine distance between their TF*IDF features is less than 0.7. As we would like to visualize more relations between nodes, 0.85 is set as the threshold to build edges.

4 Microblog Retrieval

This section introduces the similar microblog retrieval method which is helpful when user is interacting with the graph layout. For example, user may get interested in some part of data in the graph layout, then it is necessary to show contents of these microblogs and other similar microblogs in real-time to users. It is a natural idea to use TF*IDF feature as keys to create hash tables. However, it is a tough task to hash TF*IDF feature directly, because the dimension is too high to achieve efficient indexing. To overcome this, we propose to use Minhash features as keys to index similar items with high recall and then TF*IDF features are used to filter out the dissimilar items. By using Minhash based locality sensitive hashing (LSH) scheme, this method can achieve near-real-time performance.

A p-stable distributions based LSH algorithm [11] is adopted here to retrieve similar documents which costs about 0.15 s for each query in a dataset containing about 1.5 million microblogs. 10 hash tables are created offline and p is set to 2. The detailed steps for indexing a microblog document are summarized as follows:

Retrieve Minhash features by LSH.

Compute hamming distances with Minhash features between the seed and the retrieved microblogs. Keep the microblog if the corresponding distance is less than 20.

Compute cosine distances with TF*IDF features between the seed and the kept microblogs. Keep the microblog if the corresponding distance is less than 0.7.

We show some experimental results to demonstrate the effectivity of the proposed microblog retrieval method using randomly selected seeds. Because of the sparsity of microblog data, some of the seeds do not have similar microblogs or just have a small number of similar microblogs. So they are omitted and we just consider seeds which contain more than 20 similar microblogs. The observations we get are as follows.

Firstly the recall score of Minhash features is close to 1 which means that most Minhash features are retrieved that are similar to seeds. Secondly the recall of TF*IDF features is vary from different seeds. The lowest recall score is about 70 % and the highest one is about 90 %. Numbers of similar microblogs for different seeds are different, so it make little scenes to calculate the average recall score. Here 70 % to 90 % is acceptable for our visualization application.

5   Graph Based Microblog Visualization

Our proposed method is used for visualizing large scale textual data without much priori knowledge. It is applied in a microblog data set in this paper, but this method can be used in other kinds of data sets. A multilevel force-directed layout method [15] is applied to get a graph layout result to show the distributions of the input dataset. It helps users to have a deeper understanding of the data efficiently. In above sections, some steps of our proposed graph based sensemaking method for large scale microblog data have been introduced. In this section, we will combine all of these steps and show a framework of the proposed method. An easy to use interface is provided for users to interact with the generated graph layout.

5.1   The Framework

The proposed visualization framework is shown in Fig. 2. The inline part of the method has been discussed in Sect. 4, so we just show the offline part below.

The input is a large set of textual data. Firstly, preprocess method, feature extraction, and duplicates removal are conducted. Then a small set of data is sampled from the filtered large set of data. Limited by $O(n^2)$ space complexity in multi-level graph layout method, we cannot use all of data to construct a graph and run the graph layout algorithm. Samples are not randomly selected among all the data. Only data that contains more information is considered and microblogs which contains less words are

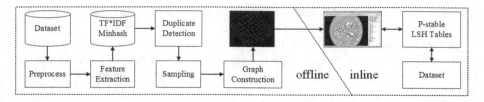

Fig. 2. The framework of our proposed graph based visualization method. The offline part shows steps that contain preprocess, feature extraction, duplicates removing, sampling and edge construction. The inline part shows the retrieval process of the proposed method.

ignored. Each sampled microblog is presented by a node, and edges are built between them. After the graph has been constructed, a multi-level graph layout method is used to get a layout. An easy coarsening strategy is also applied. In each level, a maximal matching is found by randomly selecting edges in the graph as Walshaw [15] did. This strategy is simple but good enough because it must be a sparse graph in large scale data. We use a (0, -2)-energy model [13] for layout which makes clusters more obvious. The quad-tree and the Barnes-Hut algorithm are also applied to speed up the layout process.

5.2   The Interface

We have developed an interface to show the layout result of our graph based visualization method. With this interface, users can observe a graph layout and interact with it at the same time.

The graph layout is shown as the mesh mode where a heat map is presented to show the data density. The score of each grid in the mesh is derived from nearby nodes with their weights calculated by a Gaussian function. We use a 7-color heat map to highlight the cluster in the graph layout.

It is easy to learn the information of the corresponding microblogs by interacting with a mouse. Firstly key words can be shown as a list ranked by their frequencies. Then similar microblogs can be fast retrieved by selecting nodes in one interesting area. The retrieval method has been discussed in Sect. 4, and this indexing operation is near-real-time. Thirdly user can interact with the graph layout by moving and scaling it. When scaling the layout, the mesh size for the heap map changes automatically. The mesh size will increase when it scales up, and verse vice. Finally user can change the date interval, and nodes in the layout will appear or disappear according to the selected date interval. The heat map and the retrieved microblogs is updated according to the selected interval, too.

6   Experimental Results and Discussion

This section gives the details of two experiments to validate the feasibility of the proposed method. Our experiments are conducted on a Brand-Social-Net dataset [16] which contains 3,000,000 microblogs collected from Sina Weibo[1]. There are about

---

[1] www.weibo.com

150,000 brand related microblogs that belong to 100 different brands. Different experiments are performed by selecting different data for graph layout. In the first experiment microblogs are randomly selected from the data set. We mainly show some details of the visualization method and the overview of the dataset. In the second experiment microblogs are randomly selected from the retrieved data using some brand related keywords. Graph layout results are carefully observed and some patterns and knowledge are learnt from it. All experiments are running on a PC which has an Intel i3 2.4 GHz processer and 4G memory.

6.1 The Experimental Result for Randomly Selected Data

We collect microblog documents from the dataset. After the preprocess step, Jieba[2] is used to segment words in these documents. There are about 22,000 words in the dictionary after the words filter step. Then TF*IDF and Minhash features are extracted as text features from the segmented documents. The next step is to remove duplicate documents. One microblog for each group of duplicates is kept, and the weight of the kept microblog is set as the number of its duplicates. The default weight for the rest microblog is set to 1. There are about 1,500,000 distinct documents left after removing all the duplicates. About 6,000 microblogs are randomly selected from these microblogs and construct a graph which has 6,000 nodes and 38,000 edges with the threshold of 0.85. It takes about 15 min to generate the graph layout by the multilevel algorithm. Finally we use the developed interface to show the graph layout result.

We can clearly see that four topics are most interesting to people: constellations, philosophy stories and advertisement (see Fig. 3). It is interesting to see that London Olympic Games is also a big cluster. So we scale the graph layout to see more about this topic. Detail information shows that these data are advertisements. The result is reasonable because these data is collected from 6/1/2012 to 7/31/2012 and the Olympic Games is held at 7/28/2012.

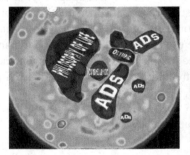 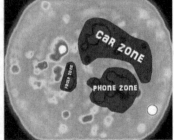

Fig. 3. The left figure shows the visualization result of randomly selected data where some areas have been tagged with their topics. Here we have tagged the top 4 topics in the layout graph. The right figure gives the visualization result of brand related data. The 'car zone', the 'phone zone' and the 'fault zone' are tagged in the graph.

---

[2] A Chinese word segmentation python library.

6.2   The Experimental Result for Brand Related Data

This experiment shows the visualization result generated by a set of brands related data from the dataset. We select some keywords for each brand and use them to retrieve microblogs using the inverted index method. There are about 7,000 microblogs are randomly selected from the retrieved microblogs to construct a graph which contains about 67,000 edges. It takes about 30 min to generate a graph layout with the multi-level graph layout method. Compared to the graph in Sect. 6.1 there are more edges in this graph, but this graph layout looks sparser. The reason is that there are no huge topics like constellation but many small topics. From the developed interface we can find some interesting phenomena. Firstly positions of some low level topics are close in the layout, and they gather themselves as some high level topics. For example, microblogs of each car brand gather as a cluster, and all these clusters are located in a big area, resulting in a 'car zone' generated in the graph layout. Similarly, a 'phone zone' can be found in the layout, too (see Fig. 3). Secondly we can observe events intuitively from the interface. Though there is no precise definition for an event in social networks, it must have the attribute that the number of the related microblogs increases in a small time interval. Then our developed interface can help user to find information that has this attribute. The heat map changes as we change the time interval, and it is easy to catch the changing of clusters.

7   Conclusion

In this paper we present a graph based sensemaking method for large social network data. In the offline part a multi-level force-directed graph layout algorithm is used to generate a 2D visualization result for a group of textual data which are sampled from a large microblogs dataset. In the online part we present a fast LSH based indexing method for short document retrieval. Also an interface which can easily observe the visualization result in different scale levels is provided. Some graph layout examples show that the proposed method can help user to have a deeper understand of a large scale dataset. There are some directions could be further investigated in the future. Firstly the proposed method can not only be used for static data, but also dynamic data flow; Secondly, as we use sampling method to achieve the dynamic interaction of scaling in real-time, a kind of hierarchy structure which is pre-calculated in offline process could be applied instead.

Acknowledgements. This work was partially supported by National Natural Science Funds of China (61173104, 61472059, 61428202).

References

1. Doyle, M., Smeaton, A.F., Bermingham, A.: TriVis: visualising multivariate data from sentiment analysis (2014)
2. Gansner, E., Hu, North, S.: Visualizing streaming text data with dynamic maps (2012). arXiv preprint arXiv:1206.3980

3. Paulovich, F.V., Minghim, R.: Hipp: A novel hierarchical point placement strategy and its application to the exploration of document collections. IEEE Trans. Visual. Comput. Graph. **14**(6), 1229–1236 (2008)
4. James, A., Van Ham, F., Krishnan, N.: Ask-graphview: A large scale graph visualization system. IEEE Trans. Visual. Comput. Graph. **12**(5), 669–676 (2006)
5. Gansner, E.R., Hu, Y., Kobourov, S.: GMap: visualizing graphs and clusters as maps. In: 2010 IEEE Pacific Visualization Symposium (PacificVis). IEEE (2010)
6. Gretarsson, B., et al.: Smallworlds: visualizing social recommendations. Comput. Graph. Forum **29**(3), 833–842 (2010). Blackwell Publishing Ltd
7. Jing, L., Yu, X., Wan, W.: Visualization research of the tweet diffusion in the microblog network. In: 2014 International Conference on Audio, Language and Image Processing (ICALIP). IEEE (2014)
8. Changbo, W., et al.: Analyzing internet topics by visualizing microblog retweeting. J. Vis. Lang. Comput. **28**, 122–133 (2015)
9. Ren, D., et al.: WeiboEvents: a crowd sourcing weibo visual analytic system. In: 2014 IEEE Pacific Visualization Symposium (PacificVis). IEEE (2014)
10. Nan, C., et al.: Facetatlas: multifaceted visualization for rich text corpora. IEEE Trans. Visual. Comput. Graphics **16**(6), 1172–1181 (2010)
11. Datar, M., et al.: Locality-sensitive hashing scheme based on p-stable distributions. In: Proceedings of the Twentieth Annual Symposium on Computational Geometry. ACM (2004)
12. Charikar, M.S.: Similarity estimation techniques from rounding algorithms. In: Proceedings of the Thiry-Fourth Annual ACM Symposium on Theory of Computing. ACM (2002)
13. Andreas, N.: Modularity clustering is force-directed layout. Phys. Rev. E **79**(2), 026102 (2009)
14. Manku, G.S., Jain, A., Das Sarma, A.: Detecting near-duplicates for web crawling. In: Proceedings of the 16th International Conference on World Wide Web. ACM (2007)
15. Walshaw, C.: A multilevel algorithm for force-directed graph drawing. In: Marks, J. (ed.) GD 2000. LNCS, vol. 1984, pp. 171–182. Springer, Heidelberg (2001)
16. Yue, G., et al.: Brand data gathering from live social media streams. In: Proceedings of International Conference on Multimedia Retrieval. ACM (2014)

Boosting Accuracy of Attribute Prediction via SVD and NMF of Instance-Attribute Matrix

Donghui Li[1], Zhuo Su[2,4(✉)], Hanhui Li[1], and Xiaonan Luo[1,3]

[1] National Engineering Research Center of Digital Life,
School of Information Science and Technology,
Sun Yat-sen University, Guangzhou, China
[2] School of Advanced Computing, Sun Yat-sen University, China, Guangzhou
suzhuo3@mail.sysu.edu.cn
[3] Shenzhen Digital Home Key Technology Engineering Laboratory, Shenzhen, China
[4] Collaborative Innovation Center of High Performance Computing,
National University of Defense Technology, Changsha, China

Abstract. Attribute-based methods for image classification have received much attentions in recent years due to the high-level or human-specified nature of attributes. Given a new image, attribute-based methods can predict its category by exploiting the attribution representation of the given image. However, the foundation of attribute-based methods is predicting attributes precisely, which is still a difficult problem in real world applications. Therefore, in this paper, we propose an Attribute Prediction boosting framework with Matrix Factorization techniques (APMF) to boost the accuracy of attribute prediction. APMF explores the potential relationships of instances and attributes by utilizing the singular value decomposition (SVD) and non-negative matrix factorization (NMF). A series of experiments show that our APMF achieves better attribute prediction accuracy than the state-of-the-art methods.

Keywords: Image classification · Matrix factorization · Singular value decomposition · Non-negative matrix factorization · Attribute prediction

1 Introduction

The idea of attribute was first introduced by Farhadi et al. [1]. Intuitively, given a particular object, attributes are the properties of that object. Semantic attributes could be parts ("has paw"), Shape ("lean"), materials ("furry"), and discriminative description ("birds have it but fish do not"). Later, relative attributes [2] were studied to enable richer textual descriptions for new images, which are more precise for human interpretation in practice. Further, relative

This research is supported by the Natural Science Foundation of China (NSFC) (No. 61320106008, 61370186), NSFC-Guangdong Joint Fund (No. U1135005), and the Shenzhen Technology R&D Program for Basic Research (No.JCYJ2013 0401160945590).

© Springer International Publishing Switzerland 2015
Y.-S. Ho et al. (Eds.): PCM 2015, Part II, LNCS 9315, pp. 466–476, 2015.
DOI: 10.1007/978-3-319-24078-7_47

attributes were used for refining retrieved results of image search systems [3]. One of the valuable property of attribute-based methods is to tackle the problem of object classification in the case of zero-shot learning, when training and test classes are disjoint [4,5]. Jayaraman et al. [6] introduced a multi-task learning approach for using semantics to guide attribute learning, which overcame misleading training data correlations, and successfully learned semantic visual attributes to help in the recognition and discovery of unseen object categories. Different tasks with attribute recognition were benefited from their method.

Since attribute learning is significant to high-level tasks, attribute relationships have attracted more and more attention in the recent years. Turakhia et al. [7] identified which attributes in an image are more dominant than others by modelling attribute dominance. Han et al. [8] encouraged highly correlative attributes to share a common set of relevant low-level features. However, in Turakhia's and Han's work, they only explored the relationships among attributes. They did not consider that how an instance is related to other instances and what's the relationships between attributes and instances. Li et al. [9] proposed the relative forest algorithm to achieve more accurate attribute prediction. Note that their work just focus on relative attribute prediction, but ignores the potential relationship between instances and binary attributes. Not difficult to understand, binary attributes are necessary in lots of cases, such as "has pads", "has tail". Therefore, the problem of boosting attribute prediction accuracy for attribute-based methods is still unsolved and we need to utilize the potential relationship between instances and attributes to tackle this problem.

Recently, matrix factorization is applied to recommendation system to realize the latent factor models [10] between users and items. As demonstrated in [10,11], matrix factorization models are superior to classical nearest-neighbour solutions for producing product recommendations. The reason is that it allows the incorporation of additional information such as implicit feedback, temporal effects, and confidence levels. In this paper, matrix factorization is used as an

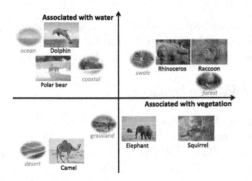

Fig. 1. An illustration for potential relationship models space. It describes both instances (the rectangular ones) and attributes (the oval ones) in the unified space. In this example we just use two dimensions ("Associated with vegetation" and "Associated with water") for demonstration purpose.

effective way to boost the precision of the attribute prediction, which discovering and utilizing the potential relationship models between instances and attributes.

There are many traditional methods for matrix factorization, such as Triangular Factorization (TF), QR Factorization and Singular Value Decomposition (SVD) [12]. Some new methods for matrix factorization are also presented, such as Non-negative Matrix Factorization (NMF) [13], and Latent Factor Model (LFM) [10]. SVD conveys important geometrical and theoretical insights about linear transformations due to its interesting and attractive algebraic properties [12]. In recent years, SVD has been broadly applied to computer vision and recommendation system [14–16]. Zhang et al. [15] utilized discriminative k-SVD for dictionary learning. Sali [14] maked use of SVD for predicting movie ratings. NMF is different from other methods by its non-negativity constraint. This allows a parts-based representation because only the addition is allowed in additive combinations [17].

In this paper, we will explore the relationships among instances, among attributes, as well as the relationships between instances and attributes to boost attribute prediction accuracy. After using attribute classifiers to make attribute prediction for the test images, we apply the proposed framework (Attribute Prediction boosting framework based on Matrix Factorization techniques (APMF)) with the predicted Instance-Attribute matrix in two ways. One is to make use of SVD to detect groupings. The other is to take the advantages of NMF, since all the predicted attribute probabilities are non-negative. Given a new instance from the test set, APMF allows us to find the instance a certain number of neighbours from the training set in the "reduced"potential relationship models space (see Sects. 3.2 and 3.3), whose attributes patterns are similar to that of the given instance. Enhanced attribute prediction for the image are made by utilizing the ground truth attribute labels of its own neighbours. Figure 1 shows the intuition of APMF in two dimensions form.

The advantages of our proposed APMF for attribute-based image classification system are that when we employ the "reduced" potential relationship models space by SVD and NMF models

- Potential relations among instances, and among attributes could be located.
- Potential relations between instances and attributes could be located.
- The located relations could be used to boost attribute prediction accuracy.

2 Preliminaries to Matrix Factorization

In this section, we will introduce two matrix factorization techniques, SVD and NMF, which play an important role in our method. In the next section, we will address the problem of how we utilize SVD and NMF to boost the accuracy of attribute prediction.

2.1 Singular Value Decomposition Model

Generally, singular value decomposition (SVD) is significant if the matrix has a large dimensions, sometimes up to tens of thousands or even larger. Given an arbitrary matrix $A \in R^{m \times n}$, SVD produces two orthogonal matrices $U \in R^{m \times m}$ and $V \in R^{n \times n}$, along with a diagonal matrix $S \in R^{m \times n}$, such that

$$A = USV^{\mathbf{T}}, \tag{1}$$

where S is rectangular with the same size of A and its diagonal entries, i.e. $S_{ii} = \sigma_i$, are in descending order. The singular values of A means the positive ones of σ, and the left and right singular vectors of A means the columns of U and V respectively. SVD is of significant importance in several applications of linear algebra, such as the calculation of the generalized inverse of a matrix and computing the eigen value decomposition of a matrix product $A^{\mathbf{T}}A$. Further more, SVD also can be utilized as a numerically reliable estimation of the effective rank of the matrix.

2.2   Non-negative Matrix Factorization Model

We call a matrix as non-negative matrix if all entries of the matrix are non-negative. Non-negative matrix factorization (NMF) algorithm has been demonstrated to be able to learn parts of faces and semantic features of text [17]. NMF is different from other matrix factorization methods by using non-negativity constraint, which only allows additive, not subtractive combinations. Given a non-negative matrix B, NMF finds the two non-negative matrix factors W and H such that

$$B \approx WH. \tag{2}$$

As mentioned before, only no negative entries are allowed in the matrix factors W and H. If B is size of $m \times n$, then W and H can be size of $m \times r$ and $r \times n$, respectively. Usually r is smaller than m or n, so that W and H are smaller than the original matrix B, resulting in a compressed version of the original data matrix. The difference between SVD and NMF is that the entries of W and H are all non-negative. The non-negative constraint of W and H makes it possible that a object can be decomposed into a non-negative linear combination of its different parts. How NMF learns a parts-based representation [17] is just benefited from the non-negative linear combination. In addition, both W and H possess special physical meanings to distinguish NMF from other decomposition methods. Each column of W is treated as a vector and W is treated as a set which composed of r vectors. In this way, each column of H is an approximation of the projection of the corresponding column of B onto the linear space formed by the vectors of W.

3   Boosting Accuracy of Attribute Prediction Based on Matrix Factorization

Now we will make a detailed explain for how to incorporate SVD and NMF into the attribute prediction step of attribute-based image classification task. Our proposed framework is presented in Fig. 2.

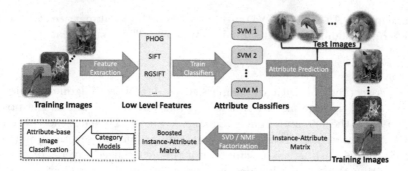

Fig. 2. The pipeline of our method. Given the training set, low-level features of images are extracted firstly. Then a SVM classifier [18] is trained for each attribute using the extracted features. After performing attribute prediction, SVD and NMF models are applied to the Instance-Attribute matrix to boost the attribute prediction accuracy. Content in the red dotted box is a category model based on our method for image classification (Color figure online).

3.1   Baseline Algorithm

The baseline algorithm means that we use the traditional manipulation steps of attribute-based image classification to perform attribute prediction. Given a training set $(x_1, x_2, ..., x_{n_1}) \in X_{N_1}$ of size N_1, a test set $(x_1, x_2, ..., x_{n_2}) \in X_{N_2}$ of size N_2, and a user-specified attributes space $(a_1, a_2, ..., a_m) \in A^M$, we define M attribute classifiers $(f_1, f_2, ...f_m) : X^d \rightarrow A^M$, that map the test images to the attributes space A^M, where d is the dimension of low level features for images.

Note that, after getting the attribute classifiers, we not only use them to predict the probabilities of having attributes in images from the test set (X_{N_2}), but also those of images from the training set (X_{N_1}), as shown in Fig. 2. Denote the original predicted Instance-Attribute matrix by $I \in R^{N \times M}$, which is generated from the baseline algorithm with $N = N_1 + N_2$. Row n of I is the instance n, column m of I means the attribute m, and the element I_{nm} means the predicted probability value of attribute m for instance n.

Next, we will describe the steps of cascading SVD and NMF with the baseline algorithm to achieve better accuracy of attribute prediction.

3.2   Boosting Accuracy of Attribute Prediction by SVD

We want to discover the potential relationship models from I, which indicate the relationships among instances, attributes and the potential models. If two instances are similar (e.g. monkey and chimpanzee), then their attributes patterns should also be analogous. Thus SVD is utilized to achieve this goal as the first method we present in this paper to boost the accuracy of the resulting attribute prediction from the baseline algorithm. We refer this method as APMF-SVD. To prevent our methods from overfitting, we divide the rows of I into two parts. The first part is used as training instances and the second part is used as test instances. But we still put them together to proceed the following steps and still denote it as I for convenience.

The first step of APMF-SVD is to compute the SVD of I and obtains the three following matrices which satisfy

$$I = USV^{\mathrm{T}}. \tag{3}$$

Note that the produced three matrices are also divided into training and test parts, respectively. In Eq. 3, matrix $U \in R^{N \times N}$ tells us how the instances are related to the potential models; $V \in R^{M \times M}$ demonstrates the correlations between attributes and the potential models, and $S \in R^{N \times M}$ shows us the weights of the potential models in descending order. Relationships among instances, attributes and potential models can be discovered in this potential model space. For example, a raccoon is much more like a squirrel than a rhinoceros since their corresponding row vectors in U are more closed. Similarly, attribute "ocean" is more relative with attribute "coastal" than attribute "desert", since their corresponding row vectors in V are more closed. In order to remove the redundant potential models which are useless for boosting the accuracy of attribute prediction, we need to effectively reduce the dimension of I. More specifically, we select only k diagonal entries from matrix S to obtain $S_k \in R^{k \times k}$ while keeping more than 95 % information of S. Correspondingly, we get the reduced $U_k \in R^{N \times k}$ and $V_k^{\mathrm{T}} \in R^{k \times M}$. These reduced three matrices form the so called "reduced" potential model space. Let's denote the "reduced" I as I_k, which is an approximation of I and expressed as

$$I_k = U_k S_k V_k^{\mathrm{T}}, \tag{4}$$

In Eq. 4, the n-th row in the second part of I is the projection of instance n in that "reduced" potential model space.

The second step of APMF-SVD is to enhance the attribute prediction of test instances. Note that, any learning-based classification technique can be used here to achieve this goal, but we prefer to use KNN method [19] because it is easy to apply and bring out the effectiveness of the proposed APMF framework. We build up a KD-Tree to store the reduced Instance-Attribute matrix. This data structure allows us to access to each instance and its neighbours in the "reduced" potential model space effectively. Then, for each instance in the test set, we refine its attribute prediction result by using the majority voting of its neighbours in the training set.

3.3   Boosting Accuracy of Attribute Prediction by NMF

Just like APMF-SVD, the second method to boost the accuracy of the resulting attribute prediction from the baseline algorithm is to utilize NMF technique. For each instance in the test set, it also finds the corresponding instance-neighbourhood in the "reduced" potential model space to get its boosted attribute prediction. Shortly, we refer this algorithm as APMF-NMF.

APMF-NMF is different with APMF-SVD in two steps, one is the factorization step and the other is the finding instance-neighbourhood step. In the first one, APMF-NMF computes the NMF of I to obtain only two non-negative entries factors W and H, which should be satisfied that

$$I \approx WH, \tag{5}$$

where $W \in R^{N \times r}$, $H \in R^{r \times M}$, and r is the selected dimension for W and H. Note that, the "reduced" potential model space of APMF-NMF is formed by W and H. Empirically, we chose the value of r through experiments. Secondly, finding the neighbours for instances in the test set is performed on the W space instead of the U_k space.

4   Experiments

In this section, we demonstrate the effectiveness of the proposed APMF-SVD and APMF-NMF methods through some experiments. Note that for all the following experiments, the attribute prediction accuracy is measured by the area under ROC curve (AUC). All the experiments were tested on PC with Intel i3-2100M 3.10 GHz CPU, 4 GB DDR3 Ram, and MATLAB R2013a.

4.1   Data Set and Parameters Tuning

The "Animals with Attributes" (AwA) data set [4] is used in our experiments, which contains more than 30000 images from 50 classes and each class is denoted with 85 human-named attributes. The 6 kinds of low-level features provided by AwA are cascaded as our image features. When training the attributes SVM classifiers, we use the radial basis kernel function with g=0.01 and c=10.

To get the best performance of the two proposed methods, tuning the parameters that used in the methods is necessary. For k used in APMF-SVD, we chose its value by constraining that I_k keeps more than 95 % information of I. For r used in APMF-NMF, the tuning results are shown in Fig. 3(a). We set $r = 20$ to save time and memory. When r gets larger, the accuracy almost remains unchanged (maximum increasement of AUC is 0.5 %). Denote the number of neighbours for each instanceas as $|Nei|$, the adjusting result of different values of $|Nei|$ is shown in Fig. 3(b). Again, for the sake of computational cost, we set $|Nei| = 100$. When $|Nei|$ is greater than 100, the AUC curve keeps stable (the best one improves less than 1 %).

4.2   Experimental Results

Figure 4 shows the result of different partitions of the training/test data set. As we can see, for all the divisions of training/test set, APMF-SVD and APMF-NMF get a much better accuracy of attribute prediction than the baseline algorithm, which effectively proves the validity of the proposed APMF framework.

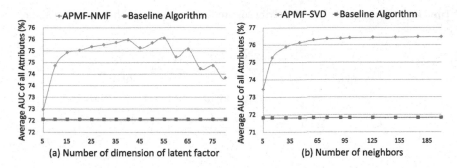

Fig. 3. It is necessary to tune the parameters to get the best performance of APMF framework. (a) shows the average AUC values of all 85 attributes corresponding to the different values of r. (r is the selected dimension of potential model space in APMF-NMF). (b) gives the average AUC values of all 85 attributes corresponding to the different number of neighbours ($|Nei|$) for each test instance.

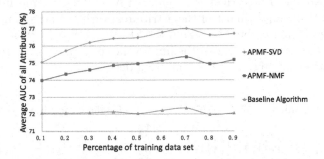

Fig. 4. Result of different partitions between training/test data set ("1" means the whole data set).

The discovered potential relationship models by APMF-SVD and APMF-NMF can explain the prediction of attributes by characterizing both instances and attributes on. As mentioned at the beginning, for attributes, the discovered potential models can measure obvious dimensions such as how closed are the attributes with water, or the cover of vegetation of the attributes, and so forth; The discovered potential models can also measure less well-defined dimensions such as mutual exclusion of attributes; What's more, it can measure completely uninterpretable dimensions. For instances, each potential model measures the probability of the instance having the attributes that score high on that potential models. We can discover and utilize relationships among instances, attributes and potential models for boosting accuracy of attribute prediction by effectively factorizing the Instance-Attribute matrix.

In order to demonstrate the effectiveness of APMF more convincingly, we also give the average AUC values of 10 times cross validation experiments, showed in Table 1. Here, we use 80 % of the data set as training set, and test on the left more than 6000 instances. From Table 1, we can see that the accuracy of attribute prediction has been greatly increased by utilizing matrix factorization techniques with roughly 8 % for the largest increasement.

Table 1. Average accuracy of attribute prediction from 10 times cross validation experiments.

	Baseline Algorithm (SVD)	APMF-SVD
Average AUC	71.94%	79.38%
	Baseline Algorithm (NMF)	APMF-NMF
Average AUC	72.00%	76.11%

Figure 5 gives the quality of individual attribute predictors combined with APMF-SVD and APMF-NMF methods. Note that within the APMF frame work, for some attributes (e.g. attributes "plankton" and "skimmer") can be learned perfectly with AUC=0.97 and AUC=0.98, respectively. And for all attributes the performance of the proposed methods is much better than random (AUC=0.5). Note also, that the performance of APMF-SVD is much better than the baseline algorithm for most of the attributes, with the largest increasement of 12.58 % for attribute "bipedal". So does APMF-NMF, with 8.55 % rising for attribute "vegetation". This shows the importance of exploring the relationships among instances, attributes and potential models.

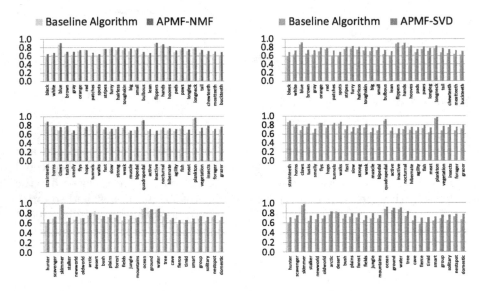

Fig. 5. Quality of individual attribute predictors combined with APMF-SVD and APMF-NMF methods, as measured by AUC.

5   Conclusion

In this paper, we propose the APMF framework to boost the accuracy of attribute prediction, which is based on SVD and NMF models. The potential model space indicates the relationships among instances, attributes and potential models. By discovering and utilizing the potential model space, we demonstrate our APMF-SVD and APMF-NMF methods have superior performance on boosting the attribute prediction accuracy. Once we get a good prediction of attributes we could deal with the attribute-based image classification tasks to obtain a better object recognition or scene understanding.

Although we have a better accuracy of attribute prediction than the baseline, it takes too much time to get the boosted attribute prediction for the instances of the test set. Thus, in future work, we will consider utilizing fast parallel algorithms to perform matrix factorization and to find neighbours of test instances for improving efficiency. In addition, we will try the Latent Factor Model to improve the prediction performance.

References

1. Farhadi, A., Endres, I., Hoiem, D., Forsyth, D.: Describing objects by their attributes. In: IEEE Conference on Computer Vision and Pattern Recognition, CVPR 2009, IEEE (2009)
2. Parikh, D., Grauman, K.: Relative attributes. In: IEEE International Conference on Computer Vision (ICCV) 2011, pp. 503–510. IEEE (2011)
3. Kovashka, A., Parikh, D., Grauman, K.: WhittleSearch: Image search with relative attribute feedback. In: IEEE Conference on Computer Vision and Pattern Recognition, CVPR 2012, IEEE (2012)
4. Lampert, C.H., Nickisch, H., Harmeling, S.: Learning to detect unseen object classes by between-class attribute transfer. In: IEEE Conference on Computer Vision and Pattern Recognition, CVPR 2009, pp. 951–958. IEEE (2009)
5. Suzuki, M., Sato, H., Oyama, S., Kurihara, M.: Image classification by transfer learning based on the predictive ability of each attribute. In: Proceedings of the International MultiConference of Engineers and Computer Scientists, vol. 1 (2014)
6. Jayaraman, D., Sha, F., Grauman, K.: Decorrelating semantic visual attributes by resisting the urge to share. In: Computer Vision and Pattern Recognition, CVPR 2014 (2014)
7. Turakhia, N., Parikh, D.: Attribute dominance: What pops out. In: Computer Vision (ICCV) (2013)
8. Han, Y., Wu, F., Lu, X., Tian, Q., Zhuang, Y., Luo, J.: Correlated attribute transfer with multi-task graph-guided fusion. In: Proceedings of the 20th ACM International Conference on Multimedia, pp. 529–538. ACM (2012)
9. Li, S., Shan, S., Chen, X.: Relative forest for attribute prediction. In: Lee, K.M., Matsushita, Y., Rehg, J.M., Hu, Z. (eds.) ACCV 2012, Part I. LNCS, vol. 7724, pp. 316–327. Springer, Heidelberg (2013)
10. Koren, Y., Bell, R., Volinsky, C.: Matrix factorization techniques for recommender systems. Computer 42(8), 30–37 (2009)
11. Takács, G., Pilászy, I., Németh, B., Tikk, D.: Matrix factorization and neighbor based algorithms for the netflix prize problem. In: Proceedings of the 2008 ACM Conference on Recommender Systems, pp. 267–274. ACM (2008)

12. Kalman, D.: A singularly valuable decomposition: the SVD of a matrix. College Math J. **29**, 2–23 (1996)
13. Lee, D.D., Seung, H.S.: Algorithms for non-negative matrix factorization. In: Advances in Neural Information Processing Systems, pp. 556–562 (2001)
14. Sali, S.: Movie rating prediction using singular value decomposition. In: Machine Learning Project Report by University of California, Santa Cruz (2008)
15. Zhang, Q., Li, B.: Discriminative K-SVD for dictionary learning in face recognition. In: 2010 IEEE Conference on Computer Vision and Pattern Recognition (CVPR), pp. 2691–2698. IEEE (2010)
16. Rajwade, A., Rangarajan, A., Banerjee, A.: Image denoising using the higher order singular value decomposition. Pattern Anal. Mach. Intell. IEEE Trans. **35**(4), 849–862 (2013)
17. Lee, D.D., Seung, H.S.: Learning the parts of objects by non-negative matrix factorization. Nature **401**(6755), 788–791 (1999)
18. Chang, C., Lin, C.: LIBSVM: A library for support vector machines. Intell. Syst. Technol. ACM Trans. **2**(3), 27:1–27:27 (2011)
19. Min-Ling, Z., Zhou, Z.: ML-KNN: A lazy learning approach to multi-label learning. Pattern Recogn. **40**(7), 2038–2048 (2007)

Fatigue Detection Based on Fast Facial Feature Analysis

Ruijiao Zheng[1], Chunna Tian[1(⊠)], Haiyang Li[1], Minglangjun Li[1], and Wei Wei[2]

[1] State Key Laboratory of Integrated Services Networks,
School of Electronic Engineering, Xidian University, Xi'an 710071, China
chnatian@xidian.edu.cn
[2] School of Computer Science and Engineering,
Northwestern Polytechnical University, Xi'an 710072, China

Abstract. A non-intrusive fatigue detection method based on fast facial feature analysis is proposed in this paper. Firstly, the facial landmarks are obtained by the supervised descent method, which automatically tracks the faces and fits the facial appearance very fast and accurately. It covers facial landmarks over a wide range of human head rotations. Then the aspect ratios of eyes and mouth are computed with the coordinates of the detected facial feature points. We interpolate and smooth those aspect ratios by a forgetting factor to deal with the occasionally missing detection of facial features. Thirdly, the degrees of eye closure and mouth opening are evaluated with two Gaussian based membership functions. Finally, the driver fatigue state is inferred by several IF-Then logical relationships by evaluating the duration of eye closure and mouth opening. Experiments are conducted on 41 videos to show the effectiveness of the proposed method.

Keywords: Fatigue monitoring · Driver drowsiness · Facial feature localization

1 Introduction

Driver fatigue resulting from sleep deprivation or sleep disorder is one of the major causes inducing the increase of traffic accidents on today's roads. Thus, the automatic detection of the driver fatigue early enough to warn the driver may save a significant amount of lives and personal suffering. However, general standards for fatigue are unavailable. In this context, it is crucial to use new technologies to monitor drivers and to measure their fatigue level during the entire driving process. Main fatigue measurements developed in recent years are based on the running status of vehicles, the driver's physiological features or the driver's visual features.

The running status of the vehicle such as the sudden changes of steering wheel angle [1], instable speed and special lane changing mode etc. usually reflects the drowsiness of the driver. Since fatigue is one of the physiological states of human being, the drowsiness of driver can also be monitored by the biomedical signals such as pulse rate, heart rate, electromyography signal [2] and electroencephalogram

© Springer International Publishing Switzerland 2015
Y.-S. Ho et al. (Eds.): PCM 2015, Part II, LNCS 9315, pp. 477–487, 2015.
DOI: 10.1007/978-3-319-24078-7_48

(EEG) data [3] etc. Jung et al. monitored the driver's fatigue levels by evaluating the HRV (heart rate variability) sensed by an embedded electrocardiogram (ECG) sensor with electrically conductive fabric electrodes on the steering wheel of a car [4]. Nambiar et al. classified the power spectrum of the driver's HRV obtained from ECG signals to detect the onset of driver drowsiness [5]. Yang et al. detected driver drowsiness through observing blink's pattern and other driving-relevant eye movements of the driver recorded by electrooculography [6]. Murugappan et al. analyzed the frequency bands of EEG signal to classify the driver's drowsiness level with subtractive fuzzy classifier [7]. Usually, the devices to obtain the biomedical signals are expensive and intrusive. The stability of the obtained signals is prone to recede by noise in the real applications.

With the popularity of low-cost cameras, the video of drivers can be obtained non-intrusively. The drowsiness level of the driver can be determined by their visual facial features. Different symptoms for driver drowsiness monitoring have been introduced such as eye closure, blinking, yawning, fixed gaze direction [8, 9] etc. According to the comparisons among those commonly used detecting and evaluating techniques for motoring driver fatigue, the PERCLOS [10] (percentage of eyelid closure over the pupil over time) is the most closely correlated with driver fatigue in the study of simulating driving instrument. The driver is considered as drowsy when more than 15 % of time that the driver's eyes are closed (e.g. 80 % or more) in a unit time interval. Hong et al. [10] detected human face by the Adaboost classifiers [11]. The eyes in the video are tracked based on CAMSHIFT algorithm. Eye states were estimated using a complexity function with a dynamic threshold. The drowsiness state of driver is determined by PERCLOS. Hardeep et al. detected fatigue of the driver by the blinking and closure of the driver's eyes and issued a timely warning. This approach requires the camera to focus on the area around the driver's eyes to improve the detection speed [12]. Coetzer et al. showed that AdaBoost is the most suitable eye classifier among artificial neural networks, support vector machines and AdaBoost based classifiers for a real world driver fatigue monitoring system [13]. Zhang et al. localized the eye region by AdaBoost classifier. In [14], driver fatigue is detected by calculating PERCLOS based on eye closure determined by the lattice degree of nearness of Fourier descriptor. Zhang et al. used drowsiness features extracted from the tracked eyes of the driver in the video. Six measures such as PERCLOS, blink frequency, maximum close duration, average opening level, opening velocity and closing velocity are combined using Fisher's linear discriminant functions to classify the drowsiness state [15]. More recently, driver drowsiness detection is realized by facial expression analysis method. Tadesse et al. analyzed the facial expression of the driver through Hidden Markov Model (HMM) based dynamic modeling to detect drowsiness [16]. Ling et al. adopted the Gabor filter on facial images, which are divided into multiple sub-blocks to learn a discriminative local structured dictionary via the Fisher discrimination dictionary learning method, for efficient features describing drowsiness [17]. Moreover, Anumas et al. implemented a system which combined with facial features and biomedical signals using a fuzzy classifier to infer the level of driver fatigue [18].

The ultimate goal of this research is to develop a camera-based driver fatigue monitoring system, centered around the tracking of driver's eyes, since the eyes provide the most information with regards to fatigue. The precise of facial feature extraction is important for assessing drowsiness. Therefore, it is important to explore innovative technologies to monitor the driver visual attention. For the test videos, we initialize the facial position using Adaboost based face detector. Then, we use supervised descent method (SDM) [19] to fit facial landmarks on eyes, nose, mouth, eyebrows accurately. These facial landmarks are used to initialize the landmarks in the next frame. When the current frame cannot detect face image or locate facial landmarks correctly, the next frame will be re-detected with Adaboost detector. Our method defines visual fatigue based on detected facial landmarks.

2   Facial Landmark Detection and Tracking by SDM

Given an image $\mathbf{I}$, we assemble the coordinates of all facial landmarks detected in the last subsection as an initial coordinate vector $\mathbf{t}_0$. The refinement of the locations of those landmarks is realized by the SDM method [19]. We denote the true location of facial landmarks as $\mathbf{t}^*$ and the refined location of facial landmarks as $\mathbf{t}_0 + \Delta\mathbf{t}$. In order to achieve a robust representation against illumination, we use SIFT as the feature of different landmarks. Denote $\mathbf{h}(\mathbf{t})$ as a SIFT operator function on facial landmarks $\mathbf{t}$. e.g. $\mathbf{h}(\mathbf{t}_0)$ is the SIFT feature extracted from $\mathbf{t}_0$ in image $\mathbf{I}$. Denote $f(\mathbf{t})$ as the distance between the SIFT features extracted from $\mathbf{t}$ and $\mathbf{t}^*$, which is $f(\mathbf{t}) = \|\mathbf{h}(\mathbf{t}) - \mathbf{h}(\mathbf{t}^*)\|_2^2$.

Suppose $\mathbf{t}^*$ is known for both training and testing images, the refinement of facial landmarks is modeled by minimizing the following function over $\Delta\mathbf{t}$.

$$f(\mathbf{t}_0 + \Delta\mathbf{t}) = \|\mathbf{h}(\mathbf{t}_0 + \Delta\mathbf{t}) - \mathbf{h}(\mathbf{t}^*)\|_2^2 \tag{1}$$

With Taylor's theorem, $f(\mathbf{t}_0 + \Delta\mathbf{t})$ is represented by

$$f(\mathbf{t}_0 + \Delta\mathbf{t}) \approx f(\mathbf{t}_0) + \mathbf{J}_f(\mathbf{t}_0)^\mathrm{T}\Delta\mathbf{t} + \frac{1}{2}\Delta\mathbf{t}^\mathrm{T}\mathbf{H}(\mathbf{t}_0)\Delta\mathbf{t} \tag{2}$$

where $\mathbf{J}_f(\mathbf{t}_0)$ and $\mathbf{H}(\mathbf{t}_0)$ are the Jacobian and Hessian matrices of function $f(\mathbf{t})$ evaluated at $\mathbf{t}_0$. We drop $\mathbf{t}_0$ from $\mathbf{J}_f(\mathbf{t}_0)$ and $\mathbf{H}(\mathbf{t}_0)$ to simplify the notation in the following. "T" denotes transposition operation. The refinement of facial landmarks always needs multiple iterations to converge to $\mathbf{t}^*$. $\Delta\mathbf{t}_i$ denotes the i-th update for $\Delta\mathbf{t}$. The first update equation of $\mathbf{t}$ is obtained by differentiating $f(\mathbf{t}_0 + \Delta\mathbf{t})$ with respect to $\Delta\mathbf{t}$ and setting it to zero, which results in Eq. (3).

$$\Delta\mathbf{t}_1 = \mathbf{H}^{-1}\mathbf{J}_f \tag{3}$$

With the chain rule, $\mathbf{J}_f$ can be further represented as $\mathbf{J}_f = 2\mathbf{J}_h^\mathrm{T}(\mathbf{h}(\mathbf{t}_0) - \mathbf{h}(\mathbf{t}^*))$ by differentiating Eq. (1) with respect to $\Delta\mathbf{t}$ [19], where $\mathbf{J}_h$ is a simplicity representation of Jacobian matrix from SIFT operator function $\mathbf{h}(\mathbf{t})$ evaluated at $\mathbf{t}_0$. Then the update Eq. (3) evolves as

$$\Delta t_1 = -2H^{-1}J_h^T(h(t_0) - h(t^*)) \tag{4}$$

With Δt_1, a new facial landmark location $t_1 = t_0 + \Delta t_1$ can be obtained, where t_1 represents the facial landmark location after the first iteration. It can be used together with t^* to find the new update item for Δt in the following iteration. The i-th update for Δt is given as follows.

$$\Delta t_i = t_i - t_{i-1} = -2H^{-1}J_h^T(h(t_{i-1}) - h(t^*)) \tag{5}$$

However, the true location of facial landmarks t^* is only known for the training image, while unknown for the test image. It leads Eqs. (4) and (5) cannot be used to refine facial landmark locations for the test data. In addition, SIFT operator is not differentiable, thus numerical approximations are very computationally expensive to calculate the Jacobian and Hessian matrix in Eqs. (4) and (5).

To address this problem, a new method named as SDM is adopted in [19] to refine the location facial landmarks iteratively. Noticing that Δt_1 can be regarded as projecting $h(t_0) - h(t^*)$ onto the row vectors of R_0 if $R_0 = -2H^{-1}J_h^T$ in Eq. (4), SDM models the update of Δt as a linear regression model without using t^* to enhance its generalization performance for the test image. Δt_1 is modeled as

$$\Delta t_1 = R_0 h(t_0) + b_0 \tag{6}$$

Where b_0 is a bias item to approximate the result of $-R_0 h(t^*)$, R_0 and b_0 are parameters learned from training images. Similarly, the i-th update for Δt is modeled as Eq. (7) in [19].

$$t_i = t_{i-1} + \Delta t_i = t_{i-1} + R_{i-1}h(t_{i-1}) + b_{i-1} \tag{7}$$

Parameters including descent directions $\{R_0\}$ and biases $\{b_i\}$ are estimated with the training images for the SDM method. R_0 and b_0 are learned with a Monto Carlo sampling method, R_i and bias b_i for $i \geq 1$ are learned iteratively. We refer the readers to [19] for details. The learned parameters are then used to infer the location of facial landmarks in test images. The facial landmarks detected by SDM are illustrated in Fig. 1.

3 Extraction and Definition of Visual Fatigue Rules

The facial appearance of a person in fatigue or in the onset of fatigue can usually be characterized by closing eyes slowly and yawning frequently. For fatigue detection, the facial features around the eyes and mouth include enough information to capture the fatigue patterns. Thus, we focus on the analysis of facial features around the eyes and mouth in this research. The aspect ratios of eyes and mouth are considered as the evaluation features. The states of eyes and mouth are estimated through the membership function defined based on the experience.

3.1   Acquisition and Processing of Feature Data

According to the results of SDM algorithm, we can get 49 facial landmarks (See Fig. 1). We calculate the aspect ratios of eyes and mouth according to the facial landmarks in each video. The facial landmarks around each eye are shown in Fig. 2. They can provide the percentage of openness of eyes. The height of each eye is represented by the distance between the upper-eyelid and the lower-eyelid. The width of the eye is represented by the distance between two corners of each eye. The aspect ratio of each eye is calculated by Eq. (8). The degree of eye closure is determined by the smaller aspect ratio of two eyes. The aspect ratio of mouth can be calculated in a similar way, which is illustrated in Fig. 3. It can be used to estimate the openness of mouths. The height of the mouth is represented by the distance between the upper-lip and the lower-lip. The width of the mouth is represented by the distance between the corners of the mouth.

In real application, the experiments show that it would bring error information for the fatigue recognition event, if the rotation angle of the drive is too large, the focus of attention of the driver may be dismissed, which leads to increase the risk of a catastrophic event. In such a case, the human face usually cannot be detected. To solve this problem, we put forward a solution to distinguish this case in two aspects. On the one hand, the proposed system will give an alarm to remind the driver not to distract if the facial landmarks cannot be located by SDM in the continuous period of more than 2 s; On the other hand, for the short time turning of the driver head happens occasionally during driving, we interpolate the missed aspect ratio caused by this case. Then we smooth the extracted features with a forgetting factor to avoid the short-term noise happened occasionally. As a result, we can extract more integrated and more reliable features because of the time continuity of the object in videos.

$$eyeRatio = \frac{d(a, a')}{d(b, b')} \tag{8}$$

 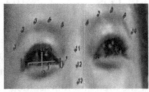 

Fig. 1. Facial landmarks detected by SDM

Fig. 2. Illustration of the eye aspect ratio

Fig. 3. Illustration of the mouth aspect ratio

3.2   Metrics for Fatigue Description

We define the metrics based on the membership function (MF), PERCLOS-like and percentage of mouth yawn (PMY) to describe the feature of fatigue. The details are given as follows.

(a) MF metric

A membership function represents the degree of truth as an extension of valuation, which maps the aspect ratios into the range of [0, 1]. The function should be simple, convenient, fast, efficient and effective. We build two membership functions based on Gaussian distribution: A simple Gaussian curve and a two-sided composite of two different Gaussian curves. The two functions are gaussmf and gauss2mf. The membership functions of eyes and mouth are as shown in Fig. 4 (a) and (b), respectively.

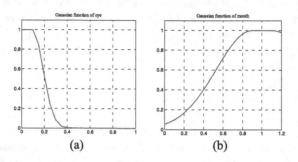

Fig. 4. The membership functions of eye and mouth features

(b) PERCLOS-Like and PMY Metrics

The duration of eye closure demonstrates the fatigue to some extent. PERCLOS [13] has been a scientifically supported measure of drowsiness associated with slow eye closure, which is defined as the proportion of time that a subject's eyes are closed over a specified period. Combining with the P80 criterion, when the percentage of time that the eyes are 80 % or more occluded in a unit interval is greater than 15 %, it is considered drowsy. It can be defined as Eq. (9).

$$\text{PERCLOS} = \frac{\sum_{i=0}^{k} P(i)}{N} * 100\% \tag{9}$$

where N = fps * t, fps represents the number of frames per second, t is the detection time, N is the total number of frames in t. P(i) represents the frames of the i-th eye closure. k represents the time of eye closure in t. In this paper, we take the aspect ratios as the features instead of the area and hierarchically label the degrees of eye and mouth determined by the membership function, so we call it PECLOS-like feature.

The aspect ratio of eye is in the range of [0.625, 1], when the eyes are open. If it is less than 0.625, it indicates the eyes are closed. The smaller the value is, the higher the degree of eye closure. We can get the hierarchical label in 4 grades based on Eq. (10). In addition, the present major researches show that the eye closure duration is about 0.2 and 0.3 s. If more than 0.5 s, the risk of a traffic accident jumps greatly. So we define the duration of PERCLOS-like feature as Eq. (11).

$$eye\,Label = \begin{cases} 1 & if \text{ eye Ratio} \geq 0.9 \\ 2 & if \ 0.7 \leq \text{eye Ratio} < 0.9 \\ 3 & if \ 0.5 \leq \text{eye Ratio} < 0.7 \\ 4 & if \text{ eye Ratio} < 0.5 \end{cases} \qquad (10)$$

$$P(i) = \begin{cases} fs & if \text{ eye Label} = 1 \ \text{and} \ \Delta t \geq \tau_{eye} \\ 0 & otherwise \end{cases} \qquad (11)$$

Similarly, studies have shown that the duration of yawning is about 6 s. We set the membership function of mouth as two-side gauss curve shown in Fig. 4(b). Moreover, if the duration of apparent yawning is more than 1.5 s, a traffic accident is more likely to happen. So we define the PMY. Then, we separate the hierarchical label to 4 grades. The yawning duration of PMY feature is defined in the same way.

Based on the analysis of video frames, we consider it as eye closure if the first grade continuous more than 0.5 s as shown in Eq. (11). Similarly, we consider it as yawning if the first grade continuous more than 1.5 s. Then the duration can be computed.

3.3   Driver Fatigue Computation Rules

At this stage, the visual features of eye closure and mouth opening are used for fatigue determination. It is important to set a reasonable warning threshold. The thresholds in our system are obtained through experience by analyzing of the extracted features of videos. We give three rules to determine the fatigue as following. If the percentage of eye closure at grade one more than 15 %, then we consider it as drowsy. If the percentage of yawning at grade one more than 10 %, then we consider it as drowsy. If the percentage of eye closure at grade 1 and 2 more than 18 % and the percentage of yawning at grade 1 and 2 more than 13 %, then we consider it as drowsy.

4   Experimental Results and Analysis

In the experiments, we have 41 videos of human being under sober condition, and simulated drowsy condition, which also include two real driving videos. In this dataset, persons in 25 videos are fatigue, persons in 16 videos are not fatigue. The first frame of each sequence with the detected facial landmarks is shown in Fig. 5. The parameters of each video are given in Table 1. The fatigue videos are obtained through a Logitech camera to capture driver's videos. This algorithm is implemented using MATLAB platform along with camera and is well suited for real world driving conditions since it can be non-intrusive by using a video camera to detect changes. The system is tested under the environment of Intel(R) Core(TM) i3-3220 CPU and 8 G RAM.

Since the fatigue should be determined over a period of time, we use a sliding window over time to obtain the video frames in 30 s as a period. After processing the new data based on the determination rules, the system outputs the driver fatigue state. It takes about 140 ms for face location and eye detection per frame to output the results timely. When the proposed method is applied to these 41 videos, the system performs well with only two videos inaccurate. To evaluate the performance of the proposed method, we use false alarm ratio (FAR) and detection ratio (DR) as the evaluation metrics. These two metrics are evaluated on the video sequence, which last 30 s. The FAR and DR are averaged for each video, which are shown in Table 1.

In [18], Anumas et al. implemented a system using facial features extracted by active shape models and HRV data. The inference parameters were combined using a fuzzy classifier to obtain the level of driver fatigue. We compare our method with the method in [18] performed on facial data. Our system is slightly better according to the results in Table 1.In this paper, a real time machine vision-based system is proposed for the detection of driver fatigue which can detect the driver fatigue and can issue a warning early enough to avoid an accident. The results are robust and promising even

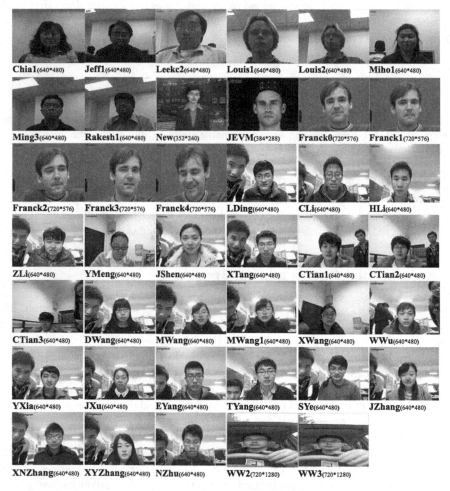

Fig. 5. The first frame of each sequence is shown with marked key feature points in green (Color figure online)

Table 1. Comparison with fuzzy method

Video-Name	Duration	fps	Frames	Video state	Fuzzy [18]	Membership	DR(%)	FAR (%)
chia1	31	14	475	normal	normal	drowsy	77.19	100 + #[1]
jeff1	40	14	605	normal	normal	normal	99.4	#
leekc2	30	14	454	normal	normal	normal	98.9	#
louis1	42	14	633	normal	normal	normal	99.98	#
louis2	31	14	474	normal	normal	normal	100	#
miho1	33	14	505	normal	drowsy	normal	100	#
ming3	39	14	585	normal	normal	normal	100	#
rakesh1	31	14	475	normal	drowsy	normal	99.17	#
New	31	15	479	normal	normal	normal	100	#
JEVM	32	25	805	normal	normal	normal	100	0
franck0	33	30	1000	normal	normal	normal	100	0
franck1	33	30	1000	normal	normal	normal	100	#
franck2	33	30	1000	normal	drowsy	drowsy	100	#
franck3	33	30	1000	normal	normal	normal	99.01	0
franck4	33	30	1000	normal	normal	normal	100	#
LDing	120	30	3600	drowsy	drowsy	drowsy	75.71	30.51 + #
Cli	120	30	3600	drowsy	drowsy	drowsy	95.18	0 + #
HLi	120	30	3600	drowsy	drowsy	drowsy	76.91	0
ZLi	120	30	3600	drowsy	drowsy	drowsy	98.79	8.51
YMeng	91	30	2740	drowsy	drowsy	drowsy	76.66	31.63
JShen	120	30	3600	drowsy	drowsy	drowsy	58.12	0 + #
Xtang	120	30	3600	drowsy	drowsy	drowsy	79.29	17.64
CTian1	60	30	1800	drowsy	drowsy	drowsy	99.65	0
CTian2	106	30	3190	drowsy	drowsy	drowsy	93.71	17.53
CTian3	60	30	1800	drowsy	drowsy	drowsy	83.37	1.78
DWang	120	30	3600	drowsy	drowsy	drowsy	73.56	0 + #
MWang	120	30	3600	drowsy	drowsy	drowsy	77.97	35.66
MWang1	120	30	3600	drowsy	drowsy	drowsy	99.72	0.10
XWang	120	30	3600	drowsy	drowsy	drowsy	92.93	15.65
WWu	120	30	3600	drowsy	drowsy	drowsy	88.65	28.96
YXia	120	30	3600	drowsy	drowsy	drowsy	99.79	4.77
JXu	120	30	3600	drowsy	drowsy	drowsy	98.17	18.06 + #
EYang	120	30	3600	drowsy	drowsy	drowsy	80.53	34.00
TYang	120	30	3600	drowsy	drowsy	drowsy	99.79	25.20
SYe	120	30	3600	drowsy	drowsy	drowsy	93.34	29.03
JZhang	120	30	3600	drowsy	drowsy	drowsy	95.12	26.39
XNZhang	120	30	3599	drowsy	drowsy	drowsy	94.73	54.87
XYZhang	120	30	3600	drowsy	drowsy	drowsy	99.83	7.22 + #
Nzhu	120	30	3600	drowsy	drowsy	drowsy	92.93	15.65
WW2	109	30	3302	normal	drowsy	normal	100	#
WW3	58	30	1769	drowsy	drowsy	drowsy	83.98	0

[1] # denotes that FAR cannot be computed in some special cases, where the groundtruth or the value of measurement indicates that the person in video is not fatigue.

when the driver is wearing glasses. The experiments conducted on the two real driving videos also prove the robustness of our method, which is better than the method in [18]. However, such a system may require further evaluations. The next improvement of the driver fatigue monitoring system will be the increase of the accuracy of the threshold which can gain from more real-world driving videos.

References

1. Takei, Y., Furukawa, Y.: Estimate of driver's fatigue through steering motion. In: IEEE International Conference on Systems, Man and Cybernetics, pp. 1765–1770 (2005)
2. Hostens, I., Ramon, H.: Assessment of muscle fatigue in low level monotonous task performance during car driving. J. Electromyogr. Kinesiol. **15**(3), 266–274 (2005)
3. King, L.M, Nguyen, H.T, Lal, S.K.L.: Early driver fatigue detection from electroencephalography signals using artificial neural networks. In: 28th Annual International Conference of the IEEE on Engineering in Medicine and Biology Society, pp. 2187–2190 (2006)
4. Jung, S.-J., Shin, H.-S., Cshung, W.-Y.: Driver fatigue and drowsiness monitoring system with embedded electrocardiogram sensor on steering wheel. IET Intel. Transport Syst. **8**(1), 43–50 (2014)
5. Nambiar, V.P., Khalil-Hani, M., Sia, C.W., Marsono, M.N.: Evolvable block-based neural networks for classification of driver drowsiness based on heart rate variability. IEEE International Conference on Circuits and Systems (ICCAS), pp. 156–161(2012)
6. Ebrahim P., Stolzmann W., Yang, B.: Eye movement detection for assessing driver drowsiness by electrooculography. In: 2013 IEEE International Conference on Systems, Man, and Cybernetics (SMC), pp. 4142–4148 (2013)
7. Murugappan, M., Wali, M.K., Ahmmad, R.B., Murugappan, S.: Subtractive fuzzy classifier based driver drowsiness levels classification using EEG. In: International Conference on Communications and Signal Processing (ICCSP), pp. 159–164 (2013)
8. Batista, J.: A drowsiness and point of attention monitoring system for driver vigilance. In: IEEE Conference on Intelligent Transportation Systems, pp. 702–708 (2007)
9. Wahlstrom E, Masoud O, Papanikolopoulos, N.: Vision-based methods for driver monitoring. In: IEEE Conference on Intelligent Transportation Systems, vol. 2, pp. 903–908(2003)
10. Hong T, Qin, H.: Drivers drowsiness detection in embedded system. In: IEEE International Conference on Vehicular Electronics and Safety, pp. 1–5 (2007)
11. Viola, P., Jones, M.: Rapid object detection using a boosted cascade of simple features. In: Proceedings IEEE Conference on Computer Vision and Pattern Recognition, pp. 511–518, Hawaii, USA (2001)
12. Singh, H., Bhatia, J.S., Kaur, J.: Eye tracking based driver fatigue monitoring and warning system. In: 2010 India International Conference on Power Electronics (IICPE), pp. 1–6 (2011)
13. Coetzer, R.C., Hancke, G.P.: Eye detection for a real-time vehicle driver fatigue monitoring system. In: IEEE Intelligent Vehicles Symposium (IV), pp. 66–71 (2011)
14. Zhang, S., Liu, F., Li, Z.: An effective driver fatigue monitoring system. International Conference on Machine Vision and Human-Machine Interface, pp. 279–282 (2010)
15. Zhang, W., Cheng, B., Lin, Y.: Driver drowsiness recognition based on computer vision technology. Tsinghua Sci. Technol. **17**(3), 354–362 (2012)

16. Tadesse E., Sheng, W., Liu, M.: Driver drowsiness detection through HMM based dynamic modeling. In: IEEE International Conference on Robotics and Automation, pp. 4003–4008 (2014)
17. Ling, Z., Lu, X., Wang, Y., Zhou, Y., Wang, G., Li, J.: Local sparse representation for driver drowsiness expression recognition. In: Chinese Automation Congress (CAC), pp. 733–737 (2013)
18. Anumas S., Kim, S-.C.: Driver fatigue monitoring system using video face images & physiological information. In: International Conference on Biomedical Engineering, pp. 125–130 (2011)
19. Xiong, X., De la Torre, F.: Supervised descent method and its applications to face alignment. In: Proceedings of IEEE Conference on Computer Vision and Pattern Recognition, pp. 532–539, Portland, OR (2013)

A Packet-Layer Model with Content Characteristics for Video Quality Assessment of IPTV

Qian Zhang[1], Lin Ma[2(✉)], Fan Zhang[1], and Long Xu[3]

[1] School of Information and Control Engineering,
Xi'an University of Architecture and Technology, Xi'an, China
[2] Huawei Noah's Ark Lab, Hong Kong, Hong Kong
forest.linma@gmail.com
[3] Key Laboratory of Solar Activity, National Astronomical Observatories,
Chinese Academy of Sciences, Beijing, China

Abstract. Due to the lightweight measurement and no access to the media signal, the packet-layer video quality assessment model is highly preferable and utilized in the non-intrusive and in-service network applications. In this paper, a novel packet-layer model is proposed to monitor the video quality of Internet protocol television (IPTV). Apart from predicting the coding distortion by the compression, the model highlights a novel loss-related scheme to predict the transmission distortion introduced by the packet loss, based on the structural decomposition of the video sequence, the development of temporal sensitivity function (TSF) simulating human visual perception, and the scalable incorporation of content characteristics. Experimental results demonstrate the performance improvement by comparing with existing models on cross-validation of various databases.

Keywords: Video quality assessment · Temporal sensitivity function (TSF) · Packet-layer model · Content characteristics

1 Introduction

With the increasing popularity of communication through IP network in recent decades, it has witnessed an extensive expansion of its applications such as Internet access, Internet protocol television (IPTV), and voice-over-IP (VoIP). Since the IPTV deliveries the television services through internet protocol suite over a packet-switched network, the perceived service quality may be degraded by data compression before transmission, as well as channel distortion during transmission. Therefore, the assessment of quality of service (QoS) or quality of experience (QoE) is highly demanded, in order to monitor the video quality and make the services meet the users' expectation.

The video quality can be predicted based on the compressed bitstream or the video signal itself. Compared with the quality assessment methods performed in the video signal domain, the bitstream-based video quality assessment methods allow the light demand of computational resources, which is highly preferable for non-intrusive and in-service networked video services, such as IPTV. Such bitstream-based quality assessment methods are expected to show good agreements with human visual perception,

Y.-S. Ho et al. (Eds.): PCM 2015, Part II, LNCS 9315, pp. 488–496, 2015.
DOI: 10.1007/978-3-319-24078-7_49

which are believed to provide good QoS or QoE for users. According to the different levels of accesses to the bitstream and availability of information, the bitstream-based quality assessment models can be categorized into three types, which are the parametric model, packet-layer model, and bitstream-layer model [1], respectively. The packet-layer model solely utilizes the packet header information to predict the video quality. It thus reveals more insights of the content characteristics than the parametric models which only use a few general parameters on the sequence level, and involves less computation load than the bitstream-layer model where media related payload information is necessary in addition to the packet header. Therefore, the packet-layer model is extensively investigated with the development of amount of standardization activities. For example, a packet-layer model called P.NAMS [2] has been standardized as a Recommendation in ITU-T SG12 for the quality assessment of IPTV.

Since packet-layer models allow the access to the packet header of video streams, content-dependent characteristics, such as spatial and/or temporal complexities, can be extracted to a certain extent, which can facilitate the performance improvements of the model prediction ability. Garcia et al. [3] proposed to extract the loss-related features, which describe the spatial-extent and duration of the loss by considering "frame-layer" information. Literature in [4] proposed a novel video quality monitoring model by estimating the spatio-temporal complexity and exploiting the interaction between content features. The model in [5] takes into account the video content, using an objective estimation of the spatio-temporal activity. To reveal the visible artifacts and perceived quality, the efficient loss-related parameters are proposed in [6] by estimating the error propagation in both spatial and temporal domain.

In this paper, we propose a quality assessment model with loss-rated features exploiting the content characteristics in a finer level of the video stream. The contributions lie in the following three aspects.

(1) A structural decomposition method is proposed to segment the video sequence in a coarse-to-fine manner.
(2) A temporal sensitivity function (TSF) is proposed to depict the human visual perception on the temporal complexity of the video content.
(3) The proposed loss-related feature incorporates the TSF in a scalable manner, which is combined with coding-related features to formulate the quality prediction model.

The remainder of the paper is presented as follow. The proposed packet-layer quality assessment model is described in Sect. 2. The experimental settings and results are shown in Sect. 3. In Sect. 4, we summarize our model with its advantages.

2 Packet-Layer Video Quality Assessment Model

2.1 Framework

The framework for the proposed packet-layer video quality assessment model is illustrated in Fig. 1. The input of the model is the encoded bitstream and the output is the predicted value indicating the video perceptual quality. By analyzing the packet header with the encrypted payload, the information such as bit-rate, frame-rate, packet loss rate, can be obtained. Also based on the packet header information, some features, which are closely related with video quality distortion and human perception, can be further computed. Such features are extracted to represent the compression-related and loss-related distortions. Finally, with such extracted features, the compression and loss degradation are estimated simultaneously to predict the perceptual quality of the video, where the parameters of the quality prediction model are obtained from the training samples.

2.2 Structural Decomposition

The sequence-level features such as bit-rate, frame-rate and packet loss rate are commonly used in the packet-layer quality assessment models. However, these features can only provide a general and crude reflection, while quality may vary on different video content even with the same features. Even for one video sequence with same features, the perceptual quality may vary dramatically with the content complexity changes. Thus the content-dependent characteristics incorporated with the sequence structural decomposition should be taken into consideration to improve the model effectiveness.

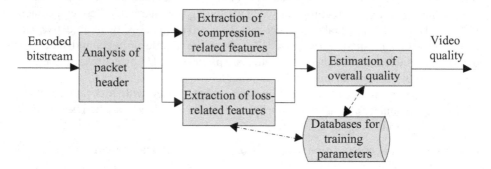

Fig. 1. The framework of the proposed packet-layer video quality assessment model

The video coding technique encodes a sequence with specific configurations before transmission, for example "IPBB" GOP structure with a GOP length of 25 in Fig. 2. The inter-frame coding modes, specifically the "P" and "B" frames, are usually employed to remove temporal redundancies. According to the coding structure, a video sample can be structurally decomposed from the coarse "SEQUENCE" level to the "GOP" level, and then the finest "GROUP" level, as shown in Fig. 2. Generally speaking, the temporal complexity may vary between GOPs, while it may keep more consistent in a same GROUP. The reason is that the GROUP consists of only a few frames, which present the same content with a large probability. In order to structurally decompose the sequence in different levels, the frame type estimation algorithm in [6] is adopted, where the "scene-cut frame" and "non scene-cut I frame" are not differentiated and denoted as I frame for simplicity. As such, the frame type is estimated, based on which the GOP and GROUP structures are identified. Consequently, the GOP-level and GROUP-level features are further computed for the quality assessment model.

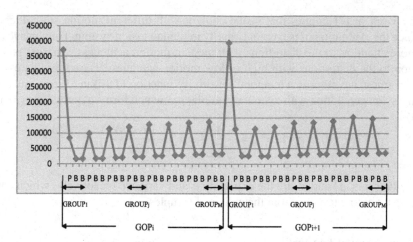

Fig. 2. Frame type estimation and sequence decomposition: y-axis is frame bytes and x-axis is the estimated frame type. A sequence is decomposed into N GOPs, and one GOP is further decomposed into M GROUPs, which starts from the I/P frame till the next I/P frame.

2.3 Temporal Sensitivity Function

As mentioned before, the perceived video quality is not only predicted by the extracted features but also dependent on the video content. The temporal complexity reflects the acuteness of temporal changes of a video sequence, which is usually high for sequences with high motions. Generally, the motion vectors can be employed to measure the motion activities [8]. However, it is not available in the packet-layer model where the payload is encrypted. Yang et al. [9] proposed a measurement of temporal complexity using the ratio of number of bit encoding P and I frames in "IPPP" GOP structure. Inspired by this, we propose a temporal complexity measurement in the GOP and GROUP level, which are calculated in (1) and (2) respectively:

$$\alpha^i_{GOP} = R_{PB}/R_I \tag{1}$$

$$\alpha^j_{GROUP} = \begin{cases} R_{PB}/R_I & if\ GRUOP_j\ contains\ I\ frame \\ R_B/R_P & otherwise \end{cases} \tag{2}$$

where α^i_{GOP} and α^j_{GROUP} are the measurement of temporal complexity in the i-th GOP and j-th GROUP respectively. R_I is the number of bits coding I frame, and R_{PB} is the average number of bits coding each frame (including the P and B frames) in the same decomposition level. R_P is the number of bits coding each P frame, and R_B is average number of bits coding each B frame in the same GROUP. If α tends to be large, each B/P frame tends to consume more bits for compression, which means that the motion estimation can hardly remove the temporal redundancies. Therefore, a high temporal complexity with fast motion is expected in the video sequence. For small α value, a low temporal complexity with smooth motion is expected.

The work in [7] implies that visual content with different motion strengthes will influence the evaluation of perceptual quality in human vision system. Human eyes will pay more attention to the distortions with median motion acuteness, whereas pay less attention to the distortions with high or low motion acuteness. Based on this psycho-physical study, we propose a temporal sensitivity function (TSF) in (3) to simulate the human vision perception with temporal complexity, which forms a normal distribution:

$$TSF(\alpha) = N(\mu, \sigma^2) = \frac{1}{\sqrt{2\pi}\sigma} exp\left(-\frac{(\alpha - \mu)^2}{2\sigma^2}\right) \tag{3}$$

where $\mu(0 < \mu < 1)$ and σ^2 are the expectation and variance receptively, which can be obtained by offline training from the collected samples.

2.4 Loss-Related Feature

An efficient loss-related method named ALAE [6] is proposed by estimating the error propagation in spatio-temporal domain. ALAE is the averaged loss artifact extension (LAE), which is calculated for each frame as the sum of initial artifact caused by the loss in the current frame and propagated artifact caused by the loss in reference frames. By considering the incorporation of sequence structural decomposition in Sect. 2.2 and human vision perception reflected by temporal complexity in Sect. 2.3, an improved parameter named as averaged loss artifact extension with human vision perception (ALAEhvp) is proposed to predict the loss-related distortion:

$$ALAEhvp = \frac{\sum_{i=1}^{N} TSF\left(\alpha_{GOP}^i\right) \times \frac{\sum_{j=1}^{M} TSF\left(\alpha_{GROUP}^j\right) \times LAE_{sum}^j}{\sum_{j=1}^{M} TSF\left(\alpha_{GROUP}^j\right)}}{\sum_{i=1}^{N} TSF\left(\alpha_{GOP}^i\right) * (f * \sqrt{s})} \tag{4}$$

where LAE_{sum}^j is the sum of LAE in the j-th GROUP. M is number of GROUPs in i-th GOP. N is number of GOPs in the sequence. The LAE is weighted by temporal complexity which reflects the importance of human perception and is accumulated at GROUP, GOP and SEQUENCE levels. The quality value is then averaged by the number of frames f and a function of number of slices per frame S. TSF is employed to calculate the human visual perception to the video temporal complexity at the GROUP and GOP levels.

2.5 Video Perceptual Quality Prediction

Considering a video sequence may be corrupted simultaneously by compression and transmission distortion, the overall quality prediction model is capable of predicting the video quality combining the coding artifacts with loss artifacts, which can be obtained by a logistic function [6]:

$$V_q^N = \frac{1}{1 + a * Br^b * ALAEhvp^c} \tag{5}$$

In (5), Br is the bit-rate used to model coding artifacts and $ALAEhvp$ is used to model slicing channel artifacts. a,b,c are the constants obtained from curve-fitting using a least square fitting method through training databases. V_q^N is the normalized mean opinion score (NMOS) within [0,1], which is transformed from MOS by the linear mapping. It should be noticed that the overall model can be reduced to predict the compression distortion only by setting $c = 0$.

Table 1. The database configuration: Df: display format (p-progressive; i-interlace); Br: bitrate (Mbps); Fr: frame rate (fps); Ns: no. of the slices per frame.

	Training	Validation
Df	1080p/i,720p,576i	1080p/i,720p,576i,480i
Br	15,9,7,6,2.5,2,1,0.5	15,7,6,5,4,3,3.5,2.5,2,1.5,0.5
Fr	50,30,25	60,50,30,25
Ns	1, 18, 68	1, 15, 18, 34, 45, 68

3   Experimental Results

3.1   Experimental Setting

In order to develop the model and conduct the experimental comparison, 5 training databases are built for training the parameters μ and σ^2 in Eq. (3) and determining the coefficients a,b,c in Eq. (5), and 6 validation databases for testing the model performance. Each database contains 8 video contents with 10 second duration of high dimension (HD) or standard dimension (SD). The hypothetical reference circuits (HRCs) are encoded by H. 264 with several different GOP structures of fixed or adaptive GOP length. The packet-loss-concealment (PLC) mode is slicing and packet-loss-duration is random or burst. The configurations of training and validation databases are summarized in Table 1. The testing environment is conformed to ITU-R BT.500 [10] and the subjective test is performed using the absolute category rating with hidden reference method in ITU-T Rec. P.910 [11]. The MOS value per HRC is the averaged rating values from 24 subjects. The subjective quality of the video content is categorized into 5 scale: 1 (very annoying), 2 (annoying), 3 (slightly annoying), 4 (perceptible but not annoying), 5 (imperceptible). The subjects are required to watch the sequence and provide their opinions ranging from 1 to 5.

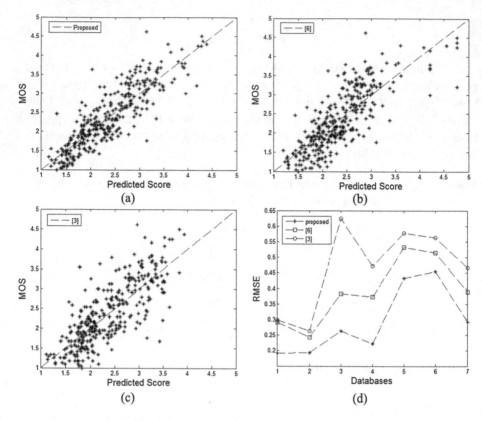

Fig. 3. Experimental results and comparisons

3.2 Experimental Results

To demonstrate the performance of the proposed quality assessment model, we carry out the experiments and make the comparisons with two related models in [6, 3]. The results are shown in Fig. 3. Figure 3(a)–(c) illustrate the predicted quality scores versus MOS values from the subjective rating, by the proposed model, the model in [6], and the model in [3], respectively. It can be observed that the data points are more closely distributed around diagonal line in the proposed model (Fig. 3(a)) than those in [6] (Fig. 3(b)) and in [3] (Fig. 3(c)), which highlights the superiority of the proposed method to the other two models. Figure 3(d) illustrates the root mean square error (RMSE) between the predicted and perceived quality using our model, the model in [6], and the model in [3]. For evaluation, the smaller the RMSE value, the better is the performance of the quality metric. The RMSE value generated by our model is outperformed by other models in all the databases from index 1-6, and clearly better in the mean value in index 7, which demonstrates its good performance. From the experimental results, we believe that our model outperforms the model in [3], because of the introduction of content-related characteristics, which are not considered in [3]. Furthermore, the outperformance over model in [6] lies in the fact that in the proposed model, the temporal complexity is used to reflect the human vision perception by the TSF and incorporated into assessment model based on structural decomposition, which is not taken into account in [6]. It also should be noticed that the content characteristics, the GOP structure, bit-rate, frame-rate, number of the slices per frame, and frame-type are considered in the calculation of model parameters. The trained one set of coefficients is sufficiently used in other 6 databases for cross-validation, which shows the generalization of the proposed mode as well.

4 Conclusion

In this paper, a packet-layer model for video quality assessment is proposed in non-intrusive and in-service network applications, such as IPTV. Apart from predicting the compression distortion, the model mainly focuses on the development of a loss-related feature which is capable of predicting the slicing type distortion during video transmission. The novelty of the feature extraction is the incorporation of human vision perception with content characteristics. The contributions lie in the following three aspects: (1) based on the essential principle of video coding technique, a video stream is structurally decomposed from coarse "SEQUENCE" to "GOP", and finest "GROUP" level, which makes the content characteristics of frames within the same level exhibit highly consistency; (2) a TSF of the video temporal complexity is proposed, which is then used to reflect the human vision perception; (3) the human vision perception is incorporated into the loss-related features in a scalable manner. The performance comparison with related existing methods in the experimental results demonstrates the superiority of the proposed model.

Acknowledgement. The work described in this paper was partially supported by Young Scientist Foundation QN1304, and Talent Technology Foundation RC1349 of Xi'an University of Architecture and Technology, and Natural Science Basic Research Plan in Shaanxi Province of China (Program No. 2014JM2-6127), and National Natural Science Foundation of China under Grant 61202242.

References

1. Yang, F., Wan, S.: Bitstream-based quality assessment for networked video - a review. IEEE Commun. Mag. **50**(11), 203–209 (2012)
2. P.NAMS (Parametric non-intrusive assessment of audiovisual media streaming quality). Recommendation in ITU-T SG12. http://www.itu.int/ITU-T/workprog/wp_item.aspx?isn=6441
3. Garcia, M.N., Raake, A.: Frame-layer packet-based parametric video quality model or encrypted video in IPTV services. In: International Workshop on Quality of Multimedia Experience (QoMEX) (2011)
4. Liao, N., Chen, Z.: A packet-layer video quality assessment model with spatiotemporal complexity estimation. EURASIP J. Image Video Process. **5**(5), 1–13 (2011)
5. Joskowicz, J. Ardao, J.C.L.: A general parametric model for perceptual video quality estimation. In: Proceedings of Communication Quality and Reliability, Vancouver, June 2010

6. Zhang, Q., Zhang, F., Ma, L.: Packet-layer model for quality assessment of encrypted video in IPTV services. In: APSIPA Annual Summit and Conference (2013)
7. Stocker, A.A., Simoncelli, E.P.: Noise characteristics and prior expectations in human visual speed perception. Nat. Neurosci. **9**, 578–585 (2006)
8. Xu, L., Ma, L., Ngan, K.N., Lin, W., Weng, Y.: Visual quality metric for perceptual video coding. In: IEEE Visual Communications and Image Processing (2013)
9. Yang, F.Z., Song, J.R., Wan, S., Wu, H.R.: Content adaptive packet layer model for quality assessment of networked video services. IEEE J. Sel. Topics Signal Process. **6**(6), 672–683 (2012)
10. ITU-R Rec. BT.500-11, Methodology for the subjective assessment of the quality of television pictures, International Telecommunications Union, Technical report (2000)
11. ITU-T Recommendation P. 910, Subjective video quality assessment methods for multimedia applications. Available Online: http://www.videoclarity.com/PDF/

Frame Rate and Perceptual Quality for HD Video

Yutao Liu[1]([✉]), Guangtao Zhai[2], Debin Zhao[1], and Xianming Liu[1]

[1] Department of Computer Science,
Harbin Institute of Technology, Harbin, China
{liuyutao2008, xmliu.hit}@gmail.com, dbzhao@hit.edu.cn
[2] Institute of Image Communication and Information Processing,
Shanghai Jiao Tong University, Shanghai, China
zhaiguangtao@gmail.com

Abstract. The frame rate (FR) of a video plays an important role in affecting the perceptual video quality. Most studies about the effect of FR on the video quality mainly focused on low frame rate, e.g. less than 30 frames per second (fps), at low resolutions like CIF or QCIF. As the video frame rate and resolution advance, we reconsider this issue and investigate the relationship between frame rate and the perceptual video quality under high frame rate and high resolution. In this paper, we discuss the impact of frame rate on the perceptual quality of High Definition (HD) video with high frame rates (up to 120 fps) considered. Firstly, we design and conduct subjective experiment to construct the video dataset, which includes video sequences at different frame rates and the corresponding mean opinion scores (MOS) which represent the perceptual video quality. Based on the MOS results, we analyze how perceptual video quality changes as frame rate varies among different video sequences and propose some meaningful findings. The video dataset will be made publicly available. We deem that this study will enrich video quality assessment and benefit the development of high frame rate and high definition video business.

Keywords: Video quality assessment · Subjective experience · High frame rate · High definition

1 Introduction

It is well known that frame rate plays an important role in affecting the perceptual quality of a video. As a matter of fact, videos with different frame rates lead to different subjective experience for viewers. Another occasion is that special videos, like computer games or 3D videos, with improper frame rate may cause visual discomfort or visual fatigue that is harmful to people's health. Therefore, frame rate acts as an important video attribute which indeed affects the perceptual quality and study on the frame rate impact on the video quality occupies a necessary part of video quality assessment (VQA). Previous work on this point mainly assessed the video quality under several given frame rates, which rarely exceeded 30 frames per second (fps). However, as video frame rate advances, videos that own higher frame rates than 30 fps,

Y.-S. Ho et al. (Eds.): PCM 2015, Part II, LNCS 9315, pp. 497–505, 2015.
DOI: 10.1007/978-3-319-24078-7_50

like 60 fps, 120 fps (called high frame rate), have come into being nowadays. It can also be imagined that high frame rate will get another developing trend and become a new selling point in the movie industry. Therefore, it is needed to know how perceptual video quality changes as the frame rate arrives at higher values than 30 fps that we have already been used to.

Previous works covering the impact of frame rate on the perceptual video quality have been conducted as follows: the work in [1] concerned three factors that affect the video quality which are spatial, temporal and amplitude resolutions and proposed functions to model the relationships between influential factors and the video quality. Zhai et al. in [2] investigated cross-dimensional perceptual quality assessment where frame rate was taken as an important parameter and presented some interesting observations. From another aspect, negative impact of frame dropping on video quality was investigated in [3]. In [4], the authors performed subjective experiments by varying the video's frame rate and quantization parameters and found the gap between PSNR and subjective experimental results. Based on this observation, they formed a more accurate subjective quality metric which combines frame rate, motion speed and PSNR to bridge the gap and got desired performance. Variable frame rate was also studied in [5] for the low bit rate video rate control scenarios.

Generally, the above works all take frame rate as an influential factor that will affect the perceptual video quality severely. However, the frame rates or resolutions of the videos employed in their study like 7.5 fps or CIF (352×288) are relatively low, which will limit the applicability when it comes to video applications in high frame rate and high definition.

In this paper, we investigate how visual quality changes as frame rate varies, particularly choosing the high frame rate and high definition video sequences as our test materials. Firstly, subjective experiments are performed to construct the test video dataset containing video sequences at different frame rates and the corresponding mean opinion scores (MOS), which represent the videos' perceptual quality. Then we make sufficient statistical analysis on the obtained MOS values and present some meaningful findings. The video dataset of this work will be made publicly available for the research community. We expect that this study will enrich video quality assessment and benefit the development of high frame rate and high definition video business.

The rest of the paper is organized as follows. Section 2 details the design of the subjective experiment. The experimental results and analysis are presented in Sect. 3. At last, we conclude this paper and point out the future work in Sect. 4.

2   Subjective Experiment

2.1   Experiment Materials

In this work, high frame rate and high definition videos were employed as our test materials for the subjective experiment. Specifically, the video sequences we adopted are from Ultra Video Group [6], which are "Beauty", "Bosphorus", "HoneyBee", "Jockey", "ReadySetGo", "ShakeNDry" and "YachtRide" respectively, with their snapshots shown in Fig. 1. The original frame rate of these video sequences is 120 fps

Fig. 1. Snapshots of the source video sequences (from left to right, top to bottom: "Beauty", "Bosphorus", "HoneyBee", "Jockey", "ReadySetGo", "ShakeNDry", "YachtRide").

and the resolution is HD of 1080 p (1920 × 1080). The spatial and temporal activities of these sequences cover a wide range, as we show in Fig. 2, where we utilized intra-frame variance and the mean of absolute inter-frame difference to represent the sequences' spatial and temporal activities respectively.

In our subjective experiment, we concerned the video frame rate at four different degrees, which are 15 fps, 30 fps, 60 fps and 120 fps respectively. The frame rate around 15 fps seems to be the threshold of humans' satisfaction level [7]. 30 fps is the frame rate that we have get used to. 60 fps and 120 fps are the emerging high frame rates. To emulate video sequences of lower frame rate than the original frame rate of 120 fps, we down-sampled the video sequences on the time dimension and refilled the missing frames with their previous frames, the same as the method adopted in [8]. Figure 3 gives an illustration to generate low frame rate videos, with the horizontal axis representing time dimension, N fps video sequence above and N/2 fps video sequence below.

2.2 Environment Setup

We arranged the subjective experiment environment under the test conditions suggested in ITU-R BT.500-12 [9]. The illumination in the test room keeps low in order to avoid disturbance from other irrelevant light sources. The distance between the subject and the display is set about three times the height of the display. The display's refresh rate is 120 Hz and the resolution is 1920 × 1080. It should be noted that the size of the display is 22.9". Both of the display refresh rate and resolution are set the same as that of the test video sequences, which guarantees normal playing of the test materials.

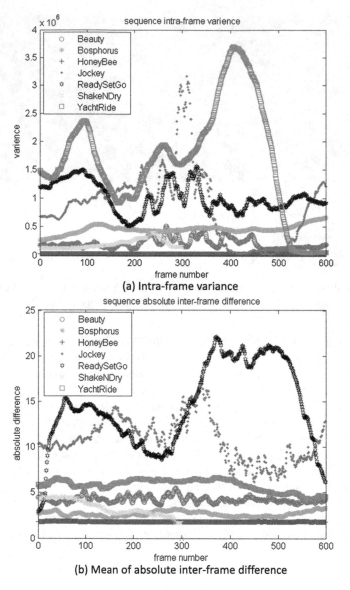

Fig. 2. Intra-frame variance and absolute inter-frame difference for the sequences. (a) Intra-frame variance (b) Mean of absolute inter-frame difference

2.3 Experiment Design

In our subjective experiment, we invited 20 inexperienced college students with normal vision or corrected to normal vision as our subjects for the test. In the process of the experiment, subjects watch the video sequences as usual and rate the overall video quality after each video sequence played. The scores rated by the subjects represents the perceptual quality for the sequences. The rating standard we adopted are 0-20,

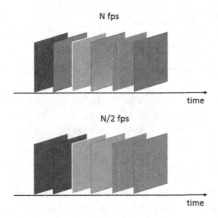

Fig. 3. The illustration of generating low frame rate video sequence.

20-40, 40-60, 60-80, 80-100, which stand for the video quality level of "bad", "poor", "fair", "good" and "excellent" respectively. The test video sequences are played randomly with frame rate unknown to the subjects. To guarantee the reliability of the rating procedure, each video sequence was played 3 times continuously. The subject is allowed to give a continuous score in [0,100]. Single stimulus evaluation method [9] without comparison to a reference version is employed during the whole test.

3   Experimental Results and Analysis

3.1   MOS Results

After the subjective experiment, we got 20 raw scores rated in [0,100] from 20 subjects for each video sequence. The 20 raw scores represent the perceptual quality of the video sequence. However, different viewers have different psychological feelings to the same video sequence so that they may give different scores. For example, when watching the same video sequence of bad quality, one subject may rate 0, while the other subject may rate 15. Therefore, the obtained scores given by different subjects may drop into different intervals in [0,100]. To compare uniformly, we firstly normalized each viewer's scores into a unified interval [0, 1] by linear transform. Specifically, we found the maximum and minimum scores assigned by each subject. Then all the scores given by this subject can be normalized via:

$$n\_score = (score - min)/(max - min) \tag{1}$$

where n_score represents the normalized score, $score$ represents the raw score, min is the minimum score and max is the maximum score. By performing score normalization one subject by one subject, the raw scores were all normalized into [0, 1] and different subjects' experience can be compared uniformly. Then we averaged all the normalized scores for each sequence to get its final MOS.

The results are shown in Table 1.

Table 1. MOS results

	Beauty	Bosphorus	HoneyBee	Jockey	ReadySetGo	ShakeNDry	YachtRide	AVG
@15 fps	0.2142	0.0940	0.1259	0.0541	0.1789	0.4573	0.3283	0.2075
@30 fps	0.7886	0.7398	0.5368	0.6017	0.6739	0.7387	0.6837	0.6805
@60 fps	0.8177	0.8402	0.8057	0.8427	0.8878	0.8594	0.9731	0.8609
@120 fps	0.8621	0.8901	0.8908	0.9093	0.9399	0.8822	0.9553	0.9042

3.2 ANOVA on MOS

To statistically verify that frame rate affects the perceptual video quality, we performed one-way analysis of variance (ANOVA) on the obtained MOS results. Specifically, we divided MOS values into four groups according to the four frame rates. For example, group 1 contains all the MOS values of 15 fps. Then ANOVA was performed on the four MOS groups. The ANOVA results are listed in Table 2, as we can see the first column is the Sum of squares of the four groups. The second column is the degrees of freedom which is defined as the number of groups minus 1. The third column is the mean squares calculated by dividing the Sum of squares by the Degrees of freedom. The fourth column is F statistic and the last column gives the p-value. It can be observed that F statistic is much greater than 1 and the p-value is almost 0, which shows statistical significance, namely the MOS variance coming from different groups takes the dominating place and then verifies that frame rate affects the perceptual video quality severely.

Table 2. Results of ANOVA on MOS

Sum of squares	Degrees of freedom	Mean squares	F statistic	p-value
2.13597	3	0.71199	89.02	3.8288×10^{-13}

3.3 Model Validation

The authors in [1] investigated the impact of spatial, temporal and amplitude resolution on the perceptual quality for compressed videos and proposed a video quality assessment model with considering these three factors. In this work, the spatial and amplitude resolution are fixed and we only focused on the temporal resolution (frame rate) effect, which in [1] was modelled by:

$$MNQT(t) = \frac{1 - e^{-\alpha_t \left(\frac{t}{t_{\max}}\right)^{\beta_t}}}{1 - e^{-\alpha_t}} \tag{2}$$

where $MNQT(t)$ refers to the perceptual quality when the temporal resolution is t, t_{max} is 120 fps here, β_t is set 0.63 according to [1], α_t controls the dropping rate as temporal resolution decreases, more details can be referred to [1].

We examined this model on our video dataset and calculated the correlation coefficient between the MOS values and the predicted MOS values obtained by Eq. (2).

The correlation coefficient result is 0.81, while the average correlation coefficient is 0.95 in [1]. It should be noted that the max frame rate of the sequences tested in [1] is 30 fps and the max frame rate of our test sequences is 120 fps. Therefore, the results proves subjective experience changes differently as the video frame rate arrives at high values.

3.4   MOS Results Analysis

We show the MOS results visually in Fig. 4. It can be observed that the horizontal axis in each subfigure refers to frame rate and the vertical axis represents the value of MOS which reflects the overall perceptual quality of the video sequence. The red circles show the final results of our subjective experiment. The vertical bar means 95 % confidence interval (CI). Additionally we list the average lengths of 95 %

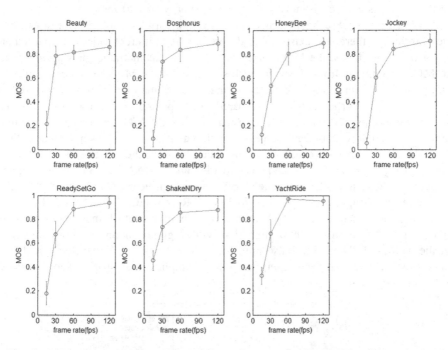

Fig. 4. MOS of the test sequences at different frame rates with 95 % confidence interval.

CI at different frame rates in Table 3. As we can see in Fig. 4, the overall changing trend of the lines keeps much similar, namely, as the frame rate grows, the MOS values increase for all the test sequences, which means the perceptual quality of all the videos improves as frame rate increases. In addition, The MOS values are low at 15 fps, most are less than 0.2, and the corresponding 95 % confidence intervals are also small, that reflects the quality of videos at 15 fps is poor and the viewers can easily sense the inferiority so that give relative low scores. However, the MOS values at 30 fps increase quickly, most exceed 0.6, which means the frame rate of 30 fps satisfies the subjects basically and the subjects tend to give much higher scores than that at 15 fps. While the 95 % confidence intervals at 30 fps are generally larger than that at 15 fps, which says

Table 3. Average lengths of 95 % CI at different frame rates

	@15 fps	@30 fps	@60 fps	@120 fps
95 % CI length	0.1551	0.2353	0.1357	0.1132

there exists much more uncertainty to assess the video quality at 30 fps compared to 15 fps. It is interesting to find that the MOS values at 60 fps are obviously higher than MOS values at 30 fps. It should be noted that our subjective experiment is single stimulus without playing different frame rate videos for comparison, which implies that the perceptual video quality can improve at high frame rate that exceeds 30 fps we've get used to. Similarly, the perceptual quality continuously improves when the frame rate reaches 120 fps, while the growth rate slows down and the MOS improvements are less than that from 30fps to 60fps. All the confidence intervals at 120 fps are also small and the subjec ts can consistently sense the superiority of 120 fps.

Although the MOS changing trend as frame rate varies for all the videos exhibits much similarity, there still exists some differences for different videos. In particular, among different videos, the MOS increments from adjacent frame rates are different like MOS increments from 30 fps to 60 fps. We show the MOS increment results in Fig. 5, the first group at 15-30 fps refers to the video frame rate at 30 fps compared to 15 fps. The second group is 60 fps compared to 30 fps and the third group is 120 fps compared to 60 fps. It can be clearly seen that most of the MOS increments in the first group exceeds 0.3, which means perceptual video quality at 30 fps is much better than that at 15 fps, as mentioned before. In the second group, the increments of "Honey-Bee", "Jockey", "ReadySetGo" and "YachtRide" are above 0.2, while the increments of "Beauty", "Bosphorus" and "ShakeNDry" are not significant. Combined with Fig. 2, the spatial and temporal energies of "Jockey", "ReadySetGo" and "YachtRide" are higher than other sequences, which indicates high frame rate brings about obvious perceptual quality improvement of the videos with high spatial and temporal energies. Yet the quality of "HoneyBee" also improves a lot, in Fig. 2, "HoneyBee" yields the lowest spatial and temporal energies, which implies the spatial and temporal

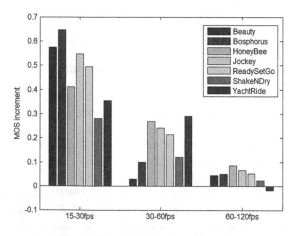

Fig. 5. The MOS increment results.

complexities of this video are the lowest. Therefore, we infer that subjects can also tell the perceptual difference of this kind of videos with simple scene and slight temporal changes as the video frame rate varies. While in the third group, the MOS of all the videos increases a little when the frame rate arrives at 120 fps.

4 Conclusion

In this paper, we focused on the question of how frame rate affects the perceived video quality. Specially, we employed the emerging high frame rate and high definition videos as our video dataset. Subjective experiments were designed and conducted to assess the video quality under different frame rates. We performed thorough statistical analysis on MOS and presented some meaningful findings. In the future, we intend to extend our work from two aspects. Firstly, we will enlarge our high frame rate and high definition video dataset by including more videos like sports videos or computer game videos. Secondly, we will consider more factors that may have influence on the perceptual video quality as frame rate varies, like changing the video resolution or introducing some artifacts to the video.

Acknowledgement. This work is supported by the Major State Basic Research Development Program of China (973 Program 2015CB351804), the National Science Foundation of China under Grants 61300110.

References

1. Yen-Fu, Ou, Xue, Yuanyi, Wang, Yao: Q-STAR: a perceptual video quality model considering impact of spatial, temporal, and amplitude resolutions. IEEE Trans. Image Process. **23**(6), 2473–2486 (2014)
2. Zhai, G., Cai, J., Lin, W., Yang, X., Zhang, W., Etoh, M.: Cross-dimensional perceptual quality assessment for low bit-rate videos. IEEE Trans. Multimedia **10**(7), 1316–1324 (2008)
3. Lu, Z., Lin, W., Choong Seng, B., Sadaatsu, K., Ong E., Yao, S.: Perceptual quality evaluation on periodic frame-dropping video. In: IEEE International Conference on Image Processing, 2007 ICIP 2007. vol. 3, pp. III – 433-III - 436, September 16 2007-October 19 2007
4. Feghali, R., Speranza, F., Wang, D., Vincent, A.: Video quality metric for bit rate control via joint adjustment of quantization and frame rate. IEEE Trans. Broadcast. **53**(1), 441–446 (2007)
5. Song, H., Kuo, C.-C.J.: Rate control for low-bitrate video via variable-encoding frame rates. IEEE Trans. Circuits Syst. Video Technol. **11**(4), 512–521 (2001)
6. Ultra Video Group. http://ultravideo.cs.tut.fi/#main
7. Chen, J.Y.C., Thropp, J.E.: Review of low frame rate effects on human performance. IEEE Trans. Syst. Man Cybern. Part A Syst. Hum. **37**(6), 1063–1076 (2007)
8. Zinner T., Hohlfeld O., Abboud O., Hossfeld T.: Impact of frame rate and resolution on objective QoE metrics. In: 2010 Second International Workshop on Quality of Multimedia Experience (QoMEX), pp. 29-34, 21 –23 June 2010
9. Recommandaton ITU-R BT.500-12, Methodology for the subjective assessment of the quality of television pictures (2009)

No-Reference Image Quality Assessment Based on Singular Value Decomposition Without Learning

Jonghee Kim, Hyunjun Eun, and Changick Kim[✉]

Korea Advanced Institute of Science and Technology (KAIST), Daejeon, Korea
{jonghee.kim,hj.eun,changick}@kaist.ac.kr

Abstract. Recently no-reference image quality assessment (NR-IQA) methods take advantages of machine learning techniques. However, machine learning approaches need a number of human scored images and cause database dependency. In this paper, we propose a simple NR-IQA method that can estimate quality of distorted images without learning, producing comparable performance to learning based approaches. We employ singular value decomposition (SVD) since we have observed that singular values are commonly affected by various distortions. In detail, a decreasing rate of singular values is highly correlated to a degree of distortions regardless of their type. From the observation, our approach utilizes the decreasing rate of singular values to model a simple and reliable NR-IQA method. Experimental results show that the proposed method has reasonably high correlation to human scores. And the proposed method can secure simplicity and database independence.

Keywords: No-reference image quality assessment · Singular value decomposition · Training-free

1 Introduction

With the spread of digital imaging devices and social networking services, a huge amount of images are captured and transmitted effortlessly. During the processes, images can be easily distorted by compression, noise, and so on. Such distortions may affect perceived visual quality of images and lower user satisfaction level. Image quality assessment (IQA) can be employed to ensure the service quality. IQA is classified into full-reference, reduced-reference, and no-reference approaches depending on an amount of prior information of source images. However, in a real scenario, full- and reduced-reference methods can not be used since an end-user only acquires distorted images. Therefore, we focus on no-reference image quality assessment (NR-IQA) rather than other approaches.

Recently NR-IQA methods take advantage of machine learning techniques since it is easy to generate a model with human scored images. In [1,2], Moorthy and Bovik propose a two-step framework for NR-IQA. At the first step, a type of distortion is classified. Once the type of distortion is identified, the general

© Springer International Publishing Switzerland 2015
Y.-S. Ho et al. (Eds.): PCM 2015, Part II, LNCS 9315, pp. 506–515, 2015.
DOI: 10.1007/978-3-319-24078-7_51

NR-IQA problem is regarded as the distortion-specific problem. Then, they utilize natural scene statistics (NSS) based on sub-band coefficients of the wavelet transform to determine the quality. Saad *et al.* propose an NR-IQA method based on NSS of DCT coefficients in [3]. In [4], Mittal *et al.* propose an NR-IQA approach which is performed in the spatial domain. They exploit spatial statistics such as normalized image patches to model NSS in the spatial domain. In [5], Ye *et al.* propose an unsupervised feature learning approach based on Bag of Visual words frameworks. They perform unsupervised clustering to learn a dictionary with distorted images, and then determine quality of an image by soft assignment coding and max pooling methods. Furthermore, recent approaches utilize common machine learning frameworks such as sparse representation [6] and convolutional neural networks [7].

Although machine learning based approaches have achieved high performance, the approaches require a number of human scored images to keep generality of models. However, constructing the database needs a huge amount of effort in spite of their usability. Furthermore, the databases often consist of distorted images stemmed from a small number of reference images at a few degradation levels due to limited manpower.

There are several methods that aim to deal with these problems. In [8], Mittal *et al.* propose a method that does not require human scores. They conduct probabilistic latent semantic analysis (pLSA) on the quality-aware visual words that are based on NSS features. Then, the similarity between the distribution of an unseen image and the learned distribution by using pristine images yields a quality measure. Although they do not utilize human scores for learning, they use distorted images from LIVE [9]. It can not make sure that the method work well on other databases. In [10], Xue *et al.* propose a method that utilizes neither human scores nor images from databases. They generate a new image set with few quality levels of distortions. The images are obtained from other sources to remove database dependecy. Then, they perform quality-aware clustering (QAC) to learn a set of centroids on each quality level. The distance between each patch of images and the centroids determines a score of distorted images. Although they do not exploit images in existing databases, distorting procedure is identical to what has done in the databases. Therefore, database independence is not fully achieved.

To remove the database dependency, in this paper, we propose a method that estimates image quality without any type of learning while producing comparable perfomance with learning based methods. The database dependency can be divided into dependency on *contents* and *type of distortion*. To remove the dependency of contents, singular value decomposition (SVD) [11] is suitable since SVD well represents characteristics of images regardless of their contents by adaptively changing its basis vectors while other transformation methods such as DFT, DCT, and DWT adopt fixed bases [12]. In addition to its outstanding representation ability, we observe that a decreasing rate of singular values is highly correlated to the degradation level without regard to the type of distortion. Due to this capability of removing the database dependency, we propose an NR-IQA method that estimates quality of images based on the decreasing rate of singular values.

Experimental results show that the proposed method has reasonably high correlation to visually perceived quality. In addition, the proposed method has low computational complexity and complete database independence that can secure generality of the method.

The rest of this paper is organized as follows. In Sect. 2, our proposed method is described in detail with observation of SVD. Then, our experimental results and comparison to state-of-the-arts are presented in Sect. 3. Finally, Sect. 4 gives a conclusion of this paper.

2   Proposed Method

2.1   Relationship Between Singular Values and Type of Image Distortions

Singular value decomposition (SVD) [11] decomposes an image into several basis images and corresponding singular values. In other words, an image is decomposed into left singular vector matrix U and right singular vector matrix V and the corresponding diagonal singular value matrix Σ, i.e.,

$$
\begin{aligned}
I &= U\Sigma V^T, \\
U &= [u_1 \ u_2 \ ... \ u_r], \\
\Sigma &= diag(\sigma_1 \ \sigma_2 \ ... \ \sigma_t), \\
V &= [v_1 \ v_2 \ ... \ v_c],
\end{aligned}
\tag{1}
$$

where I is an image whose size is $r \times c$, t is $min(r,c)$, u_i and v_j are column vectors, and σ_k is a singular value ($i = 1, 2, ..., r, j = 1, 2, ..., c, k = 1, 2, ..., t$). The singular values are sorted in a descending order, i.e., $\sigma_1 \geq \sigma_2 \geq ... \geq \sigma_t \geq 0$. And, by using singular vectors and singular values, images can be represented as

$$
I = \sum_k \sigma_k u_k v_k^T.
\tag{2}
$$

Here, singular vectors u_k and v_k constitute basis images whereas singular value σ_k corresponds to their significance. And basis images corresponding to dominant singular values can be considered as bases with low-frequency and vice versa since low-frequency components stand for rough structure of images while high-frequency components hold details.

Distortions like Gaussian blur make loss of detail in the image as shown in Fig. 1(a)-(f). Losing details causes decrease of weak singular values corresponding to high-frequency components. Therefore, the decreasing rate in singular values is easily affected by blur. This phenomenon can be observed straightforwardly by cumulative sum of singular values. The kth normalized cumulative sum of singular values $C(k)$ can be defined as

$$
C(k) = \frac{\sum_{i=1}^{k} \sigma_i}{\sum_{j=1}^{t} \sigma_j},
\tag{3}
$$

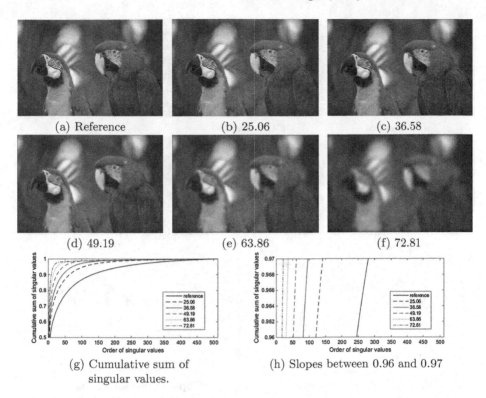

(a) Reference   (b) 25.06   (c) 36.58

(d) 49.19   (e) 63.86   (f) 72.81

(g) Cumulative sum of   (h) Slopes between 0.96 and 0.97
singular values.

Fig. 1. Example of images distorted by Gaussian blur and their cumulative sum of singular values. Difference mean opinion socre (DMOS) of each image is presented at the below of each image. The higher DMOS corresponds to the higher degree of Gaussian blur.

where t is a total number of sinular values.

As shown in Fig. 1 (g), $C(k)$ depends on the degree of blur. In detail, the singular values corresponding to high-frequency components will be close to zero since the image loses details as a degree of blur increases. Namely, cumulative sum $C(k)$ will increase faster when k is small. The characteristic is previously observed and employed for detecting blurred region in an image. Su *et al.*, in [13], utilize the few largest singular values to identify whether the region is blurred or not. In the same context, the characteristic can also be used to determine the degree of blur. Furthermore, JPEG and JPEG2000 distortions may share similar characteristic with blur distortion since they also reduce high-frequency components to compress images. It will be illustrated in Sect. 2.2 by demonstrating that blur, JPEG, and JPEG2000 distortions show similar tendencies.

Note that white noise distortion shows different characteristic compared to distortions by blurring as shown in Fig. 2(a)-(f). While blurring removes high-frequency components, white noise distortion adds every frequency components equally. Therefore, increasing rate of $C(k)$ would be close to 1 as a degree of noise increases. It can be verified by cumulative sum $C(k)$ as shown in Fig. 2(g).

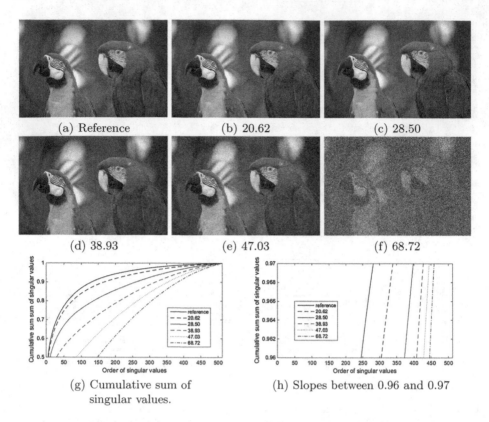

(a) Reference   (b) 20.62   (c) 28.50

(d) 38.93   (e) 47.03   (f) 68.72

(g) Cumulative sum of
singular values.

(h) Slopes between 0.96 and 0.97

Fig. 2. Example of images distorted by white noise and their cumulative sum of singular values. DMOS of each image is presented at the below of each image. The higher DMOS corresponds to the higher degree of white noise.

2.2   Quality Metric Based on Singular Values

As we observe, blur and white noise distortions show conflicting characteristics. It is the main problem to construct a general NR-IQA method without classification of distortion types. Nevertheless, fortunately, blur and white noise distortions commonly affect saturation of $C(k)$. Blur distortion lets $C(k)$ saturate early, and white noise distortion makes $C(k)$ hard to saturate as shown in Figs. 1(g) and 2(g). Therefore, $C(k)$ may increase rapidly before saturation when blur occurs. And, in the case of white noise, the increasing rate of $C(k)$ will be close to 1 as the degree increases. Namely, the increasing rate of $C(k)$ will increase in near saturation since the rate is usually smaller than 1. Thus, when distortions occur regardless of its type, the increasing rate of $C(k)$ may be bigger than usual in the vicinity of saturation as shown in Figs. 1(h) and 2(h).

By observing natural images, we carefully assume that saturation happens where $C(k)$ is around 0.97. Then, we utilize the increasing rate before 0.97 for IQA. More precisely, we use the increasing rate from 0.96 to 0.97 of $C(k)$.

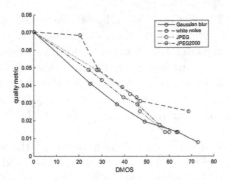

Fig. 3. Quality metric of Gaussian blur, white noise, JPEG, and JPEG2000 distortion. The results are presented with corresponding DMOS.

In Figs. 1(h) and 2(h), the increasing rate of $C(k)$ is accelerated as the degree of distortion increases. The increasing rate from 0.96 to 0.97 of $C(k)$ can be represented as

$$r = \frac{0.01}{C^{-1}(0.97)/t - C^{-1}(0.96)/t}, \qquad (4)$$

where $C^{-1}(p) = k$ and $C(k) = p$. Since the numerator of (4) is fixed and the denominator varies, we simply use the denominator of (4) for quality metric instead. Thus, proposed quality metric is

$$q = \frac{C^{-1}(0.97) - C^{-1}(0.96)}{t}. \qquad (5)$$

Proposed quality metric shows reliable correlation to DMOS as shown in Fig. 3. As DMOS increases, the proposed metric decreases without regard to the type of distortions. Although we do not show $C(k)$ of JPEG and JPEG2000 distortions due to the shortage of space, it can be verified that JPEG and JPEG2000 distortions share the same characteristics as shown in Fig. 3.

3 Experimental Results

3.1 Protocol

We validate the proposed method by its capability of predicting subjective human scores. For comparison to state-of-the-arts, we utilize public databases such as LIVE [9], CSIQ [14], and TID2008 [15]. The databases contain distorted images and corresponding mean opinion score (MOS) or difference mean opinion score (DMOS).

The LIVE database consists of 779 distorted images degenerated from 29 reference images with 5 distortions at various levels. The distortions are JPEG2000 compression (JP2K), JPEG compression, white noise (WN), Gaussian blur (GB), and fast fading. The CSIQ database includes 900 distorted images degraded from 30 original images by 6 types of distortions at 6 levels. The distortions are JP2K, JPEG, WN, GB, global contrast decrements, and pink noise. The TID2008 database is composed of 25 pristine images and their distorted counter parts by deteriorating with 17 types of distortions at 4 levels. The database contains 4 types of distortions commonly used in both LIVE and CSIQ databases and others. We only consider these 4 common distortions (JP2K, JPEG, WN, GB) as in many previous works [2,10].

Performance is evaluated by two common measures: the Spearman rank order correlation coefficients (SROCC) which measures the prediction monotonicity and the Pearson correlation coefficients (PCC) which is related to the prediction linearity. If both measurements are close to 1, it means that a prediction method shows high correlation to human scores.

3.2   Performance Comparisons

We tabulate our prediction results in terms of SROCC and PCC on LIVE, CSIQ, and TID2008 databases in Tables 1, 2, and 3. Since there is no training-free method, we compare our results to pLSA [8] and QAC [10] which are NR-IQA methods which perform learning without human scores. Although the methods do not utilize human scores, the methods exploit distorted images from the LIVE database or arbitrarily degenerated images. And full-reference image quality assessment (FR-IQA) metrics such as PSNR, SSIM, FSIM are used as

Table 1. Quality assessment results on LIVE database.

	FR NR	SROCC					PCC				
		JP2K	JPEG	WN	GB	ALL	JP2K	JPEG	WN	GB	ALL
proposed	NR	.837	.866	.942	.932	.850	.823	.831	.900	.928	.833
pLSA [8]	NR	.850	.880	.800	.870	.800	.870	.900	.870	.880	.790
QAC [10]	NR	.851	.940	.961	.909	.886	.838	.933	.924	.906	.861
PSNR	FR	.895	.880	.985	.783	.875	.873	.865	.979	.775	.858
SSIM	FR	.961	.976	.969	.952	.948	.893	.928	.958	.888	.829
FSIM	FR	.972	.983	.965	.971	.969	.902	.907	.909	.909	.865

Table 2. Quality assessment results on CSIQ database.

	FR NR	SROCC					PCC				
		JP2K	JPEG	WN	GB	ALL	JP2K	JPEG	WN	GB	ALL
proposed	NR	.907	.943	.848	.930	.837	.931	.927	.836	.933	.872
QAC [10]	NR	.870	.913	.862	.848	.863	.882	.938	.874	.844	.877
PSNR	FR	.936	.888	.936	.929	.922	.927	.790	.944	.908	.846
SSIM	FR	.961	.954	.897	.961	.933	.897	.917	.804	.869	.862
FSIM	FR	.969	.965	.926	.973	.962	.907	.903	.764	.884	.880

Table 3. Quality assessment results on TID2008 database.

	FR	SROCC					PCC				
	NR	JP2K	JPEG	WN	GB	ALL	JP2K	JPEG	WN	GB	ALL
proposed	NR	.935	.881	.688	.891	.843	.925	.882	.636	.887	.862
QAC [10]	NR	.889	.898	.707	.850	.870	.878	.924	.720	.850	.838
PSNR	FR	.936	.825	.876	.918	.934	.881	.868	.942	.927	.836
SSIM	FR	.960	.935	.817	.960	.902	.947	.947	.758	.891	.893
FSIM	FR	.976	.926	.857	.953	.953	.956	.931	.783	.907	.929

Table 4. SROCC comparison between the proposed method and state-of-the-arts.

BLIINDS-II [3]	Training ratio			DIIVINE [2]	Training ratio			CORNIA [5]	Training ratio		
	80 %	50 %	30 %		80 %	50 %	30 %		80 %	50 %	30 %
LIVE	.943	.920	.897	LIVE	.895	.877	.795	LIVE	.953	.941	.928
CSIQ	.900	.883	.847	CSIQ	.870	.825	.784	CSIQ	.885	.871	.861
TID	.898	.831	.769	TID	.893	.790	.713	TID	.899	.881	.868
Average	.916	.885	.848	Average	.885	.837	.772	Average	.915	.901	.889
BRISQUE [4]	Training ratio			BIQI [1]	Training ratio			proposed method	Training ratio		
	80 %	50 %	30 %		80 %	50 %	30 %		0 %		
LIVE	.956	.941	.909	LIVE	.843	.799	.748	LIVE	.850		
CSIQ	.909	.886	.863	CSIQ	.760	.721	.672	CSIQ	.837		
TID	.909	.870	.823	TID	.844	.751	.678	TID	.843		
Average	.927	.904	.872	Average	.812	.759	.703	Average	.844		

Table 5. Computation time of NR-IQA methods.

	BLIINDS-II [3]	DIIVINE [2]	CORNIA [5]	BIQI [1]	QAC [10]	proposed method	BRISQUE [4]
time (s)	82.3	18.3	2.01	1.14	0.28	0.13	0.08

references. Note that we can not list the results of pLSA on CSIQ, and TID2008 databases since the authors did not show the results. For the LIVE database, performance of the proposed method is better than that of pLSA and less effective than that of QAC. In addition, in terms of PCC, the proposed method presents better results than SSIM. For the CSIQ database, in terms of PCC, the proposed method shows comparable results to QAC and superior results to PSNR and SSIM. Finally, for the TID2008 database, the proposed method is preferable to QAC and PSNR in terms of PCC. In summary, in terms of SROCC, the proposed method is less effective than QAC. Note that better performance of QAC is attained at the expense of machine learning methods which need a training procedure. However, in terms of PCC, the proposed method is comparable to QAC and sometimes better than even FR-IQA metrics although we do not perform any type of learning.

For rich comparison, we also compare our results to methods that exploit human scores for learning in Table 4. (We use the results from [10] since they tuned other methods to get the best performances with various ratio of training samples.) In Table 4, performances of state-of-the-arts are tested under different ratio of training samples to analyze an effect of training. As shown in Table 4, as ratio of training samples decreases, prediction performances of learning-based methods drop 0.026 to 0.113 in terms of SROCC. It shows that machine learning based methods can be affected by a quantity of data, which is not a desirable property of NR-IQA. However, the proposed method can ignore an effect of database so that it can secure generality. In addition, the proposed method is comparable to DIIVINE [2] and BLIINDS-II [3] at some ratio of training samples. Moreover, the proposed method shows superior results to BIQI [1] without regard to the ratio of training samples. Consequently, experimental results demonstrate that the proposed method achieves high correlation to human scores without learning compared to learning-based methods (Table 5).

3.3 Computational Complexity

Computational complexity is one of significant properties of NR-IQA since it is often applied to real-time systems. To compare the computational complexity, we measure the elapsed time to return score. Therefore, it includes predicting time occupied in machine learning methods which can affect the computation time significantly. The time is measured by using codes in MATLAB provided by the authors. In Table 4, we tabulate computation time of state-of-the-arts and the proposed method. We run the methods 100 times, and take median to get reliable results. Computation time of the proposed method is faster than state-of-the-arts except [4]. It shows that the proposed method has low complexity.

4   Conclusion

In this paper, we have presented an simple NR-IQA method that does not exploit any type of learning to remove database dependency. We utilize singular values to estimate quality of an image based on the observation that the decreasing rate in singular values are highly correlated to a degree of distortions rather than a type. Experimental results show that the proposed metric has reasonably high correlation to human scores without regard to the specific type of distortion. Furthermore, performance of the proposed method is comparable to some of state-of-the-arts based on machine learning frameworks. At last, the method has low computational complexity so that it can be applied to real time systems.

References

1. Moorthy, A.K., Bovik, A.C.: A two-step framework for constructing blind image quality indices. Signal Process. Lett. IEEE **17**(5), 513–516 (2010)
2. Moorthy, A.K., Bovik, A.C.: Blind image quality assessment: from natural scene statistics to perceptual quality. Image Process. IEEE Trans. **20**(12), 3350–3364 (2011)

3. Saad, M., Bovik, A., Charrier, C.: Blind image quality assessment: a natural scene statistics approach in the DCT domain. Image Process. IEEE Trans. **21**(8), 3339–3352 (2012)
4. Mittal, A., Moorthy, A.K., Bovik, A.C.: No-reference image quality assessment in the spatial domain. Image Process. IEEE Trans. **21**(12), 4695–4708 (2012)
5. Ye, P., Kumar, J., Kang, L., Doermann, D.: Unsupervised feature learning framework for no-reference image quality assessment. In: 2012 IEEE Conference on Computer Vision and Pattern Recognition, pp. 1098–1105 (2012)
6. He, L., Tao, D., Li, X., Gao, X.: Sparse representation for blind image quality assessment. In: 2012 IEEE Conference on Computer Vision and Pattern Recognition, pp. 1146–1153 (2012)
7. Kang, L., Ye, P., Li, Y., Doermann, D.: Convolutional neural networks for no-reference image quality assessment. In: 2014 IEEE Conference on Computer Vision and Pattern Recognition, pp. 1733–1740 (2014)
8. Mittal, A., Muralidhar, G.S., Ghosh, J., Bovik, A.C.: Blind image quality assessment without human training using latent quality factors. Signal Process. Lett. IEEE **19**(2), 75–78 (2012)
9. Sheikh, H.R., Wang, Z., Cormack, L., Bovik, A.C.: LIVE Image Quality Assessment Database Release, vol. 2 (2005)
10. Xue, W., Zhang, L., Mou, X.: Learning without human scores for blind image quality assessment. In: 2013 IEEE Conference on Computer Vision and Pattern Recognition, pp. 995–1002 (2013)
11. Kalman, D.: A singularly valuable decomposition: The SVD of a matrix. College Math J. **27**, 2–23 (1996). Citeseer
12. Narwaria, M., Lin, W.: SVD-based quality metric for image and video using machine learning. Syst. Man Cybern. Part B: Cybern. IEEE Trans. **42**(2), 347–364 (2012)
13. Su, B., Lu, S., Tan, C.L.: Blurred image region detection and classification. In: Proceedings of the 19th ACM International Conference on Multimedia, pp. 1397–1400. ACM (2011)
14. Larson, E.C., Chandler, D.M.: Most apparent distortion: full-reference image quality assessment and the role of strategy. J. Electron. Imaging **19**(1), 011006–011006 (2010)
15. Ponomarenko, N., Lukin, V., Zelensky, A., Egiazarian, K., Carli, M., Battisti, F.: TID2008-a database for evaluation of full-reference visual quality assessment metrics. Adv. Mod. Radioelectron. **10**(4), 30–45 (2009)

An Improved Brain MRI Segmentation Method Based on Scale-Space Theory and Expectation Maximization Algorithm

Yuqing Song[1(✉)], Xiang Bao[1], Zhe Liu[1], Deqi Yuan[2],
and Minshan Song[1]

[1] School of Computer Science and Communication Engineering,
Jiangsu University, Zhenjiang, China
157448323@gg.com, bx425bob@163.com
[2] New District Branch of Zhenjiang First People's Hospital, Zhenjiang, China

Abstract. Expectation Maximization (EM) algorithm is an unsupervised clustering algorithm, but initialization information especially the number of clusters is crucial to its performance. In this paper, a new MRI segmentation method based on scale-space theory and EM algorithm has been proposed. Firstly, gray level density of a brain MRI is estimated; secondly, the corresponding fingerprints which include initialization information for EM using scale-space theory are obtained; lastly, segmentation results are achieved by the initialized EM. During the initialization phase, restrictions of clustering component weights decrease the influence of noise or singular points. Brain MRI segmentation results indicate that our method can determine more reliable initialization information and achieve more accurate segmented tissues than other initialization methods.

Keywords: Expectation Maximization algorithm · Initialization information · Scale-space theory · Fingerprints · Brain MRI segmentation

1 Introduction

Magnetic resonance imaging (MRI) is an imaging method which can obtain the picture of soft parts inside a patient's body using a powerful magnetic field. It has been widely applied for its advantage of being noninvasive and no known health hazards.

Clustering methods have been widely applied in MRI segmentation. Features including gray level and textural features are commonly used for segmentation. It is universally acknowledged that pixels within one cluster are homogeneous and such clusters always have specific medical significance. Different clustering approaches have been proposed to model the data. Centroid models including K-means [1] represent the groups as mean vectors; distribution models [2] such as the Gaussians Mixtures Models (GMM) model the generative distributions of the data by statistics and probabilities; graph clustering models like Markov clustering algorithm (MCL) [3] organize datasets on the basis of the edge structure of the observations; the agglomerative and divisive algorithms [4] which use connectivity models focus on the distance

© Springer International Publishing Switzerland 2015
Y.-S. Ho et al. (Eds.): PCM 2015, Part II, LNCS 9315, pp. 516–525, 2015.
DOI: 10.1007/978-3-319-24078-7_52

connectivity. Distribution model-based approach is often used in MRI segmentation, it assumes that the data set follows a mixture model of probability distributions so that a mixture likelihood approach like the Expectation Maximization (EM) algorithm may be used [5]. But EM algorithm has obvious a deficiency that the quality of segmentation depends highly on initial values [6] including the number of clusters of the whole dataset, clustering center, standard deviation value and weight of each component. Many methods have been proposed to overcome the shortcomings. K-means [1] is a common method to acquire the initialization information, but unknown number of clusters makes it difficult to achieve good results. Particle Swarm Optimization (PSO) [7] is also applied in the initialization acquisition, but it needs large amount of time so that it is not feasible in practical application. Melnykov [8] acquired the initial mean vector by choosing a higher concentration of the points in the neighborhood space. However, most EM variant methods have not settled the problem of number of clusters. For estimating the number of clusters, Information criterion including Akaike's Information Criterion (AIC), Bayesian Information Criterion (BIC) and Minimum Description Length (MDL) [9, 10] have been widely used. However, these information criteria methods imply consideration of different numbers of components ranging from 1 to K [11]. In fact, these methods will add much computation burden. Xie [11] proposed a method related to characteristic function(CF). This method regards the log-CFs(LCF) as the convergence function of the GMM, the component number can be obtained when the real and imaginary parts of its LCF become stable, it can address the problem of over-fit without any priori; clustering evaluation called variance ratio criterion (VRC) based on the between-cluster and within-cluster scatter matrix was proposed by Calinski and Harabasz [12], larger value means better clustering performance; The Davies-Bouldin criterion proposed by Davies [13] is based on the ratio of within-cluster and between-cluster distances, better clustering results always have smaller values; The silhouette value [14] is a measure of how similar that point is to points in its own cluster, the higher value, the better.

In this paper, a method in EM initialization process based on scale-space theory [15–17] is proposed. In the experiment, we compare the obtained number of clusters with that estimated by other criterions and perform segmentation experiments with different EM initialization methods in real brain MRI. It can be inferred that the obtained number of clusters is valid from both medical and statistical significance. Experimental results on real MRI justify the potentiality of proposed method in terms of similarity measure between segmentation results and segmentation groundtruth.

2   GMM Based EM Algorithm

Gaussian distribution is the most common choice for component density function and has been commonly used in MRI segmentation. The quality of a given set of parameters $\theta_h = \{(\alpha_h, \mu_h, \sum_h), h = 1, 2 \ldots k\}$ is determined by how well the corresponding mixture model fits the data. This is quantified by the log-likelihood of the data given the mixture model. The method using maximum likelihood estimation (ML) for parameter estimation [5] is commonly known as EM algorithm, which is an iterative procedure to get the local maxima of log-likelihood. The detailed process of maximum likelihood estimation with EM algorithm is displayed in [12]:

3   Density Estimation and Initialization for EM

3.1   Kernel Density Estimation

Kernel density estimators are smoother and it can also converge to the true density faster. Kernel density estimators [18] is defined given $x_1, x_2, \ldots x_n$, a kernel K and a bandwidth h. In this paper, K is assumed to be Gaussian curve and bandwidth h is obtained by The Normal Reference Rule [18].

3.2   Initialization Based on Scale-Space Theory

Yuille [17] suggested a possible method in decomposing a signal into sums of normal distributions. For a continuous signal $I(x)$, Witkin [15] defined its scale-space image to be

$$F(x, \sigma) = I^*G \tag{1}$$

Where "*" denotes convolution, G is assumed to be Gaussian which can be written as:

$$G(x, \sigma) = \frac{1}{\sqrt{2\pi}\sigma} e^{-\frac{x^2}{2\sigma^2}} \tag{2}$$

So scale-space image can be written as:

$$F(x, \sigma) = \int I(\zeta) \frac{1}{\sqrt{2\pi}\sigma} e^{-\frac{(x-\zeta)^2}{2\sigma^2}} d\zeta \tag{3}$$

We define $E(x, \sigma)$ as:

$$E(x, \sigma) = \frac{d^2}{dx^2} F = \frac{d^2}{dx^2} \int I(\zeta) \frac{1}{\sqrt{2\pi}\sigma} e^{-\frac{(x-\zeta)^2}{2\sigma^2}} d\zeta \tag{4}$$

We can obtain $x - \sigma$ image by $E(x, \sigma) \Rightarrow 0$, $x - \sigma$ image is constructed by plotting locations of zero-crossings of the second derivative of F. Yuille [17] proved that $x - \sigma$ image captured all the information which is needed to represent the signal, so the $x - \sigma$ image can be called fingerprints [15–17]. Figure 1(a) is the fingerprint of a normal probability density function $I(x)$ which follows $N(\mu, \delta)$, zero-crossings in the second derivative of E are the positions where curvature of E changes sign. Sections of the signal over which $E < 0$ are convex, while sections over $E > 0$ are concave [17], zero-crossing where changes from $E > 0$ to $E < 0$ is labeled a lower turning point, while zero-crossing where changes from $E < 0$ to $E > 0$ is labeled an upper turning point. Carlotto [16] proved that a funnel between a lower turning point and an upper

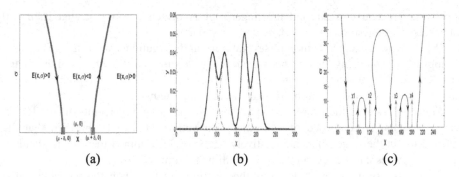

Fig. 1. (a) Fingerprints of a normal probability density function,

Table 1. Estimating initialization information for EM algorithm

---

Step 1: Estimating kernel density P by Section 3.1

Step 2: Obtaining $x-\sigma$ image by Section 3.2

Step 3: Obtaining n possible centers and standard deviation values from $x-\sigma$ image when $\sigma = 0$, labeling centers as $c_1, c_2, ... c_n$, standard deviation values as $s_1, s_2, ... s_n$

Step 4: Recording the density values of possible centers as $P_1, P_2, ... P_3$, obtaining initialization information for EM by Rule 1

Rule 1:

Input: Density values of possible centers $P_1, P_2, ... P_3$, estimated centers $c_1, c_2, ... c_n$, standard deviation values as $s_1, s_2, ... s_n$, $T_1 = 90\%, T_2 = 10\%$

Output: Estimated component weights, centers and standard deviation

$$\omega_i = \sum_{j=c_i-3s_i}^{c_i+3s_i} P_j (i = 1, 2, ... n) , \text{SUM} = \sum_{i=1}^{n} \omega_i , P_{component}[i] = \frac{\omega_i}{SUM} (i = 1, 2, ... n)$$

Sort $P_{component}[i]$ by descending order listed as PP

if($\sum_{i=1}^{n_1} PP[i] > T_1$ or $PP[n_1] < T_2$)

estimated cluster number= n_1

estimated clustering centers & standard deviation values: the corresponding values whose labels are in front of n_1 in the list PP

estimated weight by $\widehat{\omega}_i = \sum_{j=c_i-3s_i}^{c_i+3s_i} P_j (i = 1, 2, ... n_1)$

---

turning point could determine a normal distribution, and the intersection points in the fingerprints when $\sigma \approx 0$ could reconstruct the original signal, its width is approximately equal to 2δ, and the fingerprint is centered at $x = \mu$. That is, $\sigma \approx 0$ intersects the fingerprint at $(\mu - \delta, \sigma)$ and $(\mu + \delta, \sigma)$.

When there is a superposition of several normal distributions. $x - \sigma$ image can also be applied for analyzing the original signal, we go through all the funnels in the image successively so that information including the number of clusters, component centers, standard deviation values can be shown as the same method.

Figure 1(c) shows the interpretation of the corresponding fingerprints of Fig. 1(b) in which a signal is composed of 4 normal distributions. It can be inferred that the information in the fingerprints are relatively accurate. Determined means and standard deviation values can be regarded correct within the limit of error.

As is discussed above, scale-space theory helps us to obtain the valuable initial information μ_k, Σ_k. For weight of each component α_k, we take consideration into Gaussian Empirical Rule [19] that 99.7 % of the features fall within 3 standard deviations of the mean in a normal distribution, so that we can obtain the estimated weight of each component by the estimated probability values between $(\mu_k - 3^*\Sigma_k, \mu_k + 3^*\Sigma_k)$. The determined modes are not the true Gaussian components necessarily because of the influence of noise or singular points, so we should delete the modes which have less contribution to reconstruction of the original signal, so we should consider the estimated weights of estimated modes. Our method to determine the initial information by scale-space theory in the EM algorithm is as follows in Table 1.

4   Experimental Results

In this experiment, we apply brain MRI shown in Fig. 2 from one patient, MRI and their segmentation groundtruth are from http://brainweb.bic.mni.mcgill.ca/brainweb/. It is assumed that kernel density of gray level in the brain MRI follows GMM. First, we compare the number of clusters based on some classical criterions such as Silhouette criterion, Calinski-Harabasz criterion, Davies-Bouldin criterion and BIC. Second, we compare the segmentation results with scale-space based EM initialization method with other methods including K-means, PSO [7] and Melnykov's method [8]. The segmentation groundtruth including main tissues such as cerebral spinal fluid (CSF), gray

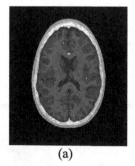   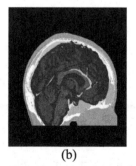

(a)   (b)

Fig. 2. original brain MRI

matter, white matter, muscles skin and marrow are selected to evaluate the segmentation results. In this paper, we illustrate the segmentation process by analyzing Fig. 2 (a), all the experiment are carried out in Matlab 7.10.0(R2010a) from a Lenovo computer whose CPU is Intel (R)Core(TM) @ 2.93 GHz, RAM 1.96 GB.

4.1 Determining the Number of Clusters

Many methods have been proposed to estimate the number of clusters in a dataset. In this paper, we estimate the number for the brain MRI segmentation by 4 classical criterions. Silhouette criterion and Calinski-Harabasz criterion obtain the best number at the maximum criterion values while Davies-Bouldin criterion and BIC are the opposite. Different criterions of Fig. 2(a) are displayed in Fig. 3 from which we can infer that the number of clusters by the Silhouette criterion and Calinski-Harabasz criterions is 7, while Davies-Bouldin criterion and BIC minimize when the number is 9.

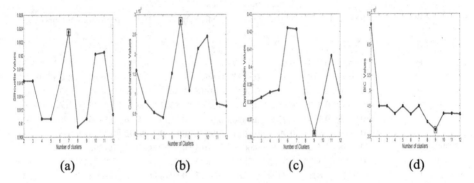

(a)	(b)	(c)	(d)

Fig. 3. Number of clusters estimated by different criterions of Fig. 2(a), (a) Silhouette criterion, (b) Calinski-Harabasz criterion, (c) Davies-Bouldin criterion, (d) BIC

Our method of estimating the initial information for EM algorithm of Fig. 2(a) is displayed above. Figure 4(a) is the estimated kernel density of gray level in the brain MRI and Fig. 4(b) is the fingerprints of estimated density. As is displayed, we can estimate 11 components in the first step from the fingerprints, and the estimated centers are also shown in Fig. 4(a). Then we delete the 4th and 9th component out of the results by algorithm shown in Table 2, because they have relatively small-scale component proportion. So the ultimate number of clusters is 9, the corresponding component weights, centers and standard deviation can be regarded as initial information for EM algorithm. The estimated number of clusters is valid from the point of statistical significance, and it is similar to the 4 classical criterions discussed above. For the medical significance, the number is also reliable, for the main tissues in the brain MRI includes cerebral spinal fluid (CSF), gray matter, white matter, muscles skin, marrow, fat, vessels, skull and background.

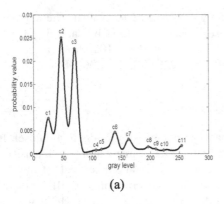
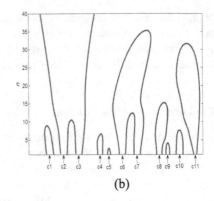

(a) (b)

Fig. 4. (a) Kernel density of brain MRI in Fig. 2(a), (b) Fingerprints of (a)

Table 2. P,R,F criterions of different tissues by different methods in Fig. 5

Test tissue	Evaluation index	Initialization methods									
		EM(rand)		EM(Kmeans)		EM(PSO)		EM(Melnykov)		EM(ours)	
		a	b	a	b	a	b	a	b	a	b
CSF	P	0.2284	1	1	1	0.5017	1	1	0.5367	1	1
	R	0.9574	0.5137	0.6783	0.5137	0.9551	0.5137	0.6783	0.8765	0.6783	0.5137
	F	0.3688	0.6787	0.8083	0.6787	0.6578	0.6787	0.8083	0.6657	0.8083	0.6787
white matter	P	0.9816	0.3324	0.9816	0.8916	0.9816	0.8916	0.9816	0.4091	0.9816	0.8916
	R	0.8666	0.9621	0.8666	0.5302	0.8666	0.5302	0.8666	0.6826	0.8666	0.5302
	F	0.9206	0.4941	0.9206	0.6650	0.9206	0.6650	0.9206	0.5116	0.9206	0.6650
gray matter	P	0.6799	0.4091	0.9971	0.9611	0.9971	0.8569	0.9542	0.4323	0.8533	0.9611
	R	0.9998	0.4826	0.8013	0.6659	0.8013	0.6947	0.8163	0.7252	0.9996	0.6659
	F	0.8094	0.4428	0.8885	0.7867	0.8885	0.7673	0.8799	0.5429	0.9204	0.7867
marrow	P	0.9240	0.8086	0.9306	0.8086	0.9227	0.8048	0.9188	0.8086	0.9239	0.8347
	R	0.6820	0.6278	0.6240	0.6278	0.6922	0.6278	0.6922	0.6278	0.6813	0.7592
	F	0.7847	0.7068	0.7470	0.7068	0.7910	0.7054	0.7895	0.7068	0.7843	0.7952
muscle skin	P	0.9746	0.9995	0.9746	0.9995	0.9746	0.9995	0.9746	0.9995	0.8545	0.9995
	R	0.5458	0.6548	0.5458	0.6548	0.5458	0.6548	0.5458	0.6548	0.7503	0.6548
	F	0.6997	0.7912	0.6997	0.7912	0.6997	0.7912	0.6997	0.7912	0.7990	0.7912

4.2 Initialization Impact on Segmentation Results

In the segmentation experiments, we compare our initialization method with several common methods on MRI. The numbers of clusters of both MRI in Fig. 4 are set 9 based on the methods discussed above. We set $\alpha_k = 1/9, (k = 1, 2, \ldots 9)$, $\mu_k = 255/8^*(k - 1), (k = 1, 2, \ldots 9)$, $\sum_k = \sqrt{\Sigma/9}, (k = 1, 2, \ldots 9)$ where Σ is the standard deviation value of the whole data set in the first method, and call EM(rand) for short; K-means, PSO and Melnykov's method are used for initialization in the second, third and fourth methods respectively. We choose main tissues including cerebral spinal fluid (CSF), gray matter, white matter, muscles skin and marrow for evaluation with segmentation groundtruth. The segmentation results are displayed as follows in Fig. 5. The tissues in Fig. 5 from top to down are CSF, white matter, gray matter, marrow and muscles skin. The first column of Fig. 4 displays the segmentation groundtruth of different tissues. Segmentation results by EM(rand), EM(K-means), EM (PSO), EM(Melnykov) and our method are shown in the next columns. In Fig. 5(a), it can be seen that EM(rand) does not separate CSF and gray matter very well, they are tangled together causing recognition error; CSF segmented by EM(K-means) is not satisfactory for it is mixed with other tissues; Gray matter and muscle skin by the proposed method is more similar to segmentation groundtruth than other method. In Fig. 5(b), it can be inferred that EM(rand) has tangled white matter with gray matter; CSF and gray matter segmented by EM(Melnykov) are not satisfactory; muscle skin by our method achieve best results compared with groundtruth.

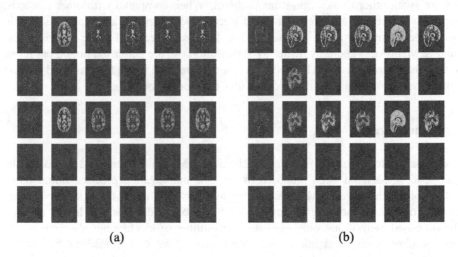

Fig. 5. (a) segmentation results of Fig. 2(a), (b) segmentation results of Fig. 2(b)

In order to evaluate the segmentation results using different methods, we compare the segmented tissues with segmentation groundtruth from point of objective view. We choose two basic criterions which are Precision Rate and Recall Rate, and can be called P and R for short respectively [20]. P reflects the separating capacity of segmentation of different tissues, while R reflects the recognition capability from the same tissue. They are calculated by:

$$P_i = \frac{A_i}{B_i} \tag{5}$$

$$R_i = \frac{A_i}{C_i} \tag{6}$$

Where P_i, R_i, $i = 1, 2, \ldots n$ represent the Precision Rate and Recall Rate in the i-th cluster after segmentation, A_i represents the number of overlapped pixels between the segmented tissue and the corresponding segmentation groundtruth, B_i represents the number of pixels of the segmented tissue, C_i represents the number of pixels of the corresponding segmentation groundtruth. We also consider the F criteria which can be regarded as an integrated index given P and R, and it can be calculated as:

$$F_i = \frac{2^* P_i^* R_i}{P_i + R_i} \tag{7}$$

The larger value of P,R,F, the better results of the segmentation. Table 2 below reflects the P,R,F in different tissues by different methods.

In Fig. 2(a), it can be concluded that the proposed method gets almost largest F values of all tissues in the experiment except marrow, F value of marrow segmented by EM (PSO) is a bit larger than that of our method. For CSF, although Recall rates of EM (rand) and EM (PSO) in are larger, while the Precision rates and F values are smaller, mixtures with other tissues cause this problem. When compared with other methods, the proposed method achieves best F and Recall rates in gray matter and muscle skin although a bit smaller Precision rates. In Fig. 2(b), recall rates of white matter by EM (rand & Melnykov) and gray matter by EM(Melnykov) are better than ours, but it can be seen that these tissues are tangled with other tissues. All in all, the proposed method can segment more reliable and precise issues than other methods.

5 Conclusion

In this paper, to settle the problem of classical EM, a novel method based on the scale-space theory has been proposed in the EM initialization process. Initial information of GMM based EM includes the number of clusters, component centers, standard deviations and weights, among which number of clusters is of higher priority. The proposed method not only estimates the number of clusters but also obtains the corresponding component information. Furthermore, we take consideration into the influence of noise or singular points making the estimation more reliable and convincing. Experimental results indicate the correctness and credibility of estimated the number of clusters, comparison with segmentation groundtruth shows the superiority of the proposed initialization method. In fact, EM (K-means), EM (PSO) and EM (Melnykov) also need the number of clusters which cannot be obtained in advance.

Acknowledgement. The paper is supported by the following fund projects: The National Natural Science Foundation of China (61402204); The Natural Science Foundation of Jiangsu Province (BK20130529); Research Fund for Advanced Talents of Jiangsu University (14JDG141); Science and Technology Project of Zhenjiang City (SH20140110); China Postdoctoral Science Foundation (Project No. 2014M551324); Special Software Development Foundation of Zhenjiang City (No. 201322); Science and Technology Support Foundation of Zhenjiang City(Industrial) (GY2014013).

References

1. Masroor Ahmed, M., Mohammad, D.B., Masroor Ahmed, M., et al.: Segmentation of brain MR images for tumor extraction by combining kmeans clustering and perona-malik anisotropic diffusion model. J. Bus. Educ. **1**, 27–34 (2008)

2. Zeger, L.S.M.S.L.: A smooth nonparametric estimate of a mixing distribution using mixtures of gaussians. J. Am. Statist. Assoc. **91**(435), 1141–1151 (2012)
3. Brandes, U., Gaertler, M., Wagner, D.: Engineering graph clustering: models and experimental evaluation. J. Exp. Algorithmics **12**, 1–5 (2007)
4. Cimiano, P., Hotho, A., Staab, S.: Comparing conceptual, divisive and agglomerative clustering for learning taxonomies from text. In: Proceedings of Eureopean Conference on Artificial Intelligence Ecai Including Prestigious Applicants of Intelligent Systems Pais (2004)
5. Dempster, A.P., Laird, N.M., Rubin, D.B.: Maximum likelihood estimation from incomplete data via the EM algorithm (with discussion). J. Royal Statist. Soc. Ser. B **39**, 1–38 (1977)
6. Yang, M.S., Lai, C.Y., Lin, C.Y.: A robust EM clustering algorithm for Gaussian mixture models. Pattern Recogn. **45**(11), 3950–3961 (2012)
7. Abraham, A., Das, S., Roy, S.: Swarm intelligence algorithms for data clustering. In: Soft Computing for Knowledge Discovery and Data Mining, pp. 279–313 (2008)
8. Melnykov, V., Melnykov, I.: Initializing the EM algorithm in Gaussian mixture models with an unknown number of components. Comput. Statist. Data Anal. **56**(6), 1381–1395 (2012)
9. Zhou, X., Wang, X., Dougherty, E.R.: Gene selection using logistic regressions based on AIC, BIC and MDL criteria. New Math. Nat. Comput. **1**(1), 129–145 (2005)
10. Hansen, M., Yu, B.: Bridging AIC and BIC: an MDL model selection criterion. In: Proceedings of IEEE Information Theory Workshop on Detection, Estimation, Classification and Imaging, vol. 63 (1999)
11. Xie, C., Chang, J., Liu, Y.: Estimating the number of components in Gaussian mixture models adaptively for medical image. Optik – Int. J. Light Electron. Opt. **124**(23), 6216–6221 (2013)
12. Calinski, T., Harabasz, J.: A dendrite method for cluster analysis. Commun. Statist. **3**(1), 1–27 (1974)
13. Davies, D.L., Bouldin, D.W.: A cluster separation measure. IEEE Trans. Pattern Anal. Mach. Intell. **PAMI-1**(2), 224–227 (1979)
14. Rouseeuw, P.J.: Silhouettes: a graphical aid to the interpretation and validation of cluster analysis. J. Comput. Appl. Math. **20**(1), 53–65 (1987)
15. Witkin, A.P.: Scale-space filtering: A new approach to multi-scale description. In: IEEE International Acoustics, Speech, and Signal Processing
16. Carlotto, M.J.: Histogram analysis using a scale-space approach. IEEE Trans. Pattern Anal. Mach. Intell. **9**(1), 121–129 (1987)
17. Yuille, A.L., Poggio, T.A.: Scaling theorems for zero crossings. Pattern Anal. Mach. Intell. IEEE Trans. **1**, 15–25 (1986)
18. Wasserman, L.: All of Nonparametric Statistics. Springer Texts in Statistics (2006)
19. Kisku, D.R., Rattani, A., Gupta, P., et al.: Offline signature verification using geometric and orientation features with multiple experts fusion. In: 2011 3rd International Conference on Electronics Computer Technology (ICECT), pp. 269–272. IEEE (2011)
20. Yeo, C., Ahammad, P., Ramchandran, K.: Coding of image feature descriptors for distributed rate-efficient visual correspondences. Int. J. Comput. Vis. **94**(3), 267–281 (2011)

User-Driven Sports Video Customization System for Mobile Devices

Jian Qin[1], Jun Chen[1,2]($\boxtimes$), Zheng Wang[1], Jiyang Zhang[1], Xinyuan Yu[1], Chunjie Zhang[3], and Qi Zheng[1]

[1] National Engineering Research Center for Multimedia Software, School of Computer, Wuhan University, Wuhan 430072, China
{jamesqin,chenj,wangzwhu,bopzjy,bluesyu,zhengq}@whu.edu.cn
[2] Collaborative Innovation Center of Geospatial Technology, Wuhan 430079, China
[3] School of Computer and Control Engineering, University of Chinese Academy of Sciences, Beijing 100190, China
zhangcj@ucas.ac.cn

Abstract. In this paper, we have implemented a user-driven sports video customization system, aiming to provide interesting video clips for mobile users according to their personalized preferences. In this system, we use the web-casting text to detect events from sports video and generate rich content description. In particular, the video clock time on the scoreboard is recognized for the purpose of aligning these events from web-casting text to sports video clips. The proposed extended-hidden Markov model (extended-HMM) is proved to be able to recognize the clock time precisely. To save mobile web traffic, an optional function based on the proposed event based video encoding approach is embedded in the system. Compared with traditional encoding approach, this approach provides bitrate saving of about 34 % while the quality of frames which users are interested in keeps the same. Both quantitative and qualitative experiments have been conducted to prove the proposed approaches' effectiveness.

Keywords: User-driven · Extended-HMM · Event based encoding · Sports video customization · Mobile devices

1 Introduction

With the rapid development of mobile devices and network technology, growing individuals tend to use mobile devices rather than computer to enjoy the video services [1]. Among all types of videos, sports video is quite popular with mobile users. Equipped with such multimedia-enable mobile devices, sports fans can freely enjoy sports videos at anytime in anyplace.

However, due to users' various appetites, not only match videos but also highlight collections made by studio professionals cannot meet every audience's demands. For example, some one is interested in seeing Yao Ming's slam dunks, while another one may only want to watch Kobe Bryant's jump shots.

© Springer International Publishing Switzerland 2015
Y.-S. Ho et al. (Eds.): PCM 2015, Part II, LNCS 9315, pp. 526–535, 2015.
DOI: 10.1007/978-3-319-24078-7_53

In such condition, providing personalized video customization which offers users their most desired video clips is of great importance. To realize this scenario, the source video has to be tagged with detailed semantic labels [2]. For task of sports video analysis, the content based method which adopts various low-level features such as audio, visual are commonly used [3]. However, due to the semantic gap between low-level features and high-level concepts, it is difficult to extract detailed semantics we need precisely by using these methods. Fortunately, a valid approach combining a video with corresponding web-casting texts has been proposed by some researchers [4,5]. In [5], Xu et al. adopted the timestamp overlapped on the sports video to bridge web-casting texts. Furthermore, based on this work, Liang et al. [6] raised a user-participant multi-constraint 0/1 knapsack problem and implemented a mobile personalized sports video customization system. Thus, the time recognition is one of key tasks of sports video analysis [7]. To recognize time-stamp overlapped on the sports video, template matching method is generally adopted in related work [4–6]. However, this easily leads to incorrect recognition due to the noise disturbance. Moreover, time-stamp in sports video changes periodically. Taking this periodic rule into consideration, incorrect recognition will be decreased extremely. Hence, we mainly investigate this time recognition based on its periodic rule and propose an extended-HMM based method for this task.

In addition, we observe that sports events extracted from the sports video always last for some duration. For example, one goal in a soccer match may contain the long time ball pass, shot and cheers. Meanwhile, users commonly not only want to watch the whole process for the goal, but they also tend to pay more attention to the several seconds around the moment when goal actually takes place and require high-quality video for the several seconds. Moreover, traditional encoding method in related work requires relatively high average bitrate to meet users' watch experience. Considering these factors, we also investigate encoding approach to meet users' watch experience with a relatively low average bitrate.

In this paper, we develop a sports video customization system for mobile devices based on web-casting text and broadcast video. The main contributions of our work can be summarized as follows: Firstly, we propose a novel algorithm for sports video time recognition to reach nearly 100 % time-stamp recognition accuracy; Secondly, we also provide an event based video encoding method which can not only ensure user experience but also save more user's resource especially the mobile web traffic; Thirdly, we design a user-friendly and elegant client interface to facilitate mobile users enjoying their favorite sports video with diverse preferences.

2   System Framework

Based on above idea, we propose a C/S framework for the user-driven sports video customization system, as shown in Fig. 1. It consists of two major components: the multimedia server and the mobile client. For the multimedia server, the sports video and its corresponding web-casting text are aligned to detect and label the events in the original sports video. For example, we know that

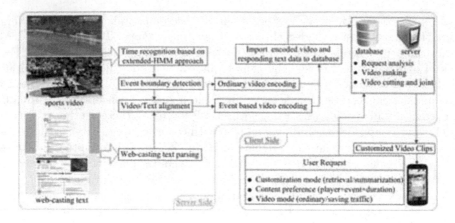

Fig. 1. System framework

Ronaldo goals at 18:33 by parsing web-casting text, moreover, we annotate corresponding video with match time-stamp by recognizing its clock on overlaid scoreboard. Then not only the video clip of this event can be segment by utilization of this time-stamp and event boundary detection, but also it is annotated with rich semantic labels from web-casting text. We present these semantic labels with MPEG-7 description. Based on above methods, we can construct a multimedia database to store the sports video data and their corresponding MPEG-7 description. The mobile client allows the user to customize his/her favorite sports video clips with personalized preference on video content. An event based video encoding is carried out for our labeled source videos to provide two kinds of sports videos for user to choose.

2.1   Semantic Annotation of Sports Video

As the web-casting text webpage[1] consists of different regions, we firstly extract different region of interest (ROI) based on their unique features. Then the keywords related to the events are defined. After that, we use keyword matching approach [5] to extract the match content descriptions.

Meanwhile, video analysis algorithm is also conducted on the corresponding sports video. Firstly, we detect and recognize the clock time on the scoreboard by our proposed extended-HMM method (detail about this can be seen in Sect. 3). After detecting the event moment in the video, we need further identify a temporal range containing the whole event clips. In our observation, we found that due to the common production rule in sports video broadcasting, event evolvement process is always accompanied by a group of specific shots transition, different broadcast sports videos usually have similar structure for an event. Hence, we can model these shots transition patterns in sports video, and use it to detect the event start/end boundaries effectively. The technique details can be referred to [4,8].

---

[1] http://www.bbc.com/sport/0/football/25285050.

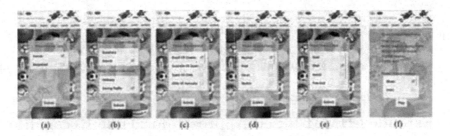

Fig. 2. The client application on the Huawei G610 phone. (a) choose sports type; (b) choose both query mode and video code mode; (c) choose match; (d) choose player; (e) choose event; (f) confirm choice and input viewing time.

2.2 Video Encoding

In our system, we provide two video encoding approaches for video sequence. One traditionally adopts x264[2] with a fixed target bitrate. The other changes bitrate according to the semantic content in our sports event (technical detail can be seen in Sect. 4). Videos encoded by these two methods are stored in our multimedia server.

2.3 Personalized Customization for Mobile Devices

Since users are usually interested in watching the video content that matches his/her preference, once certain mobile user sends a request, our server will analyse this request and rank for the video clips by considering both user's query and every event's importance synthetically. We can formulate this consideration as below:

$$I(E_i) = \lambda \cdot I_m(E_i) + (1 - \lambda) \cdot I_u(E_i) \tag{1}$$

where $I_m(E_i)$ is event importance on its corresponding match, and $\lambda(0 \leq \lambda \leq 1)$ is the fusion parameter controlling the weights on event influence on the match $I_m(E_i)$ and its semantic consistency to the user's request $I_u(E_i)$. With parameter λ, two customization modes can be treated in a unified manner with different fusion parameters. When λ is approximating to 0, event importance is mainly decided by user preference, thus only semantically consistent events can be assigned higher importance score and presented to the user. While in the case of λ approximating to 1, event's significance degree largely depends upon the match itself, hence the selected events can reflect the global situation of the match. For more details about $I_m(E_i)$ and $I_u(E_i)$, readers can refer to [6].

2.4 User Client Design

In order to facilitate user's customization conveniently, we have designed a friendly graphic user interface based on android operating system. By using

---
[2] http://www.videolan.org/developers/x264.html.

the application, mobile users can easily submit their personalized customization requirements and watch video clips sent from server. In addition, if user choose video mode of saving traffic (see Fig. 2(b)), server will send video clips encoded by our event based video encoding approach (technical detail can be seen in Sect. 4).

3   Extended-HMM Based Approach for Time Recognition

As we observe that the clock digits change periodically, we can propose an integrated approach which makes full use of its important temporal pattern to recognize the game time precisely. Firstly, we segment the static overlaid region using a static region detection approach which mainly consists of calculation of the variance map [9]. By monitoring the periodic changes of the clock digit, we can locate the digits area. Then the template of the 0–9 digits are captured in the SECOND digit region as binary images. Finally, we use those digit templates to recognize the game time by template matching method initially and verify these result by our proposed extended-HMM method. Here we mainly present extended-HMM method.

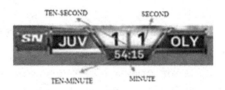

Fig. 3. Digital clock on scoreboard

Traditional template matching method used in [5,6] only recognize the digit by matching the digit template, this doesn't make full use of the periodic rule existing in sports video clock, and may lead to mistaken recognition. Take soccer game for example (see Fig. 3), not only single digit changes periodically but also there are mutual relations among these changes. Once the game starts, while the SECOND digit changes from "9" to "0" in two continuous frames, the TEN-SECOND digit increases by 1, or changes from "5" to "0". Similarly, while the TEN-SECOND digit changes from "5" to "0" in two continuous frames, the MINUTE digit increases by 1, or changes from "9" to "0". While the MINUTE digit changes from "9" to "0" in two continuous frames, the TEN-MINUTE digit increases by 1. In this paper, we propose two novel methods to improve the performance of recognizing video time. One just models four digits as four independent hidden Markov models [10] which just use periodical change of single digit. The other models these digits as four correlative hidden Markov models which takes change relationship of adjacent digits into consideration. We named the first method HMM method, the second method extended-HMM method.

3.1   Introduction to Hidden Markov Model

The hidden Markov model (HMM) is represented by a finite set of states, each of which is associated with a probability distribution [10]. Transitions among the states are governed by a set of probabilities called transition probabilities. In a particular state an outcome or observation can be generated according to the associated probability distribution. For details about HMM, readers can refer to [10].

3.2   HMM Method

As shown in Fig. 3, the SECOND digit character changes from 0, 1, 2, $\cdots$ to 9 every second. In the same way, the TEN-SECOND digit changes once every 10 s, while the MINUTE digit every one minute and the TEN-MINUTE digit every 10 min. Based on above knowledge, four digits of clock are treated as four independent hidden Markov models. For every digit on the clock, we regard ground truths as HMM's states, and template matching results as HMM's distinct observation symbols. Ground truth and template matching result are taken every second. Take SECOND digit as an example, N (the number of states in the model) equals 10, M (the number of distinct observation symbols) also equals 10. We use unsupervised learning to obtain HMM's complete parameter set $\lambda = (A, B, \pi)$, here A is the set of state transition probabilities, B is the observation probability distribution, and π represents the initial state distribution.

3.3   Extended-HMM Method

HMM method just uses an independent digit pattern change to verify the result. So we modify the Viterbi algorithm to make use of correlative transition pattern of the digits on the clock. Take TEN-SECOND digit for example, when SECOND digit changes from "9" to "0", TEN-SECOND digit will increase by 1, or change from "5" to "0" correspondingly. Namely, when state of SECOND digit transfers from state 9 to state 0, state of TEN-SECOND digit will transfer to the next state with 100 % probability. This constraint also exists between TEN-SECOND and MINUTE, MINUTE and TEN-MINUTE. So we take this constraint into consideration to modify the recursion of Viterbi algorithm as below (take TEN-SECOND for example, others are similar):

Initialization:

$$\delta_1(i) = \pi_i b_i(o_1), \ i = 1, 2, \cdots, N \tag{2}$$

$$\psi_1(i) = 0, \ i = 1, 2, \cdots, N \tag{3}$$

Recursion, for t=1,2,$\cdots$,T-1

$$\psi_{t+1}(i) = arg \max_{1 \leq j \leq N} [\delta_t(j) a_{ji}], \ 1 \leq i \leq N \tag{4}$$

$$\delta_{t+1}(i) = \max_{1 \leq j \leq N} [\delta_t(j) a_{ji}] b_i(o_t) +$$

$$c_1 \cdot I(i_t^* = q_9^*, i_{t+1}^* = q_0^*) \cdot F([\psi_{t+1}(i) + 1]\%6 = i)$$
$$+c_2 \cdot I(i_t^* \neq q_9^* || i_{t+1}^* \neq q_0^*) \cdot F(\psi_{t+1}(i) = i),$$
$$1 \leq i \leq N \tag{5}$$

Termination:

$$P^* = \max_{1 \leq i \leq N} [\delta_T(i)] \tag{6}$$

$$i_T^* = arg \max_{1 \leq i \leq N} [\delta_T(i)] \tag{7}$$

Path backtracking:

$$i_t^* = \psi_{t+1}(i_{t+1}^*), \quad t = T - 1, T - 2, \cdots, 1 \tag{8}$$

where $\delta_{t+1}(i)$ is the prob. of highest prob. path ending in state i at t+1 for model of TEN-SECOND digit, π_i is the initial state distribution, $b_i(o_t)$ is the observation probability. a_{ji} is the state transition probability. I(x) is the indicator function. i_t^* is the inferred state at t for model of SECOND digit by Viterbi algorithm. q_i^* is i^{th} state in the set of hidden states of SECOND digit, $\psi_{t+1}(i)$ is the state at t in the highest prob. path ending in state i at t+1. $F(exp)$ here means that $F(exp)$ is 1 if expression exp is true, otherwise it is -1. $c_1, c_2 > 0$ is the restraint coefficient, and we use $c_1 = c_2 = 10$ in our implementation empirically.

4   Event Based Video Encoding Approach

An event in the sports match often lasts for some duration. Compared with the whole time length, we can observe that users tend to pay more attention to the several seconds around the moment when event actually takes place. Moreover, the event moment also can be precisely located by our video annotation (see Fig. 1). Considering this factor, we make video bitrate change with the event's evolution to save user's mobile web traffic. Moreover, this mode is optional in the client end. We adjust the video bitrate according to temporal distance between current frame and reference frame namely event moment. Bitrate at the event moment should be highest among these frames in this situation. This approach can ensure high-definition for the event moment as well as decrease consumption of network resource. Video bitrate r_t at the certain moment t can be defined as below:

$$r_t = R \cdot e^{-(t-t_{em})^2/c^2} \tag{9}$$

where R is the average bitrate of traditional approach, t_{em} is the moment when the event actually occurs. We denote the starting moment of temporal range which contains the whole event evolvement process as t_{st}, the ending moment of this temporal range as t_{en}, and define the minimum of the bitrate after video encoding with this approach as $minR$, which appears either at the starting or at the ending of the event video clip. Here c is parameter which ensure equality $\min[R \cdot e^{-(t_{st}-t_{em})^2/c^2}, R \cdot e^{-(t_{en}-t_{em})^2/c^2}] = minR$. Thus, c can be presented as below:

$$c = \frac{\max[(t_{em} - t_{st}), (t_{en} - t_{em})]}{\sqrt{In(R/minR)}} \qquad (10)$$

In our implement and experiment, R equals 3000 kb/s and $minR$ is set to $0.3 \cdot R$ empirically. We only change bitrate every 100 frame (25 frames per second) in a event for consideration of implement's efficiency.

5   Experimental Results

In this section, we present our experimental results for time recognition and video encoding. In addition, our system use soccer matches of FIFA 2014 and NBA 2014 basketball matches as our data source and the corresponding web-casting texts are from ESPN for basketball and BBC for soccer.

5.1   Time Recognition

In our project, we use Visual C++ to implement the algorithm. As long as the clock time digits appear on the video, and last for more than 10 seconds, we could locate the score board and recognize the video clock time in real time. For HMM method and extended-HMM method, we use buffer to store data and reach purpose of quasi real-time recognition. In addition, because basketball matches have pause and scoreboard may disappear, we detect these case and use extended-HMM for every continuous video clips. The proposed approach in this paper and methods in [6,9] has been tested on 8 different matches which contains five soccer matches of FIFA 2014 and three NBA 2014 basketball matches from different television stations. Results are shown in the Table 1. Here accuracy is computed by following formula:

$$accuracy = \frac{r}{s} \qquad (11)$$

where r is seconds of recognizing every digit on the clock correctly at the same time, s is match's time length in second format.

Table 1. Time clock recognition accuracy(%). gn: game number (1–5: soccer video, 6–8: basketball video). tm: template match method in [6]; tmv: template match method with verification in [9]; hmm: our proposed HMM method; ehmm: our proposed extended-HMM method.

gn	1	2	3	4	5	6	7	8
tm	95.2	95.4	95.6	94.4	94.1	93.7	94.3	93.1
tmv	98.1	98.1	98.4	97.9	97.9	96.6	96.4	95.7
hmm	98.1	97.9	98.4	98.1	97.8	96.7	96.6	95.8
ehmm	**99.7**	**99.8**	**99.8**	**99.7**	**99.5**	**98.2**	**98.1**	**98.0**

Through the experiment, we can draw some conclusions. Within the acceptable time consumption, recognition rate of extended-HMM method is significantly higher than template match method, this is because our extended-HMM
method makes full use of the fact that digits of time appear in sequence like
0,1,2,3··· and transformational relationship between the digits. In the case of
similar time consumption, extended-HMM method also outperforms template
match method with verification method in [9], which is because verification
method in [9] cannot handle some cases. For example, when time 05:19 is recognized as 05:18 incorrectly, because this recognized result 05:18 is near to adjacent
results in this type of figure like Fig. 9 in [9], this result 05:18 cannot be taken
as an isolated point to the other results and this error cannot be verified by
using method in [9]; But it can be verified by using extended-HMM method
because this method can detect the fact that it doesn't conform to the rule of
time transition sequence and verifies this error.

5.2   Video Encoding

In this experiment, we compare event based video encoding approach with traditional approach. For comparison, we use video sequence extracted from soccer
video, at 1280*720 pixels resolution, 25 frames per second (fps), and 1000 frames.
We set a target bitrate of 3000 kb/s in traditional approach, but change target
bitrate according to the event moment in the proposed approach. As we can
see from the Fig. 4(b), we find that proposed approach provides bitrate saves of
33.7 % over traditional approach. Meanwhile, from the Fig. 4(a), the quality of
frames which users pay attention to keeps the same.

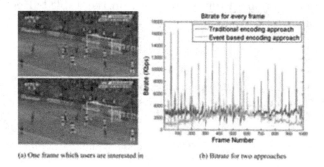

(a) One frame which users are interested in   (b) Bitrate for two approaches

Fig. 4. (a) The 500th frame when the player actually goals. Above frame is encoded by
traditional approach while below frame is encoded by proposed approach. (b) Bitrate
for these two encoding approaches. In addition, discrete and extreme peaks in (b)
correspond to I frames.

6   Conclusions

In this paper, we proposed a sports video customization system for mobile users.
Compared with previous work, the system can provide precise results of cus-

tomized video for users with our proposed extended-HMM based approach. In addition, a novel video encoding approach based on sports event is raised to save users' mobile web traffic.

Acknowledgement. The research was supported by the National Nature Science Foundation of China (61231015, 61172173, 61303114, 61170023), the Technology Research Program of Ministry of Public Security (2014JSYJA016), the National High Technology Research and Development Program of China (863 Program, 2015AA016306), the EU FP7 QUICK project under Grant Agreement (PIRSES-GA-2013-612652), the Major Science and Technology Innovation Plan of Hubei Province (2013AAA020), the Internet of Things Development Funding Project of Ministry of industry in 2013 (No. 25), the China Postdoctoral Science Foundation funded project (2014M562058), the Specialized Research Fund for the Doctoral Program of Higher Education (20130141120024), the Nature Science Foundation of Hubei Province (2014CFB712), the Fundamental Research Funds for the Central Universities (2042014kf0025, 2042014kf0250, 2014211020203), the Scientific Research Foundation for the Returned Overseas Chinese Scholars, State Education Ministry ([2014]1685).

References

1. Wang, Z., Meng, D.: Personalising sports events viewing on mobile devices. In: ICSESS, pp. 710–713 (2013)
2. Liang, C., Xu, C., Lu, H.: Personalized sports video customization using content and context analysis. Int. J. Digit. Multimed. Broadcast. **2010**, 1–20 (2010)
3. Miyauchi, S., Hirano, A., Babaguchi, N., Kitahashi, T.: Collaborative multimedia analysis for detecting semantical events from broadcasted sports video. In: ICPR, pp. 1009–1012 (2002)
4. Xu, C., Wang, J., Wan, K., Li, Y., Duan, L.: Live sports event detection based on broadcast video and web-casting text. In: ACM Multimedia, pp. 221–230 (2006)
5. Xu, C., Wang, J., Lu, H., Zhang, Y.: A novel framework for semantic annotation and personalized retrieval of sports video. TMM **10**(3), 421–436 (2008)
6. Liang, C., Jiang, Y., Cheng, J., Xu, C., Luo, X., Wang, J., Fu, Y., Lu, H., Ma, J.: Personalized sports video customization for mobile devices. In: Boll, S., Tian, Q., Zhang, L., Zhang, Z., Chen, Y.-P.P. (eds.) MMM 2010. LNCS, vol. 5916, pp. 614–625. Springer, Heidelberg (2010)
7. Yu, X.: Localization and extraction of the four clock-digits using the knowledge of the digital video clock. In: ICPR, pp. 1217–1220 (2012)
8. Bertini, M., Cucchiara, R., Bimbo, A., et al.: Object and event detection for semantic annotation and transcoding. In: ICME, pp. 421–424 (2003)
9. Bu, F., Sun, L.-F., Ding, X.-F., Miao, Y.-J., Yang, S.-Q.: Detect and recognize clock time in sports video. In: Huang, Y.-M.R., Xu, C., Cheng, K.-S., Yang, J.-F.K., Swamy, M.N.S., Li, S., Ding, J.-W. (eds.) PCM 2008. LNCS, vol. 5353, pp. 306–316. Springer, Heidelberg (2008)
10. Rabiner, L.R.: A tutorial on hidden markov models and selected applications in speech recognition. In: Processings of the IEEE, pp. 257–286 (1989)

Auditory Spatial Localization Studies
with Different Stimuli

Tao Zhang[⊠], Shuting Sun, and Chunjie Zhang

School of Electronic Information Engineering, Tianjin University, Tianjin, China
zhangtao@tju.edu.cn

Abstract. Many localization studies have tested the ability of auditory spatial localization in humans. However, broadband noise sources, such as Gaussian white noise and pink noise, were usually chosen as stimuli and the distribution is sparse. In this paper, an intuitive systematic subjective evaluation method is proposed. Subjective used a laser pointer to indicate the perceived direction accurately. Except the Gaussian white noise stimuli, the auditory localization performance was also tested with 1 kHz pure-tone stimuli ranging from − 45° to + 45° in the horizontal plane. In Experiment 1, stimuli is the Gaussian white noise and is distributed with a spacing of 10° to verify that the method is accurate and suitable for the localization research. In Experiment 2 and 3, the distribution of the speakers turns closer with each other. Stimuli are the Gaussian white noise and 1 kHz pure-tone separately. All experiment results are presented and compared with other studies.

Keywords: Subjective evaluation · Auditory localization · Pointing methods · Stimuli

1 Introduction

Nowadays, 3D multimedia has been used in movies and television shows successfully. Compared with the visual field, the research of spatial audio is much less developed. 3D Audio System, also known as Spatial Audio System, could provide audience realistic, immersive and compelling reproduction audio. Because the study of the auditory localization performance can make a great contribution to the development of spatial audio technology, many researchers committed to finding a suitable subjective evaluation method for auditory localization.

To investigate auditory localization, the subject is required to specify the perceived auditory direction. These methods could be classified into two types, one is distinguishing between two source directions, and the other is pointing to the perceived direction directly. Feng Wu et al. [1] used the former method. Subject is required to distinguish between two sound sources located at symmetrical position each time until the results meet the MAA (Minimum Audible Angle). This method can get the MAA accurately in a fixed direction. But the workload will be particularly large. The second indication method is used widely. In Wightman et al. [2] system, subject is asked to express the azimuthal and elevation angle they perceived in verbal way. In general this method is easy to implement, but not intuitive. The description of perceived direction

© Springer International Publishing Switzerland 2015
Y.-S. Ho et al. (Eds.): PCM 2015, Part II, LNCS 9315, pp. 536–543, 2015.
DOI: 10.1007/978-3-319-24078-7_54

has a significant effect on the result. Another simple graphical method requires the subject to draw the perceived direction onto a piece of paper or a screen [3]. However, since head moving includes dynamic localization cues, it is sensible to prevent the subjects from moving their head during listening. So the graphical method is not suitable. Gilkey et al. [4] introduced a method called GELP-method. Subject indicates the perceived direction of the sound by pointing at a 20-cm-diam spherical model. This comfortable method covers the whole space and provides fast responses. However, training is required and a small spherical is needed to project the real spherical. System error will be introduced, so the accuracy is reduced. Pointer methods, which include hand-point [5, 6, 11, 12], head-point [6–8], and eye-point [9, 10], are natural and intuitive methods. Pointing by head and eye gestures is restricted by human physiology. Some eye-point methods even used a sclera search coil to record eye movements [9], which is considerable discomfort for the subject. The necessary cornea anaesthetization requires the presence of a medical professional. Usually, additional head or eye movement is needed in experiments, which makes the calibration of the system and date logging become difficult. A hand-point method is simple to learn and has a high accuracy. But the method is limited mainly to the frontal sector for the head or body is not allowed to move.

Many tests have been done to study the ability of localization in three dimensional space as well as to compare the pros and cons among different methods. The stimuli in these tests were usually broadband noise sources with a sparse distribution. Seeber [11] chose the Gaussian white noise as the targets sound and the speakers were spanned an angle of 50° left to 50° right with a spacing of 10°. Schleicher et al. [10] used the pink noise impulses and all speakers were centered around the subject's seat in steps of 15° from + 45° to − 45°. The acoustic stimuli used in Populin's [9] experiments consisted of 0.1– 25 kHz broadband 150 ms noise bursts, and the speakers were almost uniformly distributed with a spacing of 10 at the eye level. Lewald and Ehrenstein [12] used band pass filtered noise with bandwidth 1–3 kHz from + 22° to − 22° The narrowband sources was seldom used as the stimuli, therefore, the knowledge of human auditory spatial localization performance is not as much as localization performance with those broadband sources.

In this paper, a systematic method for the research of auditory localization is designed. The method for the investigation of localization utilizes a laser pointer for pointing the perceived auditory direction. The method attempts to make the experimental operation more intuitive and the errors introduced by system is reduced. At the same time, complicated calibration operations can be avoided. The whole system is subjects friendly. Based on the proposed method, three experiments are carried out. Experiment 1 tries to prove that the method is effective and accurate for the localization research in audiology. Experiment 2 and 3 test the ability of sound localization with broadband source (Gaussian white noise which is more natural) and narrowband source (1 kHz pure-tone to which human hearing is more sensitive) separately.

2 Experimental Setup

When studying the spatial hearing, a spheroidal model is frequently used. As is shown in Fig. 1(a), different planes in space are contained in this model. The horizontal plane intersects the median plane at the origin. Based on the assumed model shown in

Fig. 1(a), a special semi-spherical experiment equipment is designed and produced to measure the perceived direction in 3D localization experiments. The sketch of the true equipment is shown in Fig. 1(b) (Fig. 1(b) shows half of the semi-spherical).

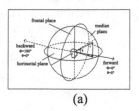 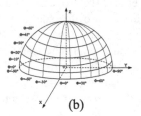

(a)   (b)

Fig. 1. (a) Spheroidal model used in psychoacoustic experiments. (b) The sketch of the real equipment used in experiments.

The rings are designed as detachable. Therefore, the whole equipment is convenient to install and move as required. The radius of the spherical is 1.5 m. The system is easy to calibrate. The hight of the chair can be adjusted. The equipment is designed to provide a basic platform for the subjective listening test.

The whole equipment is installed in an acoustically treated room of size L * W * H = 6.2 m * 5.3 m * 2.9 m. A number of exactly equal loudspeakers, with perfect directivity are choose as point sound source. The speakers can be divided into two categories, one is the real speaker which is really sounding during the experiment, the other is the confused speaker without really sounding. All speakers are mounted on the surface of the equipment one after another. In the horizontal plane, azimuth is + 90° for the direction directly to the left of the subject and − 90° for the symmetry direction and the angle of the sound source varies only in the horizontal plane from − 45° to + 45° in both experiments. As the diameter of one speaker is 1.8 cm, a total number of 130 speakers can localized on the horizontal plane from the angle − 45° to + 45° in theory. In fact, for statistical purpose, 90 speakers are mounted on the equipment uniformly and each speaker represents 1° in the horizontal direction. It should be noted that subjects do not know the exist of the "confused speakers". They consider that all speakers are the "real speakers". During experiments, each real speaker is played randomly and separately. The real speakers will be placed with different intervals on depend. In Experiment 1, the array of nine real loudspeakers is mounted on the horizontal plane with an angular spacing of 10° in the range from − 40° to + 40°. In Experiment 2 and 3, the intervals become smaller and the number of real speakers increases. In theory, the interval between two neighboring real speakers should be less than MAA. Real loudspeakers azimuth angles are shown in Fig. 2 (only the real speakers are presented), uneven speaker placement is based on spatial hearing symmetry theory. The level of each stimulus is adjusted to produce 80 dB SPL at the listening point. The sound pressure responses of all loudspeakers are measured at the listening position. Pick one real speaker to play sound randomly and the sounding time lasts 1 s each time.

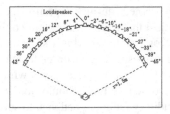

Fig. 2. The real speakers mounted diagram in experiment 2 and 3.

As the sounding cannot be "seen" by subject, subject can directly point the perceived direction onto one speaker. In this way, the experiment does not have to be carried in the dark environment. Both the experiments are carried out under the natural conditions.

3 Method and Subjects

The method used in subjective evaluation of auditory localization should be intuitive to handle and easy to learn for the subject. The response bias introduced by the method should be small. As the head movements are typically not allowed, the formerly hand-point method is generally limited mainly to the frontal sector. In order to break through this bottleneck, head of subject is kept still when listening, but is free when pointing. After each listening, subject points to the perceived direction with a laser pointer. Then turn the head back to the original position. The head fixtures ensure that head of subject can accurately return to the calibration location each time.

Block diagram of the new systematic subjective evaluation method for auditory localization experiments is shown in Fig. 3.

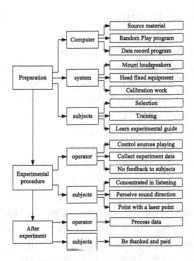

Fig. 3. Block diagram of the subjective evaluation method.

Some matters should be noted during experiments. Lead the subject to the semi-spherical instrument, adjust the height of the chair so that the midpoint line of two ears can coincide with the cross, and the subject's ear is at the same level with the horizontal plane; Fix the head of subject; In order to avoid any discrepancy due to handedness, subjects are asked to hold the laser point with both hands; After subjects point, the pointed angle is recorded; Subject should not listen to a sound source for a long time in order to avoid auditory fatigue; Feedback to subject of pointing accuracy is forbidden after they point to a direction.

Experiment 1: 10 subjects participate in the experiment 1, aged 22–28, five females, five males. All subjects are the naive subjects without any experience with localization tests and have normal audiometric thresholds. Stimulus used in Experiment 1 is the Gaussian white noise. Nine real speakers are distributed on the horizontal plane from − 40° to + 40° with a uniform interval of 10°. A total number of 90 test items are carried out in random order to each subject per session. Therefore, there are 10 test items in every real speaker position. Each stimulus sounding lasts for 1000 ms. The playback level is adjusted to 80 dB SPL sound pressure level at the listening position. The purpose of Experiment 1 is to verify that the pointing method used in this paper is a simple, intuitive and accurate method in the study of auditory localization.

Experiment 2: To investigate the influence on the localization as the sound sources increase, Experiment 2 is proposed. The subjects, procedure and stimuli of Experiment 2 are identical to Experiment 1, except for the fact that the number of real speakers increases to 19 in the range of − 45° to + 45° in the horizontal plane, as is shown in Fig. 2, and the test items for each real speaker position become 5.

Experiment 3: The goal of Experiment 3 is to estimate human locating ability for the narrowband signal. Therefore, the subjects, distribution of the real speakers, procedure in Experiment 3 are in accord with Experiment 2. Only the stimuli in Experiment 3 is the 1 kHz pure-tone source.

4 Data Processing and Results

A point sound source produces an auditory event, which is spread out to a certain degree in space. That means the auditory system processes less spatial resolution than that achieved using physical measuring techniques. The localization errors are calculated by subtracting the real speaker angles from the pointed angles. According to the conventions of Heffner and Heffner [13], the localization bias is signed mean of the localization errors and the localization precision, sometimes also called localization blur, is the standard deviation of the localization errors.

Figure 4 displays the mean and standard deviation of localization errors for the all 10 subjects in Experiment 1. The maximum absolute mean value is 2.25°, appears in the real position of − 40°. The 95 % confidence interval is within the scope of 5°. The result of pointed method used in this paper seems not as good as the method used in Schleicher et al. [10] research, which utilized eye movements for acoustic source localization tests and achieved a very accurate result with the absolute error below 1.5° and a variance not exceeding 0.5°. However, it should be attention that in Schleicher et al. study, head of the subject is unrestrained during the listening experiments. As the head movement is of great help to the precise positioning, it is reasonable that the localization result in Schleicher et al. study is more accurate than our research. This localization data is more agreement with the results of Seeber [11], who carried out one localization experiment using a laser spot which is randomized in initial position. Compared with the experimental results, the subjective evaluation method proposed in this paper, which with simple operation and intuitive pointing way, is a more suitable and accurate method for the localization research in auditory.

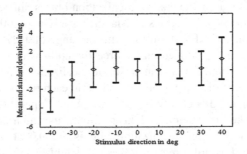

Fig. 4. Mean and standard deviation of localization error for experiment 1.

Figure 5 shows the localization bias and localization precision from the stimulus direction obtained in two experiments with different types of sound sources, Experiment 2 with Gaussian white noise and Experiment 3 with the 1 kHz source. Data with Experiment 2 are represented with () and data with Experiment 3 are represented with (). For better distinction, the data of both experiments are shifted horizontally. Due to the subjects can see the distribution of the speakers, the data around the boundary are inaccurate. Therefore, data for angle 42° and − 45° are rejected during calculation.

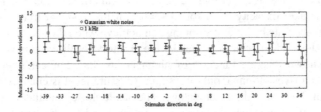

Fig. 5. Mean and standard deviation of localization error for experiment 2 and experiment 3. exp. 2: stimuli with Gaussian white noise, exp. 3: stimuli with 1 kHz.

By increasing the number of the real speakers in the range of ± 45°, the overall trend of localization performance can be better investigated. For the Gaussian white noise, the absolute error is below 2.1° with a variance not exceeding 5°, this result is keeping with Experiment 1, in which the distribution of the real speakers is more sparse. The sound source at 0° is localized at 1.36°, with a small variation range of ± 0.71°. In fact, bilateral symmetry is showed with the localization blur in Experiment 2. According to the localization blur, the localization can be roughly divided into 3 intervals: angle within

± 8°, locating in this area has the highest localization precision with an average localization blur of ± 0.89°; angle within [8°, 20°] and its symmetrical angle, the average
localization blur is ± 1.58°; angle within [20°, 39°] and its symmetrical angle, the
average localization blur is ± 2.17°. On the whole, the trend of localization performance
shows a high agreement with the most localization research which used white noise
stimuli, like Seeber [11] and Blauert [14]. Though in Blauert's research, the localization
blur at 0° is ± 3.6°, it seems that the result in our experiment is more accurate. However,
Blauert also pointed that the last time of the stimuli will have influence on the direction
localization. Thus, it is reasonable that the localization blur is different between the two
experiments since duration was 100 ms in Blauert's study and 1000 ms in this paper.

For the localization of 1 kHz stimuli within the range of ± 27°, the result shows a
high agreement between the pointed and source direction with a maximum absolute
localization bias of 2.52° and a minimum absolute localization bias of 0.24°. However,
compared with the white noise, the localization blur at each real sound source direction
with 1 kHz in Experiment 3 is clearly larger than that in Experiment 2. From Fig. 5 it
can be seen that for the localization of 1 kHz stimuli, the minimum localization blur
occurs in the forward direction (localization blur is ± 1.98° at 0°). With increasing
displacement from the forward direction toward the right or left, the localization blur
increases. The localization blur attains between two and four times its values for the
Gaussian white noise.

Localization bias increases abruptly at the source direction of − 33° (4.6°) and − 39°
(7.16°), as is shown in the profiles for Experiment 3. We analyze the reason of this
abnormal phenomenon as follow: speakers are placed within the range of ± 45° in the
horizontal direction and this visual information is available for subjects, therefore,
when perceives the sound source direction like − 39°, the perceived direction may
appear at − 50°, but the pointed direction will indicate at − 45° in the actual. For this
reason, localization bias increases abruptly at the border. Frankly speaking, this is a
defect of experimental design, it can be fixed by expanding the distribution of speakers.

Based on the results of Experiment 2 and 3, we can arrive at the conclusion human
hearing has a weaker capacity for a sound source of just a signal frequency, and relatively speaking, accurate hearing positioning can achieve when the sound source is white
noise. The conclusion had been explained by the auditory localization mechanism, it has
already been known that the function of pinna is just like a linear filter. When narrowband signals arrives at pinna, only the sound pressure level is changed by the linear
filter. Timbre of the source does not change. The white noise contains many frequency
components. So the frequency spectrum will be changed by the transfer function of
pinna. For this reason, human hearing has a higher capacity of auditory localization for
sources with broadband signals like white noise, human normal speech, music and etc.,
but a lower capacity of auditory localization for sources with narrowband signals.

5   Conclusion and Discussion

A systematic subjective evaluation method for the investigation of auditory localization
experiments is presented in this paper. The major advantages of the proposed method
lie in good stability as head is fixed and high accuracy of pointing. Subject utilizes a

laser point to indicate the perceived direction, which is intuitive to handle and easy to learn. It turns out that the subjective evaluation method proposed in this paper is a suitable and accurate method for the localization research in auditory. To study the auditory spatial localization for both broadband signal and narrowband signal, Gaussian white noise and 1 kHz pure tone are used as stimuli separately, with 19 pointed source distributing within the range of ± 45° in the horizontal direction. The result shows that localization on the broadband source is more accurately compared with narrowband source. The localization blur attains between two and four times its values for the Gaussian white noise. For both stimuli, the minimum localization blur occurs in the forward direction and with increasing displacement from the forward direction toward the right and left, the localization blur increases.

References

1. Feng, W., Gao, X., Li, Z.-H., et al.: Devising and initial realization of testing hearing system for sound location. J. Fourth Mil. Med. Univ. **22**(7), 656–658 (2001)
2. Wightman, F.L., Kistler, D.J.: Monaural sound localization revisited. J. Acoust. Soc. Am. **101**(2), 1050–1063 (1997)
3. Mason, R., Ford, N., Rumsey, F., et al.: Verbal and nonverbal elicitation techniques in the subjective assessment of spatial sound reproduction. J. Audio Eng. Soc. **49**(5), 366–384 (2001)
4. Gilkey, R.H., Good, M.D., Ericson, M.A., et al.: A pointing technique for rapidly collecting localization responses in auditory research. Behav. Res. Methods Instrum. Comput. **27**(1), 1–11 (1995)
5. Tabry, V., Zatorre, R.J., Voss, P.: The influence of vision on sound localization abilities in both the horizontal and vertical planes. Front. Psychol. **4**, 1–7 (2013)
6. Majdak, P., Goupell, M.J., Laback, B.: 3-D localization of virtual sound sources: effects of visual environment, pointing method, and training. Atten. Percept. Psychophys. **72**(2), 454–469 (2010)
7. Ashby, T., Mason, R., Brookes, T.: Head movements in three-dimensional localization. In: Audio Engineering Society Convention 134. Audio Engineering Society (2013)
8. Minnaar, P., Pedersen, J.A.: Evaluation of a 3D-audio system with head tracking. In: Audio Engineering Society Convention 120. Audio Engineering Society (2006)
9. Populin, L.C.: Human sound localization: measurements in untrained, head-unrestrained subjects using gaze as a pointer. Exp. Brain Res. **190**(1), 11–30 (2008)
10. Schleicher, R., Spors, S., Jahn, D., et al.: Gaze as a measure of sound source localization. In: Audio Engineering Society Conference: 38th International Conference: Sound Quality Evaluation. Audio Engineering Society (2010)
11. Seeber, B.: A new method for localization studies. Acustica **88**(3), 446–449 (2002)
12. Lewald, J., Ehrenstein, W.H.: Auditory-visual spatial integration: a new psychophysical approach using laser pointing to acoustic targets. J. Acoustic. Soc. Am. **104**(3), 1586–1597 (1998)
13. Heffner, H.E., Heffner, R.S.: The sound-localization ability of cats. J. Neurophysiol. **94**(5), 3653–3655 (2005)
14. Blauert, J.: Spatial Hearing: The Psychophysics of Human Sound Localization. MIT press, Cambridge (1997)

Multichannel Simplification Based on Deviation of Loudspeaker Positions

Dengshi Li[1,3], Ruimin Hu[1,2]($\boxtimes$), Xiaochen Wang[1,2],
Shanshan Yang[1], and Weiping Tu[1,2]

[1] National Engineering Research Center for Multimedia Software,
School of Computer, Wuhan University, Wuhan, China
reallds@126.com, {hrm1964,clowang,yangssgood,echo_tuwp}@163.com
[2] Research Institute of Wuhan University in Shenzhen, Shenzhen, China
[3] School of Computer and Mathematics, Jianghan University, Wuhan, China

Abstract. People hope to achieve a good impression of three-dimensional (3D) spatial sound with fewer loudspeakers at home. The present method simplified the amount of loudspeakers based on the minimum area enclosed by loudspeakers while maintaining the sound pressure at the origin. However, it doesn't consider the distortion of a sound field within some specified regions since people always use two ears to listen. In this paper, we exploit that the distortion will be affected by the deviation of positions among loudspeakers and the selection of loudspeaker positions is redefined to obtain the weighting coefficients of each multichannel simplified system. For each multichannel simplified system, simulation result indicates that the distortion within the region of head generated by the proposed method is not more than that generated by the present method. Subjective evaluation shows the proposed method is slightly better in terms of sound localization.

Keywords: Spherical harmonics · Sound field reproduction · Multichannel simplification system

1 Introduction

While 3D spatial sound field can be accurately reproduced in physics over a predetermined spatial region, the listener inside the region will experience a realistic reproduction of the original sound field. The typical methods are wave field synthesis (WFS) [1] and Higher-order Ambisonics (HOA) [2,3]. Based on the Kirchhoff-Helmholtz integral equation and Huygens principle, WFS represents a sound field in the interior of a bounded region of the space by the continuous secondary sources, arranged on the boundary of that region. HOA is to achieve the

---

R. Hu—This work is supported by National Nature Science Foundation of China (No. 61231015, 61201169, 61201340), National High Technology Research and Development Program of China (863 Program) No. 2015AA016306, Science and Technology Plan Projects of Shenzhen (ZDSYS2014050916575763), the Fundamental Research Funds for the Central Universities (No. 2042015kf0206).

Y.-S. Ho et al. (Eds.): PCM 2015, Part II, LNCS 9315, pp. 544–553, 2015.
DOI: 10.1007/978-3-319-24078-7_55

(a)   (b)

Fig. 1. Locations of loudspeakers. (a) Loudspeaker $\mathbf{y}$ in the original sound field; (b) Three loudspeakers $\mathbf{y}_1$, $\mathbf{y}_2$ and $\mathbf{y}_3$ in the reproduced sound field.

similar effects by reconstructing spherical harmonics of the pressure field within the given region. However, these reproduced methods employ large number of loudspeakers, which can reach several hundreds or even more. Simultaneously, practical techniques for spatial sound reproduction have achieved considerable progress. Considering the perceptive characteristic of human, NHK laboratory proposed 22.2 multichannel sound system by 24 loudspeakers (including two low-frequency effect loudspeakers), which could be arranged optionally in the theater rather than at home.

In order to satisfy people's demand for 3D spatial sound in the family environment, Prof. Ando introduced a multichannel simplification method [5] based on minimum area among loudspeakers of the reproduced sound field while maintaining the sound pressure at the listening point (i.e., the center of head). That was, each loudspeaker in the original sound field should be regarded as a phantom source generated by three loudspeakers with minimum enclosed area. Signals of 22.2 multichannel sound system without two low-frequency effect channels could be converted successfully into those of 10- and 8-channel sound systems, resulting in a good sense of spatial sound impressions testified by the subjective experiment evaluations. However, the natural question is whether the distortion of sound field within some specified regions, such as the region of man's head, can be decreased with the area enclosed by three loudspeakers shrinking.

Therefore, our target in this paper is to develop some fundamental performance limits for the distortion of reproducing a sound field within a given region in free space (i.e., the effect of reverberation is ignored). Specifically, when using three loudspeakers to reproduce the sound field incident from a loudspeaker, we show that the distortion of reproduced sound field within a given region is not related to the area of three loudspeakers, but to the relative positions among these loudspeakers. Compared with the results of Ando's method, the proposed method can maintain lower distortion within the region of head.

2   Reproduction Based on Area Among Loudspeakers

Considering a point-source field produced by a loudspeaker which is located at $\mathbf{y} = (\sigma, \varphi, \theta)$, shown in Fig. 1(a), where σ is the distance to the origin, ϕ is the azimuthal angle, and θ is the elevation angle. Assuming that the reproduced field generated by three loudspeakers whose locations are $\mathbf{y}_1 = (\sigma_1, \varphi_1, \theta_1)$,

$\mathbf{y}_2 = (\sigma_2, \varphi_2, \theta_2)$ and $\mathbf{y}_3 = (\sigma_3, \varphi_3, \theta_3)$ (shown in Fig. 1(b)), the original and reproduced sound fields at an arbitrary observation point $\mathbf{x} = (\gamma, \phi, \vartheta)$ are

$$S(\mathbf{x}; k) = \frac{e^{-ik|\mathbf{y}-\mathbf{x}|}}{|\mathbf{y}-\mathbf{x}|}; S'(\mathbf{x}; k) = \sum_{l=1}^{3} w_l(k) \frac{e^{-ik|\mathbf{y}_l-\mathbf{x}|}}{|\mathbf{y}_l-\mathbf{x}|} \qquad (1)$$

where (1) the symbol $i = \sqrt{-1}$ is used to denote the imaginary part of a complex number; (2) $k = 2\pi\lambda^{-1} = 2\pi f c^{-1}$ is the wave number (c is the speed of wave propagation, f is the frequency and λ is the wavelength); and (3) a complex-valued frequency-dependent weighting function $w_l(k)$ is applied to the lth loudspeaker, $l = 1, 2, 3$.

Suppose that (1) all the loudspeakers are located on a sphere whose radius equals to σ, i,e., $|\mathbf{y}| = |\mathbf{y}_l| = \sigma, \forall l$; and (2) the receiving point $\mathbf{x}$ satisfies $|\mathbf{x}| < |\mathbf{y}|, |\mathbf{x}| < |\mathbf{y}_l|$, where $\mathbf{y}$ is the location of the loudspeaker in the original sound field and $\mathbf{y}_l$ is the location of the lth loudspeaker in the reproduced field. Ando's simplification method introduced that the reproduced sound field which is at the origin $\mathbf{o}$ could be maintained, i.e., $S(\mathbf{o}; k) = S'(\mathbf{o}; k)$. Considering that people used two ears to listen, Ando's method indicated that the positions of the three loudspeakers in the reproduced field satisfy the following conditions [5]:

1. a spherical triangle formed by the directions of the three loudspeakers includes the direction of $\mathbf{y}$;
2. the spherical area of the triangle is the minimum among those satisfying 1.

That was, when there were many different spherical triangles satisfying 1, Ando's method considered that the minimum spherical area of the triangle could guarantee the minimum error within people's head. In order to analyse Ando's method, the least-squares error within some specified region χ, represented as a dark grey shaded area (shown in Fig. 1(b)), is defined as [3]

$$\varepsilon(\gamma; k) = \frac{1}{4\pi} \int \frac{|S(\mathbf{x}; k) - S'(\mathbf{x}; k)|^2}{|S(\mathbf{o}; k)|^2} d(\hat{\mathbf{x}}) \qquad (2)$$

where (1) the reproduced region χ is defined as being bounded by a sphere of some specified radius, centered on the origin; (2) $S(\mathbf{o}; k)$ is the sound field which is at the origin $\mathbf{o}$; and (3) integration is taken over the unit sphere. Assuming that people's head is a normal sphere with R as the radius , we consider that the center of head is located at the origin $\mathbf{o}$ and both ears are symmetric with YOZ plane.

When the frequency and the radius of some specified region are given, Table 1 indicates that (1) two different spherical triangles with the same areas have different least-squares errors within the region of head; and (2) the three loudspeakers can reproduce larger least-squares error within the region of head when the area of spherical triangle enclosed by them is smaller. Therefore, we should exploit the the correct relationship between the distortion of sound field within a given region and the positions of three loudspeakers in the reproduced field.

Table 1. The least-squares errors in the reproduced field ($f = 1000\,\mathrm{Hz}$, $\sigma = 200\,\mathrm{cm}$, $R = 8.5\,\mathrm{cm}$)

Loudspeaker location in original field	Loudspeaker locations in reproduced field			Spherical area of triangle(cm²)	Least-squares error $\varepsilon(R; k)$
$\mathbf{y}$	$\mathbf{y}_1$	$\mathbf{y}_2$	$\mathbf{y}_2$		
$(\sigma,0,0)$	$(\sigma,0,\pi/2)$	$(\sigma,\pi/3,\pi/6)$	$(\sigma,7\pi/6,\pi/6)$	33026	0.10863
	$(\sigma,\pi/3,\pi/6)$	$(\sigma,2\pi/3,\pi/2)$	$(\sigma,3\pi/2,\pi/6)$	33026	0.10771
$(\sigma,0,0)$	$(\sigma,2\pi/3,\pi/6)$	$(\sigma,\pi,\pi/6)$	$(\sigma,11\pi/6,\pi/2)$	33431	0.22712
	$(\sigma,0,\pi/2)$	$(\sigma,\pi/2,\pi/6)$	$(\sigma,4\pi/3,\pi/6)$	44948	0.08405

3 Analysis of Reproduced Sound Field

3.1 Least-Squares Errors Within the Reproduced Region

Let the lth loudspeaker $\mathbf{y}_l = (\sigma, \phi_l, \theta_l)$ be a point source. It produces a sound field within a source-free region at an arbitrary receiving point $\mathbf{x} = (\gamma, \varphi, \vartheta)$, which can be expressed as a linear combination of these spherical harmonics [2],

$$T_l(\mathbf{x}; k) = \frac{e^{-ik|\mathbf{y}_l - \mathbf{x}|}}{|\mathbf{y}_l - \mathbf{x}|} = -ik4\pi \sum_{n=0}^{\infty} \sum_{m=-n}^{n} h_n^{(2)}(k\sigma) j_n(k\gamma) \overline{Y_n^m(\hat{\mathbf{y}}_l)} Y_n^m(\hat{\mathbf{x}}) \quad (3)$$

where $j_n(\cdot)$ and $h_n^{(2)}(\cdot)$ are the first kind spherical Bessel function and second kind spherical Hankel function of the nth degree respectively.

The spherical harmonics are defined as

$$Y_n^m(\hat{\mathbf{x}}) = \sqrt{\frac{(2n+1)}{4\pi} \frac{(n-|m|)!}{(n+|m|)!}} P_n^{|m|}(\cos\vartheta) e^{im\varphi} \quad (4)$$

where $P_n^m(\cdot)$ is the associated Legendre function (which reduces to the Legendre function for $m = 0$). The subscript n is referred to as the *order* of the spherical harmonic, and m is referred to as the *mode*. For each order n, there are $2n + 1$ modes (corresponding to $m = -n, ..., n$).

The least-squares error $\varepsilon(\gamma; k)$ in (2) can be derived as

$$\varepsilon(\gamma; k) = (k\sigma)^2 \sum_{n=0}^{\infty} (2n+1)(j_n(k\gamma))^2 |h_n^{(2)}(k\sigma)|^2 X_n. \quad (5)$$

The loudspeaker position deviation X_n in (5) is defined as

$$X_n = 1 + \sum_{l=1}^{3} (w_l(k))^2 - 2\sum_{l=1}^{3} w_l(k) P_n(\cos\beta_l) + 2w_1(k)w_2(k) P_n(\cos\beta_{12}) \\ + 2w_1(k)w_3(k) P_n(\cos\beta_{31}) + 2w_2(k)w_3(k) P_n(\cos\beta_{23}) \quad (6)$$

where (1) β_{uv} denotes the angle between $\hat{\mathbf{y}}_u$ and $\hat{\mathbf{y}}_v$, $u, v \in \{1, 2, 3\}$; and (2) β_l denotes the angle between $\mathbf{y}$ and $\hat{\mathbf{y}}_l$, $l \in \{1, 2, 3\}$. The theoretical derivation of (5) is given in the Appendix A.

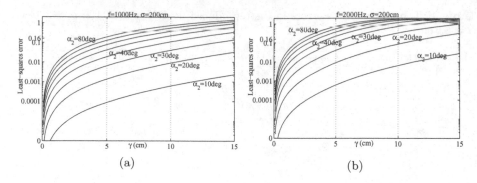

Fig. 2. Least-squares errors of reproduced sound field as a function of γ for various angles $\alpha_2 \in \{10deg, 20deg, 30deg, 40deg, 50deg, 60deg, 70deg, 80deg\}$ with no incremental azimuthal and elevation angle ($\delta = 0$ and $\Delta = 0$).

3.2   Effect of Position Deviation Among Three Loudspeakers

From the least-squares error $\varepsilon_2(\gamma; k)$ in (5), we seek to discuss the angles $\beta_l, l = 1, 2, 3$ and $\beta_{uv}, u, v \in 1, 2, 3$ in the loudspeaker position deviation X_n (shown in (6)). At first, we assume that $\beta_{12} = \beta_{23} = \beta_{31} = \alpha_1, \alpha_1 \in [0, 2\pi/3)$ and $\beta_1 = \beta_2 = \beta_3 = \alpha_2, \alpha_2 \in [0, \pi/2)$. And then, the coordinate transformation F^* is used to translate the loudspeaker positions of the current coordinate system into those of the new coordinate system. The new coordinate system will be defined as: (1) the origin $\mathbf{o}'$ is fixed on $\mathbf{o}$; (2) the positive z'-axis is oriented along the direction of $\mathbf{y}$; (3) the positive y'-axis is oriented along the direction of the vector product of $\mathbf{y}$ and $\mathbf{y}_1$; and (4) the positive x'-axis is oriented along the half space containing $\mathbf{y}_1$. That is, $F^*(\mathbf{y}, \mathbf{y}_1, \mathbf{y}_2, \mathbf{y}_3)^T = (\mathbf{y}', \mathbf{y}'_1, \mathbf{y}'_2, \mathbf{y}'_3)^T$, where $\mathbf{y}' = (\sigma, 0, 0), \mathbf{y}'_1 = (\sigma, 0, \alpha_2), \mathbf{y}'_2 = (\sigma, 2\pi/3, \alpha_2), \mathbf{y}'_3 = (\sigma, 4\pi/3, \alpha_2)$.

Now, we start to move $\mathbf{y}'_2$ on the same latitude with a incremental azimuthal angle δ firstly (i.e., $\mathbf{y}'_2 = (\sigma, 2\pi/3 + \delta, \alpha_2), 0 \le \delta < \pi/3$), and then move on the same longitude with a incremental elevation angle Δ, as a consequence, the location of $\mathbf{y}'_2$ will become $(\sigma, 2\pi/3 + \delta, \alpha_2 + \Delta), 0 \le \Delta < \pi/2 - \alpha_2$. Hence, the angles $\beta_{uv}, u, v \in 1, 2, 3$ in (6) will be derived as

$$\cos \beta_{31} = \cos \alpha_1 = \cos^2 \alpha_2 - \frac{\sin^2 \alpha_2}{2},$$
$$\cos \beta_{12} = \cos(\alpha_2 + \Delta) \cos \alpha_2 - \frac{1}{2} \sin \alpha_2 \sin(\alpha_2 + \Delta)(\cos \delta + \sqrt{3} \sin \delta),$$
$$\cos \beta_{23} = \cos(\alpha_2 + \Delta) \cos \alpha_2 - \frac{1}{2} \sin \alpha_2 \sin(\alpha_2 + \Delta)(\cos \delta - \sqrt{3} \sin \delta).$$

The weighting coefficients in (6) will be calculated by [5].

Let σ be a constant. Suppose that the incremental azimuthal and elevation angle are null (i.e., $\delta = 0$ and $\Delta = 0$), the least-squares error $\varepsilon(\gamma; k)$ is plotted according to different frequencies generated by the loudspeakers (shown in Fig. 2 with $f = 1000\,\text{Hz}$ and $f = 2000\,\text{Hz}$ as the examples). Figure 2 shows that for any given angle α_2, the least-squares error $\varepsilon(\gamma; k)$ decreases monotonically when γ is below a certain value. i.e., $\varepsilon(\gamma; k) \le \varepsilon(\gamma_0; k), \forall \gamma < \gamma_0$. Thus, the field is also

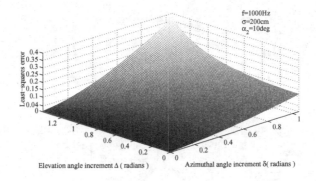

Fig. 3. Least-squares errors of reproduced sound field for various azimuthal angles δ and elevation angles Δ under some specified conditions: $f = 1000\,\text{Hz}$, $\sigma = 200\,\text{cm}$ and $\alpha_2 = 10deg$.

accurately reproduced at all points within the sphere if the angle α_2 is sufficient to accurately reproduce the original sound field on a sphere of radius γ_0.

Next, we change the incremental azimuthal angle δ and elevation angle Δ under some specified conditions, such as $f = 1000\,\text{Hz}$, $\sigma = 200\,\text{cm}$ and $\alpha_2 = 10deg$. From Fig. 3, for a given elevation angle increment Δ, the least-squares error $\varepsilon(\gamma; k)$ increases monotonically with the increase of azimuthal angle increment δ. And likewise, for a given azimuthal angle increment δ, the least-squares error $\varepsilon(\gamma; k)$ increases monotonically with the increase of elevation angle increment Δ. Furthermore, when the incremental azimuthal angle is not more than δ_0 (i.e., $\delta \leq \delta_0 = 0.6$ shown in Fig. 3), the least-squares error is less than 16.1%[1] for all elevation angel increments Δ. When the incremental azimuthal angle is more than δ_0 and elevation angle is more than Δ_0 (i.e., $\Delta \geq \Delta_0 = 0.6$), the least-squares error will increase sharply with the increase of the elevation angle increment Δ.

Therefore, the selection of three loudspeaker positions will be redefined as:

1. a spherical triangle formed by the directions of three loudspeaker includes the direction of $\mathbf{y}$;
2. the deviation of three loudspeaker directions from the direction of $\mathbf{y}$ is minimized among those satisfying 1;
3. each shape of the spherical triangle among those satisfying 2 approaches that of an equilateral spherical triangle.

As for 2, the average deviation angle $\bar{\alpha}$ of unit vector in the direction is defined as $\bar{\alpha} = \frac{1}{3} \sum_{l=1}^{3} \beta_l$ where the minimum average deviation angle is used to satisfy condition 2. As for 3, the standard deviation ρ is defined as $\rho = \sqrt{\frac{(\beta_{12} - \bar{\beta})^2 + (\beta_{23} - \bar{\beta})^2 + (\beta_{13} - \bar{\beta})^2}{3}}$ where $\bar{\beta} = \frac{\beta_{12} + \beta_{23} + \beta_{13}}{3}$ and the minimum standard deviation is used to satisfy condition 3.

---

[1] [3] introduced that the sound field could be reproduced accurately when least-squares error was less then 16.1%.

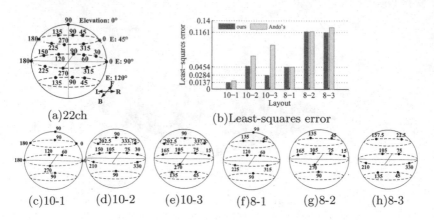

Fig. 4. Layouts for experiments and least-squares errors of reproduced sound field for six layouts (shown in c-h) with $f = 1000\,\text{Hz}$ and $R = 8.5\,\text{cm}$ as an example.

4 Experiments

4.1 Simulation

Free-field source conditions are assumed and the sound field resulted from the loudspeaker is a spherical wave. Let the center of head be located at the origin. Distance from each loudspeaker to the origin is 200 cm, i.e., $\sigma = 200$. Distance from each ear to the center of head is 8.5 cm, i.e. $R = 8.5$. Original sound field is produced by 22-channel audio system (shown in Fig. 4(a)) with 1000 Hz (i.e., $k = 0.1848$) as an example and the reproduced fields are produced by three layouts of 10 channels and three layouts of 8 channels (shown in Fig. 4(c-h)) which were introduced in [5]. Compared to Ando's method, the proposed method uses the redefined selection of three loudspeaker positions in the reproduced field.

Figure 4(b) shows the least-squares error $\varepsilon(R; k)$ of six layouts. It indicates that the least-squares error $\varepsilon(R; k)$ of layout 8-1 obtained by Ando's method is equal to that obtained by the proposed method since the loudspeaker selections of Ando's and our method happen to be the same, which is similarly happened to layout 8-2. However, the rest of the least-squares errors of Ando's method are more than those of our proposed method. Moreover, the least-squares error of layout 10-3 is 0.0284 obtained by our proposed method, which is 68.5 % smaller than that obtained by Ando's method. Suppose that the three-dimension least-squares error $\varepsilon(R; k)$ is reduced to two-dimension least-squares error, i.e., the least-squares error of measuring plane is defined as $\epsilon(\gamma; k) = \int \frac{|S(\mathbf{x};k) - S'(\mathbf{x};k)|^2}{\pi \gamma^2 |S(\mathbf{o};k)|^2} d(\hat{\mathbf{x}})$. Table 2 shows the simulations of the sound fields reproduced by layout 10-3 which are generated by Ando's and our method respectively within the measuring plane at z = − 4 cm, z = 0 cm and z = 4 cm (f = 1000 Hz, R = 8.5 cm, the encircled region is the region of head). From Table 2, the least-squares errors $\epsilon(\gamma; k)$ of our method are 61.9 %, 58.2 % and 62.5 % smaller than those of Ando's method at $z = -4\,\text{cm}$, $z = 0\,\text{cm}$, and $z = 4\,\text{cm}$ respectively.

Table 2. Simulations of the sound fields reproduced by layout 10-3 within the measuring plane at $z = -4\,\mathrm{cm}$, $z = 0\,\mathrm{cm}$ and $z = 4\,\mathrm{cm}$ ($f = 1000\,\mathrm{Hz}$, $R = 8.5\,\mathrm{cm}$, encircled region is the region of head)

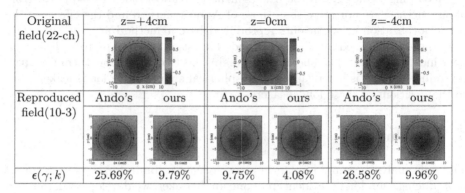

Original field(22-ch)	z=+4cm		z=0cm		z=-4cm	
Reproduced field(10-3)	Ando's	ours	Ando's	ours	Ando's	ours
$\epsilon(\gamma; k)$	25.69%	9.79%	9.75%	4.08%	26.58%	9.96%

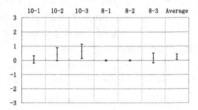

Fig. 5. Comparative Mean opinion score (CMOS) of sound localization between the proposed method and Ando's method (95 % confidence limits).

4.2   Subjective Experiment

Subjective experiments had been done by the RAB paradigm [7] with two stimuli (A and B) and a third reference (R). The reverberation time at 500 Hz in the soundproof room was 0.18 s and background noise was 30 dB (A). In this experiment, R was the original 22-channel white noise [6] with 10 s, one stimuli (A or B) was the sound of 10-(or 8-) channel system by Andos method in [5] and the other one was the sound of the same channel system generated by the proposed method. Subjects were asked to compare the differences of the sound localization between A and B relative to R and give scores according to the continuous seven-grade impairment scale. Subjects were 20 students all of whom major in audio signal processing and they had accepted the training of subjective experiment before.

Figure 5 shows the Comparative Mean Opinion Score (CMOS) of sound localization given by the subjects with 95 % confidence limits. From Fig. 5, the CMOS of sound localization reproduced by layout 8-1 and 8-2 are about zero since the loudspeaker selections of Ando's and our method happen to be the same. However, the CMOS of layout 10-1, 10-2, 10-3 and 8-3 are 0.064, 0.436, 0.645 and 0.173 respectively, which are higher than zero and the average CMOS is 0.25.

It also indicates that the effects of the sound localization reproduced by our method are slightly better than those reproduced by Ando's method.

According to the experiment results mentioned above, the least-squares errors of our method are not more than those of Ando's method and the effects of the sound localization of six layouts reproduced by our method are slightly better than those reproduced by Ando's method. That is, in order to reproduce a sound field incident from a loudspeaker, the main effect of distortion within a given reproduced region is not related to the spherical area of three loudspeakers, but to the positions of these loudspeakers.

5   Conclusion

Through analyzing some fundamental performance limits for the distortion of reproducing sound field incident from one loudspeaker, we develop a relationship between the positions among three loudspeakers of the reproduced field, the size of the reproduced region, the frequency range and the desired reproduction accuracy. On this basis, the selection of loudspeaker positions is redefined to lower the distortion within the region of head. As an example, the redefined selection of loudspeaker position was used to obtain the weighting coefficients of 10- and 8-channel sound systems from the original 22.2 multichannel system without two low-frequency effect channels. Although the least-squares errors of layout 8-1 and 8-2 within head obtained by Ando's method happened to be equal to those obtained by the proposed method, the distortion of the rest layouts obtained by our method are lower than those obtained by Ando's method respectively. Moreover, compared to subjective evaluation of Ando's method, our proposed method is slightly better for sound localization.

A   Appendix

Derivation of (5) The original sound field at $\mathbf{x}$ generated by a loudspeaker $\mathbf{y}$ is represented as $S(\mathbf{x}; k) = -ik4\pi \sum_{n=0}^{\infty} \sum_{m=-n}^{n} h_n^{(2)}(k\sigma)j_n(k\gamma)\overline{Y_n^m(\hat{\mathbf{y}})}Y_n^m(\hat{\mathbf{x}})$. The reproduced field at $\mathbf{x}$ generated by three loudspeakers is $S'(\mathbf{x}; k) = -ik4\pi \sum_{n=0}^{\infty} \sum_{m=-n}^{n} h_n^{(2)}(k\sigma)j_n(k\gamma) \times \sum_{l=1}^{3} w_l(k)\overline{Y_n^m(\hat{\mathbf{y}}_l)}Y_n^m(\hat{\mathbf{x}})$. The squared error over the unit sphere is $\int |S(\mathbf{x}; k) - S'(\mathbf{x}; k)|^2 \, d\hat{\mathbf{x}} = (4\pi k)^2 \sum_{n=0}^{\infty} |j_n(k\gamma)|^2 |h_n^{(2)}(k\sigma)|^2 \times \sum_{m=-n}^{n} \left| Y_{nm}(\hat{\mathbf{y}}) - \sum_{l=1}^{3} w_l(k)Y_{nm}(\hat{\mathbf{y}}_l) \right|^2$, which follows from the orthogonality property in [2] of the spherical harmonics. The addition theorem of Legendre functions states $\sum_{m=-n}^{n} Y_{nm}^*(\hat{\mathbf{y}})Y_{nm}(\hat{\mathbf{x}}) = \frac{2n+1}{4\pi}P_n(\cos\beta)$, where β denotes the angle between $\hat{\mathbf{y}}$ and $\hat{\mathbf{x}}$ [8]. Using this addition theorem with $\hat{\mathbf{y}} = \hat{\mathbf{x}}$, $P_n(\cos 0) = 1 \forall n$, gives

$$\int |S(\mathbf{x}; k) - S'(\mathbf{x}; k)|^2 \, d\hat{\mathbf{x}}$$
$$= 4\pi k^2 \sum_{n=0}^{\infty} (2n + 1)(j_n(k\gamma))^2 |h_n^{(2)}(k\sigma)|^2 [1 + \sum_{l=1}^{3} (w_l(k))^2 - 2 \sum_{l=1}^{3} w_l(k) P_n(\cos \beta_l)$$
$$+ 2w_1(k)w_2(k) P_n(\cos \beta_{12}) + 2w_1(k)w_3(k) P_n(\cos \beta_{31}) + 2w_2(k)w_3(k) P_n(\cos \beta_{23})].$$

And then, $|S(\mathbf{o}; k)|^2 = \dfrac{e^{-ik|\mathbf{y}_4|}}{|\mathbf{y}_4|} \dfrac{e^{ik|\mathbf{y}_4|}}{|\mathbf{y}_4|} = \dfrac{1}{|\mathbf{y}_4|^2} = \dfrac{1}{\sigma^2}$, thus completing the derivation.

References

1. Ahrens, J., Spors, S.: Applying the ambisonics approach to planar and linear distributions of secondary sources and combinations thereof. Acta Acustica United with Acustica **98**(1), 28–36 (2012)
2. Zhang, W., Abhayapala, T.: Three dimensional sound field reproduction using multiple circular loudspeaker arrays: functional analysis guided approach. IEEE Trans. Audio Speech Lang. Process. **22**(7), 1184–1194 (2014)
3. Kennedy, R.A., Sadeghi, P., Abhayapala, T.D., Jones, H.M.: Intrinsic limits of dimensionality and richness in random multipath fields. IEEE Trans. Sig. Process. **55**(6), 2542–2556 (2007)
4. Hamasaki, K., Matsui, K., Sawaya, I., Okubo, H.: The 22.2 multichannel sounds and its reproduction at home and personal enviroment. In: Audio Engineering Society Conference: 43rd International Conference: Audio for Wirelessly Networked Personal Devices. Audio Engineering Society (2011)
5. Ando, A.: Conversion of multichannel sound signal maintaining physical properties of sound in reproduced sound field. IEEE Trans. Sig. Process. **19**(6), 1467–1475 (2011)
6. Sawaya, I., Oode, S., Ando, A., Hamasaki, K.: Size and shape of listening area reproduced by three-dimensional multichannel sound system with various numbers of loudspeakers. In: Audio Engineering Society Convention 131. Audio Engineering Society (2011)
7. Recommendation, ITU-R BS. 1284–2: General Methods for the Subjective Assessment of Sound Quality. International Telecommunications Union (2002)
8. Colton, D., Kress, R.: Inverse Acoustic and Electromagnetic Scattering Theroy. Springer-Verlag, New York (1997)

Real-Time Understanding of Abnormal Crowd Behavior on Social Robots

Dekun Hu[1,3], Binghao Meng[1], Shengyi Fan[2], Hong Cheng[1(✉)],
Lu Yang[1], and Yanli Ji[1]

[1] Center for Robotics, School of Automation Engineering,
University of Electronic Science and Technology of China, Chengdu, China
{hudekun,hcheng}@uestc.edu.cn
[2] Ricoh Software Research Center of Beijing, Beijing, China
[3] College of Computer Science, University of Chengdu, Chengdu, China

Abstract. Perceiving the crowd behavior is very important for social cloud robots, who serve as guiders at transportation junctions. In this paper, we propose a real-time algorithm based on background modeling to detect collective motions in complex scenes. The proposed algorithm not only avoids unstable foreground extraction, but also has low computational complexity. To detect the abnormal crowd escape, we refer to the definition of the moving energy of patches and use the energy histogram of the patches to effectively and accurately represent the crowd distribution information in the crowd scenes. We have applied the proposed algorithm to the real surveillance videos which contain the aggregation and dispersion events. The experimental results show the significant outperformance of the proposed algorithm in comparison to the-state-of-the-art approach.

Keywords: Crowd behavior analysis · Social robots · Motion energy of patches · Background modeling

1 Introduction

With the development of the robot technology, cloud computing technology and Internet of things technology, cloud service robots have been used in various places, such as airports, bus stops, shopping malls, subway stations, *etc.* They can serve as the guiders for people to take traffic tools or the assistants of shopping. Moreover, they can serve as the safety guards by detecting the abnormal events in real-time. Real-time crowded scene analysis is a very difficult task for social robots due to the inherent complexity and vast diversity such as illumination changes, low resolution, scene depth, camera position, *etc.* With the development of computer vision technologies, various vision based approaches have been proposed to detect abnormal events in surveillance scenes, background modeling [15], sparse representation [4], object tracking [19], face recognition [20] and people counting [10], are considered as the fundamental elements that compose

© Springer International Publishing Switzerland 2015
Y.-S. Ho et al. (Eds.): PCM 2015, Part II, LNCS 9315, pp. 554–563, 2015.
DOI: 10.1007/978-3-319-24078-7_56

an intelligent surveillance system for the anomaly detection. In this paper, we aim to detect the abnormal crowd behaviors [7,9,12,16] based on the computer vision technology in real-time. A variety of algorithms have been proposed to detect abnormal events in scenes, these approaches could be divided into three categories according to the scene representation: a trajectory based approaches [8,18,19,21], patch based approaches [5,6,13,18], sparse coding based approaches [2,7,11,14].

In [8], a normal dictionary set, was constructed by collecting trajectories of normal behaviors and extracting the control point features of cubic B-spline curves, which was further divided into Route sets. Sparse reconstruction coefficients and residuals of a test trajectory to the Route sets could be calculated with the trajectory sparse reconstruction analysis (SRA). A new descriptor named as Social Affinity Maps (SAM) [21] and priors over origin and destination (OD-prior) were proposed to understand the crowd behavior at the scale of million pedestrians for human mobility in crowded spaces such as city centers or train stations. In [18], a robust algorithm was proposed to detect stationary group activities and understand crowd scene. A locally shared foreground codebook and was used to shape the 3D stationary time map.

A collectiveness descriptor for crowd as well as their constituent individuals along with the efficient computation were proposed in [6]. In [5], a novel patch entropy approach to represent the crowd distribution information and the optical flow was introduced to describe the crowd speed. The Gaussian Mixture Model (GMM) over the normal crowd behaviors was used to predict the anomalies in the detecting stage. A hybrid agent system [13] was used to detect abnormal behaviors in crowded scenes, which included static and dynamic agents to observe efficiently the corresponding individual and interactive behaviors in a crowded scene. The crowd behaviors were represented as a bag of words through the integration of static and dynamic agents information.

The Social Force model [14] treated the moving patches as individuals. And their interaction forces are estimated and mapped into the image plane to obtain Force Flow for every pixel in every frame. Randomly selected spatio-temporal volumes of Force Flow are used to model the normal behaviors of the crowd. In [11], based on inherent redundancy of video structures, an efficient sparse combination learning framework was proposed for abnormal behaviors detection. It achieved high detection rates on benchmark datasets at a speed of 140~150 frames per second on average when computing on an ordinary desktop PC by MATLAB. In [2], unlike existing approaches based on sparse coding , the abnormal events detection model directly sparsely coded the motion features of the center patches with features of its surrounding patches.

In this paper, we take advantage of the distribution of the patches in the frame to simulate the distribution of the individuals in the crowd. The speed of the patches in the frame to simulate the speed of the individuals in the crowd. We proposed patch moving energy to effectively represent the crowd distribution information. As the number of frames with abnormal crowd behaviors is only a small portion of the entire video, it is obvious that abnormal crowd behavior detection is an unbalanced problem. In this paper we simultaneously use the

crowd speed and the distribution information to predict abnormal crowd dispersion behaviors. The comparison experiments conducted on surveillance dataset
validate the advantages of the proposed algorithm.

The rest of the paper is organized as follows: In Sect. 2, we introduce the procedures of proposed algorithm. In Sect. 3, we present experimental results and the
comparisons with the-state-of-art algorithm. Section 4 concludes the paper.

2   The Proposed Algorithm

2.1   Crowd Aggregation Detection

In the field of public security, the massive mass incidents are often from small
crowd gathered to evolution. Therefore, moderate scale crowd aggregation detection and its alarming are crucial to social robots for surveillance purpose. The
gathered crowd can't be measured using the algorithm based on optical flow due
to the relatively static state of crowd in a particular area within a period of time.
Hence, a crowd aggregation detection based on real background modeling and
hierarchical alarm algorithm is proposed in this paper. The algorithm process is
shown in Fig. 1 and described in following subsections.

Background Modeling. In order to model the background in the video scenes,
a robust Pixel-to-Model (P2M) background modeling and recovery approach
[17] is used in our work. Each pixel is represented by a context feature which
consists of local compressive descriptors. The novel P2M distance is employed
to classify the potential background pixels. Furthermore, the P2M distances are
also utilized to adaptively update the background model in the space of local
descriptors in a smooth and efficient way. The P2M based background recovery
can robustly reconstruct the clean background and suppress real-world noises.

As shown in Fig. 1, there are two different types of background computing
in the crowd aggregation detection: clean background modeling, which is called
real background, and the dynamic background. At time t, the dynamic background of scene based on P2M is denoted as $B_{dy}(t)$. The real background B_{real}
is computed by

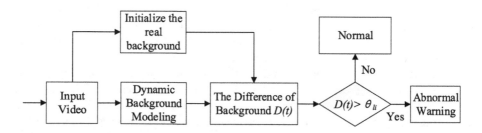

Fig. 1. The framework of the proposed crowd aggregation detection.

$$B_{real} = \frac{1}{N} \sum_{n=1}^{N} b_n, \tag{1}$$

where b_n is a random selected background image from $B_{dy}(t)$ with $t \in [0, T]$ and N is the number of the random selected background images. In practice, the real background can be selected manually for better performance.

Event Detection. One of the main characteristics of the gathered crowd is relatively static within certain areas for a period of time. In order to extract the "static" people from the scene, a $i \times j$ grid of patches is placed over every difference image of clean background and dynamic background, and the size of the patch $P_{(i,j)}$ is $m \times n$. The difference image should be binarized, thus all of pixels with value of 1 presents the "static" foreground. Part of static pixels describes the information of gathered crowd in the scene. As we use the "static" patches to represent the gathered crowd, in order to statistic the "static" patches, we denote the difference image before binarization at time t as $D(t)$, which is calculated by Eq. (2). The value of a patch $V_{P_{(i,j)}}$ is defined by the proportion of non-zero pixels in it as Eq. (3)

$$D(t) = |B_{real} - B_{dy}(t)|, \tag{2}$$

$$V_{P_{(i,j)}} = \begin{cases} 1, & if \; \sum_{i=1}^{m} \sum_{j=1}^{n} P_{i,j}(x,y) \geq \frac{m \times n}{T_p}, \\ 0, & otherwise \end{cases} \tag{3}$$

where $P_{i,j}(x,y)$ is the value of pixels in the patch $P_{(i,j)}$ of the binary difference image. And T_p is the threshold of a static patch. The value of T_p is 8 in this work.

Crowd aggregation area is composed of multiple adjacent static patches. So a weighted sliding window is used to detect the crowd aggregation. The size of the window is integer times of $m \times n$, and its sliding step length is also equal to integer times of the patch size. The value of the window is defined as

$$V_w = \sum_{a=1}^{A} \sum_{b=1}^{B} \lambda_{c1} V_{P_{(i,j)}}, \tag{4}$$

where A is integer times of m and B is integer times of n. λ_{c1} is a compensation parameter of camera calibration, which can improve the effect of the patches far from the camera. A group threshold θ_{l_i} are used to classify the different levels of crowd aggregation according to the value of V_w. An example of crowd aggregation detection is shown in Fig. 2.

2.2   Crowd Escape Detection

In real-life situations, crowd escape occurs by violent movement which is apparent as sudden speeding up, or chaotic movement in a restricted area, or movement contrasting with that of one's neighbors such as in a panic situation. In statistical

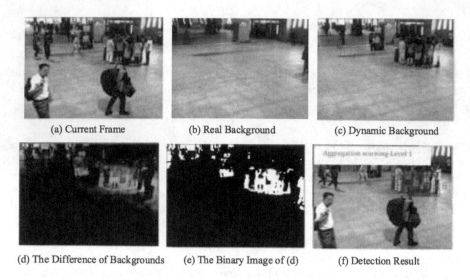

(a) Current Frame   (b) Real Background   (c) Dynamic Background

(d) The Difference of Backgrounds   (e) The Binary Image of (d)   (f) Detection Result

Fig. 2. An example of crowd aggregation detection.

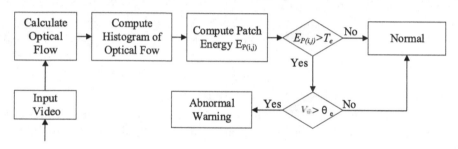

Fig. 3. The framework of the proposed crowd escape detection.

mechanics, entropy is used to measure uncertainty. The greater entropy means the higher disorder, thus the patch entropy approach is proposed to estimate the distribution of the moving patches in [6]. We refer to [6] and propose the patch energy to simulate the distribution of the pedestrians in the crowd. The main steps of the patch energy approach are summarized in Fig. 3 and described in following subsections.

Calculate the Dense Optical Flow. As the moving pedestrians are able to cause the abnormal crowd behaviors, only them need to be concerned about when we detect the crowd escape. We use the moving patches to represent the moving pedestrians in this work. The moving patches extraction stated as the following. Firstly, the velocity of every pixel is calculated by dense optical flow [3]. In order to reduce the influence of illumination change, the average optical flow of continuous several frames is extracted. The map of optical flow is shown in Fig. 4(a). Secondly, every map is divided into $M \times N$ patches. We estimate every patch's velocity with the energy of motion according to the velocity of every pixel in the patch as described in the following subsection.

Patch's Energy of Motion. Assuming the size of every patch in the map of optical flow is $X * Y$, a histogram of the patch is calculated by the different velocities of every pixel(as shown in Fig. 4(b)), every patch in the image has an energy of motion defined by Eq. (5). An example of patch energy change is shown in Fig. 4(c).

$$E_{p_{(i,j)}} = \frac{1}{2} \sum_{r=1}^{H} h_r v_r^2, \ \sum_{r=1}^{H} h_r = X * Y, \tag{5}$$

where v_r is the rth bin in the histogram and h_r is the number of pixels in the rth bin.

Moving Patches Extraction. We denote a patch as "moving patch" if the energy of motion for it is greater than the threshold T_e. In order to extract the moving patches, we compare every patch's energy of motion with the T_e in turn, the value of T_e is given by experiences in different video scenes. The value of a patch $V_{\tilde{P}_{(i,j)}}$ is defined as

$$V_{\tilde{P}_{(i,j)}} = \begin{cases} 1, & if E_{p_{(i,j)}} \geq T_e \\ 0, & otherwise \end{cases}. \tag{6}$$

Event Detection. There will be more patches involving in the escape area if there are more running directions. So a weighted sliding window is used to detect the crowd escape. The size of the window is integer times of the patch, and its sliding step length is also equal to integer times of patch size. The value of the window $V_{\tilde{w}}$ is defined as

$$V_{\tilde{w}} = \sum_{i=1}^{A} \sum_{j=1}^{B} \lambda_{c2} V_{\tilde{P}_{(i,j)}}, \tag{7}$$

where A is integer times of X and B is integer times of Y. λ_{c2}, which can improve the effect of the patches far from the camera, is a compensation parameter of camera calibration, and the value of it will be increased with the increase of the distance between the camera and the real point. If the $V_{\tilde{w}}$ is greater than the threshold θ_e, the crowd escape behavior happens in this window area. As the crowd escape behavior usually involving plenty of persons, we consider that the crowd escape behavior must happen in more than 2 adjacent window areas. In this way, a lot of false positives have been avoided. An example of crowd escape detection is shown in Fig. 4(d), and the blue boxes in the picture indicate the area where the event happens.

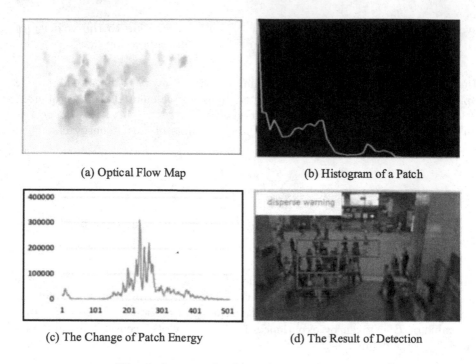

(a) Optical Flow Map

(b) Histogram of a Patch

(c) The Change of Patch Energy

(d) The Result of Detection

Fig. 4. An example of crowd escape detection.

3 Experiment Results and Analysis

3.1 Dataset

To validate the performance of the proposed algorithm, we test it on UMN and RICOH dataset in comparison to the particle entropy algorithm.

The publicly available dataset of the unusual crowd activities from University of Minnesota (UMN) [1] is used to verify the effectiveness of the proposed abnormal crowd event detection algorithm. The dataset consists of 3 different indoor or outdoor scenes with the escape events.

RICOH dataset consists of two kinds of video from two different cameras (TYZX camera and Point-Gray camera). The dataset from TYZX camera includes two dispersing events. It is a low resolution complex scene, involving more moving pedestrians, with illumination changing drastically. So it is very difficult to detection the abnormal crowd behaviors. Another kind of video from Point-Gray camera includes some crowd aggregation in 6 videos, with different view of camera, different direction of aggregation of crowd and different scale of crowd. It is also difficult to detect the event because of the serious occlusion resulting from the low installing location of the camera. The example images of RICOH dataset is shown in Fig. 5.

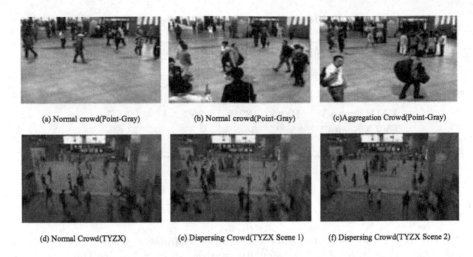

(a) Normal crowd(Point-Gray)   (b) Normal crowd(Point-Gray)   (c)Aggregation Crowd(Point-Gray)

(d) Normal Crowd(TYZX)   (e) Dispersing Crowd(TYZX Scene 1)   (f) Dispersing Crowd(TYZX Scene 2)

Fig. 5. The forward and backward estimation of optical flow.

(a)UMN Scene 1   (b)UMN Scene 2   (c)UMN Scene 3

Fig. 6. The result of crowd escape detection on UMN dataset.

3.2   Experiments

The experiment is conducted as follows. Firstly, the experiment on the crowd aggregation detection is devised on the videos from the Point-Gray camera, the average precision rate is 88% with 94% recall rate. In order to compare our algorithm with the state-of-the-art particle entropy algorithm, we conduct the experiments on UMN and TYZX dataset for crowd escape detection. Figure 6 shows some results on UMN Scenes.

The experiment for crowd escape is conducted secondly. Table 1 shows the quantitative comparisons to the particle entropy algorithm in the UMN dataset and TYZX. The precision and recall rate of our algorithm are much better than the particle entropy except the precision on UMN scene2. And the best result even achieves 100% precision rate and 82% recall rate. Hence, the proposed algorithm can significant outperform the state-of-the-art particle entropy algorithm on most tested datasets.

Table 1. The result of comparison.

Method Dataset	Our method		The particle entropy [6]	
	precision	recall	precision	recall
UMN Scene 1	99.1%	78.1%	98.1%	75.3%
UMN Scene 2	83.5%	51.1%	96.4%	37.3%
UMN Scene 3	100%	82%	96%	57.5%
TYZX Scene 1	85.3%	35.8%	9.3%	13.6%
TYZX Scene 2	94.3%	58.3%	7.6%	11.4%

4   Conclusions

In the future, robots will play more and more important roles in our life. As one of the crucial role of public security guards, they can sense the abnormal crowd behaviors and activate alarm and evacuate the stream of people, thus reducing the occurrence of public events. In this paper, we propose a novel crowd aggregation detection algorithm based on background modeling firstly. The algorithm can make grading warning to crowd congestion in public security. Secondly, another energy of moving approach is proposed to represent the crowd distribution information. The experimental results on RICOH dataset show the good performance of the proposed approach. Specially, our algorithm is robust to illumination changes, low resolution, scene depth and camera position. The experiments conducted on publicly available dataset showed the effectiveness of the approach and that our algorithm outperforms the state-of-the-art particle entropy algorithm. In the future work, the thresholds in our algorithm will be self-adaptive to avoid the complicated manual modulation for different surveillance scenes.

Acknowledgment. The authors would like to thank all the reviewers for their insightful comments and special thanks to the RICOH (Beijing) for their dataset, financial and technical support. This work was partially supported by NSFC (No.61305033, 61273256, NO.61305043), Fundamental Research Funds for the Central Universities (ZYGX2013J088, ZYGX2014Z009) and SRF for ROCS, SEM. Youth Research Funds of Chengdu university (20805066)

References

1. Unusual crowd activity dataset of university of minnesota. http://mha.cs.umn.edu/proj_events.shtml#crowd
2. Alahi, A., Ramanathan, V., Fei-Fei, L.: Socially-aware large-scale crowd forecasting. In: CVPR. IEEE (2014)
3. Brox, T., Bruhn, A., Papenberg, N., Weickert, J.: High accuracy optical flow estimation based on a theory for warping. In: Pajdla, T., Matas, J.G. (eds.) ECCV 2004. LNCS, vol. 3024, pp. 25–36. Springer, Heidelberg (2004)

4. Cheng, H., Liu, Z., Yang, L., Chen, X.: Sparse representation and learning in visual recognition: theory and applications. Signal Process. **93**(6), 1408–1425 (2013)
5. Cho, S.H., Kang, H.B.: Abnormal behavior detection using hybrid agents in crowded scenes. Pattern Recogn. Lett. **44**, 64–70 (2014)
6. Gu, X., Cui, J., Zhu, Q.: Abnormal crowd behavior detection by using the particle entropy. Optik **125**, 3428–3433 (2014)
7. Kratz, L., Nishino, K.: Anomaly detection in extremely crowded scenes using spatio-temporal motion pattern models. In: CVPR. IEEE (2009)
8. Li, C., Han, Z., Ye, Q., Jiao, J.: Visual abnormal behavior detection based on trajectory sparse reconstruction analysis. Neurocomput. **119**, 94–100 (2013)
9. Liao, Z., Yang, S., Liang, J.: Detection of abnormal crowd distribution. In: Int'l Conference on Green Computing and Communications & Int'l Conference on Cyber, Physical and Social Computing. IEEE/ACM (2010)
10. Liu, X., Tu, P.H., Rittscher, J., Perera, A., Krahnstoever, N.: Detecting and counting people in surveillance applications. In: AVSS. IEEE (2005)
11. Lu, C., Shi, J., Jia, J.: Abnormal event detection at 150 FPS in MATLAB. In: ICCV. IEEE (2013)
12. Mahadevan, V., Li, W., Bhalodia, V., Vasconcelos, N.: Anomaly detection in crowded scenes. In: CVPR. IEEE (2010)
13. Mehran, R., Oyama, A., Shah, M.: Abnormal crowd behavior detection using social force model. In: CVPR. IEEE (2009)
14. Tang, X., Zhang, S., Yao, H.: Sparse coding based motion attention for abnormal event detection. In: ICIP. IEEE (2013)
15. Thijs, G., Lescot, M., Marchal, K., Rombauts, S., De Moor, B., Rouze, P., Moreau, Y.: A higher-order background model improves the detection of promoter regulatory elements by gibbs sampling. Bioinform. **17**(12), 1113–1122 (2001)
16. Wang, B., Ye, M., Li, X., Zhao, F., Ding, J.: Abnormal crowd behavior detection using high-frequency and spatio-temporal features. Mach. Vis. Appl. **23**(3), 501–511 (2012)
17. Yang, L., Cheng, H., Su, J., Li, X.: Pixel-to-model distance for robust background reconstruction. In: TCSVT PP (2015)
18. Yi, S., Wang, X., Lu, C., Jia, J.: L0 regularized stationary time estimation for crowd group analysis. In: CVPR. IEEE (2014)
19. Yilmaz, A., Javed, O., Shah, M.: Object tracking: a survey. CSUR **38**(4), 13 (2006)
20. Zhao, W., Chellappa, R., Phillips, P.J., Rosenfeld, A.: Face recognition: a literature survey. CSUR **35**(4), 399–458 (2003)
21. Zhou, B., Tang, X., Zhang, H., Wang, X.: Measuring crowd collectiveness. Pattern Anal. Mach. Intell. **36**(8), 1586–1599 (2014)

Sparse Representation Based Approach for RGB-D Hand Gesture Recognition

Te-Feng Su[1], Chin-Yun Fan[2], Meng-Hsuan Lin[2],
and Shang-Hong Lai[1,2(✉)]

[1] Department of Computer Science,
National Tsing Hua University, Hsinchu, Taiwan
{tfsu,lai}@cs.nthu.edu.tw
[2] Institute of ISA, National Tsing Hua University, Hsinchu, Taiwan
aril2292000@gmail.com, mayrall021@livemail.tw

Abstract. In this paper, we present a new algorithm for RGB-D hand gesture recognition by using multi-attribute sparse representation enforced with group constraints. Firstly, the hand region is segmented from the background according to the depth information. Then, we process all gesture-performing hand region images with PCA to reduce the feature dimension. To obtain a more accurate and discriminative representation, a multi-attribute sparse representation is employed for hand gesture recognition from different view angles. The multiple attributes for a gesture image can be represented by individual binary matrices to indicate the group properties for each gesture. Then, these attribute matrices are incorporated into the formulation of l_1-minimization in the sparse coding framework. Finally, the effectiveness and robustness of the proposed method are demonstrated through experiments on a public RGB-D hand gesture dataset.

Keywords: Gesture recognition · Sparse coding · Depth image · Attribute constraint

1 Introduction

Recognizing human gesture from images or videos has received increasing attention in computer vision mainly due to the potential applications, such as hand tracking, human-computer interaction (HCI), etc. Gesture recognition is touch-free and it can be used to substitute for keyboard and mouse, providing a new way of interaction which is more natural and unlimited, such as interacting with 3D applications, augmented reality, smart glasses. Although there has been much work on hand gesture recognition over the past decades [1–3, 5], human hand motion exhibits high degrees of freedom with large viewpoint variations and partial occlusion, which still makes the hand gesture recognition problem very challenging. In recent years, the emergence of depth cameras, such as Kinect sensor and Intel Creative Senz3D camera, has opened new possibilities for acquiring depth information. Following the popularity of depth sensor, depth information based hand gesture recognition gains considerable attention (Fig. 1).

© Springer International Publishing Switzerland 2015
Y.-S. Ho et al. (Eds.): PCM 2015, Part II, LNCS 9315, pp. 564–570, 2015.
DOI: 10.1007/978-3-319-24078-7_57

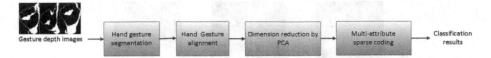

Fig. 1. Flowchart of the proposed hand gesture recognition system

There are quite some researches focused on vision-based hand tracking and gesture recognition. A brief review on the use of depth for hand tracking and gesture recognition is provided in [4], which focused on pervious approaches for depth-based gesture recognition. The two common approaches for hand gesture recognition are the model-based approach and appearance-based approach. The recognition of hand gesture highly depends on the adopted features, which are used to detect and segment the hand from the captured RGBD image. The accuracy will be increased if hand region can be well presented by adopted features. Skin color [7–9] is an important feature for detecting and segmenting human hands. Bretzner et. [7] proposed to represent hand postures as multi-scale colour image features at different scales. Dardas and Georganas [10] proposed to detect skin color region and use face subtraction techniques to remove face. However, color-based methods are doomed to fail when the color of background or other objects are similar to the skin color. To simplify the task, additional tools are applied in some methods like colored marker [6]. However, they take a high setup costs which limits development and make these kinds of methods hard to be applied to different applications. In [10], they achieved ideal performance of hand gesture detection and recognition by using the bag-of-features approach in conjunction with the multiclass support vector machine (SVM) and a grammar that generates gesture commands to control an application. However, full DOF hand pose information is restricted by their method.

Recently, sparse coding-based methods [11–13] have been developed to improve the accuracy of classification. Sparse coding is a technique which reconstructs a target instance by a sparse linear combination of dictionary atoms. The coefficients in the sparse representation are determined by imposing an l_1-norm minimization, which is commonly referred to as the lasso problem [14, 15]. Wright et al. 11 exploited the entire training set as the dictionary and converted the query face image into a sparse linear combination of dictionary atoms. By imposing l_1-norm constraint on the coefficients, their sparse representation classification (SRC) method achieved promising results on face recognition.

In this paper, we present a multi-attribute sparse coding approach for gesture recognition. Firstly, the hand region is segmented according to its depth different from background. Then, we process all gesture-performing hand region images with PCA to reduce the dimension. To obtain a more accurate and discriminative representation, a multi-attribute sparse representation is applied for hand gesture recognition. Different rotation angles can be represented by individual attribute mask to describe group property which enforces basis selection from groups with the same rotation as best as possible. Then, these attribute matrices are incorporated into the formulation of l_1-minimization (Fig. 2).

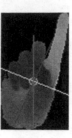

Fig. 2. Some examples of the depth image alignment for hand gesture recognition: blue lines are eigenvectors corresponding to the largest eigenvalues, yellow lines are the second eigenvectors and red lines are the y-axis in each depth image (Color Figure online).

2 Proposed Method

2.1 Hand Gesture Segmentation and Alignment

In each class of hand pose, images with different translations, scales and rotations should be aligned to remove those factors that may reduce the accuracy of recognition. The depth images with corresponding masks are firstly segmented to a set of images that only the hand regions in the images remain, and the hand segmentation result is represented by a mask. The mask is a binary image with 1 representing the pixel in the hand region and 0 otherwise.

After the hand segmentation, images are furthering aligned by calculating the two eigenvectors of all the 3D pixel coordinates in the hand region. The eigenvector with the largest eigenvalue represents the main orientation of the deformation due to hand pose. We rotate the whole image according to the angle θ between the primary eigenvector and image y-axis to align the principle axis of hand poses for all images to the same direction. Cropping the hand regions from the depth images by appropriate thresholds can remove the translation factor, and all images are resized to a fixed resolution for reducing the scaling effect. In addition, to make depth value invariant, mean value of hand depth image should be subtracted from each aligned image. Finally, PCA is used to reduce the dimensionality of the gesture images.

2.2 Multi-attribute Sparse Coding for Gesture Recognition

Sparse coding is usually employed to represent a signal or feature vector as a linear combination of very few atoms in a dictionary. For this instance, only a small number of weight coefficients will be non-zero, and thus the resulting coefficient vector is sparse. Consider a feature vector $y \in R^m$, and a finite training set of signals $\mathbf{X} = [\mathbf{x}_1, \ldots, \mathbf{x}_n]$ in $R^{m \times n}$, the sparse representation problem can be formulated as:

$$\min_{\alpha} \|\mathbf{y} - \mathbf{X}\alpha\|_2^2 + \lambda \|\alpha\|_1 \tag{1}$$

where λ is a regularization parameter that weights the l_1-norm regularizer and controls the sparsity of coefficient vector $\boldsymbol{\alpha}$. This convex optimization problem is also referred to as Lasso. Furthermore, the group property is also introduced into the sparse representation to obtain more sparsity in the group level.

Chiang et al. [16] proposed to utilize the attribute properties in the data to enhance the discriminative power of data prediction. Suppose that there a set of n training data represented by $\mathbf{X} = [\mathbf{x}_1,\ldots,\mathbf{x}_n] \in R^{m \times n}$, and the associate attribute property for the training data is encoded by the attribute matrix $\mathbf{A}_k \in R^{c_k \times n}$, for the k-th attribute, $k = 1,$ $\ldots,K$. The attribute matrix $\mathbf{A}_k$ is given by

$$\mathbf{A}_k = \begin{bmatrix} a_{11} & a_{12} & \cdots & a_{1p} \\ a_{11} & a_{22} & \cdots & a_{2p} \\ \vdots & \vdots & \ddots & \vdots \\ a_{m_k 1} & a_{m_k 2} & \cdots & a_{m_k p} \end{bmatrix}, \; a_{ip} = \begin{cases} 1, & \text{if } x_p \in \text{class } i, \\ 0, & \text{otherwise,} \end{cases} \tag{2}$$

where i is the class index, $i = 1,\ldots,c_k$. Each column n in $\mathbf{A}_k$ corresponds to the data $\mathbf{x}_n$. The number of rows of $\mathbf{A}_k$ is determined by the total number of different attribute values for the k-th attribute. In this paper, gesture class is considered as the attributes used for gesture recognition.

By adding all attribute matrices as group constraints into the sparse representation, the formulation is given as follows:

$$\min_{\hat{\boldsymbol{\alpha}}} ||\hat{\boldsymbol{\alpha}}||_1 \quad \text{subject to} ||\hat{\mathbf{X}}\hat{\boldsymbol{\alpha}} - \hat{\mathbf{y}}||_2 < \varepsilon$$

$$\hat{\mathbf{X}} = \begin{bmatrix} \mathbf{X} & \mathbf{0}_{N \times m_1} & \cdots & \mathbf{0}_{N \times m_k} \\ \mathbf{A} & -\mathbf{I}_{m_1 \times m_1} & \cdots & \mathbf{0}_{m_1 \times m_k} \\ \vdots & \vdots & \vdots & \vdots \\ \mathbf{A}_k & \mathbf{0}_{m_k \times m_1} & \cdots & -\mathbf{I}_{m_k \times m_1} \end{bmatrix} \tag{3}$$

$$\hat{\boldsymbol{\alpha}} = \begin{bmatrix} \boldsymbol{\alpha} \\ \boldsymbol{\alpha}_{A_1} \\ \vdots \\ \boldsymbol{\alpha}_{A_k} \end{bmatrix}, \; \hat{\mathbf{y}} = \begin{bmatrix} \mathbf{y} \\ \mathbf{0}_{m_1 \times 1} \\ \vdots \\ \mathbf{0}_{m_k \times 1} \end{bmatrix}$$

The coefficient vectors $\boldsymbol{\alpha}, \boldsymbol{\alpha}_{A_1}, \ldots, \boldsymbol{\alpha}_{A_K}$ should be sparse due to the l_1 minimization in the objective function. Note that $\mathbf{A}_K \cdot \boldsymbol{\alpha} = \boldsymbol{\alpha}_{A_K}$ is called group constraint because this constraint enforces non-zeros coefficient to occur at few specific atoms in $\boldsymbol{\alpha}$. In order to make $\boldsymbol{\alpha}_{A_K}$ also sparse, group constraint enforces the data prediction mainly from the same type for the k-th attribute.

For most classification methods based on sparse coding technique, the classification is determined by the minimal reconstruction error. In our implementation, we use the tool package cvx for solving (http://cvxr.com/cvx/) the convex optimization. The decision of gesture classification $\hat{c}$ is determined by the minimal reconstruction error, which can be formulated as:

$$\hat{c} = arg \min_p \left\| \mathbf{y} - \mathbf{X}\delta_p(\boldsymbol{\alpha}) \right\|_2 \tag{4}$$

where $\delta_p(\boldsymbol{\alpha})$ selects the sparse coefficient in $\boldsymbol{\alpha}$ from class p. Based on our framework, the target instance is encoded by a compact representation which benefits from group constraint to improve the accuracy of gesture recognition.

3   Experimental Results

In this section, we show the experimental results by applying the proposed method to Large RGB-D Extensible Hand Gesture Dataset (LaRED) datasets, recorded with an Intel's newly-developed short range depth camera. This dataset is composed of a color image, a depth image, and a mask of the hand region. The LaRED dataset contains 27 gestures in 3 different orientations which are performed by 10 subjects. The corresponding hand gesture and 300 images are collected from each subject with the Intel camera providing a pair of synchornized color and depth images. Meanwhile, a mask image is given to localize the boundary of the gesture-performing hand. Figure 3 shows some sample depth images extracted from LaRED dataset.

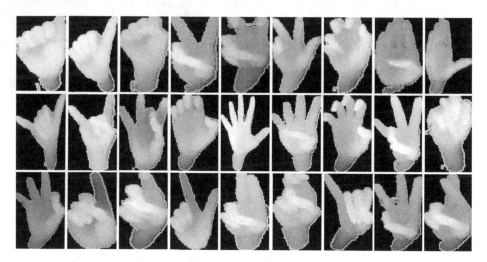

Fig. 3. Example depth images from LaRED dataset.

Table 1. Recognition accuracy (%) of LaRED dataset among different methods.

Method	Measure	Acc(%)
SVM	80 % training	72.84
SRC[11]	80 % training	76.37
Ours	80 % training	84.83

In our experimental setting, we select 7 subjects (5 men and 2 women) and randomly pick 5 gesture depth images for the 3 viewpoints for each class, and there are totally 2268 gesture depth images in all classes to evaluate the performance of our proposed method. For the selected gesture images, we select 80 % data examples for training, and the remaining data examples for testing. In addition, two types of attributes are exploited in our experiment: gesture class and view angle. Table 1 shows the recognition accuracies for the proposed method and two different methods. The proposed algorithm can achieved the 84.83 % gesture recognition accuracy in this experiment. Compared to SRC [11] and SVM classifiers, the proposed algorithm provides significant improvement with the same setting of the experiment.

4 Conclusions

In this paper, we propose a multi-attribute sparse coding based algorithm for hand gesture recognition from depth images. Different attributes can be employed as group constraints which enforce basis selection from groups. Based on our framework, the target instance is encoded by a compact representation which benefits from imposed group constraints. We demonstrate the effectiveness of our proposed algorithm on two facial expression datasets. Under the same experimental setting, the proposed algorithm provides significant improvement compared with the other methods for gesture recognition in our experiments.

Acknowledgements. This work was partially supported by Ministry of Science and Technology in Taiwan under the project MOST 103-2218-E-007-017-MY3.

References

1. Pavlovic, V., Sharma, R., Huang, T.: Visual interpretation of hand gestures for human-computer interaction: A review. IEEE Trans. Pattern Anal. Mach. Intell. **19**(7), 677–695 (1997)
2. Erol, A., Bebis, G., Nicolescu, M., Boyle, R., Twombly, X.: Vision based hand pose estimation: A review. Comput. Vis. Image Underst. **108**(1-2), 52–73 (2007)
3. de La Gorce, M., Fleet, D., Paragios, N.: Model-based 3d hand pose estimation from monocular video. IEEE Trans. Pattern Anal. Mach. Intell. **33**(9), 793–1805 (2011)
4. Suarez, J., Murphy, R.R.: Hand gesture recognition with depth images: a review. In: IEEE RO-MAN (2012)
5. Gustus, A., Stillfried, G., Visser, J., Jorntell, H., van der Smagt, P.: Human hand modelling: kinematics, dynamics, applications. Biol. Cybern. **106**(11-12), 741–755 (2012)
6. El-Sawah, A., Georganas, N., Petriu, E.: A prototype for 3-D hand tracking and gesture estimation. IEEE Trans. Instrum. Meas. **57**(8), 1627–1636 (2008)
7. Bretzner, L., Laptev, I., Lindeberg, T.: Hand gesture recognition using multi-scale colour features, hierarchical models and particle filtering. In: International Conference on Automatic Face Gesture Recognition (2002)

8. McKenna, S., Morrison, K.: A comparison of skin history and trajectory-based representation schemes for the recognition of user specific gestures. Pattern Recogn. **37**(5), 999–1009 (2004)
9. Imagawa, K., Matsuo, H., Taniguchi, R., Arita, D., Lu, S., Igi, S.: Recognition of local features for camera-based sign language recognition system. In: International Conference on Pattern Recognition (2000)
10. Dardas, N.H., Georganas, N.D.: Real-time hand gesture detection and recognition using bag-of-features and support vector machine techniques. IEEE Trans. Instrum. Measur. **60**(11), 3592–3607 (2011)
11. Wright, J., Yang, A.Y., Ganesh, A., Sastry, S.S., Ma, Y.: Robust face recognition via sparse representation. IEEE Trans. Pattern Anal. Mach. Intell. **31**(2), 210–227 (2009)
12. Zafeiriou, S., Petrou, M.: Sparse representations for facial expressions recognition via l1 optimization. In: IEEE Computer Vision and Pattern Recognition Workshops (2010)
13. Liu, W., Song, C., Wang, Y.: Facial expression recognition based on discriminative dictionary learning. In: International Conference on Pattern Recognition (2012)
14. Tibshirani, R.: Regression shrinkage and selection via the lasso (Series B). J. Royal Statist. Soc. **58**, 267–288 (1996)
15. Osborne, M.R., Presnell, B., Turlach, B.A.: On the lasso and its dual. J. Comput. Graphical Stat. **9**, 319–337 (1999)
16. Chiang, C.-K., Su, T.-F., Yen, C., Lai, S.-H.: Multi-attribute sparse representation with group constraints for face recognition under different variations. In: IEEE International Conference on Automatic Face and Gesture Recognition, Shanghai, China, April 2013

Eye Gaze Correction for Video Conferencing Using Kinect v2

Eunsang Ko, Woo-Seok Jang, and Yo-Sung Ho[(⊠)]

School of Information and Communications,
Gwangju Institute of Science and Technology (GIST),
123 Cheomdangwagi-ro, Buk-gu, Gwangju 500-712, Republic of Korea
{esko,jws,hoyo}@gist.ac.kr

Abstract. In video conferencing, eye gaze correction is beneficial for effective communication. In this era, video conferencing at homes using laptops is straightforward. In this paper, we propose an eye gaze correction method with a low-cost simple setup using Kinect v2. Our method detects an ellipse that connects edge points of the face after identifying several feature points within the face using Kinect v2 SDK. Then, we apply a 3D affine transform that allows eye gaze correction using camera space points that are acquired from depth information. Thus, in the preprocessing step, an ellipse model should be extracted when the user gazes the camera and display, respectively. Also, we fill holes that are caused by the affine transform using color inpainting. As a result, we produced a natural eye gaze-corrected image in real-time.

Keywords: Eye gaze correction · Eye contact · Video conferencing · Kinect v2

1 Introduction

When we are talking with other people, eye contact provides important information. Eye contact is looking at each other's eyes at the same time. By contacting other's eyes, we can guess what they are thinking, or whether they are interested. Thus, eye contact is significant in video conferencing. Video conferencing is one of telecommunication technologies which allows people communicate in two or more locations simultaneous video transmission. In these days, as the telecommunication technologies have grown, many people use the video conferencing in their homes as well as work places. In particular, people who is personal user use free video conferencing program such as Skype using their laptop or webcam. However, general video conferencing system occurs lack of eye contact due to the disparity between the locations of the subject and the camera [1]. These lack of eye contact cause unnatural communication and negative signs to other users.

For solving the lack of eye contact, various eye gaze correction methods have been proposed. One method is a remote collaboration system based on a semi-transparent see-through display. This method creates an experience where local and remote users are seemingly separated only by a vertical sheet of glass [2]. However, this method cannot be widely used due to expensive customized hardware. In this paper,

© Springer International Publishing Switzerland 2015
Y.-S. Ho et al. (Eds.): PCM 2015, Part II, LNCS 9315, pp. 571–578, 2015.
DOI: 10.1007/978-3-319-24078-7_58

we propose an eye gaze correction method using Kinect v2. The proposed method can be used cost-efficient and simple.

The proposed method bases a gaze correction approach using a single Kinect v1 [1]. Kinect v1 is composed of a color camera and a depth camera. They produce images at 640 × 480 resolution. However, the depth image of Kinect v1 is inaccurate with silhouette of the color image due to structure of Kinect v1. Thus, Kuster's method applies smoothing and filling holes on the depth image using Laplacian smoothing. Then, they make a novel view where the gaze is corrected. This is accomplished by applying a rigid transform that allows eye gaze correction. A matrix of the rigid transform is computed only once during the calibration stage. Next, they extract a user's face to track facial feature points in the original color image. They compute 66 feature points along the chin, nose, eyes and eyebrows. Then, they apply a seam optimization that is an optimal stencil to cut the face from a transformed image using the feature points. As a result, they generate a natural result of eye gaze correction by extending the seam of face by 5–10 pixels. Also, they solve a problem of flickering artifacts in a sequence result of eye gaze correction by optimizing the face tracker vertices. Their method runs at about 20 frames per second (fps) on a consumer computer.

2   Proposed Method

2.1   System Design

We use 27 inches display monitor and Kinect v2. Kinect v2 is combined color camera with depth camera. In Fig. 1(a), the color camera is on the left of Kinect v2, and the depth camera is on the center of Kinect v2 [3]. As we use depth information for eye gaze correction, we align the depth camera with center of the display monitor. Then we set Kinect v2 on the bottom of the display monitor. And, we raise an angle of Kinect v2 to face to the user's face for effective eye gaze correction, but the user's spine must be shown in the color image for face detection using Kinect v2 SDK. Thus, moderate the angle of Kinect v2 and distance between the users and Kinect v2 are required when the user sets the proposed system. Figure 1(b) displays an example of the proposed eye gaze correction system configuration.

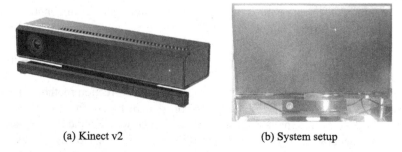

(a) Kinect v2   (b) System setup

Fig. 1.   Kinect v2 and system configuration

2.2 Preprocessing

Preprocessing comprises four steps: down-sampling a color image, detecting face feature points, estimating a 3D affine transform matrix that allows eye gaze correction, and creating face mask that is applied 3D affine transform. Although the preprocessing steps are many, some are repeatedly processed each frames, the others take only once in a few seconds by users.

The color camera of Kinect v2 has a resolution of 1920 × 1080 pixels, and the depth camera has a resolution of 512 × 424 pixels. For mapping the color image to depth information, we down-sample the color image using Kinect v2 SDK. We produce eye gaze-corrected image using the down-sampled color image. Thus, the eye gaze-corrected image has a resolution of 512 × 424 pixels. Then, we limit depth value between 450 and 900 for reducing depth noise and subtracting background. The depth value of Kinect v2 means a distance between the object and the camera. The depth camera of Kinect v2 can capture the depth value from 450 mm. Figure 2(a) and (b) shows a real-depth map and the down-sampled color image, respectively.

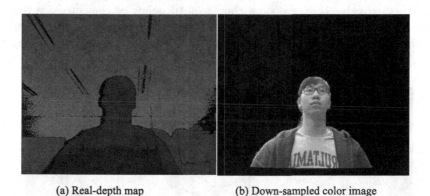

(a) Real-depth map

(b) Down-sampled color image
(depth limit: 450-900)

Fig. 2. Depth map and down-sampled color image of Kinect v2

(a) 1347 points of face feature   (b) Sampled 10 points in 1347 points

Fig. 3. Face feature point detection using Kinect v2 SDK

Second is the detecting face feature points. For detecting the face feature points in Kinect v2 SDK, it needs body tracking data. Thus, the user's upper body like head, neck, shoulder and spine must be shown in field of view of Kinect v2 for detecting skeleton data of body. Once skeleton data is detected, 1347 points of face feature are tracked if face is in the field of view of Kinect v2. And, once tracking of the face feature points are started, Kinect v2 continuously tracks the face feature points in the color image. The face feature points are represented with camera space point. We can display the face feature points on the down-sampled color image by converting camera space point to depth image point using Kinect v2 SDK. Figure 3(a) exhibits a result of detected 1347 points of face feature. However, most of face feature points are convergent in the user's eye, nose and mouth. Thus, we sample 10 points in the 1347 points of face feature for decreasing complexity and improving accuracy of eye gaze correction. Figure 3(b) shows the sampled 10 points in the user's forehead, cheek bone, jaw and chin.

Third, for estimating the matrix of 3D affine transform, we need the face feature points of two models when the user gazes the camera and display, respectively. Figure 4(a) and (b) display an example image when the user gazes the camera and display, respectively. Thus, we store couple of 10 points of camera space that are camera model and display model, respectively. Then, we estimate the optimal matrix of 3D affine transform that converts the display model to the camera model using the

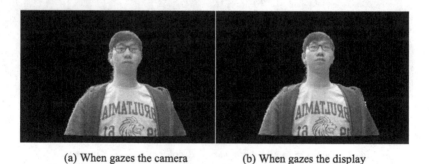

(a) When gazes the camera   (b) When gazes the display

Fig. 4. An example image when the user gazes the camera and display

(a) Ellipse fitting   (b) Ellipse mask

Fig. 5. An example image of fit ellipse and created mask

random sample consensus (RANSAC) algorithm for eliminating outliers. We can get converted camera space points by multiplying the matrix of 3D affine transform by original camera space points.

Finally, we create a mask that is applied eye gaze correction using the matrix of 3D affine transform. The mask is ellipse area that connects the sampled 10 face feature points. For finding an optimal ellipse model using the 10 face feature points, we use ellipse fitting function within OpenCV that algorithm [Fitzgibbon95] is used [4, 5]. The mask can be largely changed by coordinates of the sampled face feature points in the user's forehead, cheek bone, jaw and chin, respectively. Thus, the users can adjust the coordinates of the face feature points within Kinect v2 SDK to generate the optimal ellipse mask. Then, we draw the ellipse mask on an empty black image using a simple OpenCV flag [6]. Figure 5(a) and (b) exhibit an example image of fitting ellipse and creating the filled ellipse mask.

2.3   Eye Gaze Correction

The proposed eye gaze correction method converts a camera space point of Kinect v2. X, Y, and Z of the camera space point means yaw, pitch, and depth, respectively. Thus, we have to change the Y value of the camera space point to make eye gaze correction. Figure 6(a) and (b) simulate a face model that is original model and decreased model by 0.1 the Y value of the camera space point, respectively. The matrix of 3D affine transform that we estimated in the processing step allows eye gaze correction by changing the camera space point. The Y value of the camera space point is more changed than the X or Z value. Equation 1 shows the changing camera space point by multiplying the matrix of 3D affine transform by original camera space point. The matrix of 3D affine transform is 3 by 4, m means an element of the transform matrix. The X, Y, and Z denote the original camera space point, and the $X`$, $Y`$, and $Z`$ represent the converted camera space point. When multiplying the matrix by the original camera space point, we make the original camera space point to a homogeneous matrix.

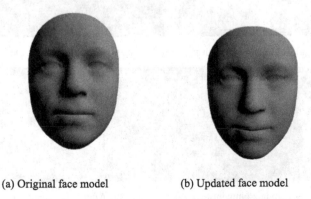

(a) Original face model   (b) Updated face model

Fig. 6. Simulation result of eye gaze correction

$$\begin{bmatrix} X' \\ Y' \\ Z' \end{bmatrix} = \begin{bmatrix} m_{11} & m_{12} & m_{13} & m_{14} \\ m_{21} & m_{22} & m_{23} & m_{24} \\ m_{31} & m_{32} & m_{33} & m_{34} \end{bmatrix} \begin{bmatrix} X \\ Y \\ Z \\ 1 \end{bmatrix} \quad (1)$$

To apply eye gaze correction, we convert a depth point that is in the face mask to a camera space point using Kinect v2 SDK. The camera space point can be converted using a pair both the depth image point and depth value of the depth point. This camera space point is corrected to make eye contact using the 3D affine transform. The corrected camera space point is converted to depth image point. Thus, y coordinates of the depth point are increased by depth value of the depth point.

2.4   Color Inpainting

Figure 7(a) shows an example of eye gaze correction result. There are many holes due to a pixel rounding error when applying the 3D affine transform. Also, there is one large hole in the user's forehead that represents eye gaze correction is applied well. Figure 7(b) displays an error map of the result of eye gaze correction. For removing the error, we fill hole using color inpainting using [Telea04] method [7]. However, the large hole causes an unnatural result after color inpainting. Thus, we create a color inpainting mask by updating the face mask. Figure 7(c) and (d) represent the updated color inpainting mask and a result of color inpainting, respectively.

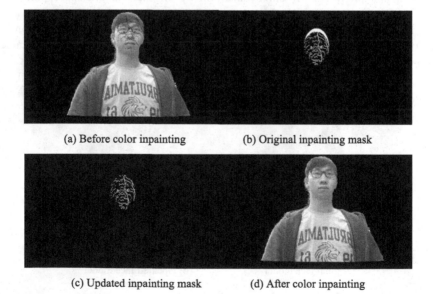

(a) Before color inpainting   (b) Original inpainting mask

(c) Updated inpainting mask   (d) After color inpainting

Fig. 7. A result of color inpainting

3 Experiment Result

To evaluate our proposed eye gaze correction method, we capture several results of eye gaze correction when the user watches around the monitor. Figure 8(a)–(c) display the results of eye gaze correction, respectively. Figure 8(a) is a result of eye gaze correction when the user watches center of the monitor, and Fig. 8(b)–(c) are results of eye gaze correction when the user watches left and right of the monitor, respectively. As watching the results, there are three problems. First, there is a depth hole in area of the user's glasses due to reflected light. Second, there is boundary noise due to time-of-flight (ToF) depth error. The boundary noise occurs while the color image is

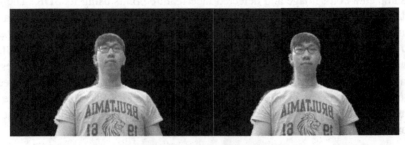

(a) When the user gazes center of the monitor

(b) When the user gazes left edge of the monitor

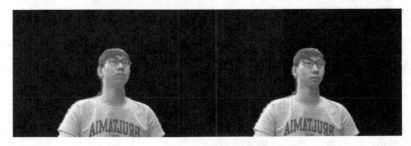

(c) When the user gazes right edge of the monitor

Fig. 8. Experiment result of eye gaze correction

down-sampled in the processing step. Finally, a result of eye gaze correction is out of shape if the user's depth value is largely changed like shifting the user's position.

We processed the proposed eye gaze correction method on a desktop computer that uses a CPU: Intel Core i7 4960X @ 3.6 GHz. To achieve processing of the eye gaze correction method in real-time, we handled a multi-thread scheduling when capturing the color and depth image, down-sampling the color image, applying eye gaze correction. As a result, a result of eye gaze-corrected is generated at a rate of 28 fps.

4 Conclusion

In this paper, we proposed an eye gaze correction method using Kinect v2. We down-sampled a color image and detected face feature points using its SDK. Then, we estimated the matrix of 3D affine transform that allows eye gaze correction by means of a face model as the user gazes the camera and display, respectively. Finally, we applied the 3D affine transform and color inpainting within the face mask. As a result, we acquired an eye gaze-corrected natural image in real-time. There exists depth holes and boundary noise due to the limitation of Kinect v2 depth camera. Hence, if we process a depth up-sampling using high-performance computer and GPU processing, we can generate a result of eye gaze-corrected at 1920×1080 resolution as well as solve depth holes and boundary noise of the result image.

Acknowledgement. This research was supported by Basic Science Research Program through the National Research Foundation of Korea (NRF) funded by the Ministry of Science, ICT & Future Planning (No. 2011-0030079)

References

1. Kuster, C., Popa, T., Bazin, J.C., Gotsman, C., Gross, M.: Gaze correction for home video conferencing. ACM Trans. Graphics **31**(6), 1–6 (2012)
2. Tan, K.H., Robinson, I.N., Culbertson, B., Apostolopoulos, J.: ConnectBoard: Enabling Genuine Eye Contact and Accurate Gaze in Remote Collaboration. IEEE Trans. Multimedia **13**(3), 466–473 (2011)
3. http://www.microsoftstore.com/store/mssg/en_SG/pdp/Kinect-for-Windows-v2-Sensor/productID. 299057000
4. Fitzgibbon, A.W., Fisher, R.B.: A Buyer's guide to conic fitting. In: Proceedings of the British Machine Vision Conference (BMVC), Birmingham, England, vol. 2, pp. 513–522 (1995)
5. http://docs.opencv.org/modules/imgproc/doc/structural_analysis_and_shape_descriptors. html?highlight=fitellipse#fitellipse
6. http://docs.opencv.org/modules/core/doc/drawing_functions.html?highlight=ellipse#ellipse
7. Telea, A.: An image inpainting technique based on the fast marching method. J. Graphics, GPU, Game Tools **9**(1), 25–36 (2004)

Temporally Consistence Depth Estimation from Stereo Video Sequences

Ji-Hun Mun and Yo-Sung Ho(✉)

School of Information and Communications,
Gwangju Institute of Science and Technology (GIST),
123 Cheomdangwagi-ro, Buk-gu, Gwangju 500-712, Republic of Korea
{jhm,hoyo}@gist.ac.kr

Abstract. In this paper, we propose complexity reduction method in stereo matching. For complexity reduction in video sequences, we start from generating of initial disparity information. Initial disparity information can give a restricted disparity search range when performing the local stereo matching. As an initial disparity information, we use 4 different kinds of input images. The initial disparity information types can divide into two main streams like '*calculated*' and '*given*' materials. Iterative stereo matching method, motion prediction stereo matching method and global matching result based stereo matching method are used as a '*calculated*' initial value. Captured by depth camera image is used as '*given*' information. By using those 4 different types of disparity information, we can save the time consuming when performing the local stereo matching with consecutive image sequences. Results of the experiment prove the efficiency of proposed method. By using proposed local stereo matching method, we can finish all procedures within a few seconds and conserve the quality of disparity images.

Keywords: Local stereo matching · Stereo camera · Time complexity

1 Introduction

Binocular disparity is essential concept to interpret the three-dimensionality of human visual system. Because of the importance of human visual system, jointed two-frame stereo matching is one of the most considerably studied field in computer vision. Stereo matching algorithm attempts to generate the depth information from stereo image pairs captured by stereoscopic camera or multiple cameras. It is also the major concern in a wide variety of applications such as super multi-view virtual view synthesis, free view point image and three-dimensional virtual reality. Most stereo matching algorithm can be roughly classified into two categories: *global* and *local* matching methods [1].

Global stereo matching method calculate the matching problem using an energy function with data and smoothness constraint. To find out the minimized energy value, generally dynamic programing [2], graph cuts [3] and belief propagation [4] are used. Global optimization method can considerably reduce the stereo matching ambiguities. Normally matching ambiguity factors are coming from illumination variation, saturation region and texture region. Global stereo matching can generate more efficient matching result than local stereo matching method, because global stereo matching consider all of

© Springer International Publishing Switzerland 2015
Y.-S. Ho et al. (Eds.): PCM 2015, Part II, LNCS 9315, pp. 579–588, 2015.
DOI: 10.1007/978-3-319-24078-7_59

the image conditions. But this method usually has an expensive computation time due to consideration of all image pixel value for minimization. To reduce the time complexity many state-of-the-art global stereo matching methods are developed [5].

Local stereo matching basically use the window concept in matching procedure. This method compute each pixels disparity value independently over the all image region. To derive the cost function value, the matching costs are aggregated over the window region. Within the window size, minimal cost value is selected as an output of the associated pixel. The procedure of local stereo matching method basically depend on the support window user designated. Generally window size starting from 3×3 to (2n-1) ×(2n-1) size when 'n' is not zero and larger than 2. If the window size is small, then matching result has a considerably accurate disparity result in edge or boundary region. While performing the matching procedure with large window size, then conversely matching result has an accurate value on homogeneous region or similar texture region [6].

Our algorithm basically focus on the local stereo matching method due to the complexity problem in global stereo matching. Generally to solve the complexity problem when performing the local or global stereo matching, GPU process is normally used for implementation step [7]. And one of the state-of-the-art local stereo matching, adaptive supporting-weight (ADSW) [8], could deliver disparity maps close to global optimization. However, the matching method like ADSW suffer from highly computational complexity, and the following associated research attempted to accelerate it. Chang et al. [9] simplify the ADSW by using a hardware implementation. Although the previous researching for high complexity problem, but they basically use other hardware assistance like using a GPU: CUDA or onboard implementation.

In this paper, we propose a new local stereo matching algorithm in video sequences which based on the initial disparity information. Different from the previous complexity reduction research, our algorithm provide a significant solution only using a software method. As an initial disparity information we use 4 different type of disparity generation method. In consecutive image sequence, between the neighbor image frames, they has a small disparity differences. Because of that reason from the previous stereo images matching result information, current frame stereo images just consider smaller disparity range than previous stereo images disparity range.

To testify proposed algorithm, we use four different computer graphic video sequences with depth ground truth image that provided by Cambridge computer laboratory. We provide full description of the proposed local stereo matching algorithm in Sect. 2. In Sect. 3, we discuss about the experiment result of proposed algorithm. In experiment result we provide a 4 different test sequences matching results and compare the matching results with given depth ground truth image for efficiency. After showing the experiment results with analysis of time complexity, we conclude this paper in Sect. 4.

2 Temporal Domain Stereo Matching

In this section, we will discuss about temporal domain local stereo matching method for time complexity reduction. The proposed temporal domain stereo matching methods can divide into two main idea. Firstly we use general stereo matching method for initial disparity information which comes from the first frame image pairs. And the other method use given depth information which captured by ToF depth camera or Kinect depth camera etc. Because we basically use general local stereo matching method for several proposed method, we will briefly explain about that method.

2.1   Local Stereo Matching

Generally used local stereo matching method basically use pre-designated window size. The basic local stereo matching use restricted disparity search range. To find out the most similar pixel value between image pairs, local stereo matching restrict the disparity search range. Figure 1 show general local stereo matching procedure.

In Fig. 1, each consecutive image frame pairs has different disparity result. But between that results it just has small different disparity value. We focus on that point and will apply this properties in temporal domain stereo matching method.

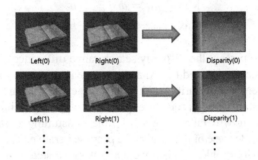

Fig. 1.   General stereo matching procedure

2.2   Iterative Stereo Matching

Among the consecutive image frames, disparity value has small difference between closest neighbor image frames. As stated in previous section we use this properties in iterative stereo matching method procedure. Iterative stereo matching method consider temporal domain disparity information differences. But this method need only one time general local stereo matching method on the beginning. Because as an initial disparity information we need initial stereo image matching result. Figure 2 represent the iterative stereo matching procedure.

In iterative stereo matching method, initially generated disparity information 'Disparity (0)' used for following image pairs stereo matching procedure. As mentioned in previous section, we use following image pair disparity result value and previous disparity result value differences. In this paper we apply smallest disparity searching range with difference of adding disparity sign. As represented in Fig. 2, following image pair disparity result has similar to previous frame image pair result. Iterative stereo matching method continuously used in video sequences, from frame (1) pair, frame (2) pair to frame (n) pair.

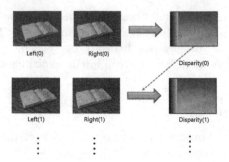

Fig. 2. Iterative stereo matching procedure

Based on the properties of previous and current disparity value differences, Eq. 1 can be derived.

$$\text{Min}_{\text{disparity}} = disp_{pre} - n$$
$$\text{Max}_{\text{disparity}} = disp_{pre} + n \tag{1}$$

In Eq. 1 value n represent the disparity search range differences in following frame image pair stereo matching procedure. And $disp_{pre}$ mean previous image pair disparity result value. When generate disparity result using image pair, disparity search range has to be defined before starting that procedure. If we use original minimum and maximum disparity range, then it will take amount of time consuming compare to restricted disparity search range. Because of that reason we propose iterative stereo matching method to solve time complexity problem. When we apply iterative stereo matching method, we can more efficiently compute the disparity value than general stereo matching method. Even we diminish the disparity search range, iterative stereo matching method can consider the significant disparity search range while performing the stereo matching.

2.3   Motion Prediction Stereo Matching

In consecutive image sequences, following frame like 'Left (0)' and 'Left (1)' has a small difference between that frames. Even human eyes hardly perceive the difference of two images, but it has a difference value in terms of image pixels. Considering motion flow for stereo matching research has been actively performed [10, 11]. In previous works about motion prediction, proximity and similarity have been considered in the computation of image disparity map. However, that information insufficient for video disparity estimation because motion cues are very important for accurate disparity calculation near edge of moving objects. We include motion flow to compute disparity value more clearly in objects edge or moving area. Figure 3 indicate that stereo matching procedure with considering motion flow in consecutive image frames.

In Fig. 3, *Moving difference* represent difference of following image frames difference result. Comparing to iterative stereo matching method, motion prediction stereo matching method added motion information. But when compute motion information, it

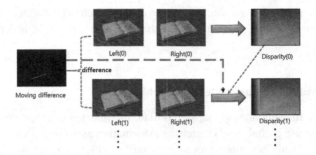

Fig. 3. Motion prediction stereo matching procedure

has different result depending on threshold value to determine 0 and 255 pixel value. Equation 2 represent the motion difference determining method.

$$Color_{diff} \geq th \quad diff_{map} = 255$$
$$Color_{diff} < th \quad diff_{map} = 0$$

(2)

If difference of following image value is bigger than threshold value then Moving difference has 255 pixel value, otherwise it has 0 pixel value. So pre-determined threshold value effect on motion estimation result. Figure 4 show different result of motion estimation of consecutive image frames.

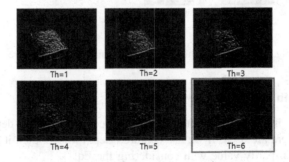

Fig. 4. Difference of motion estimation result for threshold value

Indicated in Fig. 4, as the threshold value is increased, the motion estimation results change into blurred image. Because we focusing on the object moving area or boundary region, texture region and inside of object information is unnecessary. In this paper we use fixed threshold value 6. If we use bigger threshold value, then object boundary region will be removed. Then it will spoil the stereo matching result.

The main framework same as iterative stereo matching, but to accurately compute the object boundary or moving region we use motion prediction stereo matching method. Comparing to time consuming of iterative stereo matching method, it will takes more time because of the motion predicted region. During the stereo matching, window face to prediction region, then it have to search for original minimum and maximum disparity range. As a result of that we can get more accurate disparity information.

2.4   Global Matching Based Stereo Matching

As mentioned in introduction, global stereo matching has better result than local stereo matching. So we use global stereo matching information as an initial value [12]. Global stereo matching result has more accurate disparity value compare to previous initial information, consecutive frame stereo matching result also more accurate than previous one. Figure 5 represent the global matching based stereo matching method.

Likewise previously explained proposed method it has same framework, but only initial value is different as a global matching result.

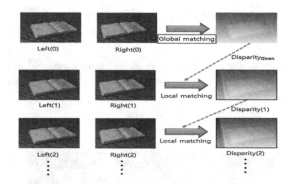

Fig. 5. Global matching based stereo matching procedure

2.5   Given Depth Based Stereo Matching

To generate depth information directly from the object, usually ToF depth camera and Kinect depth camera are used. If we use captured depth image, then it need to change depth value into disparity value with considering the Eq. 3.

$$Z_{\text{near}} = \frac{f \cdot l}{d_{max} + \Delta d'} \quad Z_{\text{far}} = \frac{f \cdot l}{d_{min} + \Delta d} \tag{3}$$

In Eq. 3, f is focal length of camera and l represent the base line of cameras. And also we already know the Z_{near} and Z_{far} value, so apply those parameter value in Eq. 3, then disparity minimum and maximum values can easily derived.

In this paper we use given computer graphics depth ground truth value, so we don't need to compute depth information to disparity value changing procedure. As like the iterative stereo matching method in Fig. 2, it also has same framework. But as an initial value it use given depth information, like a captured depth image or computer graphics ground truth image.

3   Experiment Results

We testify our proposed stereo matching method using computer graphic sequences provided by *Cambridge computer laboratory*. Test platform is a PC with Core(TM) i7-5960X 3.00 GHz CPU and 32.0 GB memory. And we use 4 different video sequence as indicated in Fig. 6: (a) *book*, (b) *street*, (c) *tanks* and (d) *temple*.

(a)   (b)   (c)   (d)

Fig. 6.   Test video sequences

All of the test sequences has same resolution as 400×300 and also depth ground truth image has same resolution. Figure 7 show provided depth ground truth images.

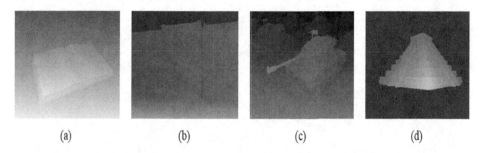

(a)   (b)   (c)   (d)

Fig. 7.   Depth ground truth images for each test sequences

To compare disparity result image and depth ground truth image, we use bad pixel rate (BPR). If difference of computed disparity value and depth ground truth value is bigger than 1, then we determine that pixel as bad pixel. And also to verify the efficiency of our algorithm compare processing time for each proposed method with general local stereo matching method.

As represented in Table 1, temporal domain information based stereo matching is 10 % faster than general stereo matching method.

Figure 8 represent bad pixel rate comparison results for each test sequences. Except for test sequence 'Street', other test sequences has highest bad pixel rate when we use plus/minus 3 disparity search range. And stereo matching result with original minimum and maximum disparity search range has similar bad pixel rate even the frame numbers increase.

Table 1. Temporal domain stereo matching methods time comparison

Matching method	Disparity range			
	Min/Max	±3	±5	±7
General	21.76(sec)	–	–	–
Iterative	–	2.43(sec)	3.77(sec)	5.12(sec)
Prediction	–	3.17(sec)	4.45(sec)	5.76(sec)
Global	–	2.09(sec)	3.50(sec)	4.86(sec)
Given	–	2.10(sec)	3.47(sec)	4.84(sec)

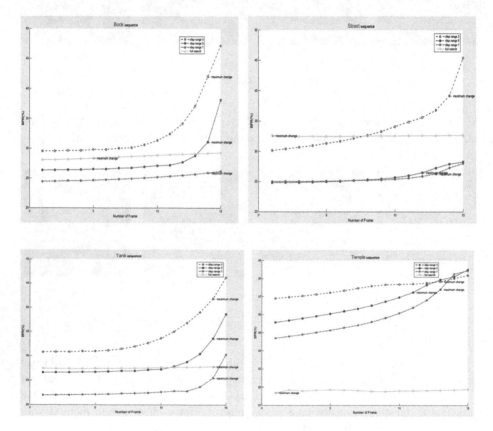

Fig. 8. BPR test for each test sequences

Proposed stereo matching results are represented in Fig. 9. When we use global matching method and given depth image as a base information, matching results has better quality than other matching methods.

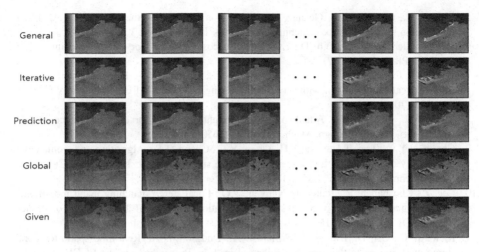

Fig. 9. Test results for sequence 'Tank'

4 Conclusion

In this paper we propose local stereo matching method in video sequences for reducing time complexity. Iterative matching, motion prediction and global matching method need 1 time general stereo matching procedure. And given depth based method directly use the temporal domain information for local stereo matching. From the experiment result we can check that proposed method is 10 % faster than general local stereo matching and wide disparity search matching has better BPR value. General local stereo matching has similar BPR value over the frame number, but proposed methods BPR values are decreased. From that result we need to refresh the reference disparity image by using minimum/maximum disparity range. We will research about that kind of problem to prevent the error propagation.

Acknowledgments. This research was supported by the 'Cross-Ministry Giga KOREA Project' of the Ministry of Science, ICT and Future Planning, Republic of Korea (ROK). [GK15C0100, Development of Interactive and Realistic Massive Giga-Content Technology]

References

1. Scharstein, D., Szeliski, R.: A taxonomy and evaluation of dense two-frame stereo correspondence algorithms. IJCV **47**(103), 7–42 (2002)
2. Bobick, A.F., Intille, S.S.: Large occlusion stereo. IJCV **33**(3), 181–200 (1999)
3. Boykov, Y., Veksler, O., Zabig, R.: Fast approximate energy minimization via graph cuts. IEEE TPAMI **23**(11), 1222–1239 (2001)
4. Meltzei, T., Yanover, C., Weiss, Y.: Globally optimal solutions for energy minimization in stereo vision using reweighted belief propagation. IEEE Int. Conf. ICCV **1**, 428–435 (2005)

5. Bleyer, M., Chambon, S., Gelautz, M.: Evaluation of different methods for using colour information in global stereo matching approaches. ISPRS Congr. **1**, 1–6 (2008)
6. Hirschmuller, H., Scharstein, D.: Evaluation of cost functions for stereo matching. IEEE Conf. CVPR **1**, 1–8 (2007)
7. Zhang, K., Lu, J., Yang, Q., Lafruit, G., Lauwereins, R., Gool, L.V.: Real-Time and accurate stereo: a scalable approach with bitwise fast voting on CUDA. IEEE Trans. CSVT **21**(7), 867–878 (2011)
8. Yoon, K.J., Kweon, I.S.: Adaptive support-weight approach for correspondence search. IEEE Trans. Pattern Recogn. Mach. Intell. **2**(4), 650–656 (2006)
9. Chang, N.Y.C., Tsai, T.H., Hsu, P.H., Chen, Y.C., Chang, T.S.: Algorithm and architecture of disparity estimation with mini-census adaptive support weight. IEEE Trans. Circ. Syst. Video Technol. **20**(6), 792–805 (2010)
10. Lee, Z., Khoshabeh, R., Juang, J., Nguyen, T.Q.: Local stereo matching using motion cue and modified census in video disparity estimation. In: Signal Processing Conference (EUSIPCO), pp. 1114–1118 (2012)
11. Richardt, C., Orr, D., Davies, I., Criminisi, A., Dodgson, N.A.: Real-time spatio-temporal stereo matching using the dual-cross-bilateral grid. In: Proceedings of the ECCV, pp. 510–523 (2010)
12. Jnag, W.S., Ho, Y.S.: Discontinuity preserving disparity estimation with occlusion handling. J. Vis. Commun. Image Representation **25**(7), 1595–1603 (2014)

A New Low-Complexity Error Concealment Method for Stereo Video Communication

Kesen Yan[1], Mei Yu[1,2], Zongju Peng[1], Feng Shao[1],
and Gangyi Jiang[1,2(✉)]

[1] Faculty of Information Science and Engineering,
Ningbo University, Ningbo 315211, China
jianggangyi@126.com
[2] National Key Lab of Software New Technology,
Nanjing University, Nanjing 210093, China

Abstract. The emerging popularity of three dimensional content has instigated the advances of stereo video coding and transmission techniques. According to the existing video compression standards, highly compressed stereo video bit streams are susceptible to transmission errors. As a consequence of the unavoidable spatial, temporal, and inter-view error propagation, the display quality is severely degraded at the receiver side. In this paper, to deal with the whole-frame loss of the right view in H.264-compressed stereo video bitstream transmission, a highly efficient perceptual whole-frame loss of right view error concealment algorithm is proposed with a fast error concealment strategy. Firstly, a fast error concealment strategy is presented, and an efficient image quality Index, gradient magnitude similarity deviation (GMSD) is extended to the concepts of temporal GMSD (TGMSD) and inter-view GMSD (VGMSD), respectively, to evaluate the perceptual quality of stereo image. Then, according to the temporal correlation of video sequences, macroblock (MB) prediction modes of the previous right-view frame are used as the MB prediction mode of the lost frame. Then, MBs in the previous frame of the lost frame are also matched, in both temporal and inter-view domains, to obtain the pixel-based TGMSD and VGMSD maps. Finally, for each MB in the lost right-view frame, its TGMSD and VGMSD values are calculated and compared to obtain the MB prediction mode, after which either motion compensation or disparity compensation can be quickly decided and used for resilience of the lost right-view frame. Experimental results show that compared with traditional error concealment algorithms, the proposed algorithm has superior subjective and objective qualities, and compared with the error concealment algorithms based on structure similarity, its error concealment time is reduced by about 40 % with almost same perceptual quality.

Keywords: Error concealment · Stereo video · Whole-frame loss

1 Introduction

With the rapid development of multimedia communication and networks, three-dimensional (3D) video technologies are developed with more vivid visual effects to audiences. 3D video have drawn extensive attention [1]. Transmission errors such as

© Springer International Publishing Switzerland 2015
Y.-S. Ho et al. (Eds.): PCM 2015, Part II, LNCS 9315, pp. 589–597, 2015.
DOI: 10.1007/978-3-319-24078-7_60

packet losses and bit errors are inevitable and more seriously, since the errors may propagate to the current frame or even the whole video streaming due to high compression efficiency for stereo video. Thus, the researches on stereo video error concealment (EC) are important for practical 3D video applications [2, 3].

Stereo video EC methods can be classified into macroblocks (MBs) loss EC and whole-frame loss EC. For resilience of MBs loss EC, Tang et al. proposed a stereo image EC algorithm based on smoothness of boundaries [4], they judged the characteristics of lost MBs and choose different EC schemes according to boundary smoothing. Zhu et al. [5] proposed a stereo video MB loss EC algorithm, combining the partition mode of the related reference MBs, in which the current MBs are divided into smooth or texture blocks for EC. Tang et al. [6] proposed a MB loss EC method based on reliable position from stereo images, and they used a greedy strategy to estimate the disparity of the lost MB for EC. Zhou et al. utilized the temporal and inter-view correlations of stereo video to choose the temporal or inter-view EC scheme for error recovery based on the vitality strength of the damaged block [7].

For the case of a whole-frame loss, no spatial information is available and the lost information can only be recovered by the temporal and inter-view information. Some whole-frame loss EC methods for stereo video had been proposed recently. Li et al. [8] proposed a stereo video whole-frame loss EC algorithm based on structure similarity. According to temporal structure similarity and inter-view structural similarity, the prediction mode of each MB is obtained as the prediction mode of corresponding MB in lost frame, and motion compensation prediction or disparity compensation prediction is used for recovery. Bilen et al. uses redundancy, disparity and motion information of decoded frames to estimate the lost frame. The temporal correlation of disparity vector and the inter-view correlation of motion vector are employed for error resiliency [9]. Chen et al. proposed an EC algorithm based on frame difference of disparity prediction, using the global disparity and local disparity to make varying region detection with the lost content [10]. Lin et al. [11] improved an whole frame error concealment method for multiview video with depth. The error pixels are first concealed by the pixels from another view, and the light differences between views are also considered.

The existing whole-frame loss EC methods of stereo video mainly considered the temporal and inter-view correlations, rarely took the problems of the restored images' visual perceptual quality and EC complexity into account. In this paper, considering the human subjective perception of videos and EC complexity, we propose a low complexity EC algorithm for whole-frame loss of right-view in stereo video. Firstly, gradient magnitude similarity deviation (GMSD) is extended to the concepts of temporal GMSD (TGMSD) and inter-view GMSD (VGMSD), respectively. TGMSD and VGMSD are used to judge the MB prediction modes of the previous right-view frame. Secondly, according to the different prediction modes, motion or disparity compensation prediction is used for error recovery of the current frame respectively. In addition to the traditional temporal and inter-view correlation, the method also considers the human eye subjective perception of video image structure and the complexity of EC, thus the objective quality is improved and the recovery effect is closer to the human eye perception.

2 Stereo Video EC Method Based on GMSD

In stereo video transmission, Fig. 1 shows the IPPP prediction structure of stereo video coding. In Fig. 1, the left-view is encoded by the coding standard h.264/AVC independently. The right-view is encoded by combining motion-compensation prediction (MCP) with disparity-compensated prediction (DCP). When right-view video image at time t is lost, there is no available information of the current frame for EC. So, temporal and inter-view correlations of stereo video sequences have to be used for EC. The detailed flow chart of the proposed method is shown in Fig. 2. In order to evaluate the

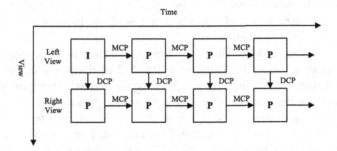

Fig. 1. IPPP prediction structure of stereo video coding.

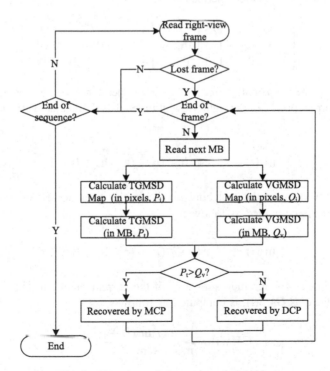

Fig. 2. Flowchart of the proposed algorithm.

perceptual quality of stereo video combining with temporal and inter-view correlation effectively, the concept of temporal highly efficient perceptual quality index (TGMSD) and inter-view highly efficient perceptual quality index (VGMSD) are presented. Then, after getting pixel-wise TGMSD and VGMSD for right-view frame at time t-1, the MB-wise TGMSD and VGMSD of the right-view frame at time t-1 are calculated so as to decide the MB prediction mode of each MB at time t-1. Corresponding to the MB at time t, MCP or DCP is used for EC respectively.

2.1   Pixel-wise TGMSD and VGMSD Map of Right-View

Assuming that left-view is received correctly, the frame of the right-view is lost. Selecting the lost video image of right-view at time t, the process of calculating the pixel-wise GMSD map as follows:

There are two kinds of MB matching: temporal matching and inter-view matching. The motion vector at time t-1 relating to at time t-2 of right-view based on motion vector extrapolation and the disparity vector of the right-view relating to the left-view at time t-1 based on disparity matching are obtained for MB matching.

A good video image quality assessment model should not only deliver high quality prediction accuracy, but also be computationally efficient. Here, GMSD [12] is used to describe image feature. Therefore, the GMSD between the right-view image and the reference frame can be used as correlation index of the corresponding pixels in the two images. Gradient operators are described by Eq. (1).

$$\mathbf{h}_x = \begin{bmatrix} 1/3 & 0 & -1/3 \\ 1/3 & 0 & -1/3 \\ 1/3 & 0 & -1/3 \end{bmatrix}, \mathbf{h}_y = \begin{bmatrix} 1/3 & 1/3 & 1/3 \\ 0 & 0 & 0 \\ -1/3 & -1/3 & -1/3 \end{bmatrix} \tag{1}$$

where $\mathbf{h}_x$ or $\mathbf{h}_y$ is the horizontal or vertical gradient operator, respectively. The gradient value of the current pixel i in the right-view at time t-1 is defined as $\mathbf{m}_c(i)$, is computed as follows:

$$\mathbf{m}_c(i) = \sqrt{(\mathbf{c} \otimes \mathbf{h}_x)^2(i) + (\mathbf{c} \otimes \mathbf{h}_y)^2(i)} \tag{2}$$

The gradient value of the matching pixel i in the reference frame at time t-1 is defined as $\mathbf{m}_r(i)$, is computed as follows:

$$\mathbf{m}_r(i) = \sqrt{(\mathbf{r} \otimes \mathbf{h}_x)^2(i) + (\mathbf{r} \otimes \mathbf{h}_y)^2(i)} \tag{3}$$

Then, the gradient magnitude similarity of the current pixel i in the right-view at time t-1 is defined as $GMS(i)$, is computed as follows:

$$GMS(i) = \frac{2\mathbf{m}_c(i)\mathbf{m}_r(i) + c}{\mathbf{m}_c^2(i) + \mathbf{m}_r^2(i) + c} \tag{4}$$

where c is a very small constant to avoid zero numerator or denominator.

In the process of calculating the pixel-wise GMSD map of the right-view, due to the gradient template size is 3×3, the gradient of the boundary pixels in right-view at time t-1 do not need to be calculated. If the image pixels of right-view at time t-1 are not boundary points and the matching pixels of the reference frame are boundary pixels, then the matching pixels are replaced by the neighbor pixels which are not in the boundary, the value of pixel-wise GMSD can be obtained with Eqs. (2), (3) and (4). In this case, reference mode judgment process of the MB is the same, so the replacement is reasonable relatively.

According to the difference between temporal and inter-view of the reference frame, after calculating the pixel-wise GMSD value of each pixel completed, the pixel-wise TGMSD map of right-view at time t-1 relating to at time t-2 and the pixel-wise VGMSD map relating to the left-view at time t-1 can be acquired, the pixel-wise TGMSD value in the pixel-wise TGMSD map is defined as P_i, the pixel-wise VGMSD value in the pixel-wise VGMSD map is defined as Q_i.

2.2   MB-Wise TGMSD and VGMSD of the Right View

Since the pixel-wise GMSD map has been acquired, the value of GMSD based on MBs can be calculated. The MB is divided into four kinds: 16×16, 8×16, 16×8, and 8×8. MB-wise GMSD with different sizes will be computed. MB-wise TGMSD is computed as shown in Eq. (2). Let (x,y) denote the MB position need to be restored, $P_i(x,y)$ denote the pixel-wise TGMSD value of the pixel i in the current MB of the right view frame at time t-1. Thus the standard deviation of pixel-wise TGMSD in the MB is denoted as $P_t(x,y)$:

$$P_t(x,y) = \sqrt{\frac{1}{N}\sum\nolimits_{i-1}^{N}\left(P_i(x,y) - \frac{1}{N}\sum\nolimits_{i=1}^{N}P_i(x,y)\right)^2} \tag{5}$$

where N donates the number of pixels in the current MB with the size of 16×16, 8×16, 16×8 and 8×8, N may be set as 256, 128, 128 or 64, respectively. Similarly, MB-wise VGMSD is computed as shown in Eq. (3). $Q_i(x,y)$ denotes the pixel-wise VGMSD value of the pixel i in the current MB of the right-view in stereo video at time t-1. The MB-wise VGMSD value of the current MB is defined as $Qv(x,y)$, and computed as follows:

$$Q_v(x,y) = \sqrt{\frac{1}{N}\sum\nolimits_{i-1}^{N}\left(Q_i(x,y) - \frac{1}{N}\sum\nolimits_{i=1}^{N}Q_i(x,y)\right)^2} \tag{6}$$

2.3   EC with Different Prediction Modes

According to the proposed algorithm in Fig. 2, the prediction mode of current MB can be determined followed by calculating the $P_t(x,y)$ and $Q_v(x,y)$ of each MB in right-view video image at time t-1. If $P_t(x,y)$ is larger than $Q_v(x,y)$, which means that the structure similarity of MBs in the right view frames between time *t*-1 and *t*-2 is strong, therefore the temporal correlation of the current MB is strong; On the other hand, means that the structure similarity of MB between right-view and left-view video image at time *t*-1 is strong, thus the inter-view correlation of current MB is strong.

As the MB prediction mode has been decided, according to the continuity and temporal correlation of image sequence, MB prediction modes of the previous right-view frame are used as the MB prediction mode of the lost frame. If the temporal correlation of current MB is strong, the lost information is recovered by motion compensation prediction, in which the motion vector from motion vector extrapolation is used for motion compensation to recover the lost information of current MB; Conversely, if the inter-view correlation of current MB is strong, the lost information is recovered by inter-view prediction, in which the disparity vector of the previous right-view frame is used for disparity compensation to recover the lost information of current MB. And so on, until all missing information of the current lost frame has been recovered.

3   Experimental Results and Discussions

To evaluate the efficiency of the proposed algorithm, the H.264/AVC reference software JM8.6 [13] is used for all experiments on a laptop with an Intel Core i3 Duo 3.40 GHz processor and 4 GB RAM. Three test sequences that cover different types and levels of scene content complexity, named as "Puppy", "Rena" and "Akko", are used to evaluate the performance of the proposed method. For each sequence, set one frame of the right-view is lost in each experiment, then restored by different algorithms. The right-views of test stereo videos are shown in Fig. 3.

(a) Rena (640×480)   (b) Puppy (720×480)   (c) Akko (640×480)

Fig. 3. Right views of the test stereo videos.

Table 1. PSNRs of the lost frame of different sequence at the same QP [dB].

Sequence	Decoding	MVE	DCP	Reference [8]	Proposed	$\Delta PSNR$
Puppy	38.05	33.04	27.91	37.08	37.16	0.08
Rena	41.87	28.02	34.83	37.53	37.55	0.04
Akko	39.90	22.76	29.90	32.85	32.87	0.02
Average	–	–	–	–	–	0.04

The proposed algorithm is compared with the motion vector extrapolation method (MVE), disparity compensation prediction method (DCP), and the algorithm of the literature [8] respectively, where objective quality is evaluated by using PSNR and algorithm complexity is weighed by using EC time.

In the case of the same QP (QP = 28), the 98th frame of "Rena", the 44th frame of "Puppy", the 12th frame of "Akko" and the 54th frame of "Door_flowers" are selected as the lost frames. The results of PSNR and decoding time are shown in Tables 1 and 2. Let $\Delta PSNR$ donates PSNR gain of the proposed algorithm compared with the competitor and $-\Delta T$ donates the algorithm complexity reduction compared with the competitor.

Table 2. Average run time (in seconds) of the lost frame of different sequence at the same QP.

Sequence	Reference [8]	Proposed	$-\Delta T(\%)$
Puppy	1.89	1.07	43.40
Rena	2.06	1.31	36.36
Akko	2.01	1.29	35.89
Average	–	–	38.55

Then, the algorithm in this paper compared with the literature [8] algorithm in the case of the same sequence and different QP. The 44th frame of "Rena" is selected as the lost frame, PSNR and EC time are compared respectively with different QP(23, 28, 33, 38), the results are shown in Tables 3 and 4. In order to further confirm the effectiveness of the algorithm, for the "Puppy" sequence, the EC performance of different lost frame under the condition of the same QP (QP = 28), the results are shown in Tables 5 and 6.

Table 3. PSNRs of the different QPs of "Rena" sequence [dB].

QP	Decoding	Reference [8]	Proposed	$\Delta PSNR$
23	40.85	38.89	39.05	0.06
28	38.05	37.08	37.16	0.08
33	35.04	34.56	34.60	0.04
38	31.94	31.71	31.73	0.02
Average	–	–	–	0.05

Table 1 shows the PSNR value of the proposed algorithm is higher than MVE and DCP for each sequence. Because MVE only takes the temporal correlation into account, and DCP only consider the inter-view correlation. The proposed algorithm considers the temporal correlation and the correlation between viewpoints, combined with the subjective perception of human eye to the stereo video, making recovery effect is closer to the human eye perception, an objective effect is very obvious. As can be observed in Tables 1 and 2, compared the algorithm in [8], there is 0.04 dB average PSNR gains with EC time reduced by 38.55 % on average. It is due to highly efficient perceptual quality index GMSD model is the optimization of SSIM model, which

Table 4. Average run time (in seconds) of the different QPs of "Rena" sequence.

QP	Reference [8]	Proposed	-ΔT(%)
23	2.02	1.22	39.91
28	1.89	1.07	43.40
33	1.87	1.03	44.74
38	1.88	1.05	44.13
Average	–	–	43.04

Table 5. The PSNR of the different lost frame of "Puppy" sequence [dB].

Frame	Reference [8]	Proposed	$\Delta PSNR$
8	35.97	35.83	-0.14
40	35.65	35.53	-0.12
78	36.11	36.20	0.09
90	37.12	37.16	0.04
Average	36.25	36.18	-0.03

Table 6. Average run time (in seconds) of the different lost frame of "Puppy" sequence [dB].

Frame	Reference [8]	Proposed	-ΔT (%)
8	2.12	1.40	33.85
40	2.07	1.33	35.55
78	2.07	1.33	35.53
90	2.03	1.31	35.49
Average	2.07	1.34	35.10

ensure the quality of objective conditions, greatly reduce the complexity of the algorithm.

Tables 3 and 4 show the performance in the same stereo video sequences with different QPs, the proposed algorithm is relatively better than the algorithm in [8] with the 0.05 dB average PSNR gains, and EC time decreased by 43.04 %. It illustrates that the algorithm not only suit for different sequences with the same QP, the recovery effect of the same sequence with different QPs is also improved.

Tables 5 and 6 show comparison results between the proposed algorithm and the algorithm in [8], the average PSNR of the proposed algorithm become -0.03 dB, and EC time decreased by 35.10 %. Instructions of the different frames of the same sequence, the proposed algorithm is better than the algorithm in [8].

4   Conclusion

Combined with the human eye subjective perception of video image gradient information, this paper proposed a highly efficient perceptual whole-frame loss of right view error concealment algorithm, in which a highly efficient perceptual quality index,

GMSD, is proposed to evaluate the perceptual quality of stereo image with TGMSD and VGMSD. MB-wise TGMSD and VGMSD are calculated according to the pixel-wise TGMSD map and VGMSD map respectively, thus the prediction mode of the MB in lost frame can be obtained by comparing to TGMSD and VGMSD of the MB. Finally, the lost information of the lost frame can be restored by motion compensation prediction or disparity compensation prediction based on the prediction mode. Experimental results show that the proposed algorithm improves the subjective and objective quality of recuperated right-view stereo video more efficiently.

Acknowledgements. This work is supported by the Natural Science Foundation of China (Grant Nos. 61271270, U1301257, 61311140262), and National Science and Technology Support Program of China (Grant No. 2012BAH67F01).

References

1. Hou, J.H., Chau, L.P., He, Y.: Compressing 3-D human motions via keyframe-based geometry videos. IEEE Trans. Circ. Syst. Video Technol. **25**, 51–62 (2015)
2. Zhou, Y., Xiang, W., Wang, G.K.: Frame loss concealment for multiview video transmission over wireless multimedia sensor networks. IEEE Sens. J. **15**, 1892–1901 (2015)
3. Lie, W.N., Lee, C.M., Yeh, C.H., Gao, Z.W.: Motion vector recovery for video error concealment by using iterative dynamic-programming optimization. IEEE Trans. Multimedia **16**, 216–227 (2014)
4. Tang, G.J., Zhu, X.C.: Error concealment algorithm for stereoscopic images based on boundary smoothness criteria. J. Nanjing Univ. Posts Telecommun. (Nat. Sci.) **33**, 45–49 (2013)
5. Zhu, X., Zhou, L., Song, X.: A JMVC-based error concealment method for stereoscopic video. In: International Conference on Computer Vision Theory and Applications, pp. 201–207 (2013)
6. Tang, G.J., Zhu, X.C., Liu, T.L.: An error concealment algorithm for stereoscopic images based on local reliable disparities. Acta Electronica Sin. **41**, 781–786 (2013)
7. Zhou, Y., Jiang, G.Y., Yu, M.: Region-based error concealment of right-view frames for stereoscopic video transmission. Comput. Electri. Eng. **38**, 217–230 (2012)
8. Li, X.D., Yu, M., Jiang, G.Y., Peng, Z.J., Shao, F.: Error concealment algorithm in stereoscopic video based on structural similarity. Opto-Electron. Eng. **41**, 60–68 (2014)
9. Bilen, C., Aksay, A., Bozdagi, A.G.: Motion and disparity aided stereoscopic full frame loss concealment method. In: Signal Processing and Communications Applications, pp. 1–4 (2007)
10. Chen, Y.B., Cai, C.H., Ma, K.K.: Stereoscopic video error concealment for missing frame recovery using disparity-based frame difference projection. In: IEEE International Conference on Image Processing (ICIP), pp. 4289–4292 (2009)
11. Lin, T.L., Chang, T.E., Huang, G.S., Chou, C.C.: Multiview video error concealment with improved pixel estimation and illumination compensation. In: International Symposium on Intelligent Signal Processing and Communication Systems (ISPACS), pp. 157–162 (2011)
12. Xue, W., Zhang, L., Mou, X., et al.: Gradient magnitude similarity deviation: a highly efficient perceptual image quality index. IEEE Trans. Image Process. **23**, 684–695 (2014)
13. Li, S.P.: Research on network faced stereoscopic video coding and transmission. Chinese Academy of Sciences, Beijing (2006)

Hole Filling Algorithm Using Spatial-Temporal Background Depth Map for View Synthesis in Free View Point Television

Huu Noi Doan$^{(\boxtimes)}$, Beomsu Kim, and Min-Cheol Hong

Soongsil University, Sangdo-Dong, Dongjak-Gu, Seoul, South Korea
{doanhuunoi, rhand41, mhong}@ssu.ac.kr

Abstract. This paper presents a new hole filling algorithm based on background modeling and texture synthesis. Depth information of hole areas is computed by using a local background estimation method. In order to exploit the correlation between frames in temporal domain, a background modeling method is used to extract reliable background reference scenes. The holes are then filled by combining the temporal background information and the estimated depth information. A modified exemplar-based inpainting method is used to fill remaining hole pixels. Experimental results demonstrate the capability of the proposed algorithm.

Keywords: Hole filling · View synthesis · 3D reconstruction · Background modeling

1 Introduction

For recent years, there has been a rapid development in 3D industry, so that it is possible to enjoy lots of 3D movies at theater with additional glasses [1]. Furthermore, many kinds of 3D display are produced with releasing movies in stereoscopic 3D format for home entertainment [2]. In order to improve the quality of depth perception, auto-stereoscopic display technology allowing watching 3D without using any glasses was proposed [3], in which auto-stereoscopic brings stereo effect by display multi-view images having slight difference of the same scene simultaneously. These views can be captured by each real camera, compressed and then transmitted to display directly. However, that is unrealistic because of the large number of viewpoints and big data size. Therefore, the depth image based rendering (DIBR) [4] was presented to generate virtual view from real view at different viewpoints.

The DIBR uses 3D warping to generate a new view, but leading to holes in the synthesized view. Small holes come from wrong depth value. On the other hand, large holes are caused by disocclusion problem which makes uncovered regions in the real view become visible in the virtual view. Due to lack of the depth information or representation texture for these holes, it is difficult to synthesize a high quality texture for these regions, especially large size of disoccluded areas. Therefore, this is still a big challenge in 3D reconstruction research field.

© Springer International Publishing Switzerland 2015
Y.-S. Ho et al. (Eds.): PCM 2015, Part II, LNCS 9315, pp. 598–607, 2015.
DOI: 10.1007/978-3-319-24078-7_61

Several methods have been presented for filling holes in the virtual view. Generally, they can be classified into two categories: depth image based approach and texture based approach. In the former, a low-pass filter was used to preprocess the depth map so that the discontinuity along the boundary between the background (BG) and the foreground (FG) regions was eliminated [5–7]. However, the low-pass filter leaded to geometric distortions, especially for the large baseline. In the latter, the holes were filled by applying averaging filter with depth guide, in which neighboring pixels having low depth were used [8]. A directional hole filling approach was proposed, in which hole pixel value was generated by using an orientation interpolation from known BG neighboring pixels detected by segmentation [9]. Two approaches based on Laplacian Pyramid were proposed to fill holes [10]. Hierarchical Hole Filling (HHF) used a lower resolution estimate of the wrapped image in a pyramid-like structure. The image sequences in the pyramid was generated through a pseudo zero canceling plus Gaussian filtering of the wrapped image. With the small size of hole, interpolation methods have no geometrical distortions and have advantage reduce artifact on the object boundary. However, with the large size of hole, the virtual image might show blurring and smoothing effect. The degradation increases as hole region increase.

Inpainting algorithms have been widely used in hole-filling problems [11, 12]. Inpainting approach can be classified as structure-based and exemplar-based methods. In structure-based methods, structure elements such as edges were diffused into hole areas. In addition, exemplar-based inpainting method has been studied to solve the disocclusion problem. It was reported that the exemplar-based inpainting leaded to promising results gains better performance [13–16]. The exemplar-based inpainting has been extend to 3D video by exploiting temporal correlation. For example, an image registration pattern of consecutive frames was used to reduce hole areas [15]. In addition, temporal stationary scene was extracted from the input video, and then the virtual view and extracted stationary scene are merged together to reduce the hole regions [14]. It successfully obtained the stationary BG pixels in case the activity of the FG object was high but fail to extract BG pixels when the activity of the FG object decreased. A Gaussian mixture model (GMM) algorithm was provided to generate the BG reference image, and then the hole pixels were compensated by this BG information [16]. It was observed that the performance was very sensitive to the initialized distribution.

This paper presents a new hole-filling algorithm based on background modeling and texture synthesis. A new adaptive depth codebook algorithm is defined to extract temporal background information. Local background information is used to obtain spatial background information. The spatial-temporal background information is merged to reduce the hole regions. Finally, an exemplar-based inpainting algorithm based on a new priority function is addressed to update the remaining hole areas.

The rest of this paper is organized as follows. In Sect. 2, a new hole-filling algorithm is presented. A new BG codebook and local BG estimation are described to extract temporal BG information and spatial BG information. In addition, a new priority function for exemplar-based inpainting is provided to fill remaining holes. Experimental results and conclusions are described in Sects. 3 and 4.

2   Proposed Hole Filling Algorithm

The proposed hole-filling method consists of four stages: BG modelling, local BG estimation, frame updating and exemplar-based hole filling, as shown in Fig. 1.

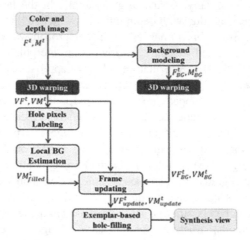

Fig. 1. Block diagram of the proposed hole filling method.

In the following, the input image is denoted as F^t and the corresponding depth map as M^t, in which t is time notation. The BG reference images are referred as F_{BG}^t and M_{BG}^t. After warping, VF^t and VM^t denote virtual images, VF_{BG}^t and VM_{BG}^t denote virtual BG reference images. The filled depth map in local BG estimation is referred as VM_{filled}^t. In addition, VF_{update}^t, VM_{update}^t represent updated virtual images.

2.1   Background Modeling

In this paper, a new background modeling (BM) approach based on [17, 18] is presented. Since almost of hole pixels lie on BG region, except for inner holes in FG region because of the wrong depth map, it is necessary to extract BG information so that hole pixels can be compensated by BG pixels existing in other previous frames.

In order to extract BG information, a new online mode BM based on codebook is proposed. The proposed BM consists of four stages: codebook construction, BG codebook extraction, temporary BG extraction, and stationary BG updating. For each frame F^t and depth M^t at time t, a codebook C^t is constructed. Based on depth value, BG codebook C_{BG}^t storing BG codewords is archived by using k-means algorithm. Temporary BG images, such as TF_{BG}^t, TM_{BG}^t are obtained and used to update stationary BG images F_{BG}^t, M_{BG}^t. Stationary BG images are used as BG reference images of the current frame.

Let X be a training sequence for a single pixel consisting of N RGBD-vectors: $X = \{x_1, x_2, x_3, \ldots, x_N\} = \{(p_1, d_1), \ldots, (p_N, d_N)\}$, where $p_n = \{R_n, G_n, B_n\}$ and d_n

denote a RGB vector and depth value. Let $C = \{c_1, c_2, c_3, \ldots c_L\}$ represent the codebook for the pixel consisting of L codewords. Each codeword consists of a vector $\bar{x}_l = (\bar{p}_l, \bar{d}_l) = (\bar{R}_l, \bar{G}_l, \bar{B}_l, \bar{d}_l)$ and a 5-tuple $aux_l = \left(I_{l,min}, I_{l,max}, D_{l,min}, D_{l,max}, f_l\right)$, in which $\bar{R}_l, \bar{G}_l, \bar{B}_l$ and $\bar{d}_l$ represent average value of each color channels and depth value. In addition, $I_{l,min}, I_{l,max}, D_{l,min}, D_{l,max}$ and f_l denote the minimum brightness, the maximum brightness, the minimum depth value, the maximum depth value and the frequency of the codeword, respectively. Then, the codebook of each pixel is constructed in similar way to conventional approach [17].

After constructing the codebook for the current frame, a list of codewords for each pixel is archived. The depth adaptive codebook using color and depth map is robust to additive noise [18]. However, due to fixed threshold, BG pixels appearing in a short-time were not collected. In order to solve the problem, k-means cluster algorithm is adopted to classify BG codewords containing low depth value with k = 2 in this work.

$$C_{BG}^t = \{c_m \in BG | 1 \leq m \leq M\}. \tag{1}$$

When the list of codewords for a pixel has only 1 codeword, this codeword is considered as BG codeword without clustering.

Temporary BG information can be extracted from current frame by matching each pixel with its corresponding BG codewords. For $x_t = (p, d) = (R, G, B, d)$ and $C_{BG}^t = \{c_m | 1 \leq m \leq M\}$, the following functions: $T_1 = colordist(p, \bar{p}_m), T_2 = brightness(I, aux_m)$ and $T_3 = disparity(d, aux_m)$, the condition $(T_1 \leq \in_1 \vee (\in_1 < T_1 \leq \in_2 \wedge T_3)) \wedge T_2 \wedge T_3$ are examined to find the best match BG pixels.

$$TF_{BG}^t(i,j) = \begin{cases} p, & matching \\ 0, & otherwise \end{cases}, TM_{BG}^t(i,j) = \begin{cases} d, & matching \\ 0, & otherwise \end{cases}, \tag{2}$$

where three functions T1, T2 and T3 were defined in [17, 18].

The temporary BG information is used for updating a stationary BG information scene F_{BG}^t and its depth map M_{BG}^t in similar way:

$$F_{BG}^t(i,j) = \begin{cases} TF_{BG}^t(i,j), & TF_{BG}^t(i,j) \neq 0 \\ F_{BG}^{t-1}(i,j), & TF_{BG}^t(i,j) = 0 \wedge disparity\left(M_{BG}^{t-1}(i,j), aux_m\right) = true, \\ 0, & otherwise \end{cases} \tag{3}$$

where $c_m = (\bar{x}_m, aux_m) \in C_{BG}^t$. Performance comparisons with BG extraction for the 99th frame of "Car Park" and "Street" video sequences are shown in Fig. 2 (video comparisons are available at https://www.youtube.com/watch?v=sLpGbrCI_c4). The results show that the BG information is well extracted with the proposed BM.

2.2   Hole Pixels Labeling and Local BG Estimation

There are still many disoccluded regions with a variety of sizes. In order to be adaptive the size of patch as well as the filling ordering, each hole pixel is classified by using

    (a)   (b)   (c)   (d)

Fig. 2. Visual comparisons of the 99th frame of "Car Park" and "Street" sequences: (a) results with the BG stationary extraction [14], (b) results with BG reference image extraction using GMM [16], (c) results with BG Modeling using Codebook [18], (d) results with the proposed BM method.

flood-fill algorithm [19]. A local BG estimation algorithm is defined to obtain approximated BG threshold for regions containing hole pixels, so that BG and FG pixels of virtual image could be discriminated to prevent from distortion phenomenon. For doing this, each rectangle bounding each hole regions is divided to many small rectangles by horizontal direction. Then, k-means cluster algorithm with k = 2 is applied to segment each small rectangle, so that two clusters are obtained. The highest value in lower depth cluster can be considered as local BG threshold for the region and therefore, the hole pixels in this area are replaced by the local BG threshold. Based on the filled virtual depth map, a local BG threshold for arbitrary rectangle Ψ_d containing hole pixels is defined as

$$\text{theshold}(\Psi_d) = \max(VM^t_{filled}(m, n) | VM^t_{filled}(m, n) \in \Psi_d \wedge VM^t(m, n) = 0). \quad (4)$$

The highest depth value of region filled by local BG estimation is used as the BG threshold for this region, thus pixels have depth value higher than this BG threshold are regarded as FG pixels. In addition, this process is used to solve the depth missing part problem, so that depth value of missing area is estimated by using neighbor information.

2.3 Frame Updating and New Exemplar-Based Inpainting

For a hole pixel in the virtual image VF^t, VM^t, but non-hole pixel in VF^t_{BG}, VM^t_{BG}. Updating equation with a depth constrain is defined as

$$VF^t_{update}(i,j) = \begin{cases} VF^t_{BG}(i,j), & VM^t_{filled}(p) - T \leq VM^t_{BG}(i,j) \leq VM^t_{filled}(i,j) + T \\ VF^t(i,j), & otherwise \end{cases}, \quad (5)$$

where the threshold T (T > 0) used to control depth matching. The missing number of updates to a BG pixel of VF^t increases as T decreases. On the other hand, the number of incorrectly updated pixels of VF^t increases as T increases. The corresponding depth map is updated in the similar way.

In order to fill remaining holes after updating step, a modified exemplar-based inpainting approach based on [13] is utilized. First, priority for all non-hole pixels lying on disoccluded area's boundary is calculated, and then similarity patches centered at the highest priority pixel are searched in known neighboring regions using energy function. For a given rectangle Ψ_p centered at pixel p for some $p \in \delta\Omega$ (see Fig. 3), the priority is defined as follows:

$$P(p) = C(p)D(p)Z(d), \qquad (6)$$

where C(p) and D (p) were defined in [13], Z(d) is a new term representing the depth term, and it is defined as

$$Z(d) = \begin{cases} 1, & d \leq threshold(\Psi_d), d \in \Psi_d \subset VM_{filled}^t \\ 0, & otherwise \end{cases}, \qquad (7)$$

where the size of Ψ_d is same to that of Ψ_p. Then, a BG threshold of rectangle Ψ_p is obtained by using Eq. (5). Equation (7) means that the proposed algorithm starts at a patch belonging to a BG region in order to prevent from geometric distortion.

According to the priority, similarity patch is searched by minimizing the energy Eq. (8). Hole pixels are filled by coping non-hole pixels from similarity patch to corresponding hole pixels in Ψ_p.

$$E = \sum_{i=1}^{K} x_i - c_i^2 + \omega \sum_{j=1}^{K_\Omega} x_j - c_j^2 \qquad (8)$$

where K is the number of original and K_Ω is the number of initialized region by using median filter [15], ω is the weighting factor for the initialized values. The filling process is repeated until there is no any hole in the virtual image. In order to reduce searching cost, all pixels in search space with depth value higher than local BG threshold are excluded and size of patch depends on size of each hole region.

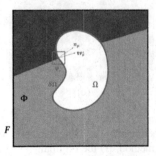

Fig. 3. Example of priority determination.

3   Experimental Results

Several experiments were conducted with various video sequences. In the set of experiments, "Ballet", "Break Dancer", "Book Arrival", and "Car Park", "Street" are described here. The proposed algorithm is compared to the depth HHF method [10], the Ming's method [14], the Koppel's method [15], and the Yao's method [16]. The peak signal-to-noise ratio (PSNR) and the structural similarity (SSIM) were used here to evaluate the performance of the algorithms.

Figure 4 shows the visual comparison (video comparisons are available at https://www.youtube.com/watch?v=1aza1DkQ7B8). It shows that the proposed method leaded to more promising results than the other methods. In particular, the enlarged version in Fig. 5 shows that the proposed method has the capability to generate more natural texture than the other competitive methods. It was observed that the depth HHF leaded to seriously blurring at boundaries between BG and FG region due to Gaussian filtering of down-sampling and up-sampling in building hierarchical structure. Also, it was observed that the Ming's method leaded to unusual textures, and the Koppel's method resulted in the distortion due to inaccurate discrimination between BG and FG using global BG threshold. In addition, the Yao's method resulted in annoying artifacts which comes from using wrong updating function. On the other hand, it was verified that the proposed method leaded to promising visual results in the boundaries between FG regions and BG regions.

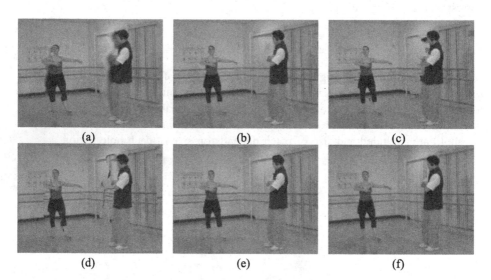

Fig. 4. Visual comparison of the 46th frame of "Ballet" with synthesizing from camera 3th to camera 4th; (a) the Depth HHF method, (b) the Ming's method, (c) the Koppel's method, (d) the Yao's method, (e) the proposed method, (f) the original.

The average PSNR comparisons of each video sequences are shown in Table 1. The results show that the DHHF method outperforms the other methods for "Car Park" and

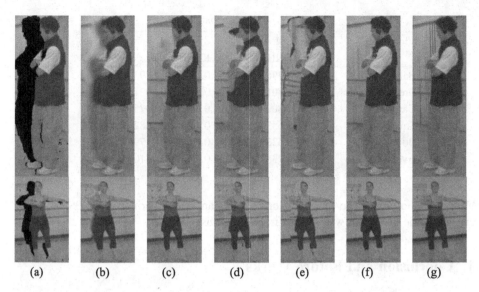

(a)   (b)   (c)   (d)   (e)   (f)   (g)

Fig. 5. Visual comparisons of zoomed and cut image of Fig. 4; (a) warped image, (b) the Depth HHF method, (c) the Ming's method, (d) the Koppel's method, (e) the Yao's method, (f) the proposed method, (g) the original.

"Book Arrival", and that the Ming's method leads to better performance than the others for "Ballet" and "Street". On the other hand, the proposed algorithm dominates the other competitive methods for "Break Dancers". It was observed that the proposed method and the Ming's leaded to more promising results as the disocclusion was more serious, and that the DHHF method resulted in better performance as it was less serious. However, the DHHF method resulted in overly blurring around boundaries between BG region and FG region. Also, the performance of the Ming's method was sensitive to the type of the disocclusion. It was verified that the proposed algorithm consistently guaranteed the promising results regardless of the type of the disocclusion. The SSIM comparisons are shown in Table 2. The results show that the proposed method and the DHHF method outperform the others. However, the DHHF method leaded to over-blurring in the boundary areas, as mentioned, since only FG information was taken into account of the synthesized texture. In addition, it was observed that the

Table 1. PSNR comparisons.

Video Seq.	Camera	PSNR				
		DHHF	Ming	Koppel	Yao	Proposed
Ballet	3 → 4	25.9432	**27.9291**	27.7211	26.0579	27.8591
Break Dancers	4 → 5	30.8381	30.5841	30.7821	29.9678	**30.9964**
Car Park	4 → 5	**27.4180**	26.1671	25.8857	26.4738	26.7490
Street	4 → 5	23.7572	**24.1040**	23.8624	23.8375	23.9836
Book Arrival	4 → 5	**24.9328**	23.5385	23.9720	24.7268	24.7964
Average		26.5779	26.4646	26.4447	26.2128	**26.8769**

Table 2. SSIM comparisons.

Video Seq.	Camera	SSIM				
		DHHF	Ming	Koppel	Yao	Proposed
Ballet	3 → 4	**0.7966**	0.7945	0.7920	0.7793	0.7959
Break Dancers	4 → 5	**0.8011**	0.7948	0.7941	0.7921	0.7954
Car Park	4 → 5	0.8020	0.8039	0.8018	0.7975	**0.8066**
Street	4 → 5	0.6530	0.6784	0.6749	0.6682	**0.6797**
Book Arrival	4 → 5	**0.6257**	0.6209	0.6220	0.6179	0.6226
Average		0.7357	0.7385	0.7370	0.7310	**0.7400**

Ming's method leaded to unsatisfactory performance, comparing to the PSNR, since the geometric distortion was serious in BG regions.

4 Conclusion and Future Works

In this paper, a new hole filling method is presented. A new BG codebook and local BG estimation are provided to extract temporal BG information and spatial BG information, respectively. Based on them, color image and depth map are updated. In addition, a hole-filling priority function for exemplar-based inpainting is provided to fill remaining holes. The experimental results demonstrate the capability of the proposed method. A new BG modeling algorithm incorporating camera tracking into view synthesis processing is under investigation. With such approach, it is expected that a more sophisticated formulation can be derived and better performance can be achieved.

Acknowledgment. This research was supported by Next-Generation Information Computing Development Program through the National Research Foundation of Korea (NRF) funded by the Ministry of Science, ICT & Future Planning (No. 2012M3C4A7032182).

References

1. Smolic, A., Kauff, P., Knorr, S., Hornung, A., Kunter, M., Muller, M., Lang, M.: Three-dimensional video postproduction and processing. Proc. IEEE **99**(4), 607–625 (2011)
2. Muller, K., Merkle, P., Tech, G., Wiegand, T.: 3D video formats and coding methods. In: IEEE International ConferenceImage Processing, pp. 2339–2392 (2010)
3. Dodgson, N.A.: Autostereoscopic 3D displays. Computer **38**(8), 31–36 (2005)
4. Fehn, C.: Depth-image-based rendering (DIBR), compression, and transmission for a new approach on 3DTV. In: SPIE Proceedings of the Stereoscopic Image Process and Rendering, pp. 93–104 (2004)
5. Zinger, S., Ruijters, D., Do, L., de With, P.H.N.: View interpolation for medical images on autostereoscopic displays. IEEE Trans. Circuits Syst. Video Technol. **22**(1), 128–137 (2012)
6. Zhang, L., Tam, W.J.: Stereoscopic image generation based on depth images for 3D TV. IEEE Trans. Broadcast. **51**(2), 191–199 (2005)

7. Lee, P.-J.: Effendi: nongeometric distortion smoothing approach for depth map preprocessing. IEEE Trans. Multimedia **13**(2), 246–254 (2011)
8. Zinger, S., Do, L., de With, P.H.N.: Free-viewpoint depth image based rendering. J. Vis. Commun. Image Represent. **21**(5–6), 533–541 (2010)
9. Doan, H.N., Nguyen, T.A., Kim, B., Hong, M.-C.: Directional hole filling algorithm in new view synthesis for 3D video using local segmentation. In: ACM International Conference Research in Adaptive and Convergent Systems, pp. 100–104, MD, USA (2014)
10. Solh, M., Alregib, G.: Hierarchical hole-filling for depth-based view synthesis in FTV and 3D video. IEEE. J. Sel. Top. Signal **6**(5), 495–504 (2012)
11. Oh, K., Yea, S., Ho, Y.-S.: Hole filling method using depth based in-painting for view synthesis in free viewpoint television and 3-D video. In: Picture Coding Symposium, pp. 1–4 (2009)
12. Marcelo, B., Guillermo, S., Vincent, C., Coloma, B.: Image inpainting. In: The 27th Annual Conference on Computer Graphics and Interactive Techniques, pp. 417–424 (2000)
13. Criminisi, A., Perez, P., Toyama, K.: Object removal by exemplar-based inpainting. In: IEEE International Conference Computer Vision and Pattern Recognition, vol. 2, pp. 721–728 (2003)
14. Ming, X., Wang, L.-H., Yang, Q.-Q., Li, D.-X., Zhang, M.: Depth-image-based rendering with spatial and temporal texture synthesis for 3DTV. EURASIP J. Image Video Process. **2013**(1), 1–18 (2013)
15. Koppel, M., Wang, W., Doshkov, D., Wiegand, T., Ndjiki-Nya, P.: Consistent spatio-temporal filling of disocclusions in the multiview-video-plus-depth format. In: IEEE International Workshop Multimedia Signal Processing, pp. 25–30 (2012)
16. Yao, C., Tillo, T., Jimin, X.: Depth map driven hole filling algorithm exploiting temporal correlation information. IEEE Trans. Broadcast. **60**(2), 394–404 (2014)
17. Kim, K., Chalidabhongse, T.H., Harwood, D., Davis, L.: Background modeling and subtraction by codebook construction. IEEE Int. Conf. Image Process. **5**, 3061–3064 (2004)
18. Fernandez-Sanchez, E.J., Javier, D., Eduardo, R.: Background subtraction based on color and depth using active sensors. Sensors **13**(7), 8895–8915 (2013)
19. Glassner, A.S. (ed.): Graphics Gems I. Academic Press, Cambridge (1990)

Pattern Feature Detection for Camera Calibration Using Circular Sample

Dong-Won Shin and Yo-Sung Ho[⊠]

Gwangju Institute of Science and Technology (GIST),
123 Cheomdan-gwagiro, Buk-gu, Gwangju 500-712, Korea
{dongwonshin,hoyo}@gist.ac.kr

Abstract. Camera calibration is a process to find camera parameters. Camera parameter consists of intrinsic and extrinsic configuration and it is important to deal with the three-dimensional (3-D) geometry of the cameras and 3-D scene. However, camera calibration is quite annoying process when the number of cameras and images increase because it is operated by hand to indicate exact points. In order to eliminate the inconvenience of a manual manipulation, we propose a new pattern feature detection algorithm. The proposed method employs the Harris corner detector to find the candidate for the pattern feature points in images. Among them, we extract valid pattern feature points by using a circular sample. Test results show that this algorithm can provide reasonable camera parameters compared to camera parameters using the Matlab calibration toolbox by hand but eliminated a burden of manual operation.

Keywords: Camera calibration · Harris corner · Feature detection

1 Introduction

The real world consists of a three-dimensional (3-D) space. However, when you capture 3-D space by a camera, it would be represented on a two-dimensional (2-D) plane. If you want to know where a point in 3-D space is placed on a 2-D plane, you should know a geometric relationship between them: a rotation and position of the camera (extrinsic parameters), a relationship between a lens and an image sensor in a camera (intrinsic parameters). In addition to the fact that we can know the relation between 2-D plane and 3-D space, a geometric relation among multi-view cameras also can be found. We call this process as the camera calibration, which find intrinsic and extrinsic parameters [1].

The camera calibration step is necessary when you deal with a geometric meaning among multiple cameras such as 3-D warping or image rectification. 3-D warping is a process to forward depth information from one view image to another view image by using camera parameters [2]. Figure 1 shows the 3-D warping process. We can forward a point p_l located in the left image to a point p_r located in the right image via a point p_w placed in the world coordinate system. It means that we can know where the point p_l is located in right image as the point p_r and vice versa.

Next, the image rectification is to align epipolar lines of the images as parallel each other [3]. It is usually performed as a pre-processing step of stereo matching. In this

© Springer International Publishing Switzerland 2015
Y.-S. Ho et al. (Eds.): PCM 2015, Part II, LNCS 9315, pp. 608–615, 2015.
DOI: 10.1007/978-3-319-24078-7_62

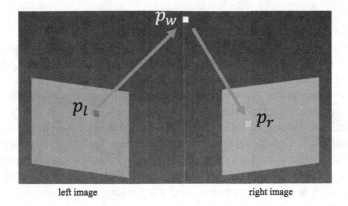

left image   right image

Fig. 1. 3-D warping

ideal camera setup, all cameras have the same intrinsic and extrinsic parameters except for the translation vector. In case of the translation vector, it is set to be had the same distance to the adjacent camera along the horizontal direction. Figure 2 shows the image rectification process.

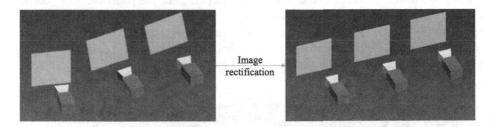

Image rectification

Fig. 2. Image rectification

Although the camera calibration is a necessary process in multi-view geometry, it is very time-consuming since a human needs to mark the points in every image by their hand. For example, in case of the camera calibration using a 4 × 6 planar grid pattern, we need to select four corner points of a pattern rectangle in a single image. If we have 5 cameras and capture 10 images at each camera, we need to choose 5 × 10 × 4 = 200 points by our hands. It is very inefficient and would be a big burden when the number of cameras and images increases. In order to relieve this inconvenience, we studied on an automatic pattern feature detection for camera calibration.

Recently, many researches in this area have presented lots of methods. Arturo *et al.* employed Hough transform to detect lines in a pattern image, and then choosing the intersections between horizontal and vertical lines as a pattern feature point [4]. J. Chu *et al.* proposed a chessboard corner detector based on image physical coordinates with subpixel precision, which can obtain the subpixel in last one step and decrease the computation complexity [5]. Lastly, M. Fiala *et al.* designed a discriminable marker-based calibration system. This calibration method employs an array of fiducial marker to find the correspondences among images [6].

In order to improve the efficiency of the camera calibration without a manual manipulation, we propose a circular sample with Harris corner detector. Our proposed algorithm extracts corners in an image by using Harris corner detector. Among the candidates, we obtain true pattern feature points by using a circular sample which inspect pixels around extracted Harris corner points.

The paper is organized as follows. In Sect. 2, we describe a corner detection. In Sect. 3, we explain a circular sample to find true pattern feature points. In Sect. 4, we show test results of the system running under various conditions and comparisons with the widely available Caltech Matlab camera calibration toolbox in terms of accuracy. Finally, in Sect. 5, we draw conclusions and offer discussions.

2   Corner Detection

A corner is usually defined as the intersection of two edges. It has a well-defined position and can be robustly detected. On the basis of the concept of the corner, we define a pattern feature point in this research. The pattern feature point is a point that we need to find in every image to perform the camera calibration. It is located in an interior of a planar calibration pattern and the corners include the pattern feature points. Figure 3 shows the example of corners and the pattern feature points. In Fig. 3, we represent the corners as red points and pattern feature points as blue points. As we can see in this figure, corners can be extracted not only in the planar grid calibration panel but also in the persons standing on the both sides of the panel.

Fig. 3. Corners and pattern feature points

In the proposed method, first we extract the corners by using Harris corner detector [7]. Figure 4 shows the principle of Harris corner detector.

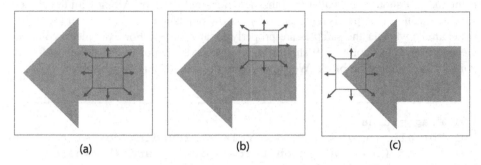

(a)   (b)   (c)

Fig. 4. Principle of Harris corner detector

When we move the window to find a corner, the corner should have large changes in any direction. Figure 4(a) shows the movement in a homogeneous region. In this case, the window does not have a change in all directions. However, in case of Fig. 4(b), the window have big changes in a horizontal direction but not a vertical direction. Finally, in case of Fig. 4(c), the window have a large amount of changes in any direction, therefore we estimate it as a corner. You can see the details of Harris corner detector in [7].

3   Circular Samples

In general, there are many corners in an image after performing Harris corner detector like Fig. 3. However, we are only interested in pattern feature points rather than all corners. Hence, we need to extract pattern feature points among all candidates. In this paper, we propose a circular sample to achieve the purpose. Figure 5 shows the diagram of the circular sample.

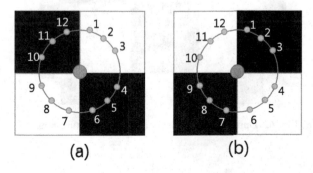

Fig. 5. Diagram of the circular sample

There are two cases of a pattern feature point which are shown in Figs. 5(a) and (b). Let's assume that the blue points in Fig. 5 are corners found by Harris corner detector. Then, we can define a circle of which center is the position of the blue point with radius r; the radius r should be smaller than the size of a single pattern. Along with the circle, we can obtain sample pixels represented by yellow points in Fig. 5. In case of Fig. 5(a), we examine whether the samples are properly located or not. For example, a color of the points 1–3 and points 7–9 should be white. Next, a color of the points 4–6 and points 10–12 should be black. We show a pseudo code of this step below.

```
01 Flag = true
02 For all samples s on a circle surrounding corner c
03      If (s is in region 1 or region 3) and (I_s ≠white)
04           Flag = false
05      If (s is in region 2 or region 4) and (I_s ≠black)
06           Flag = false
07
08 If flag = true
09      Corner c is a pattern feature point
```

Here, s means a sample point on a circle surrounding the corner and I_s is an intensity value of the sample point. In case of Fig. 5(b), we can apply this algorithm inversely.

However, right after performing this step, there are many duplicate candidates indicating a same pattern feature point, which shown in Fig. 6.

Fig. 6. Duplicated candidates for pattern feature points

In order to suppress duplicated candidates, we employ Euclidean distance among them. If a point is placed within a distance d from the other points, that point will be rejected. Otherwise, this point will be accepted.

4   Test Results

The proposed detecting method has been tested on calibration pattern sequences from five cameras placed in parallel. We captured 10 images for each camera. Figure 7 shows detected points by the proposed method.

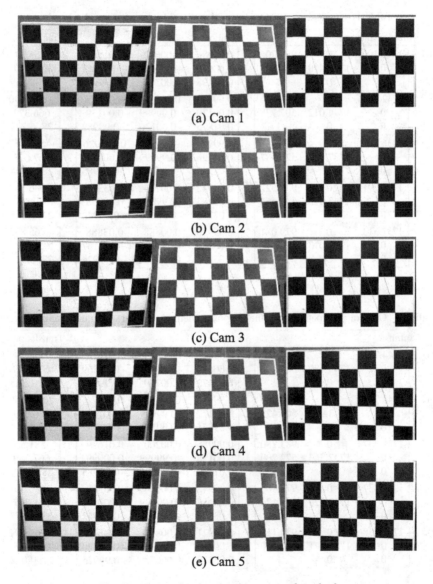

Fig. 7. Detected points by the proposed method

Here, we displayed three images for each camera and represented points as colored point and lines using cvDrawChessboardCorners() function in Opencv framework. We can find exact results of well detected points.

Next, we compared the camera parameters obtaining from Matlab calibration toolbox by hand and proposed method without a manual manipulation in Tables 1 and 2 respectively.

Finally, we show the difference between each matrix by using Euclidean norm in Table 3. we can see that the differences between camera parameters from Matlab camera calibration toolbox and the proposed method are sufficiently small to practically use.

Table 1. Camera parameters from Matlab calibration toolbox by hand

	Intrinsic matrix			Rotation			Translation
Cam1	1745.490	0	983.706	0.99986	0.00642	0.01500	−455.235
	0	1746.419	514.462	0.00646	−0.99997	−0.00267	679.347
	0	0	1	0.01498	0.00277	−0.99988	2805.469
Cam2	1742.720	0	988.246	0.99982	0.00182	0.01875	−523.490
	0	1744.012	507.778	0.00179	−0.99999	0.00139	666.373
	0	0	1	0.01875	−0.00136	−0.99982	2806.729
Cam3	1736.700	0	983.227	0.99985	0.00197	0.016926	−581.463
	0	1737.410	521.987	0.00196	−0.99999	0.00087	667.623
	0	0	1	0.01692	−0.00084	−0.99985	2801.644
Cam4	1742.291	0	977.522	0.99942	0.00184	0.03392	−686.548
	0	1743.313	511.662	0.00183	−0.99999	0.00038	670.412
	0	0	1	0.03392	−0.00031	−0.99942	2791.815
Cam5	1737.313	0	979.081	0.99920	0.00888	0.03888	−765.842
	0	1737.073	527.819	0.00889	−0.99996	−4.63137	665.155
	0	0	1	0.03888	0.00039	−0.99924	2788.822

Table 2. Camera parameters from the proposed method without a manual manipulation

	Intrinsic			Rotation			Translation
Cam1	1744.770	0	983.243	0.99987	0.00643	0.01474	−454.499
	0	1745.689	514.496	0.00647	−0.99997	−0.00269	679.270
	0	0	1	0.01472	0.00279	−0.99988	2804.305
Cam2	1742.702	0	987.739	0.99982	0.00180	0.01871	−522.660
	0	1743.945	507.428	0.00177	−0.99999	0.00124	666.949
	0	0	1	0.01871	−0.00121	−0.99982	2806.464
Cam3	1736.837	0	983.280	0.99985	0.00200	0.01699	−581.558
	0	1737.563	522.200	0.00198	−0.99999	0.00099	667.259
	0	0	1	0.01700	−0.00095	−0.99985	2801.840
Cam4	1742.837	0	977.400	0.99942	0.00184	0.03387	−686.338
	0	1743.926	511.124	0.00184	−0.99999	0.00013	671.267
	0	0	1	0.03387	−0.00007	−0.99942	2792.477
Cam5	1737.958	0	979.117	0.99921	0.00887	0.03872	−765.891
	0	1737.671	528.260	0.00887	−0.99996	0.00011	664.448
	0	0	1	0.03872	0.00022	−0.99925	2790.042

Table 3. Differences between camera parameters

	Cam1	Cam2	Cam3	Cam4	Cam5
Intrinsic matrix	0.8570	0.6170	0.2672	0.8230	0.7444
Rotation	2.6397e-04	1.5813e-04	1.3518e-04	2.4823e-04	2.2934e-04
Translation	1.3794	1.0443	0.4244	1.1014	1.4113

5 Conclusions

In this paper, we have proposed a new detection method to extract pattern feature points. From our experiments, we could obtain camera parameters with an ignorable difference with the result of Matlab calibration toolbox but eliminated the inconvenience of a manual manipulation. This proposed method is efficient when the number of camera and image increases especially in the multiview camera system. Although it can suffer from various conditions such as strong or weak illumination and a partial occlusion, it can be available under reasonable constraints. We expect that we can offer advantages of efficiency in multiview image processing and applications.

Acknowledgements. This research was supported by the 'Cross-Ministry Giga KOREA Project' of the Ministry of Science, ICT and Future Planning, Republic of Korea(ROK). [GK15C0100, Development of Interactive and Realistic Massive Giga-Content Technology]

References

1. Zhang, Z.: A flexible new technique for camera calibration. IEEE Trans. Pattern Anal. Mach. Intell. **22**, 1330–1334 (2000)
2. Mark, W., McMillan, L., Bishop, G.: Post-rendering 3D warping. In: Symposium on Interactive 3D Graphics, pp. 7–16 (1997)
3. Kang, Y., Ho, Y.: Geometrical compensation for multi-view video in multiple camera array. In: International Symposium ELMAR-2008 (2008)
4. Escalera, A., Armingol, J.: Automatic chessboard detection for intrinsic and extrinsic camera parameter calibration. Sensors 2010 **10**, 2027–2044 (2010)
5. Chu, J., GuoLu, A., Wang, L.: Chessboard corner detection under image physical coordinate. Opt. Laser Technol. **48**, 599–605 (2013)
6. Fiala, M., Shu, C.: Self-identifying patterns for plane-based camera calibration. MVA **19**(4), 209–216 (2008)
7. Harris, C., Stephens, M.: A combined corner and edge detector. In: Alvey Vision Conference, pp. 147–151 (1988)

Temporal Consistency Enhancement
for Digital Holographic Video

Kwan-Jung Oh$^{(\boxtimes)}$, Hyon-Gon Choo, and Jinwoong Kim

Electronics and Telecommunications Research Institute, Daejeon, Korea
kjoh8163@gmail.com

Abstract. Holography is an imaging technique to reconstruct wavefront information of the light scattered by real objects or a scene, allowing an observer to perceive three-dimensional (3D) images with the unassisted eye. Such 3D holographic images result from reproducing the intensity and phase of light by diffraction. This paper presents a temporal consistency enhancement for digital holographic video. The proposed temporal consistency enhancement method improves compression efficiency and visual quality by reducing the flickering artifact.

Keywords: Holographic video · Temporal consistency enhancement · CGH

1 Introduction

Holography is a technique for recording and reconstructing the amplitude and phase distributions of a three-dimensional (3D) object based on wave interference. The idea of holography was first suggested by Dennis Gabor (1948) [1], but it did not become widespread since its method was technically extremely complicated and adequate coherent source of light was unavailable. In the early of 1960s, the appearance of lasers have opened up the possibilities for the practical use of holography in radio electronics, optics, physics, and various fields of technology. The laser satisfies the properties of coherency, since it has a constant phase and a single wavelength. Holography can be better understood by comparing to photography. In photography, only intensity of light is recorded whereas both intensity and phase of light are recorded in holography. Thus, photography produces two dimensional (2D) picture whereas holography gives three dimensional (3D) picture [2].

The digital hologram is obtained by recording the interference pattern via a charge-coupled device (CCD) camera directly or by calculating the interference pattern with computer generated hologram (CGH) method. A digital hologram is represented as an image and it is an aggregation of several zone plate generated from the each object light. The hologram can be reconstructed by using spatial light modulator (SLM) which is a special device can spatially modulate the amplitude, phase or polarization of an optical wave front. To realize the 3D imaging, a huge amount of hologram data are needed but the hologram pattern has low spatial and temporal redundancies. As a result, the direct compression of hologram data with a high compression ratio is a quite challenging work [3].

In order to compress holograms, several image compression based techniques have been proposed. Yoshikawa et al. tried to reduce the hologram data based on analysis of

© Springer International Publishing Switzerland 2015
Y.-S. Ho et al. (Eds.): PCM 2015, Part II, LNCS 9315, pp. 616–622, 2015.
DOI: 10.1007/978-3-319-24078-7_63

fringe pattern [4]. They divided a fringe pattern into multiple segments and then discrete cosine transform (DCT) is applied to each segment. The transformed coefficients were encoded with moving picture compression technique. Seo et al. has upgraded this work with several 3D scanning techniques for segments [5]. Naughton et al. proposed a lossless hologram compression scheme based on conventional entropy coding such as LZ77, LZW, and Huffman coding [6]. Shortt et al. have tried to use non-uniform quantization compression technique for digital hologram [7] and Kayser et al. also proposed a compression scheme based on logarithmic quantization [8]. Ding et al. proposed a wavelet based hologram compression technique [9]. Liebling et al. proposed a Fresnelet transform which is a new basis function for multi-resolution wavelet bases for digital holography [10]. The above mentioned researches on hologram compression are mostly concentrated in holographic image compression not holographic video. However, compression on digital holographic video is important to realize a holographic video system.

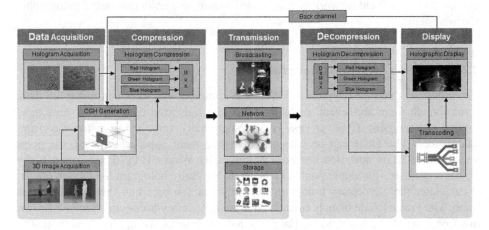

Fig. 1. Holographic video system.

To compress the certain data efficiency, the redundancy of the raw data is important as much as compression algorithms. In other words, the hologram data whether it is obtained directly or indirectly should have enough redundancy. In this paper, we propose a temporal consistency enhancement method for digital holographic video. The proposed method increases the temporal correlation of holographic video by unifying the conditions of CGH for successive frames. The efficiency of the proposed method is verified by comparing the temporal correlation and compression ratio.

2   Digital Holographic Video

2.1   Holographic Video System

The holographic video system can be designed 3D video based or hologram based approaches. We only deal the hologram based holographic video system in this paper.

The hologram based digital holographic video system is shown in Fig. 1. In this system, hologram is directly obtained or generated by CGH at the sender side. Thus, more compact receiver design is possible. The hologram data have a high dependency on holographic display parameters such as wavelength and pixel pitch. Thus, hologram based holographic video system needs a back channel to send above parameters to sender side. However, the hologram made by this way is only reconstructed perfectly by a holographic display which has same holographic display parameters. Depending on the service type, a single hologram should be compatible with several holographic displays. To solve this problem, transcoding on hologram itself have been studied. The transcoding means generation of a new hologram with a different wavelength or pixel pitch. Unfortunately, capability of transcoding is not enough to cover all holographic displays since it has a limitation to change the wavelength or pixel pitch. Thus, standardization for ranges of available wavelength and pixel pitch might be needed. In addition, compression of hologram data itself is quite challenging since hologram image has low data redundancy. Thus, high efficiency hologram coding tools are needed and broadband network is also needed to realize a hologram based holographic video system.

2.2   Computer Generated Hologram

The computer generated approach is the most common way to obtain hologram data. Even though there are several methods to acquire hologram data directly, they need limited environment. CGH process is computationally very intensive for wavefront calculations. The CGH can be categorized in two main approaches. First, the point source-based CGH approach has been proposed by Waters [11]. It assumes that the object is consists of multiple points. To generate the entire hologram for a given object, it calculates the elementary hologram for every point source and then superimposes them. The Fresnel zone plate is considered as an elementary hologram. Second, Brown and Lohmann [12] introduced the Fourier transform-based CGH approach in 1966. The Fourier transform is employed to simulate the wavefront propagation of each object plane in depth to the hologram plane. In other words, the hologram is obtained by super-positioning of the Fourier transforms of each object plane in depth [13]. In this paper, we use an amplitude hologram generated by Fourier transform hologram approach.

The result of CGH is optical field at the hologram plane consisting of real and imaginary complex values. The optical field is calculated based on Fresnel diffraction equation [14] by

$$U(x,y) = \frac{e^{i\frac{2\pi z_j}{\lambda}}}{i\lambda z_j} \sum_{j}^{N} A_j exp\left(-i\frac{\pi}{\lambda z_j} \left((x - x_j)^2 + (y - y_j)^2 \right) \right) \qquad (1)$$

where N and A_j are the total number of object points and the intensities of the object points, and λ is the wavelength of reference light. (x,y) and (x_j, y_j, z_j) are the coordinates on the CGH and the three dimensional objects, respectively.

2.3   Hologram Reconstruction

The hologram can be reconstructed both optically and numerically. The optical reconstruction uses a spatial light modulator (SLM) which is a special device can spatially modulate the amplitude, phase or polarization of an optical wavefront. The numerical reconstruction is a kind of simulation using a computer. Numerically reconstructed holographic image at distance d from SLM can be obtained by squaring the amplitude of complex field at reconstruction plane expressed as (2)

$$R(x_r, y_r) = \frac{e^{i\frac{2\pi d}{\lambda}}}{i\lambda d} \sum_{h}^{N} (cos(U_h) + 1)exp\left(-i\frac{\pi}{\lambda d}\left((x_r - x_h)^2 + (y_r - y_h)^2\right)\right) \qquad (2)$$

where N and U_h are the total number of hologram points and the values of hologram, respectively. (x_h, y_h) and (x_r, y_r) are the coordinates on the CGH and the reconstructed plane. The sampling interval at reconstruction plane was determined as pixel pitch of display panel of floating display system. However, the reconstructed images only present the intensity at the plane but not the propagation direction of light.

3   Proposed Temporal Consistency Enhancement

The raw data of hologram is a set of complex values. This complex field data are represented as a digital hologram image by quantization process. In holography, the amplitude hologram is equal to the real part of complex field data and phase hologram is set of phase values. The real part of complex value can have negative value as well as positive value and the range of phase is from $-\pi$ to π. These values are quantized into 256 levels. In amplitude hologram, the minimum value is mapped to 0 and maximum value is mapped to 255 and then remaining values are mapped between 0 and 255. This process causes a temporal inconsistency for holographic video since the minimum and maximum values of complex field can be varied for each frame. In addition, the random phase factor is multiplied to each pixel of the desired pattern in order to spread the information of each pixel over a large area of the hologram. The random phased hologram is almost same with noise pattern and it degrade both spatial and temporal consistency.

The above approach negligibly affects to the image hologram but results in flickering artifacts and low compression efficiency in holographic video. We enhance the temporal consistency of holographic video by unify the CGH conditions such as minimum and maximum values of complex field and random phase pattern for successive frames. In other words, we intentionally unify the minimum and maximum values of complex field for all frames and apply the same random phase pattern for all frames. The detailed process is described in Algorithms 1 and 2.

Algorithm 1 Unified Random Phase

1: Generate the random phase pattern for 1st frame
2: Save the generated random phase pattern
3: Calculate the CGH for 1st frame
4: Load the saved random phase pattern
5: Calculate the CGH for next frame
6: Go to Line 4 until last frame
7: End

Algorithm 2 Unified range of min/max values

1: Obtain the min/max values for 1st frame
2: Save the min/max values
3: Obtain the min/max values for next frame
4: Compare the current min/max values and saved min/max values
5: Update the saved min/max values
6: Go to Line 3 until last frame
7: Calculate the CGHs for all frames with saved min/max values
8: End

---

4 Experimental Results

The performance of the proposed temporal consistency enhancement technique was evaluated by checking the temporal correlation and compression efficiency. The temporal correlation between successive frames is measured with correlation coefficient. The correlation coefficient of two variables, sometimes called the cross correlation, is the covariance of the two variables divided by the product of their individual standard deviations as in

$$r_{x,y} = \frac{cov(X,Y)}{\sigma_X \sigma_Y} = \frac{E[(X - \mu X)(Y - \mu Y)]}{\sigma_X \sigma_Y} \tag{3}$$

where X and Y are two random variables with expected values μX and μY and standard deviation σ_X and σ_Y. $cov(X,Y)$ means covariance between X and Y. The value close to or equal to 1 means two variables have strong correlation and close to or equal to 0 means no correlation.

We conducted coding experiments with 3DV-ATM version 0.9 to verify the improvement of compression efficiency according to the proposed temporal consistency enhancement. We check the bit saving compared to conventional method, that is, no temporal consistency enhancement. We use "UndoDancer" sequence since its depth map is accurate enough to be applied to holography experiments. The coding results in (Table 1) are compared using the BDBR (Bjøntegaard delta bit rate) metric [15]. We check the following two methods based on proposed algorithms: (1) Method 1: Algorithm 1, Method 2: Algorithm 1 + Algorithm 2.

Table 1. Comparions of Cross Correlation and bit Saving

Methods	Cross correlation	Bit saving (%)
Conventional Method	0.0004	–
Proposed Method 1	0.9073	45.41 %
Proposed Method 2	0.9074	49.65 %

As show in above results, the unification of random phase leads a big improvement and min/max unification also contributes temporal consistency enhancement. The proposed methods save bits more than 40 % but shows the similar subjective quality as shown in Fig. 2. The images are numerically reconstructed with coded hologram for QP 36.

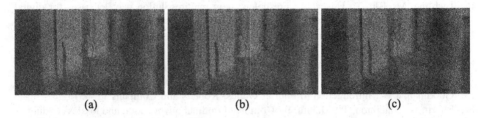

(a)   (b)   (c)

Fig. 2. Numerically reconstructed hologram images: (a) conventional method, (b) proposed method 1, (c) proposed method 2

5 Conclusion

The conventional CGH temporal inconsistency of holographic video causes flickering artifacts and low compression efficiency. In this paper, we proposed temporal consistency enhancement methods for digital holographic video. We unified some conditions such as random phase and min/max values of quantization during CGH process. In experimental results, the proposed methods achieve about 50 % bit saving compared to conventional method and show the similar subjective quality for reconstructed hologram images. In the future, we will further study on the holographic data format based on the proposed methods.

Acknowledgment. This research was supported by GigaKOREA project, (GK15D0100, Development of Telecommunications Terminal with Digital Holographic Table-top Display).

References

1. Gabor, D.: A new microscopic principle. Nature **161**(4098), 777–778 (1948)
2. Park, M., Chae, B.G., Kim, H.-E., Hahn, J., Kim, H., Park, C.H., Moon, K., Kim, J.: Digital holographic display system with large screen based on viewing window movement for 3D video service. ETRI J. **36**(2), 232–241 (2014)

3. Chung, J.K., Tsai, M.H.: Three-Dimensional Holographic Imaging. John Wiley & Sons Inc, New York (2002)
4. Yoshikawa, H., Tamai, J.: Holographic image compression by motion picture coding. Proc. of SPIE **2652**, 2–9 (1996)
5. Seo, Y.H., Choi, H.J., Kim, D.W.: 3D scanning-based compression technique for digital hologram video. Signal Process. **22**(2), 144–156 (2007)
6. Naughton, T.J., Javidi, B.: Compression of encrypted three dimensional objects using digital holography. Opt. Eng. **43**(10), 2233–2238 (2004)
7. Shortt, A., Naughton, T.J., Javidi, B.: Histogram approaches for lossy compression of digital holograms of three-dimensional objects. IEEE Trans. Imag. Proc. **16**(6), 1548–1556 (2007)
8. Kayser, D., Javidi, B., Psaltis, D.: Compression of digital holographic data using its electromagnetic field properties. Opt. Inf. Syst. III, Proc. SPIE **5908**, 97–105 (2005)
9. Ding, L., Yan, Y., Xue, Q., Jin, G.: Wavelet packet compression for volume holographic image recognition. Opt. Commun. **216**, 105–113 (2003)
10. Liebling, M., Blu, T., Unser, M.: Fresnelets: new multiresolution wavelet bases for digital holography. IEEE Trans. Image Process. **12**, 29–43 (2003)
11. Waters, J.P.: Holographic image synthesis utilizing theoretical methods. Appl. Phys. Lett. **9**(11), 405–406 (1966)
12. Brown, B.R., Lohmann, A.W.: Complex spatial filtering with binary masks. Appl. Opt. **5**, 967–969 (1966)
13. Goodman, J.W.: Introduction to Fourier Optics. Roberts and Company (2005)
14. Shortt, A., Naughton, T.J., Javidi, B.: Combined optimal quantization and lossless coding of digital holograms of three-dimensional objects. In: Three-Dimensional TV, Video, and Display V, Proceedings of SPIE vol. 6392, no. 63920A, October 2006
15. ITU-T SG16 Q.6, An excel add-in for computing Bjøntegaard metric and its evolution. VCEG-AE07, January 2007

Efficient Disparity Map Generation Using Stereo and Time-of-Flight Depth Cameras

Woo-Seok Jang and Yo-Sung Ho[⊠]

Gwangju Institute of Science and Technology (GIST),
123 Cheomdan-Gwagiro, Buk-gu, Gwangju 500-712, Republic of Korea
{jws,hoyo}@gist.ac.kr

Abstract. Three-dimensional content (3D) creation has received a lot of attention due to numerous successes of 3D entertainment. Accurate estimation of depth information is necessary for efficient 3D content creation. In this paper, we propose a disparity map estimation method based on stereo correspondence. The proposed system utilizes depth and stereo camera sets. While the stereo set carries out disparity estimation, depth camera information is projected to left and right camera positions using 3D warping and upsampling is processed in accordance with the image size. The upsampled depth is used for obtaining disparity data of left and right positions. Finally, disparity data from each depth sensor are combined. The experimental results demonstrate that our method produces more accurate disparity maps compared to the conventional approaches which use stereo cameras and a single depth sensor.

Keywords: Depth estimation · Stereo matching · Tof depth camera

1 Introduction

With the huge success of three-dimensional (3D) movies, interest in 3D entertainment systems is increasing recently. 3D multimedia applications give audiences the opportunity to experience 3D. The 3D experience comes from the left and right eyes seeing different views. Thus, audiences can perceive 3D using two separate views for the left and right eyes. For this, several data formats have been proposed. The texture plus depth approach is one of these formats. This format uses an ordinary 2D image accompanied by a depth map. The non-existing view can be synthesized by depth image based rendering (DIBR) [1]. Benefits of this format are the flexibility to render views with variable baseline and the increased compressibility of depth data due to its characteristics [2]. Thus, this format is practical for many 3D multimedia applications.

Depth information can mainly be acquired by two approaches: active and passive sensor based depth estimation methods. The former employs physical sensors, such as infrared ray (IR), laser and light pattern, to measure depth data directly. Depth cameras, structured light sensors, and 3D scanners are employed in this approach [3]. Usually, active sensors are more effective than passive sensors in terms of the quality of produced depth images. However, they produce only low resolution images and generally require expensive devices. The latter, on the other hand, indirectly estimates depth information from 2D images captured by at least two cameras [4]. Stereo matching is

© Springer International Publishing Switzerland 2015
Y.-S. Ho et al. (Eds.): PCM 2015, Part II, LNCS 9315, pp. 623–631, 2015.
DOI: 10.1007/978-3-319-24078-7_64

the most widely used passive sensor based method [5]; its advantages are low cost and flexible resolution. However, passive sensors also obtain miscalculated depth information in several types of regions.

Disparity data can be converted into depth information by using a stereo image pair in combination with triangulation [6]. Disparity data can be acquired by finding the corresponding points in other images for pixels in one image. The correspondence problem is to compute the disparity map which is a set of the displacement vectors between the corresponding pixels. For this problem, two images of the same scene taken from different viewpoints are given and it is assumed that these images are rectified for simplicity and accuracy of the problem. From this assumption, corresponding points are found in same horizontal line of two images. A disparity map acquired by stereo matching can be represented by a gray scale image. Depth of each pixel is perceived from the disparity map. The object is close to viewpoint as intensity value of a pixel in the disparity map is high.

The objective of this paper is to obtain accurate depth information using depth and stereo images. Thus, we present an accurate disparity map acquisition method through stereo correspondence. We design a disparity estimation system to strengthen the merits and make up for the weaknesses for active and passive depth sensors. The proposed method deals with fully unsolved problems by fusing and refining the depth data.

2 Problem Statement

Over the past several decades, a variety of stereo-image-based depth estimation methods have been developed to obtain accurate depth information. However, accurate measurement of stereo correspondence from natural scene still remains problematic due to difficult correspondence matching in several regions: textureless, periodic texture, discontinuous depth, and occluded areas [7]. First, since color data of the textureless and periodic texture region in left and right images are so similar each other in a wide range, correspondence matching often fails because of its ambiguity. Second, in case of the depth discontinuous region, such as the edge region, smeared color values exist, which leads to ineffective correspondence matching. Lastly, in the occluded region, some pixels may appear in one image but not in the other image; So there is no corresponding pixel. These problems can be solved for accurate depth information. Figure 1 illustrates correspondence matching problem in stereo matching.

Usually, depth cameras are more effective in producing high quality depth information than the stereo-based estimation methods. However, depth camera sensors also suffer from inherent problems. Especially, they produce low resolution depth images due to challenging real-time distance measuring systems. Such a problem makes depth cameras not practical for various applications. Figure 2 represents resolution difference between the regular color camera and depth camera.

Recent approaches of fusing active and passive sensors have shown improvements on depth quality by making up for weakness of each sensor method [8, 9]. In our work, we propose a disparity fusion method to make up for weakness of stereo-based estimation and carry out better correspondence matching by adding a depth camera to the stereo system.

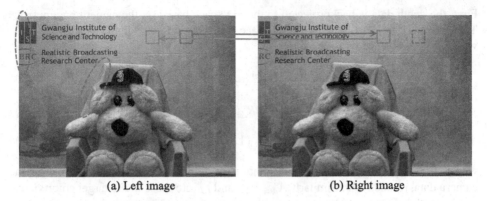

(a) Left image   (b) Right image

Fig. 1. Correspondence problem in stereo matching

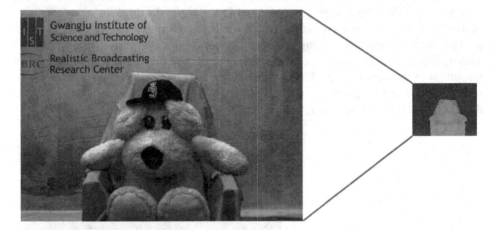

Fig. 2. Principle of Harris corner detector

3   Depth Fusion System

In general, stereo matching can be categorized into two approaches: local and global methods. Local methods are generally efficient for complexity [10]. However, these make blurred object borders and the removal of small detail at the depth discontinuity depending on the size of correlation window. In order to solve this problem, global methods have been proposed [11]. Global methods define an energy function by Markov Random Field (MRF) and optimize this function using several optimization algorithms such as belief propagation [12] and graph cut [13].

The proposed method is initially motivated by global stereo matching [14]. The information of the depth camera is included as a component of the global energy function to acquire more accurate and precise depth information. Depth camera processing is implemented as follows. (1) Depth data is warped to its corresponding position of the stereo views by 3D warping. 3D warping is composed of two processes.

First, the depth data is backprojected to the 3D space based on the camera parameters. Then, the backprojected data in the 3D space is projected to the target stereo views. 3D warping is performed as follows.

$$(x, y, z)^T = RsrcA_{src}^{-1}(u, v, 1)^T du, v + tsrc, \tag{1}$$

$$(l, m, n)^T = AdstR_{dst}^{-1}(x, y, z)^T - tdst, \tag{2}$$

$$(u', v') = (1/n, m/n) \tag{3}$$

where A_{src}, R_{src}, and t_{src} are intrinsic, rotation, and translation parameters in the depth camera data, respectively. Similarly A_{dst}, R_{dst}, and t_{dst} are those in the target color view. $d_{u,v}$ is depth value at (u,v) coordinate in the depth camera. The depth data is sent to the 3D space by Eq. (1) and projected to the target view by Eq. (2). (u', v') in (3) represents the projected coordinate to the target view. Figure 3 shows the 3D warping of depth data. (2) depth-disparity mapping is processed [15]. Due to the different representation of the actual range of the scene, correction of the depth information is required. (3) We perform joint bilateral upsampling (JBU) to interpolate the low resolution depth data. This method used high resolution color and low resolution depth image to increase the resolution of the depth image. Figure 4 illustrates JBU process. The processed information of the depth camera is applied as the additional evidence for data term of disparity fusion energy function.

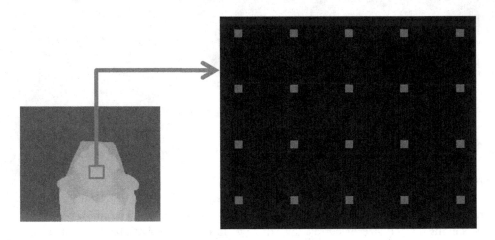

Fig. 3. 3D warping of depth camera data

Figure 5 illustrates error detection in stereo matching. To fuse both depth information, we consider error regions from both sensing. The error regions are obtained from the results of stereo matching. We applied the occlusion detection method using several constraints [7] to find the error regions.

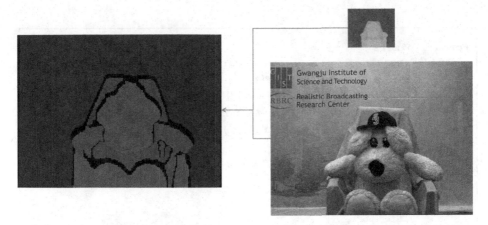

Fig. 4. Joint bilateral upsampling process

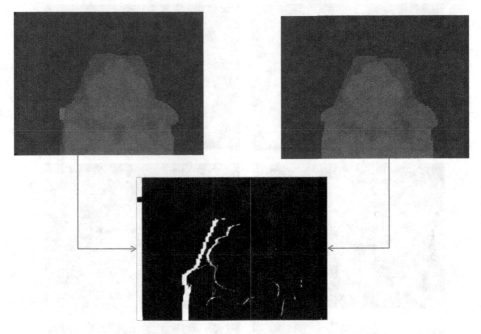

Fig. 5. Detection of error region in stereo matching

The disparity fusion energy function for accurate disparity estimation is defined as

$$E = \sum_{x,y} (Edata + Esmooth), \tag{4}$$

where E_{data} is a data term which measure the pixel similarity and E_{smooth} is the smoothness term which penalizes depth variations. The following function is applied to data term for energy function of depth estimation.

```
For all pixels
    If(JBU result ≠ hole && non-error region in stereo)
        E_data(x,y,d) = α|I_L(x,y) − I_R(x′,y′,d)| + β|d − d_up|)
    end if
    else if(JBU result == hole && non-error region in stereo)
        E_data(x,y,d) = |I_L(x,y) − I_R(x′,y′,d)|
    end else if
    else if(JBU result ≠ hole && error region in stereo)
        E_data(x,y,d) = |d − d_up|
    end else if
end for
```

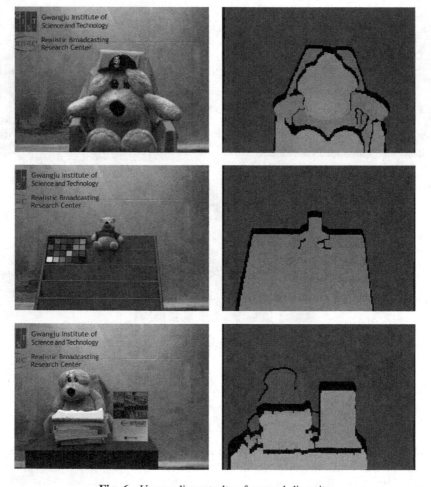

Fig. 6. Upsampling results of warped disparity

$I_L(x,y)$ is the pixel value in the left image given (x,y) coordinate. $I_R(x',y', d)$ is the matched pixel value in the right image given the disparity value at (x,y) in the left image, denoted by d. d_{up} is the upsampled disparity data obtained from the previous step. The upsampled disparity information enhances the precision and accuracy of the final depth values by allowing large depth variation.

The smoothness term is based on the degree of difference among the depth values of neighboring pixels.

$$Esmooth = \sum_{t\in N(x,y)} |d - d_t| \qquad (5)$$

$N(x,y)$ represents the neighboring pixels of the current pixel. The algorithm is hierarchically processed to acquire more accurate depth values in the textureless region.

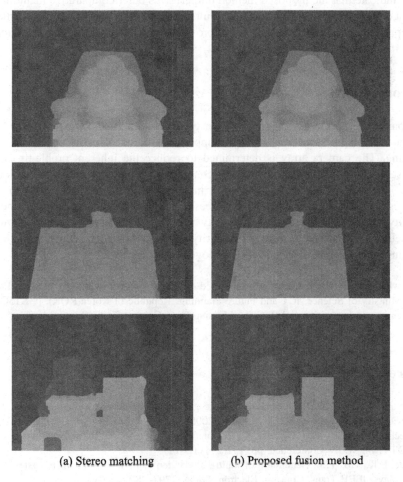

(a) Stereo matching   (b) Proposed fusion method

Fig. 7. Comparison of Depth results

Furthermore, we apply post-processing to the acquired disparity map to improve the quality [16].

4 Experimental Results

In order to evaluate the performance of the proposed method, we compare our proposed method with the stereo-image-based and depth upsampling methods. They are captured at the resolution of 1280 × 960 pixels. The depth camera employs the resolution 176 × 144 pixels. Camera calibration was applied to these data sets for the accuracy of the processes. Figure 6 shows the original image and disparity results of JBU.

Figure 7 shows the results of stereo matching and proposed fusion method. The JBU results represent that boundary is not good, but we can obtain the precision disparity data, especially in regards to homogeneous region. On the other hand, the stereo matching method cannot produce disparity detail in homogeneous region. The results indicate that the proposed method outperforms other comparative methods. The visual comparison of the experimental results demonstrates that the proposed method can represent the disparity detail and improve the quality in the vulnerable areas of stereo matching.

5 Conclusions

This paper presents a novel stereo disparity estimation method exploiting a depth camera. The depth camera is used to supplement crude disparity results of stereo matching. The camera array is determined to reduce the inherent problems of each depth sensor. We fuse the both depth data considering confidence regions of each depth sensor. The fusion depth estimation which includes refinement process overcomes the weakness in each depth sensing. This increases the precision and accuracy of the final disparity values by allowing large disparity variation. Experimental results show that our method produce more accurate disparity maps compared to conventional stereo matching especially in occlusion and homogeneous regions.

Acknowledgment. This research was supported by the 'Cross-Ministry Giga KOREA Project' of the Ministry of Science, ICT and Future Planning, Republic of Korea(ROK). [GK15C0100, Development of Interactive and Realistic Massive Giga-Content Technology]

References

1. Zhang, L., Tam, W.J.: Stereoscopic image generation based on depth images for 3DTV. IEEE Trans. Broadcast. **51**(2), 191–199 (2005)
2. Tech, G., Muller, K., Wiegand, T.: Evaluation of view synthesis algorithms for mobile 3DTV. In: 3DTV Conference, pp. 132(1–4) (2011)
3. Lee, E.K., Ho, Y.S.: Generation of multi-view video using a fusion camera system for 3D displays. IEEE Trans. Consum. Electron. **56**(4), 2797–2805 (2010)

4. Jang, W.S., Ho, Y.S.: Efficient disparity map estimation using occlusion handling for various 3D multimedia applications. IEEE Trans. Consum. Electron. **57**(4), 1937–1943 (2011)
5. Scharstein, D., Szeliski, R., Zabih, R.: A taxonomy and evaluation of dense two-frame stereo correspondence algorithms. Int. J. Comput. Vis. **47**, 7–42 (2002)
6. Hartley, R., Zisserman, A.: Multiple View Geometry in Computer Vision, 2nd edn, pp. 262–278. Cambridge University Press, Cambridge (2003)
7. Jang, W.S., Ho, Y.S.: Discontinuity preserving disparity estimation with occlusion handling. J. Vis. Commun. Image Represent. **25**(7), 1595–1603 (2014)
8. Lee, E.K., Ho, Y.S.: Generation of high-quality depth maps using hybrid camera system for 3-D video. J. Vis. Commun. Image Represent. **22**(1), 73–84 (2011)
9. Kang, Y.S., Ho, Y.S.: Generation of multi-view images using stereo and time-of-flight depth cameras. In: International Conference on Embedded Systems and Intelligent Technology, pp. 104–107 (2013)
10. Hirschmuller, H., Innocent, P.R., Garibaldi, J.M.: Real-time correlation-based stereo vision with reduced border errors. Int. J. Comput. Vis. **47**(1/2/3), 229–246 (2002)
11. Veksler, O.: Fast variable window for stereo correspondence using integral images. IEEE Comput. Soc. Conf. Comput. Vis. Pattern Recogn. **1**, 556–561 (2003)
12. Sun, J., Zheng, N.N., Shum, H.Y.: Stereo matching using belief propagation. IEEE Trans. Pattern Anal. Mach. Intell. **25**(7), 787–800 (2003)
13. Boykov, Y., Veksler, O., Zabih, R.: Fast approximate energy minimization via graph cuts. IEEE Trans. Pattern Anal. Mach. Intell. **23**(11), 1222–1239 (2001)
14. Yang, Q., Wang, L., Ahuja, N.: A constant-space belief propagation algorithm for stereo matching. In: Computer Vision and Pattern Recognition, pp. 1458–1465 (2010)
15. Kang, Y.S., Ho, Y.S.: High-quality multi-view depth generation using multiple color and depth cameras. In: International Workshop on Hot Topics in 3D, pp. 1405–1408 (2010)
16. Yang, Q., Engels, C., Akbarzadeh, A.: Near real-time stereo for weakly-textured scenes. In: British Machine Vision Conference, pp. 80–87 (2008)

Super-Resolution of Depth Map Exploiting Planar Surfaces

Tammam Tilo, Zhi Jin[✉], and Fei Cheng

Department of Electrical and Electronic Engineering,
Xi'an Jiaotong-Liverpool University (XJTLU), 111 Ren Ai Road, SIP,
Suzhou 215123, Jiangsu Province, People Republic of China
tammam.tillo@xjtlu.edu.cn
zhi.jin10@student.xjtlu.edu.cn
http://www.mmtlab.com

Abstract. Depth map, with per-pixel depth values, represents the relative distance between object in the scene and the capturing depth camera. Hence, it has been widely used in 3D applications and Depth Image-Based Rendering (DIBR) technique to provide an immersive 3D and free-viewpoint experience to the viewers. Depth maps could be generated by using software- or hardware-driven techniques. However, most generated depth maps suffer from a combination of the following shortcomings: noise, holes and limited spatial resolution. Therefore, to tackle the limited spatial resolution problem of Time-of-Flight depth images, in this paper, we present a planar-surface-based depth map super-resolution approach, which interpolates depth images by exploiting the equation of each detected planar surface. Aided with these equations the surfaces will be categorized into three groups, namely: planar surfaces, non-planar surfaces, and finally edges. For the first category the analytical equations of the planar surfaces will be used to super-resolve them, while a traditional interpolation method will be used for the non-planar surfaces, whereas, a combination of the two previous approaches will be used to up-sample edges. Both quantitative and qualitative experimental results demonstrate the effectiveness and robustness of our approach over the benchmark methods.

Keywords: Depth map · Super-resolution · Planar surface detection · Time-of-flight (Tof) camera

1 Introduction

A depth map describes the geometric relationship between objects in the scene and the capturing cameras [11] and it has been widely used in 3D video and Depth Image-Based Rendering (DIBR) based multivew and free-viewpoint video. Depth information, is usually, obtained through two major approaches: stereo or multiview matching-based methods [3,9] and depth-camera-based methods [5,13].

© Springer International Publishing Switzerland 2015
Y.-S. Ho et al. (Eds.): PCM 2015, Part II, LNCS 9315, pp. 632–641, 2015.
DOI: 10.1007/978-3-319-24078-7_65

For matching-based methods, they require at least two color images captured at slightly different viewpoints, then the joint features and areas in these captured images are analyzed to extract depth information. However, the matching-based approach may fail when no match is found between some areas in the two views, this could happen for example when some areas are occluded in one view and viewed in the other view, or for some textless areas. Depth-camera-based approach, such as Time-of-Flight (ToF) camera, can overcome most of these shortcomings. ToF depth camera obtains the depth information by calculating the phase delay of the received light with respect to the emitted light point by point. Hence, the depth map can be generated in real-time and released from texture interference. Compared with matching-based approach the depth-camera-based approach has higher accuracy. However, due to the intrinsic physical constraints of sensors, depth-camera-generated depth map, compared with traditional texture images, typically has Low Resolution (LR) (e.g. 176×144 for SR4000 [1] and 640×480 for Kinect [2]) and suffers from intrinsic noise and extrinsic environmental interference. Therefore, in order to successfully use depth maps in 3D applications, several depth map Super-Resolution(SR) techniques have been proposed to increase the spatial resolution of depth maps.

Kopf et.al proposed to upsample the LR depth map by using a Joint Bilateral Upsampling (JBU) filter. Aided by the associated High Resolution (HR) texture, the edges of upsampled depth map can be well preserved [8]. A similar but advanced joint bilateral filtering technique was proposed in [12] which iteratively refined the input LR depth map by referring the registered HR textures. On one hand, the adoption of texture information can help to get sharp depth edges. However, on the other hand, the color or lighting variations on the same areas of the texture images can cause false discontinuities in HR depth maps. Hence, the texture images need to be used in a more sophisticated way. By applying the Markov Random Fields (MRF) to super-resolve the LR depth map, *Diebel et.al* formulated the SR process as an energy minimization problem to fuse LR depth images and HR texture images. Different from Diebel's work, in [10], a nonlocal means (NLM) term was used into the MRF to preserve the edges. However, during the optimization process of MRF-based methods, the estimation errors are easily propagated into the obtained HR depth map. *Lo et.al* proposed to incorporate a texture-guided weighting factor into the MRF model for reducing the texture copying artifacts and the weighting factor was obtained based on a learning approach. Although the demonstrated results were good, the learning-based approaches usually have higher computation complexity, which might prevent their adoption for real-time applications.

Since after projection, the objects in a 3D scene can be represented by several planar surfaces with different shapes in a 2D image, each planar surface will have linearly changing depth values in the corresponding depth map and the boundaries of surfaces indicate the discontinuities of the depth values. If the equation of each surface can be obtained, the SR of LR depth map can be obtained by inserting pixels based on this equation. Therefore, the whole depth map can be classified into three categories: planar surfaces, non-planar surfaces, and edges.

In work [14], the SR of depth map relayed on the local planar hypothesis and the candidates of potential HR depth values were obtained by either linear interpolation along horizontal and vertical directions or the estimated local planar surface equations. However, since the surface equation was evaluated locally, it may be biased by the noise contained in the local pixels which later on will magnify the estimated error of generated HR depth map. Therefore, to address the above problem, in this paper, we propose to use the global analytical equations of the detected surfaces in the scene. For each of these three categories a proper up-sampling approach is proposed to exploit its intrinsic properties.

2   Planar Surface Detection

Obviously, to obtain the analytical equations of all surfaces in the scene is a hugely complicated task, if it is not impossible, thus in this paper we will only use the analytical equations which represent planar surfaces in the scene. To get these equations we will use the Depth-map Driven Planar Surface Detection (DDPSD) method which was proposed in [6], thus in the following this approach will be reviewed. In DDPSD, planar surfaces are detected by using a growing-based approach, which starts from some seed patchs. The first stage of the DDPSD algorithm detects all the valid seed patches over the whole depth map, these then will be arranged according to their planarity. In the second stage the arranged seed patches will be used sequently, so the most planar seed patch will be used first to initiate the growing mechanism, then each used seed will be expanded outwards until no new neighbor point fits into the current planar surface.

At each growing stage the linear least squares plane fitting approach is used to find the best fitting surface $a_m x + b_m y + c_m z + d_m = 0$, where (a_m, b_m, c_m) defines the estimated normal vector $\hat{\mathbf{n}}_m$ of the plane and d_m is the estimated distance from the 3D space origin, where $\mathbf{p} = (x, y, z)$ indicates an arbitrary point on the plane.

The surface equation could be represented by the function $f_m(a_m, b_m, c_m, d_m)$. This estimated equation is used to steer the growing process, and will be refined at each growing stage, so the more pixels are englobed into the surface the more accurate the equation will be. Aided with this mechanism, each seed patch grows to its maximum extent, and then the growing process will be iteratively repeated for the remaining unused seeds patches until the whole depth map is covered. In [6] it has been reported that DDPSD algorithm is able to detect semi-planar surfaces, and over-segment non-planar surfaces.

3   The Proposed Approach

In this section the proposed approach will be explained. If the analytical equation of a surface in the scene is known then it could be used to up-sample this surface, by plugging the coordinate of each missing pixel (x, y) in the surface equation to find its corresponding depth value z. So aided with this paradigm, in the proposed approach, the surfaces in the depth-scene will be categorized into

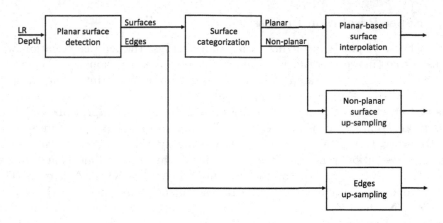

Fig. 1. The framework of the proposed depth map super-resolution method

three groups, namely: planar surfaces, non-planar surfaces, and finally edges. For each of these three categories a proper SR approach will be devised to better exploit its intrinsic properties. In the first stage of the proposed approach the DDPSD [6] method will be also used, nevertheless, other approaches for Planar Surface Detection(PSD) could be used. Since the output of the DDPSD method is mainly surfaces and the surrounding edges, thus the edges category will be simply the one outputted by the DDPSD method, whereas, in the second stage of the proposed approach the detected surfaces will be categorized into planar surfaces and non-planar surfaces. The block diagram of the proposed approach is shown in Fig. 1. The idea behind the proposed categorization mechanism is to check if the estimated planar equation fits well the measured depth values of the surface. So the pixels of each detected surface $\mathcal{S}_m$ will be plugged into the surface equation so as to evaluate their estimated depth values (i.e., $\hat{z} = \frac{-1}{c_m}(a_m x + b_m y + d_m)$, where $\mathbf{p} = (x, y)$ is a pixel belonging to the surface $\mathcal{S}_m$. Then the Mean Square Fitting Error (MSFE) is evaluated for this surface as:

$$\delta_m = \sqrt{\frac{1}{|S_m|} \sum_{\forall \mathbf{p} \in S_m} (z - \hat{z})^2} = \sqrt{\frac{1}{|S_m|} \sum_{\forall \mathbf{p} \in S_m} (z + \frac{1}{c_m}(a_m x + b_m y + d_m))^2} \quad (1)$$

This value will be compared against a threshold related to the maximum roughness value used in the planar surface detection unit, for furtehr details on the maximum roughness value please refer to [6]. If the MSFE is larger than the threshold then this indicates that this surface is non-planar or/and the estimated surface equation is not accurate. In this case, this surface will be classified as non-planar surface. Otherwise it will be classified as planar. The details of the SR approach for each of the three categories of pixels will be explained hereinafter.

3.1 Super-Resolution Process

For each set of pixels belonging to a planar surface the surface equation will be used to estimate the values of the pixels to-be-filled. Although a planar surface could be easily up-sampled by using a first-order linear interpolator, however, given that the measured depth data is affected by measurement noise, this will affect the accuracy of the interpolator-based up-sampled pixels. Thus, it is more accurate to use the estimated equation of the plane to up-sample it. To prove this property let us firstly simplify the analysis by modeling the high resolution version of the depth data using a one-dimensional vector $\bar{\mathbf{Z}}_h = [\bar{z}_0, \ldots, \bar{z}_i, \ldots, \bar{z}_{N-1}]^T$, and let us consider the case where N is even. The measured depth data by the depth-camera will be represented by the vector $\mathbf{Z}_l = \bar{\mathbf{Z}}_l + \mathbf{N}$ where $\bar{\mathbf{Z}}_l = [\bar{z}_0, \ldots, \bar{z}_{2i} \ldots, \bar{z}_{N-2}]^T$ is the low resolution, or in other words the down-sampled version of $\bar{\mathbf{Z}}_h$ where the downsampling factor is 2; the vector $\mathbf{N} = [n_0, \ldots, n_{2i}, \ldots, n_{N-2}]^T$ is a zero mean iid random process representing the measurement noise affecting depth data, and its variance is σ_n^2.

Let us represent by $\mathbf{Z}_z$ the zero-filled version of $\mathbf{Z}_l$ where the measured samples in $\mathbf{Z}_l$ have been separated by inserted zeros, i.e., $\mathbf{Z}_z = [z_0, 0, \ldots, 0, z_{2i}, 0, \ldots, z_{N-2}, 0]^T$. This version will be used as basis to generate the final super-resolved version of the vector $\mathbf{Z}_l$ by filling the zeros using some super-resolved values. If a first order linear estimator is used to estimate the zero-fill position in $\mathbf{Z}_z$ starting from their two-side neighbors then we could write that:

$$\hat{\mathbf{Z}}_h = \begin{pmatrix} 1 & 0 & 0 & 0 & 0 & \cdots \\ \alpha_1 & 0 & \beta_1 & 0 & 0 & \\ 0 & 0 & 1 & 0 & 0 & \\ 0 & 0 & \alpha_3 & 0 & \beta_3 & \\ \vdots & & & & & \ddots \end{pmatrix} \mathbf{Z}_z \qquad (2)$$

where $\hat{\mathbf{Z}}_h$ is the super-resolved version of the vector $\mathbf{Z}_l$. The variance of the estimation error of say for example z_{2i+1} could be evaluated as:

$$\sigma_e^2(2i+1) = E\left\{(\alpha_{2i+1}z_{2i} + \beta_{2i+1}z_{2i+2} - \bar{z}_{2i+1})^2\right\}$$
$$= E\left\{((\alpha_{2i+1}\bar{z}_{2i} + \beta_{2i+1}\bar{z}_{2i+2}) + (\alpha_{2i+1}n_{2i} + \beta_{2i+1}n_{2i+2}) - \bar{z}_{2i+1})^2\right\} \qquad (3)$$

if we suppose that $\bar{z}_{2i}$, $\bar{z}_{2i+1}$ and $\bar{z}_{2i+2}$ are the depth distances from a planar surface then in this case $\bar{z}_{2i+1} = \frac{\bar{z}_{2i}+\bar{z}_{2i+2}}{2}$, which means the best estimate of $\bar{z}_{2i+1}$ could be obtained when $\alpha_{2i+1} = \beta_{2i+1} = 1/2$. In this case, if we take into account the assumption that the noise $\mathbf{N}$ is a random iid process with zero mean value then (3) could be simplified as:

$$\sigma_e^2(2i+1) = \frac{1}{4} E\left\{(n_{2i} + n_{2i+2})^2\right\} = \frac{1}{2}\sigma_n^2 \qquad (4)$$

Obviously the accuracy of the up-sampling process depends on the accuracy of the depth measurement, and the error shown in (4) cannot be minimized by using a traditional interpolators (such as linear, bicubic, etc.). On the other hand, by using the planar surface estimated equation for up-sampling the interpolation error get reduced, this is because the depth measurement noise is canceled out during the estimation of the planar surface equation.

After super-resolving all planar surfaces the non-planar surfaces will be up-sampled by exploiting the local structure by using a traditional interpolator. In this paper we used Bicubic [7] interpolator for this task. Nevertheless, it is worth noticing that more advanced interpolators, such as directional-based interpolator could be used to estimate the values of the pixels to-be-filled.

Once all planar and non-planar surfaces get up-sampled then the turn comes to up-sample the edges and the remaining non-filled pixels. For each non-filled pixel its N_8 neighbors [4] and the surfaces that they belong to will be firstly identified. Then the missing pixel will be estimated by taking into account the surface category of each of its neighbour. To simplify the description of the proposed approach an example will be used hereinafter, let us suppose that (x, y) are the coordinate of one non-filled pixels, let suppose that one of its neighbor, $\mathbf{p}_i$, belong to a planar surface $\mathcal{S}_k$ then the surface equation of this surface will be used to have an estimate of the depth value at (x, y), as follows $\hat{z}_i = \frac{-1}{c_k}(a_k x + b_k y + d_k)$. If $\mathbf{p}_i$ belongs to a non-planar surface $\mathcal{S}_n$ then the same traditional approach which was used to up-sample $\mathcal{S}_n$ will be used now to extrapolate it and evaluate the depth value at (x, y). After obtaining, for each neighbor of the pixel (x, y) an estimated version of the to-be-filled pixel, these estimated versions will be fused by using a weighted average approach.

4   Experimental Results

Due to the difficulty to generate high resolution depth data that could be used as ground truth, for the proposed approach, some Computer Graphic (CG) depth images were generated to objectively assess the proposed approach. Due to the limited space, in the following we report the results for only one CG scene. Figure 2 shows a 3D saw-tooth structure, with each "tooth" having different height, whereas Fig. 2(b) shows the cross-section profile of this structure. The high resolution CG generated image is of 288×352 pixels. Then the rows and columns of the HR image were down-sampled by a factor 2 to generate its LR version.

For the DDPSD method [6] square seeds of 4×4 pixels were used, as for the threshold parameters the maximum allowed roughness, i.e., τ, was set to 3 and the threshold change speed, λ was set to 1.

The Mean Square Errors (MSE) evaluated row by row for the up-sampled saw-tooth image with respect to the HR ground truth are reported in Fig. 3 versus the row index. Two different approaches for up-sampling are shown, namely: the proposed approach, and traditional cubic approach. We used the traditional cubic approach as benchmark to make future comparisons with the proposed method straightforward. From the reported results we could notice the effectiveness of the proposed approach in recovering planar surfaces and edges. However, for small surfaces, such as those at the right side of Fig. 2(a) the DDPSD will

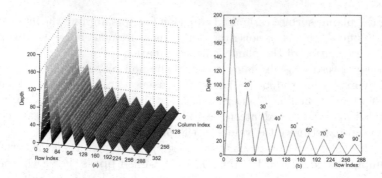

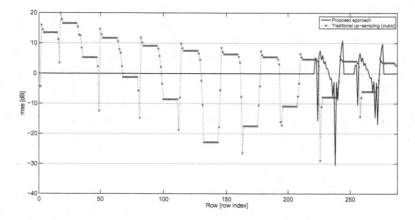

Fig. 2. (a) The 3D saw-tooth structure, each "tooth" has different height; (b) the profile of the saw-tooth structure

Fig. 3. The row by row MSE for the up-sampled saw-tooth image with respect to the HR ground truth versus the row index. Two different approaches for up-sampling are shown, namely: the proposed approach (PSNR is 47.35 dB), and traditional cubic approach (PSNR is 39.91 dB).

have some problems to estimate their equations, consequently, their HR version will have high MSE. It is worth reporting that the PSNR of the proposed SR approach is 47.35 dB versus 39.91 dB for the traditional cubic approach, a matched down/up-sampling cubic approach is also tested and its PSNR is 46.54 dB, which also confirm the superior performance of the proposed approach.

Furthermore, to visually appreciate the performance of the proposed approach, different indoor scenes were captured by using SwissRanger SR4000 depth camera [1] and then tested[1]. In the following due to the limited space we report the results for only one scene. Figure 4(a) shows the 144×176 original depth image,

---

[1] All the testing images with their setting information are available at http://www.mmtlab.com/DDPSD.ashx.

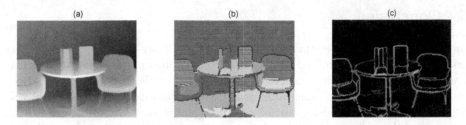

Fig. 4. Image (a) shows the original LR 144 × 176 depth image; the output of the surface categorization is shown in (b), where horizontal hatch pattern shows planar surfaces; the edges and the isolated non-filled pixels are shown in (c)

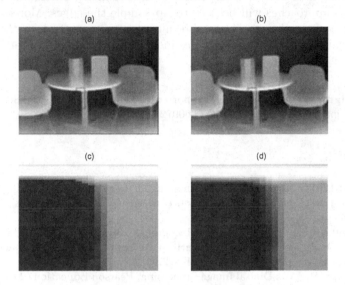

Fig. 5. The super-resolved image using: (a) proposed approach; (b) traditional interpolation approach; The delimitated area by a red box in (a) and (b) is blown-up in (c) and (d), respectively (Color figure online)

the output of the surface categorization on this image is shown in Fig. 4(b), in this figure different colors are used to represent different detected surfaces, and horizontal hatch pattern is used to show the surfaces which were categorized as planar surfaces. From this image we could observe that the proposed surface categorization approach works well, in fact non-planar surfaces such as some parts of the chair, and the room floor which although flat however is not smooth, were well identified. In Fig. 4(c) the edges and the isolated non-filled pixels are shown.

The image shown in Fig. 4(a) has been super-resolved using the proposed approach and a cubic interpolation approach, the results are shown respectively in Fig. 5(a) and (b). The delimitated area by a red box in Fig. 5(a) and (b) is blown-up in Fig. 5(c) and (d), respectively. From these cropped images we could observe the superiority of the proposed approach in recovering edges details.

5   Conclusion

In this paper, a planar-surface-based depth map super-resolution approach is proposed to tackle the limited spatial resolution problem of ToF captured depth images. Firstly, all the planar surfaces in the scene have been detected and represented by analytical equations. Referring to these equations, all the surfaces will be categorized into three groups: planar, non-planar surfaces and edges. Then, for each of these three categories, a proper up-sampling approach is applied to estimated the to-be-filled pixels. For the planar surfaces, they are upsampled by using the analytical equations, while, for the non-planar surfaces, a traditional interpolator such as bicubic interpolator is used. Finally, a combination of the two previous approaches will be used to up-sample the edges. Moreover, it can be appreciated from the experimental results that the proposed super-resolution approach achieves superior performance in comparison with traditional interpolation one.

Acknowledgments. This work was supported by the National Natural Science Foundation of China (NO. 61210006 and NO.60972085).

References

1. Internest (2014). http://www.mesa-imaging.ch/home/ Accessed 1 May 2014
2. Internest (2014). http://www.microsoft.com/en-us/kinectforwindows/ Accessed 1 May 2014
3. Bennamoun, M., Mamic, G.: Object Recognition: Stereo Matching and Reconstruction of a Depth Map. Advances in Pattern Recognition. Springer, London (2002). http://dx.doi.org/10.1007/978-1-4471-3722-1_2
4. Gonzalez, R.R.E.W.: Digital Image Processing. Pearson Education (2009). https://books.google.co.in/books?id=a62xQ2r_f8wC
5. Hansard, M., Lee, S., Choi, O., Horaud, R.P.: Time-of-flight Cameras: Principles, Methods And Applications. Springer Science & Business Media, London (2012)
6. Jin, Z., Tillo, T., Cheng, F.: Depth-map driven planar surfaces detection. In: Visual Communications and Image Processing Conference, 2014 IEEE. pp. 514–517 (December 2014)
7. Keys, R.: Cubic convolution interpolation for digital image processing. IEEE Trans. Acoustics, Speech Signal Process. **29**(6), 1153–1160 (1981)
8. Kopf, J., Cohen, M.F., Lischinski, D., Uyttendaele, M.: Joint bilateral upsampling. ACM Trans. Graph. (TOG) **26**, 96 (2007)
9. Kowalczuk, J., Psota, E., Perez, L.: Real-time stereo matching on cuda using an iterative refinement method for adaptive support-weight correspondences. IEEE Trans. Circuits Syst. Video Technol. **23**(1), 94–104 (2013)
10. Park, J., Kim, H., Tai, Y.W., Brown, M.S., Kweon, I.: High quality depth map upsampling for 3D-TOF cameras. In: 2011 IEEE International Conference on Computer Vision (ICCV), pp. 1623–1630. IEEE (2011)
11. Schwarz, S., Olsson, R., Sjostrom, M., Tourancheau, S.: Adaptive depth filtering for HEVC 3D video coding. Picture Coding Symposium (PCS) 2012, pp. 49–52 (2012)

12. Yang, Q., Yang, R., Davis, J., Nistér, D.: Spatial-depth super resolution for range images. In: IEEE Conference on Computer Vision and Pattern Recognition, 2007. CVPR 2007. pp. 1–8. IEEE (2007)

13. Zhang, Z.: Microsoft kinect sensor and its effect. IEEE MultiMedia **19**(2), 4–10 (2012)

14. Zhong, G., Yu, L., Zhou, P.: Edge-preserving single depth image interpolation. In: Visual Communications and Image Processing (VCIP), 2013. pp. 1–6. IEEE (2013)

Hierarchical Interpolation-Based Disocclusion Region Recovery for Two-View to N-View Conversion System

Wun-Ting Lin, Chen-Ting Yeh, and Shang-Hong Lai[✉]

Department of Computer Science, National Tsing Hua University, Hsinchu, Taiwan
{shiaushiauhan,yayaya80429}@gmail.com
lai@cs.nthu.edu.tw

Abstract. In this paper, we propose a novel disocclusion region recovery approach for two-view to n-view conversion system. Although the topic of view synthesis has been exhaustively studied for decades, a reliable disocclusion region recovery approach, an indispensable issue in synthesizing realistic content of virtual view, is still under research. The most common concept used for predicting these unknown pixels is inpainting-related method, which fills the disocclusion region with the information of mated exemplars in self-defined searching domain. In spite of widely taken in making up the missing values generated among the synthesis procedures, the result quality of inpainting-based approach is sensitive to the filling priority and also unstable in recovering large disocclusion region. Therefore, we propose a hierarchical interpolation-based approach to calculate the desired lost information under coarse-to-fine manner accompanied with the joint bilateral upsampling technology, applied for enlarging the estimation from small dimension to higher-resolution. Proposed hierarchical interpolation-based scheme is more robust in restoring the value of missing region and also induces fewer artifacts. We demonstrate the superior quality of the synthesized virtual views under the proposed recovery algorithm over the traditional inpainting-based method through experiments on several benchmarking video datasets.

Keywords: Disocclusion region recovery · Image inpainting · View interpolation · Novel view synthesis

1 Introduction

Video has become a popular trend in the entertainment industry in recent years, especially 3D movies. Advanced 3D display technologies allow users to experience realistic 3D effects at home. More and more 3D display applications can be found in the high-end electronic products, such as 3D LCD/LED TVs, 3D cameras, 3D camcorders, 3D laptops, 3D mobile phones, and 3D games, etc. The revolution from 2D display to 3D display has started to change many aspects of our daily life, including entertainment, communication, photography, and medical science.

© Springer International Publishing Switzerland 2015
Y.-S. Ho et al. (Eds.): PCM 2015, Part II, LNCS 9315, pp. 642–650, 2015.
DOI: 10.1007/978-3-319-24078-7_66

Fig. 1. Inpainting-based disocclusion region recovery is one of the main source of synthesized artifacts.

Nowadays, an advanced 3D display technology, autostereoscopic, allows people to watch 3D content without wearing glasses and enable user to experience 3D content of different viewpoints by changing their viewing direction toward the screen. This technology requires multi-view multimedia content, which can be obtained by either hardware multi-view camera array or software view synthesis approach.

Given a stereo pair, the target of view synthesis is to generate the content of virtual view located between these two given viewpoints. The recovery of disocclusion region, whose information is missed in both given images, is one of the vital problems in two-view to n-view conversion system. The most common and typical method to predict these unknown pixels is the inpainting-related approach [1,2,7,10], whose quality vary from the filling order and is limited from practical application due to their unstable results. Figure 1 demonstrate that inpainting-based disocclusion region recovery algorithm is one of the main source of synthesized artifacts. Lin et al. [4] imposed additional depth information to make the original exemplar-based inpainting [1] adapted to the problem in view synthesis. [13] presented a depth no-synthesis-error model which handels the depth quantization and filtering error in view synthesis.

Different from traditional inpainting-related approaches, we propose a view interpolation based algorithm for recovering the disocclusion region in a hierarchical architecture. Solh and AlRegib [9] also developed a hierarchical view synthesis scheme that fills invisible regions by low pass filtering among neighboring pixels at different resolutions. In our approach, we first iteratively downsample the synthesized image into lower resolution until all the disocclusion region are alleviated. Then, a joint bilateral upscaling scheme [3] is applied to gradually enlarge the downscaled image back to original resolution. In each upscaling procedure, the value of enlarged image is adjusted with the gradient information provided by the original image of the same resolution through bilateral filtering. With this iteratively upscaling scheme, a superior synthesized image can be constructed for the target viewpoint.

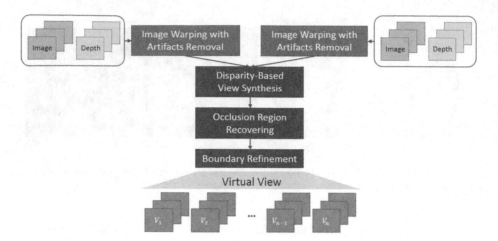

Fig. 2. Flowchart of the two-view to n-view conversion system.

2   Two-View to N-View Conversion System

Before introducing proposed disocclusion region recovery algorithm, we give a brief explanation of the two-view to n-view conversion system taken in our system. We follow the view synthesis procedures suggested by Wei et al. [11], shown in Fig. 2, except from replacing the disocclusion region recovery algorithm with proposed hierarchical interpolation-based method. After generating the required disparity maps by Yang [12], the given stereo pair, left image and right image, are both warped to the target viewpoint, which located between the given stereo pair. A disparity-based view synthesis approach is used for fusing these two warped images into a single estimation. Succeeding, a disocclusion region recovery technology, which is extensively described in the next section, is applied for filling the missing information in the disocclusion region and the directional filters, involving in eight distinct quantized orientations and ranges, are taken to reduce the jagging artifacts along the object boundary brought about by the warping process in the region has discontinuous depth maps.

3   Hierarchical Interpolation-Based Disocclusion Region Recovery

Given two warped stereo images I_0^L and I_0^R, which contain several disocclusion regions without intensity information, our target is to synthesize a refined image S_0 which fuses the message extracted from both stereo images as well as makes up the missing data in the disocclusion region. Motivated by the success of Luo et al. [5] in depth map refinement, a hierarchical architecture, illustrated in Fig. 3, is adopted to synthesize and recover the target refined image. Iteratively downsampling two warped images into lower resolution with factor-2, the

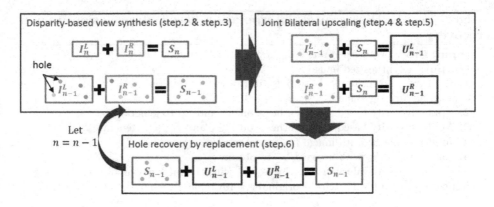

Fig. 3. Hierarchical interpolation-based disocclusion region recovery algorithm.

estimations I_n^L and I_n^R, which contain compete intensity information without lost information, are acquired after employing n times downsampling process. Next, proposed algorithm repetitively upsamples the downscaled image through disparity-based view synthesis technology and joint bilateral upscaling approach. We take I_n^L, I_n^R, which have compete information, and their finer version in upper layer I_{n-1}^L, I_{n-1}^R, which consist of some missing data, for fusing an estimation S_{n-1} in layer $n-1$.

Start with, a disparity-based view synthesis technology, recommended by Wei et al. [11], is used for constructing a fused estimation S_n and S_{n-1} in either layer n or layer $n-1$ with the warped pair I_n^L, I_n^R and I_{n-1}^L, I_{n-1}^R, respectively. Having the estimations S_n and S_{n-1}, next objective is to fill the missing information in S_{n-1} with the corresponding message located in S_n. Here, a joint bilateral upscaling approach is adopted for this work. First enlarging S_n into higher resolution, expressed in $S_n^\uparrow$, with BICUBIC approach, two warped image I_{n-1}^L, I_{n-1}^R in the same resolution with $S_n^\uparrow$ are taken for joint filtering $S_n^\uparrow$ to be U_{n-1}^L, U_{n-1}^R with bilateral equation, which is depicted in Eq. (1).

$$U_{n-1}^*(p) = \frac{1}{k_p} \sum_{q \in \Omega} f\left(\|p - q\|\right) g\left(\left\|I_{n-1}^*(p) - I_{n-1}^*(q)\right\|\right) S_n^\uparrow(q) \tag{1}$$

The function $f(x)$ and $g(x)$ in Eq. (1) are both Gaussian equation with two parameters σ_f and σ_g, controlling the weight for either position term or range term.

After constructing the estimations U_{n-1}^L, U_{n-1}^R in layer $n-1$ with two given warped image, proposed algorithm fills the required information in occlusion region of S_{n-1} by linear interpolating the values in U_{n-1}^L, U_{n-1}^R according to the ratio of viewshift, which is described in Eq. (2).

$$S_{n-1}(x) = (1 - v) * U_{n-1}^L(x) + v * U_{n-1}^R(x),$$
$$\text{for each } x \in \text{disocclusion region,} \tag{2}$$

Algorithm 1. Hierarchical interpolation-based disocclusion region recovery

---

Input: Given two warped image I_0^L, I_0^R from left or right
Output: A synthesized image S_0
1. Iteratively downsample I_0^L and I_0^R into lower resolution until I_n^L, I_n^R in layer n contain no missing value
2. Apply disparity-based view synthesis on I_n^L and I_n^R to generate S_n
3. Apply disparity-based view synthesis on I_{n-1}^L and I_{n-1}^R to generate S_{n-1}
4. Joint upscale S_n with image I_{n-1}^L to obtain U_{n-1}^L
5. Joint upscale S_n with image I_{n-1}^R to obtain U_{n-1}^R
6. Fill the missing values in S_{n-1} by Eq.(2) with the information of U_{n-1}^L and U_{n-1}^R
Repeat **step.3-6** until reach the original resolution and then output the target image S_0

---

where v stands for the ratio of viewshift between target viewpoint to left viewpoint over two given stereo pair. Finally, under the hierarchical architecture, we iteratively upsample the estimation S_n until reaching the original resolution and acquiring the final refined estimation S_0. The summary of proposed occlusion region recovery approach can be found in Algorithm 1.

4   Experimental Results

Here we conduct the experiments in three different directions. First, we compare the synthesized images under two different disocclusion region recovery approaches, either inpainting method or proposed strategy, to evaluate the improvement of our novel hierarchical interpolation-based recovery algorithm. Second, given ground truth images and disparity maps of different view points, we compare the virtual view to real view for quantitative evaluations. Third, a comparison under benchmarking video is taken for both proposed system and state-of-the approach to demonstrate the superior quality our method can achieve.

Figure 4 shows the results of both inpainting-based and interpolation-based recovery algorithm. In the left image pair, the boundary of pillar in the inpainting-based method, marked with red circle, suffers from severe aliasing artifacts while interpolation-based approach can recover a sharp and straight outline. Another example, illustrated in the right-hand side, also supports that interpolation-based recovery strategy can generate better synthesized image than traditional inpainting-based algorithm.

The Middlebury stereo dataset [8] provides images from six different viewpoints and two corresponding disparity maps. We use these two images with ground truth disparity map, view 1 and view 5, to generate virtual view 3 by proposed system. Figure 5. shows the virtual views, real views, residue images

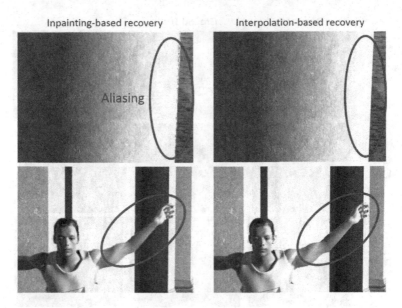

Fig. 4. Results of both inpainting-based and interpolation-based disocclusion region recovery algorithms.

and PSNR values. One drawback of proposed system is that proposed system may blur at disocclusion region, and these errorneous regions can be found at residue images.

Besides evaluating the disocclusion recovery approaches, a more general experiment on the entire two-view to n-view conversion system is also employed. We compare the proposed view synthesis system to VSRS [6] with four common benchmarking stereo video datasets. In this experiment, we randomly pick up two different viewpoints for each video and synthesize the content of middle view to calculate the average PSNR value across the whole sequence, which demonstrated in Table 1. In general, proposed system has superior quality than VSRS and some of synthesized middle images are displayed in Figs. 6 and 7.

Table 1. Average PSNR values of four selected video datasets

	VSRS	Proposed	Improvement
Lovebird1	23.2878	27.0569	3.7691
Lovebird2	23.5619	28.6482	5.0863
Bookarrival	28.1171	30.6562	2.5391
Outdoor	27.7410	32.4299	4.6889

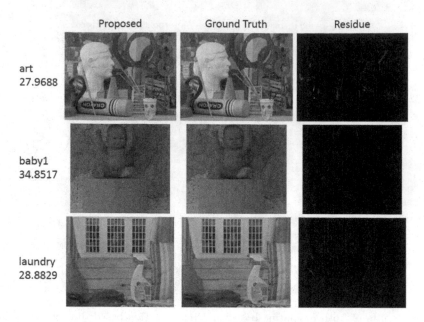

Fig. 5. Comparison of virtual view and real view for some images in Middlebury dataset, PSNR values are given under dataset names.

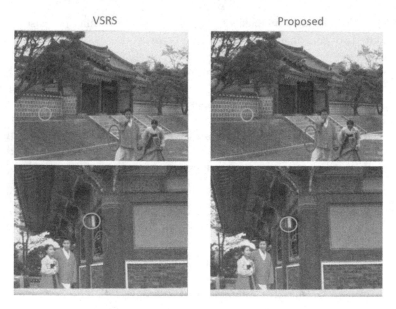

Fig. 6. Comparison of synthesized middle views by using VSRS and the proposed system on the Lovebird1 and Lovebird2 datasets.

VSRS   Proposed

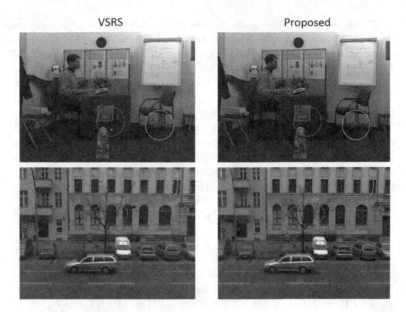

Fig. 7. Comparison of synthesized middle views by using VSRS and the proposed system on the datasets.

5   Conclusion

In this paper, we first show that the traditional inpainting-based disocclusion region recovery method is unstable in predicting the missing information of disocclusion region and in most situation becomes one of principal source of artifacts. Therefore, we proposed an interpolation-based approach, utilizing joint bilateral upscaling concept under a hierarchical architecture, to estimate the value of disocclusion region in more robust way and also induce fewer artifacts. The proposed system also runs faster than inpating-based methods when disocclusion region increases.

Acknowledgements. This work was partially supported by Ministry of Science and Technology, Taiwan, R.O.C., under the grant MOST 101-2221-E-007-129-MY3.

References

1. Criminisi, A., Perez, P., Toyama, K.: Object removal by exemplar-based inpainting. In: Proceedings of 2003 IEEE Computer Society Conference on Computer Vision and Pattern Recognition, 2003. vol. 2, pp. II-721. IEEE (2003)
2. Do, L., Zinger, S., et al.: Quality improving techniques for free-viewpoint DIBR. In: IS&T/SPIE Electronic Imaging. pp. 75240I–75240I. International Society for Optics and Photonics (2010)
3. Kopf, J., Cohen, M.F., Lischinski, D., Uyttendaele, M.: Joint bilateral upsampling. ACM Trans. Graph. (TOG). vol. 26, Article no. 96. ACM (2007)

4. Lin, S.-J., Cheng, C.-M., Lai, S.-H.: Spatio-temporally consistent multi-view video synthesis for autostereoscopic displays. In: Muneesawang, P., Wu, F., Kumazawa, I., Roeksabutr, A., Liao, M., Tang, X. (eds.) PCM 2009. LNCS, vol. 5879, pp. 532–542. Springer, Heidelberg (2009)
5. Luo, H.L., Shen, C.T., Chen, Y.C., Wu, R.H., Hung, Y.P.: Automatic multi-resolution joint image smoothing for depth map refinement. In: 2013 2nd IAPR Asian Conference on Pattern Recognition (ACPR), pp. 284–287. IEEE (2013)
6. MPEG ISO/IEC JTC1/SC29/WG11: View Synthesis Software Manual (VSRS), release 3.5 (2009)
7. Oh, K.J., Yea, S., Ho, Y.S.: Hole filling method using depth based in-painting for view synthesis in free viewpoint television and 3D video. In: Picture Coding Symposium, 2009. PCS 2009. pp. 1–4. IEEE (2009)
8. Scharstein, D., Pal, C.: Learning conditional random fields for stereo. In: IEEE Conference on Computer Vision and Pattern Recognition, 2007. CVPR 2007. pp. 1–8. IEEE (2007)
9. Solh, M., AlRegib, G.: Hierarchical hole-filling for depth-based view synthesis in FTV and 3D video. IEEE J. Select. Topics Signal Process. **6**(5), 495–504 (2012)
10. Wang, L., Jin, H., Yang, R., Gong, M.: Stereoscopic inpainting: joint color and depth completion from stereo images. In: IEEE Conference on Computer Vision and Pattern Recognition, 2008, CVPR 2008. pp. 1–8. IEEE (2008)
11. Wei, C.-H., Chiang, C.-K., Sun, Y.-W., Lin, M.-H., Lai, S.-H.: Novel multi-view synthesis from a stereo image pair for 3D display on mobile phone. In: Park, J.-I., Kim, J. (eds.) ACCV 2012. LNCS, vol. 7729, pp. 568–579. Springer, Heidelberg (2013)
12. Yang, Q.: A non-local cost aggregation method for stereo matching. In: 2012 IEEE Conference on Computer Vision and Pattern Recognition (CVPR), pp. 1402–1409. IEEE (2012)
13. Zhao, Y., Zhu, C., Chen, Z., Yu, L.: Depth no-synthesis-error model for view synthesis in 3D video. IEEE Trans. Image Process. **20**(8), 2221–2228 (2011)

UEP Network Coding for SVC Streaming

Seongyeon Kim, Yong-woo Lee, and Jitae Shin[✉]

School of Electronic and Electrical Engineering, Sungkyunkwan University,
2066, Seobu-ro, Jangan-gu, Suwon 440-746, Korea
{creative24, tencio2001, jtshin}@skku.edu

Abstract. In this paper, we propose an optimized unequal error protection (UEP) network coding for scalable video coding (SVC). First, we introduce packet-level UEP network coding schemes which encode packets with different selection probabilities according to the priorities of video layers. Secondly, we explain how to prioritize the packet selection probability using distortion degree. The distortion degree accounts for the layers as well as group of pictures (GOP). Finally, experiment results demonstrate the performance of our proposed UEP network coding compared to other existing network coding methods in terms of decoding probability and peak signal to noise ratio (PSNR).

Keywords: Unequal error protection · Network coding · Scalable video coding · Distortion degree

1 Introduction

Recently, as communication technology advances and demand of smart devices become universal, the traffic on the mobile network is rapidly increasing. It is expected that the mobile traffic occupation due to video services will be increased up to 72 % in 2019 [1]. With the advance of the smart devices and broadcast technique, a variety of video services are highly in demand, which requires that video coding techniques have compatibility in multiple terminals while providing same service. In real-time streaming service, especially, it is necessary to satisfy user's convenience by considering various devices characteristics and communication environment. For real-time streaming service, source node transmits data to a large number of users with limited feedback exchange. Random linear coding (RLC) method can address this difficulty [2–4]. In Fig. 1(a), base station transmits four packets to four user equipment (UE) and each UE cannot get the full information due to the packet loss. In this case, each UE asks retransmission for the lost packet by sending a feedback. As the number of user increases, however, such requests may burden the base station so that it is difficult for the UEs to receive the lost data. RLC can resolve this issue by sending packets linearly combined with randomly selected coefficient vector regardless of packet sequence. It is guaranteed that the users will recover the full data if they only receive the packets over a certain number. Therefore the base station does not need to know which packet should be sent; instead it only needs to send RLC coded packets. This case is illustrated in Fig. 1(b).

© Springer International Publishing Switzerland 2015
Y.-S. Ho et al. (Eds.): PCM 2015, Part II, LNCS 9315, pp. 651–659, 2015.
DOI: 10.1007/978-3-319-24078-7_67

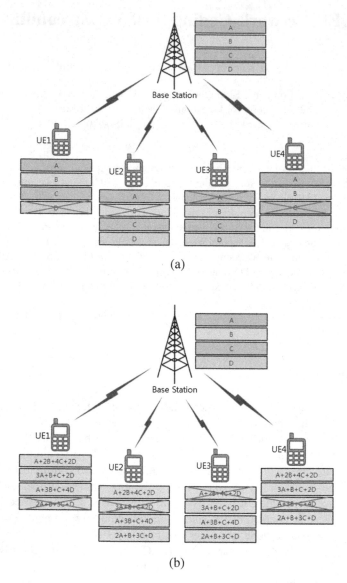

Fig. 1. Multicast scenarios: (a) Non-network coding; (b) RLC

In addition, heterogeneous networks and devices make it even more difficult because every device has different specifications which require different stream data. In order to tackle this issue, one can exploit the advantages of scalable video coding (SVC) which fully satisfy these necessities [5, 6]. By encoding base layer and enhancement layers together, SVC provides scalability of various conditions and it can also guarantees a minimum video quality in a poor network environment. The problem of SVC is that the layers have dependency on the lower layers. Even though the information of enhancement layers is completely sufficient, it is not decodable if its dependent layer information is lost during the transmission. RLC may not be well suited for transmitting SVC data since RLC does not focus on relative importance of each packet; instead it only needs to receive the packets over a certain number. To deal with this issue, one can apply unequal error protection (UEP) scheme for effective transmission of such hierarchical data. Namely, to better protect higher priority information, UEP schemes have been proposed and evaluated [6–14]. However in those UEP schemes, enhancement layers are usually vulnerable to error. On one hand, the UEP schemes intend to provide a minimum video quality on all devices by further protection of the lower layers. On the other hand, the enhancement layers are protected less so that the video quality does not reach the maximum quality in good channel condition. For that reason, with the tradeoff issue, it is very important to apply the optimal encoding method. In this paper, we proposed an UEP network coding scheme based on expanding window (EW) with optimal encoding parameters which utilizes distortion degree in SVC. First, we introduce packet-level UEP network coding schemes which encode packets with different selection probabilities according to the priorities of video layers. Secondly, we explain how to prioritize the packet selection probability using distortion degree. The distortion degree accounts for the layers as well as group of pictures (GOP). Finally experiment results demonstrate the performance of our proposed UEP network coding compared to other existing network coding methods in terms of decoding probability and peak signal to noise ratio (PSNR).

2   Proposed Scheme

2.1   Background: NOW RLC and EW RLC

There are several UEP RLC schemes to provide better protection on higher priority source data. In this paper, we introduce non-overlapping window (NOW) and expanding window (EW) RLC schemes which are typical methods of UEP RLC [14]. Figure 2 shows the UEP RLC encoding process [3].

Let $x = \{x_1, x_2, \ldots, x_K\}$ be source data consists of K packets and it is classified into L hierarchical layers. Each layer has different priority and the first layer is base layer which is the most important layer. For UEP RLC, a transmitter makes an encoded packet y_i by choosing a window which includes a certain number of packets according

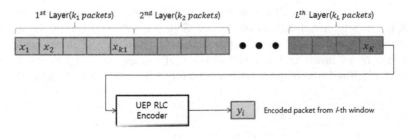

Fig. 2. UEP RLC encoding process

to the window selection probability. In NOW RLC, once a packet is included in a window, it will not be included in other windows while a packet can be repeatedly included in EW RLC. Since a packet can be repeatedly included in windows, there is more chance to be securely transmitted. EW RLC utilizes this concept to better protect higher priority data.

In Fig. 3, the difference between NOW RLC and EW RLC is described. In NOW RLC, l-th window only contains l-th layer, so each window is independent from the other layers. In case of EW RLC, l-th window consists of all other windows as well as all packets. The packets in the lower number windows are repeatedly included in the higher number windows, providing higher protection to those packets in the lower number windows. In the encoding part, window selection probability function decides windows to be used to encode a certain packets. The function is as follows:

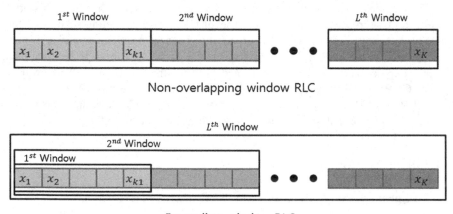

Non-overlapping window RLC

Expanding window RLC

Fig. 3. Difference between NOW RLC and EW RLC

$$f_w = \rho_1 \cdot \zeta^1 + \rho_2 \cdot \zeta^2 + \rho_l \cdot \zeta^l \qquad (1)$$

where ρ_l is the probability of selecting l-th window, ζ^l is l-th window.

2.2 Proposed UEP EW Network Coding Using Distortion Degree

In this paper, we propose UEP EW network coding using distortion degree (UEP EW-NCD). It exploits distortion degree of SVC video bitstream to set encoding parameters used for window selection probability function. In SVC bitstream, a frame is classified into l layers and a certain number of frames are grouped into N GOPs which are prioritized differently. The importance of l-th layer in n-th GOP is shown as

$$P(l, n) = \frac{\delta D_{l,n}}{\delta R_{l,n}} \qquad (2)$$

where $D_{l,n}$ indicates distortion rate and $R_{l,n}$ is bit rate of l-th layer in n-th GOP. We adopt sum of squared error (SSE) as the distortion metric of video packets [15]. Therefore we can utilize $P(l,n)$ to represent the importance of data. We use this value to set the encoding parameters of our proposed scheme.

Along with the priority, one should consider the data size as well. Since the packets in base layer are selected more frequently, the portion of packets in enhancement layers is relatively small. However, the packet size in each layer does not follow the priority, in a very extreme case, the packets in the enhancement layers do not get to be encoded in the first place. This leads to decoding fail since receiver cannot have enough amounts of enhancement layer data. To compensate for this problem, the highest window selection probability should not use its distortion degree but it should employ its data size as in Eq. 3. With this method, UEs can secure the basic video quality service as well as it provides the possibility for UEs in good channel to make sure of receiving all data, which settles the tradeoff issue. Our proposed window selection probability function is as follows:

$$f_w = \left(1 - \frac{R_3}{R_1 + R_2 + R_3}\right) \cdot \left(\frac{P(1,n)}{P(1,n) + P(2,n)} \cdot \zeta^1 + \frac{P(2,n)}{P(1,n) + P(2,n)} \cdot \zeta^2\right) + \frac{R_3}{R_1 + R_2 + R_3} \cdot \zeta^3$$

(3)

where R_l is bit rate of l-th layer. It is clear to see that the probability of selecting third window is only affected by its data size.

3   Experimental Results

First, we have compared NOW RLC and EW RLC to see which case is more appropriate in using dependent hierarchical data. In conventional EW RLC, window selection probability function only follows its data size [14]. We have demonstrated two layers case and three layers case in Figs. 4 and 5.

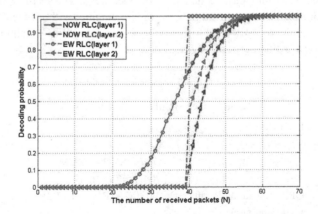

Fig. 4. Performance of NOW RLC and EW RLC in two layers case

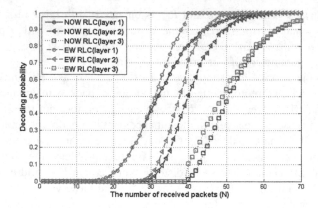

Fig. 5. Performance of NOW RLC and EW RLC in three layers case

In the simulation results, EW RLC shows better decoding probability in the lower layer and NOW RLC shows better decoding probability in the higher layers. Considering the fact that the higher layers cannot be used without their dependent lower layers, the information of the lower layers is absolutely far more significant than the higher layers. Therefore, EW RLC is better strategy than NOW RLC in case of using hierarchical data, e.g. SVC. In the simulation, we set $k_1 = 15$, $k_2 = 25$ and $f_w = 0.4 \cdot \zeta^1 + 0.6 \cdot \zeta^2$ in two layers case and $k_1 = 10, k_2 = 15, k_3 = 15$, and $f_w = 0.3 \cdot \zeta^1 + 0.4 \cdot \zeta^2 + 0.3 \cdot \zeta^2$ in three layers case where k_l indicates the number of packets in l-th window as in [14]. Henceforth we only consider EW RLC.

Next we simulated EW RLC and our proposed EW-NCD using SVC bitstream to see their decoding probability. As shown in Table 1, we used 4CIF CITY sequence encoded with three quality layers using JSVM 9.19. Base layer (BL), enhancement layer (EL) 1 and 2 were encoded with quantization parameter (QP) value of 34, 30, 26, respectively. A packet size is 500 byte and a generation is composed of 100 packets. We assumed a simple multicast scenario which includes one transmitter and multiple receivers. In addition, we assumed a packet erasure channel without feedback between the transmitter and the receivers and forward error correction (FEC) technique was used at the transmitter for this purpose.

Table 1. The simulation parameters

Parameter	Value
Video sequence	CITY
Codec	H.264 SVC – JSVM v9.19
Resolution	704 × 576 (4CIF)
QP(Quantization Parameter)	34/30/26 (BL/EL1/EL2)
FR(Frame rate)	60
Packet size	500byte
Generation size	100(12: 20: 68) packets

As we explained in Sect. 2.2, if we only use the distortion degree to get window selection probability function, the packets in enhancement layers will not be decoded since the size of the layers are relatively bigger than the others. We can see that our SVC bitstream does not have three equal data portions of layers: 12 % for BL; 20 % for EL1; and 68 % for EL2. This means that if we use the conventional window selection probability function, EL2 packets will not be decoded. To solve this issue, we used our window selection probability function and compared EW-NCD with EW RLC.

In Fig. 6, we compared our proposed EW-NCD with EW RLC. EW RLC solely uses bitrate in the window selection probability function while EW-NCD uses our window selection probability function. It is clearly seen that the receiver can decode the base layer better in EW-NCD. Even though the decoding probability of enhancement layers is better in EW RLC, they are useless if the receiver does not securely have their dependent layers.

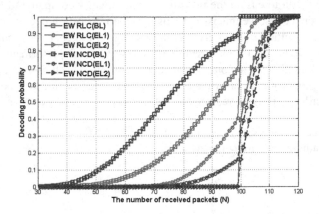

Fig. 6. The comparison of EW RLC and EW-NCD in decoding probability

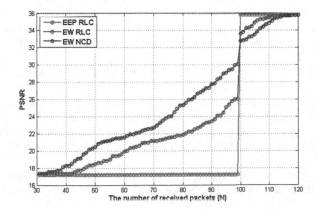

Fig. 7. The comparison of EEP RLC, EW RLC, EW-NCD in PSNR

We also simulate PSNR result with the EEP RLC, EW RLC, and EW-NCD. In Fig. 7, EW-NCD has better PSNR gain in which the number of received packets is less than 100 which guarantees better minimum quality in bad channel. Moreover, EW-NCD scheme shows almost linear PSNR property proportional to the number of received packets.

4 Conclusion

In this paper, we present UEP EW-NCD scheme with optimal encoding parameters based on distortion degree and show the performance of our proposed scheme in terms of decoding probability and PSNR. In comparisons of EEP RLC and UEP EW RLC taking into account the ratio between the data, the proposed method EW-NCD provides better video quality under relatively wide range of channel conditions by taking into consideration of distortion degree. Of course in case of UEP network coding in good channel, it is difficult to ensure the maximum video quality compared to EEP network coding. However we resolve this issue by using video distortion degree with data rate. When data is transmitted in real network, depending on the channel condition and the ratio of the redundant data, video recipients show different performance. In case of multicast, it is difficult to figure out the channel condition of each user and to manage it. Therefore the proportional performance of the proposed scheme can be much more promising in order to provide reasonable video services to users as many as possible.

Acknowledgement. This research was funded by the MSIP (Ministry of Science, ICT & Future Planning), Korea in the ICT R&D Program 2015.

References

1. Cisco Visual Network Index: Global Mobile Data Traffic Forecast Update, 2014–2019
2. Ho, T., Medard, M., Koetter, R., Karger, D.R., Effros, M., Shi, J., Leong, B.: A radnom linear network coding approach to multicast. IEEE Trans. Inf. Theor. **52**(10), 4413–4440 (2006)
3. Vingelmann, P., Fitzek, F.H., Pedersen, M.V., Heide, J., Charaf, H.: Synchronized multimedia streaming on the iphone platform with network coding. IEEE Commun. Mag. **49**(6), 126–132 (2006)
4. Lun, D.S., Medard, M., Koetter, R., Effros, M.: On coding for reliable communication over packet networks. Phys. Commun. **1**(1), 3–20 (2008)
5. Advanced Video Coding for Generic Audiovisual Services, ITU-T Rec. H.264 & ISO/IEC 14496-10 AVC, v3: 2005 Amendment 3: Scalable Video Coding
6. Schwarz, H., Marpe, D., Wiegand, T.: Overview of the scalable video coding extension of the H. 264/AVC standard. IEEE Trans. Circuits Syst. Video Technol. **17**(9), 1103–1120 (2007)
7. Nguyen, K., Nguyen, T., Cheung, S.C.: Peer-to-peer streaming with hierarchical network coding. In: IEEE International Conference on Multimedia and Expo, pp. 396–399. IEEE (2007)

8. Nguyen, A.T., Li, B., Eliassen, F.: Chameleon: adaptive peer-to-peer streaming with network coding. In: Proceedings of the IEEE on INFOCOM, pp. 1–9. IEEE (2010)

9. Chau, P., Kim, S., Lee, Y., Shin, J.: Hierarchical random linear network coding for multicast scalable video streaming. In: Proceedings of the IEEE of 2014 Annual Summit and Conference on APSIPA, pp. 1–7. IEEE (2014)

10. Ha, H., Yim, C.: Layer-weighted unequal error protection for scalable video coding extension of H. 264/AVC. IEEE Trans. Consum. Electron. **54**(2), 736–744 (2008)

11. Hom, U., Stuhluller, K., Link, M., Girod, B.: Robust internet video transmission based on scalable coding and unequal error protection. Proc. Signal Process. Image Commun. **15**(1), 77–94 (1999)

12. Maani, E., Katsaggelos, A.K.: Unequal error protection for robust streaming of scalable video over packet lossy networks. IEEE Trans. Circ. Syst. Video Technol. **20**(3), 407–416 (2010)

13. Nguyen, K., Nguyen, T., Cheung, S.C.: Video streaming with network coding. J. Sig. Process. Syst. **59**(3), 319–333 (2010)

14. Vukobratovic, D., Stankovic, V.: Unequal error protection random linear coding strategies for erasure channels. IEEE Trans. Commun. **60**(5), 1242–1252 (2012)

15. Luo, Z., Song, L., Zheng, S., Ling, N.: Raptor codes based unequal protection for compressed video according to packet priority. IEEE Trans. Multimedia **15**(8), 2208–2213 (2013)

Overview on MPEG MMT Technology and Its Application to Hybrid Media Delivery over Heterogeneous Networks

Tae-Jun Jung, Hong-rae Lee, and Kwang-deok Seo(✉)

Yonsei University, Wonju, Gangwon, Korea
jeung86@naver.com, kdseo@yonsei.ac.kr

Abstract. The MPEG has recently developed a new standard, MPEG Media Transport (MMT), for the next-generation hybrid media delivery service over IP networks considering the emerging convergence of digital broadcast and broadband services. MMT is intended to overcome the current limitations of available standards for media streaming by providing a streaming format that is transport- and file-format-friendly, cross-layer optimized between the video and transport layers, error-resilient for MPEG streams, and convertible between transport mechanisms and content adaptation to different networks. In this paper, we overview on the MPEG MMT technology and describe its application to hybrid media delivery over heterogeneous networks.

Keywords: MPEG media transport (MMT) · Video streaming · Hybrid media delivery

1 Introduction

In recent years, digital broadcasting services and IP-based multimedia services over the Internet including mobile Internet, have begun to be integrated and converged [1]. This trend is expected to continue with other multimedia services. In addition, emerging multimedia services and content are being introduced. These multimedia service developments mean that various services and content will be delivered over heterogeneous networks and that users will expect to consume such services using these networks, which depend on the availability and reach of the network at the time of consumption.

Along with this trend, there have been so many changes in multimedia service environments, such as media content delivery networks, diverse video signals, 4 K/8 K video transport systems, and various client terminals that display multi-format signals. It has become clear that the MPEG standard has been facing several technical challenges due to the emerging changes in those multimedia service environments [2]. Therefore, to address these technical challenges to existing and emerging MPEG standards, ISO MPEG developed the MPEG-H standard suite (ISO/IEC 23008) for the delivery of audio-visual information compressed with high efficiency over a heterogeneous environment. The MPEG-H suite consists of three functional areas: High Efficiency Video Coding (HEVC) [3], 3D audio [4], and MPEG Media Transport (MMT) [5]. The overall structure of MPEG-H is shown in Fig. 1.

© Springer International Publishing Switzerland 2015
Y.-S. Ho et al. (Eds.): PCM 2015, Part II, LNCS 9315, pp. 660–669, 2015.
DOI: 10.1007/978-3-319-24078-7_68

- Part 1 : Encapsulation, Delivery, Signaling
- Part 4 : MMT Reference Software
- Part 7 : MMT Conformance testing
- Part 10 : MMT FEC Codes
- Part 11 : Composition Information (CI)
- Part 13: Implementation Guidelines

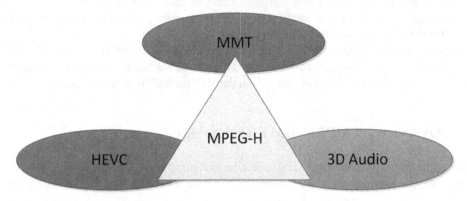

- Part 2: High Efficiency Video Coding
- Part 5: HEVC Reference Software
- Part 8: HEVC Conformance testing
- Part 12: Image File Format

- Part 3: 3D Audio
- Part 6: 3D Audio Reference Software
- Part 9: 3D Audio Conformance testing

Fig. 1. Structure of the MPEG-H standard

The HEVC standard is the successor to H.264/AVC and was jointly developed by the ISO/IEC MPEG and ITU VCEG as the second part of MPEG-H. It was designed to address essentially all existing applications of H.264/AVC and to focus on two key issues: increased video resolution and the increased use of parallel processing architectures, which have resulted from the increasing diversity of services, the growing popularity of high-definition (HD) video, the emergence of beyond-HD formats (e.g., 4 K and 8 K resolutions), and enormous network traffic caused by video applications targeting mobile devices and tablet PCs [6].

To deploy efficient solutions for the transport of HEVC video in an interoperable fashion, especially considering the recently increased demand of multimedia delivery in the heterogeneous network environment, MPEG launched in mid-2010 a new standardization work item, called MPEG Media Transport (MMT). MMT serves to address the technical challenges of existing standards due to recent changes of multimedia delivery and consumption environments and new requirements from emerging use cases and application scenarios in the area of multimedia services [7]. MMT is intended to overcome the current limitations of available standards for media streaming by providing a streaming format that is transport- and file-format-friendly, cross-layer optimized between the video and transport layers, error-resilient for MPEG streams, and convertible between transport mechanisms and content adaptation to different networks [8]. To achieve these purposes, MMT was designed with the following key considerations.

– Convergence to IP and HTML5: MMT is designed as a standard that supports complex multimedia services composed of multiple audio visual components in an environment utilizing HTML5 as a presentation layer technology and IP as an underlying delivery protocol.
– Inheritance of major features of MPEG-2 TS: MMT was designed to inherit major features of MPEG-2 TS that were identified as useful for next-generation broadcasting systems, such as multiplexing of components, easy conversion between storage and streaming formats, and so on.
– Support of hybrid delivery and multi-devices: MMT was designed to support hybrid delivery using broadcast network and the Internet for multimedia services and to support the incorporation of companion devices for additional services.

2   Overview on MPEG MMT Technology

To support efficient delivery and effective consumption of coded media data for multimedia services over packet-switched networks, including IP networks and

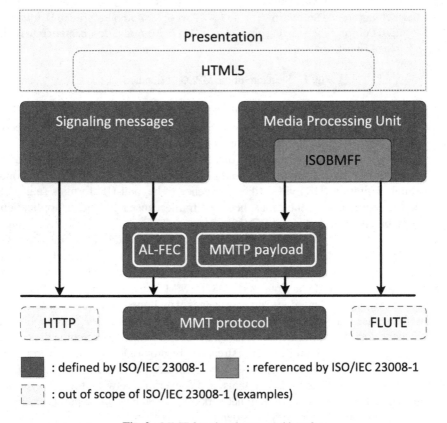

Fig. 2. MMT functional areas and interface

digital broadcasting networks, MMT defines three functional areas: encapsulation, delivery, and signaling. These areas are illustrated in Fig. 2 [5, 8].

The encapsulation functional area defines the logical structure of the media content, the package, and the format of the data units to be processed by an MMT entity and their instantiation with the ISO BMFF (Base Media File Format) as specified in ISO/IEC 14496-12 [9, 10]. It produces the media processing unit (MPU) as an output. It is designed to be self-contained so that each MPU can be completely processed at the client without requiring any further information. Defining the MPU in such a way would be beneficial to the next-generation broadcasting system in which there is no way for each client to rely on bidirectional communication networks for signaling the starting of reception and decoding of media data. The package specifies the components comprising the media content and the relationship among them to provide the necessary information for advanced delivery.

The delivery functional area defines the application layer transport protocol, including the payload format required for transferring encapsulated media data from one network entity to another. The application layer transport protocol provides enhanced features for delivery of multimedia data when compared to conventional application layer transport protocols; e.g., multiplexing and support of mixed use of streaming and download delivery in a single packet flow. The payload format is defined to enable the carriage of encoded media data, which is agnostic to media types and encoding methods.

The signaling functional area defines formats of signaling messages to manage delivery and consumption of media data. Signaling messages for consumption management are used to signal the structure of the package, and signaling messages for delivery management are used to signal the structure of the payload format and protocol configuration.

Figure 3 depicts Functional Areas that MMT covers. Figure 4 depicts the end-to-end architecture of MMT [5]. The MMT sending entity is responsible for sending the packages to the MMT receiving entity as MMTP (MMT protocol) packet flows. The sending entity may be required to gather content from content providers based on the presentation information of the package that is given by a Package provider. A package provider and content providers may be co-located. Media content is provided as an asset that is segmented into a series of encapsulated MPUs, which forms an MMTP packet flow. The MMTP packet flow of this content is generated by using the associated transport characteristics information.

An MMTP session consists of one MMTP transport flow. MMTP transport flow is defined as all packet flows that are delivered to the same destination and which may originate from multiple MMT sending entities. In the case of IP, destination is the IP address and port number. A single Package may be delivered over one or multiple MMTP transport flows. An MMTP transport flow may carry multiple Assets. Each Asset is associated with a unique packet_id within the scope of the MMTP session. MMTP provides a streaming-optimized mode (the MPU mode) and a file download mode (the GFD mode). The MPU mode supports the packetized streaming of an MPU. The GFD mode supports flexible file delivery of any type of file or sequence of files.

The packetization of an MPU that contains timed media may be performed in a MPU format aware mode or MPU format agnostic mode. In the format agnostic mode the

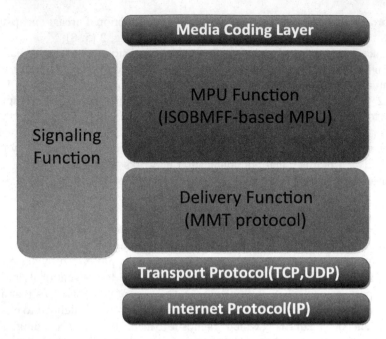

Fig. 3. Functional Areas that MMT covers

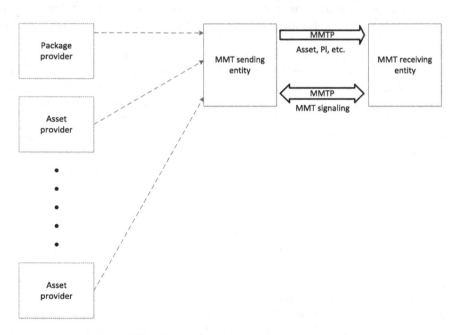

Fig. 4. End-to-end architecture of MMT

packetization procedure takes into account the boundaries of different types of data in MPU to generate packets by using MPU mode. The resulting packets shall carry delivery data units of either MPU metadata, movie fragment metadata, or MFU (media fragment unit). The resulting packets shall not carry more than two different types of delivery data units. The delivery data unit of MPU metadata consists of the 'ftyp' box, the 'mmpu' box, the 'moov' box, and any other boxes that are applied to the whole MPU. The FT (file type) field of the MMTP payload carrying a delivery data unit of MPU metadata is set to '0 × 00'. The delivery data unit of movie fragment metadata consists of the 'moof' box and the 'mdat' box header (excluding any media data). The FT field of the MMTP payload carrying a delivery data unit of movie fragment metadata is set to 0 × 01. The media data, MFUs in 'mdat' box of MPU, is then split into multiple delivery data units of MFU in a format aware way. The FT field of the MMTP payload carrying a delivery data unit of MFU is set to '0 × 02'. This procedure is described in Fig. 5. So MMT receiving entity can classify all packets as three categories. First category belongs to MPU metadata. Second category belongs to fragment metadata and third one belongs to MFU. MFU is considered as a (sub-) sample of media.

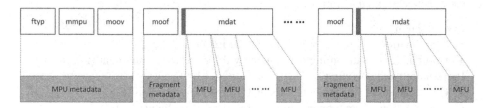

Fig. 5. Payload generation for timed media [5]

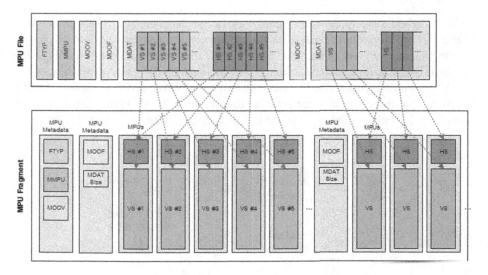

Fig. 6. MPU fragment building block

Figure 6 depicts an MPU fragment building block for packetized streaming of an MPU. This MPU fragment building block helps to prepare for the packetization of an MPU into MMTP packets. The process of creating the MMTP packet passes by two steps: generating the MMTP payload and generating the MMTP packet. This may be performed with the help of the MMT hint track.

3   Hybrid Media Delivery Based on MMT

Hybrid delivery in this Section is defined as simultaneous delivery of one or more content components over more than one different types of network. One example is that one media component is delivered on broadcast channels and the other media component is delivered on broadband networks [11]. The other example is that one media component is delivered on broadband networks and the other media component is delivered on another broadband networks [11].

Basic concept of hybrid delivery is to combine media components on different channels. However, in practice, there are several scenarios for hybrid delivery. The classification can be classified as follows.

- Live and non-live
  - Combination of streaming components (Fig. 7)
  - Combination of streaming component with pre-stored component (Fig. 8)
- Presentation and decoding
  - Combination of components for synchronized presentation (Fig. 7)
  - Combination of components for synchronized decoding (Fig. 9)
- Same transport schemes and different transport schemes
  - Combination of MMT components
  - Combination of MMT component with another-format component such as MPEG-2 TS

In the hybrid delivery services, accurate media synchronization mechanism is indispensable. In MMT, for timed media data, the presentation duration and the decoding order, and the presentation order of each AU (access unit) are signaled as part of the fragment metadata. The MPU does not have its initial presentation time. The presentation time of the first AU in an MPU is described by the presentation information (PI) document. The PI document specifies the initial presentation time of each MPU. Figure 10 depicts an example of the timing of the presentation of MPUs from different Assets that is provided

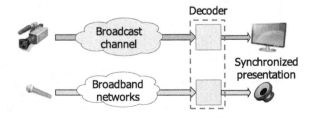

Fig. 7. Combination of streaming components for presentation

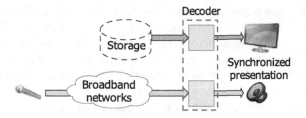

Fig. 8. Combination of streaming component with pre-stored component for presentation

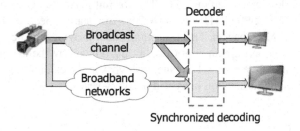

Fig. 9. Combination of components for decoding

by the PI document. In this example, the PI document specifies that the MMT receiving entity shall present MPU #1 of Asset #1 and of Asset #2 simultaneously. At a later point, MPU #1 from Asset #3 is scheduled to be presented. Finally, MPU #2 of Asset #1 and Asset #2 are to be presented in synchronization. The specified presentation time for an MPU defines the presentation time of the first AU in presentation order of that MPU. If any 'elst' box is available, the indicated offset shall be applied to the composition time of the first sample in presentation order of the MPU in addition to the presentation time provided by any presentation information for media synchronization.

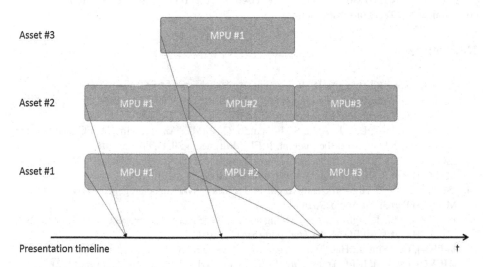

Fig. 10. Example of mapping MPUs to the presentation timeline for synchronization.

MPEG-2 TS components have timestamps based on STC. MMT components have timestamps based on UTC (Coordinated Universal Time). At an MMT compliant receiving entity, the STC based clock in MPEG-2 TS can be converted to the wall clock based on UTC by processing the Clock Relation Information (CRI). An MMT component and an MPEG-2 TS component are presented in synchronized manner since both components can share the same time domain as wall clock.

Preservation of timing relationships among packets in a single MMTP packet flow or between packets from different MMTP packet flows is another important feature of MMT. In the MMTP packet header, timestamp field of 32 bits specifies the time instance of MMTP packet delivery based on UTC. The format is the "short-format" as defined in clause 6 of IETF RFC5905, NTP version 4. This timestamp specifies the sending time at the first byte of MMTP packet. This timestamp field provides the necessary information to calculate jitter and the amount of delay introduced by the underlying delivery network during the delivery of a Package. Network jitter calculation is essential for preserving media synchronization in hybrid delivery services.

4   Conclusion

MMT is a new international standard that is intended to address emerging multimedia services over heterogeneous packet-switched networks, including IP and broadcasting networks. It succeeds MPEG-2 TS as the media transport solution for broadcasting and IP network content distribution, with the aim of serving new applications like UHD video streaming. In this paper, we overviewed the MPEG MMT technology and described its application to hybrid media delivery over heterogeneous networks.

Acknowledgement. This research was supported by the MSIP (Ministry of Science, ICT and Future Planning), Korea, under the ITRC (Information Technology Research Center) support program (IITP-2015-H8501-15-1001) supervised by the IITP (Institute for Information & communications Technology Promotion).

References

1. Aoki, S., Aoki, K., Hamada, H., Kanatsugu, Y., Yamamoto, M., Aizawa, K.: A new transport scheme for hybrid delivery of content over broadcast and broadband. In: Broadband Multimedia Systems and Broadcasting (BMSB), pp. 1–6 (2011)
2. Lim, Y., Park, K., Lee, J., Aoki, S., Fernando, G.: MMT: An emerging MPEG standard for multimedia delivery over the Internet. IEEE Multimedia **20**(1), 80–85 (2013)
3. ISO/IEC 23008-2, High efficiency coding and media delivery in heterogeneous environments – MPEG-H Part 2: High Efficiency Video Coding (HEVC) (2013)
4. ISO/IEC 23008-3, High efficiency coding and media delivery in heterogeneous environments – MPEG-H Part 3: 3D Audio (2014)
5. ISO/IEC 23008-1, High efficiency coding and media delivery in heterogeneous environments – MPEG-H Part 1: MPEG Media Transport (MMT) (2014)
6. Sullivan, G., Ohm, J., Han, W., Wiegand, T.: Overview of the high efficiency video coding (HEVC) standard. IEEE Trans. Circ. Syst. Video Technol. **22**(2), 1649–1668 (2012)

7. ISO/IEC JTC1/SC29/WG11 N11542, Use Cases for MPEG Media Transport (MMT) (2010)
8. Lim, Y., Aoki, S., Bouazizi, I., Song, J.: New MPEG transport standard for next generation hybrid broadcasting system with IP. IEEE Trans. Broadcast. **60**(2), 160–169 (2014)
9. Sodagar, I.: The MPEG-DASH standard for multimedia streaming over internet. IEEE Multimedia **18**(4), 62–67 (2011)
10. ISO/IEC 14496-12, ISO Base Media File Format (2012)
11. ISO/IEC JTC 1/SC 29 w15235, High Efficiency coding and media delivery in heterogeneous environments – MPEG-H Part 13: 2nd Edition MMT Implementation guidelines

A Framework for Extracting Sports Video Highlights Using Social Media

Yao-Chung Fan, Huan Chen[✉], and Wei-An Chen[✉]

Department of Computer Science, National Chung Hsing University,
Taichung, Taiwan
{yfan,huan}@nchu.edu.tw, maruko.and@gmail.com

Abstract. Summarizing lengthy sports video into compact highlights has many applications and plays an essential role for effective media dissemination and delivery. To perform the highlights extraction correctly and effectively is of great challenge. Extensive research efforts have been made to this problem in recent years. In practice, sports video highlights are extracted either manually or based on video content analysis schemes. The former approach is not cost effective and naturally brings the scalability concern, while the later approach suffers from high computational complexity. In this paper, we start from a novel angle to address the sports video summarization problem; we employ real-time text stream, e.g. opinion comment posts, from social media to detect events and the event semantics in live sport videos. The main idea is that one can treat the volumes of comment posts over time as a time series, and the variation of the time series, such as a spike, may reveal events in a game, which therefore can be employed to identify the important moments in the game. By aligning the events with the sports videos over time, automatically summarizing sports video may be feasible. This paper describes the implementation of this idea and reports our experience of summarizing the 2014 World Cup Video. We also evaluate our technique compared to human-generated summaries and find that the results of our technique are quite similar to the human-generated result, which demonstrate the superiority of our technique.

1 Introduction

People love sport games not only because the sport competition is entertaining, but also because it engages our mind, inspires us and touches the depths of our emotion [1]. Football, baseball, boxing, basketball, golf, ice hockey, and tennis are among the most popular sport games. However, sport games are typically with several hours long and the plays might not be always interesting. Sports video summarization, which extracts events that occurred in the lengthy sports game, is therefore of great importance. Summarizing lengthy sports video into compact video clips, also known as "highlights" has many applications and plays an essential role for effective media dissemination and delivery in facing explosive amount of video data.

© Springer International Publishing Switzerland 2015
Y.-S. Ho et al. (Eds.): PCM 2015, Part II, LNCS 9315, pp. 670–677, 2015.
DOI: 10.1007/978-3-319-24078-7_69

In practice, sports video highlights are extracted either manually or based on video content analysis schemes. The former approach is performed by manually labeling video semantics, which is impractical and expensive due to the explosive amount of video today. This approach naturally brings the scalability concern. The later approach suffers high computational complexity problem due to video contents are analyzed based on computer vision and image processing techniques. With the popularity of social platforms, people document daily life experiences, express opinions, and provide diverse observations on what they saw. Massive amounts of data are being generated on social media sites. Mining and analyzing the social data brings many interesting applications, such as detecting influenza epidemics [2] or detecting earthquake [3].

In this paper, we address the sports video summarization problem from a novel perspective: employing real-time text stream, e.g. opinion comments and posts, from social media to detect important events in live sport videos. The main idea is to treat the volumes of comment posts over time as a time series, and the variation of the time series, such as a spike, may reveal events in a sport game. By aligning the events with the sports videos over time, automatically identifying important events for summarizing sports video may be feasible. This paper describes the implementation of this idea and reports our experience of summarizing the 2014 World Cup Video. We evaluate our technique compared to human-generated summaries and find that the results of our technique are similar to the human-generated result, which demonstrates the superiority of our technique.

2 Related Work

In recent years, extensive research efforts have been made to sports video summarization. The existing approaches can be divided into two categories: analyzing by video content only and analyzing with external knowledge. In the following, we review the existing works along the two directions.

2.1 Content-Based Sport Video Summarization

Most of previous works on sports video summarization are based on employing low-level features, such as audio or visual features, extracted from video content itself. The basic idea is to extract the low level features and then performs rule-based or machine learning algorithms to detect events in sports video. For example, in [4, 5] authors employ audio features for baseball and soccer event detection. The study of using visual features for soccer event detection is addressed in [6, 7]. Textual features were utilized in [8, 9] for baseball and soccer games. However, the content-based sport video summarization all suffer from the shortcoming of heavily relying on audio/visual/textual features extracted from the video itself. The semantic gap between low-level features and high-level events are the main concern for the accuracy of the event detections.

2.2   Sport Video Summarization by External Knowledge

As mentioned, content-based sport video summarization face the computational difficulty in processing video content using computer vision and image processing techniques. Therefore, recent researches have proposed techniques by leveraging the external knowledge, such as webcast text [10, 11] or closed caption [12] to sport video summarization. For the research based on closed caption, the main idea is to make use of caption text on the video to perform sports video summarization. However, the approaches based on caption text suffer from two concerns. First, the accuracy of recognizing caption texts is affected by video quality. Second, the caption text is a simply transcript from speech to text, which may be indirectly relevant to the video content. The idea of the work using webcast is to make use of webcast text, which is a short text reporting the status for sports game and easily obtained from the web, and therefore without the problem of recognizing caption text. However, the research based on the webcast text still require low level features to detect the boundary of the events. Differing from the existing research, we start from a novel angle to address the sports video summarization problem by employing social media text stream, e.g. opinion comment posts to detect semantic events in live sport videos without the shortcoming of relying low-level features.

3   Proposed Framework

We formulate the problem to be addressed as follows: given a sequence of documents and a broadcast video of a sports game, we develop an automatic mechanism to align documents and broadcast video at both temporal and semantic levels to achieve semantic event detection in broadcast video. Along this research direction, two research challenges are expected. First, an event detection component is required, which helps to automatically identify events in a sports video. Second, an event semantic annotation component is required, which helps to annotate the semantics of the discovered events. In this section, we will introduce the proposed framework, which comprises the following three stages, (1) content retrieval and context preprocessing, (2) event detection, and (3) semantic annotation.

3.1   Content Retrieval and Context Preprocessing Stage

We first introduce the content retrieval process and context pre-processing tasks on social media for the highlight extraction. Figure 1 shows the steps to perform the social media content retrieval and context pre-processing steps. In our scheme, we implement a web crawler in python to collect the content of a given social media (using uri running on the HTTP protocol). Next in Step 2, the retrieved context is processed using regular expression to detach the tags and unrelated information. In addition, bursts statistics (such as total bursts and average burst in a minute) are analyzed. At last, a histogram of the posting bursts can be illustrated as in the bottom of Fig. 1.

Step1: Implement a web crawler (in python) to retrieve posting
       bursts on social media

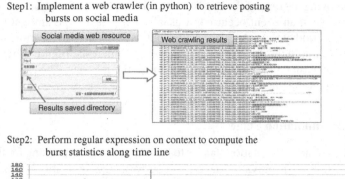

Step2: Perform regular expression on context to compute the
       burst statistics along time line

Fig. 1. Content retrieval and context preprocessing

3.2 Event Detection Stage

With the histogram of the posting bursts (as shown in Fig. 1) in the preprocessing stage, we are ready to address how to detect major events in the sport game. We propose two event detection schemes and they are designed based the on total amount of posting burst (TA) and the moving average of posting burst (MA), respectively.

In TA event detection, the events of interests are extracted in the order of total amount of bursts, which is used to identify the importance of events occurred in the sport game. The rationale behind the proposed TA scheme relies on the high correlation between the event occurrences and posting bursts. As such, high volume of posting bursts usually implies the occurrence of the event of interests. Figure 2(a) shows the top n posting bursts (e.g. n = 10) and the time index associated with those top n bursts.

The MA event detection scheme is designed based on the moving average of the total bursts. An event of interests is labeled based on the difference between the instant

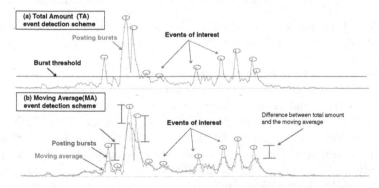

Fig. 2. Event detection schemes (a) based on total amount of bursts (top) (b) based on moving average of bursts (bottom)

burst and the moving average of the bursts. The design principle for MA event detection is to claim that when an event occurred, it usually triggers a large amount of posting bursts in a short period of time. As such, sudden bursts, compared to the moving average of bursts, usually imply the occurrence of the event of interests. Figure 2(b) shows the top n bursts and the time index associated with those top n bursts. The top-n events are assumed to occur at the time index labeled.

3.3   Semantic Annotation Stage

The purpose of semantic annotation stage is to label the semantic meaning for the events of interests identified in previous stages. To perform the semantic annotation, one popular approach is to apply the *term frequency–inverse document frequency* (TF-IDF) numerical statistic to evaluate the importance of the terms in specific contexts. TF-IDF is a widely used scheme in text mining. However, conventional TF-IDF scheme suffers too many unrelated words that are frequently present in the context. To address this issue, we introduce the user weighting concepts in addition to conventional TF-IDF scheme. As illustrated in Fig. 3, the event of interests 1 to 4 are identified by either the TA event detection scheme or the MA event detection scheme proposed in Sect. 3.2. Then a weighting for each user is computed based on the number of events in which the user has posted comments.

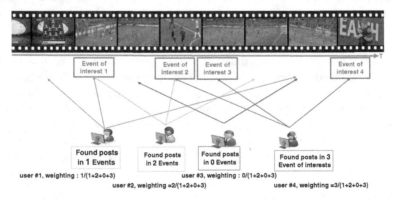

Fig. 3. Illustration of the semantic annotation scheme based on user posts found in event of interests identified in the event detection stage

Take the case shown in Fig. 3 as an example, user #1 only posts comments in one event of interests of all (i.e., in event of interest 1). Similarly, user #2, user #3 and user #4 found posts in 2, 0 and 3 events of interests respectively. As such, the weightings for user #1 to user #4 are assigned to be 1/6, 2/6, 0/6 and 3/6, respectively. Then we order the users according to the weightings. Unlike conventional TF-IDF scheme, all texts are fed into the algorithm to evaluate the importance of specific semantic terms; in our scheme only the comments posted by the first top m users (e.g., m = 3) are used to perform the TF-IDF. With the use of the first top m users, we can significantly eliminate the unrelated information that may mislead in the process of semantic annotation.

4 Performance Evaluation

4.1 Event Detection

To evaluate the performance of our proposed scheme, the 2014 Fédération Internationale de Football Association (FIFA) World Cup Finalgames are used. The matching pairs are (1) Germany vs. Brazil (2) Netherland vs. Argentina (3) Brazil vs. Netherland and (4) Germany vs. Argentina. The social media we used come from the PTT World Cup board [13–16]. The total number of posts for each game is listed in Table 1. In addition, as a benchmark, three audiences are selected to manually label the time indexes for the events of interest based on their expertise. The performance metrics we used are *precision* and *recall* and the evaluation results are reported in Table 1.

Table 1. 2014 FIFA World Cup Finals

(a) Germany vs. Argentina on 2014/7/9(8083 total posts)		
Event detection scheme	Precision	Recall
TA	0.5	0.3334
MA	0.65	0.5556
(b) Netherland vs. Argentina on 2014/7/10(8104 total posts)		
Event detection scheme	Precision	Recall
TA	0.45	0.7142
MA	0.5	0.5714
(c) Brazil vs. Netherland on 2014/7/13(4295 total posts)		
Event detection scheme	Precision	Recall
TA	0.3	0.3
MA	0.4	0.8
(d) Germany vs. Argentina on 2014/7/14(15198 total posts)		
Event detection scheme	Precision	Recall
TA	0.333	0.6667
MA	0.43	0.6667

4.2 Event Semantic Annotation

In this section, we validate the effectiveness of the event semantic annotation. In the experiments, we manually extract three types of events (score event, red card event, and yellow card) from the tested four soccer games. Totally 29 events are extracted. As proposed in the previous section, we use the keywords abstracted from opinions and comments to annotate the semantics of events. For each event, we select k keywords with highest TF-IDF weights as the annotation of the event. With the 29 events, we say there is a hit if the annotation of an event consists of the following three keywords, i.e., (score, red card, and yellow card). We employ hit rate to compare the effectiveness of the proposed methods. Figure 4 shows the experiment result with different values of k (k = 1, k = 5, k = 10, k = 20) of the compared method. From the result, we have two observations. First, when k is small, we see that the hit rate is low. Such result can be expected, since if k is small, only a few of words are employed to annotate the event, and therefore lessen the probability of hits. Second, we see that the method using the comments from power users consistently outperforms the method using all comments. Such result validates the effectiveness of the proposed power user concepts.

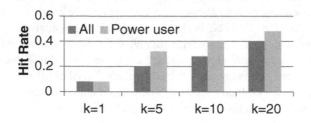

Fig. 4. Hit rate of the event annotation

5 Conclusion

Sports video summarization is an application of high value. In this study, we have introduced the steps of content retrieval process and context pre-processing tasks on social media for the highlight extraction. The contribution of this study is to identify the important events in a sport game using the proposed two event detection schemes based on the burst statistics. In particular, a novel weighting mechanism is introduced for each user to exclude those unrelated frequent terms (the key reasons for TF-IDF to get poor result). The rationale behind the proposed method relies on two facts, which makes our scheme valuable. First when the sport events occurred, people in social media tend to have high activities of message posting and replying. Secondly, the social media contents on the sport games are analyzed rather than the sport video contents themselves. Social media contents are mostly consisted of texts, and the data amount and computation complexity is much less compared to those of the video processing techniques. Results show that our proposed framework detects the events of interests more accurate and performs the semantic annotation more correctly and effectively.

References

1. Red Shannon. Why We Love Sports. Bleacher buzz, Accessed 5 December 2015. http://bleacherreport.com/articles/45904-why-we-love-sports
2. Aramaki, E., Maskawa, S., Morita, M.: Twitter catches the u: detecting inuenza epidemics usingtwitter. In: Proceedings of the Conference on Empirical Methods in Natural Language Processing, pp. 15–68 (2011)
3. Sakaki, T., Okazaki, M., Matsuo, Y.: Earthquakeshakes twitter users: real-time event detection by social sensors. In: Proceedings of WWW, pp. 851–860 (2010)
4. Rui, Y., Gupta, A., Acero, A.: Automatically extracting highlights for TV baseball programs. In: Proceedings of the ACM Multimedia, pp. 105–115 (2000)

5. Xu, M., Maddage, N.C., Xu, C., Kakanhalli, M.S., Tian, Q.: Creating audio keywords for event detection in soccer video. In: Proceedings of the IEEE International Conference Multimedia and Expo, vol. 2, pp. 281–284 (2010)
6. Gong, Y., et al.: Automatic parsing of TV soccer programs. In: Proceedings of the International Conference on Multimedia Computing and Systems, pp. 167–174 (1995)
7. Ekin, A., Tekalp, A.M., Mehrotra, R.: Automatic soccer video analysis and summarization. IEEE Trans. Image Process. **12**(5), 796–807 (2003)
8. Zhang, D., Chang, S.F.: Event detection in baseball video using superimposed caption recognition. In: Proceedings of the ACM Multimedia, pp. 315–318 (2002)
9. Assfalg, J., et al.: Semantic annotation of soccer videos: Automatic highlights identification. Comput. Vis. Image Underst. **92**, 285–305 (2003)
10. Xu, C., Zhang, Y.-F., Zhu, G., Rui, Y., Lu, H., Huang, Q.: Using webcast text for semantic event detection in broadcast sports video. Multimedia, IEEE Trans. **10**(7), 1342–1355 (2008)
11. Xu, C., Wang, J., Lu, H., Zhang, Y.: A novel framework for semantic annotation and personalized retrieval of sports video. Multimedia, IEEE Trans. **10**(3), 421–436 (2008)
12. Nitta, N., Babaguchi, N.: Automatic story segmentation of closed-caption text for semantic content analysis of broadcasted sports video. In: Multimedia information systems, p. 110
13. san70168. 2014 FIFA WORLD CUP GER 1-0 ARG Final. Pttworldcup, 9 July 2014. http://www.ptt.cc/bbs/WorldCup/M.1404846835.A.9EA.html
14. san70168. 2014 FIFA WORLD CUP NED 0-0 ARG (SF). Pttworldcup, 10 July 2014. http://www.ptt.cc/bbs/WorldCup/M.1404933803.A.AE0.html
15. san70168. 2014 FIFA WORLD CUP BRA 0-3 NED (3rd). Pttworldcup, 13 July 2014. http://www.ptt.cc/bbs/WorldCup/M.1405192187.A.3D6.html
16. san70168. 2014 FIFA WORLD CUP GER 1-0 ARG Final. Pttworldcup, 14 July 2014. http://www.ptt.cc/bbs/WorldCup/M.1405276044.A.502.html

Author Index

Printed in the United States
By Bookmasters

Printed in the United States
By Bookmasters